2nd EPI International Conference on Science and Engineering (EICSE2018)

Strengthening UIG-Based Collaborations in Anticipating the Era of Industry Disruption 4.0

IOP Conference Series: Materials Science and Engineering
Volume 676

Gowa, Indonesia
24 October 2018

Editors:

Faisal Mahmuddin
Muhammad Anshar
Intan Areni

ISBN: 978-1-7138-0928-9
ISSN: 1757-8981

Printed from e-media with permission by:

Curran Associates, Inc.
57 Morehouse Lane
Red Hook, NY 12571

Some format issues inherent in the e-media version may also appear in this print version.

Institute of Physics
Dirac House, Temple Back
Bristol BS1 6BE UK

Phone: 44 1 17 929 7481
Fax: 44 1 17 920 0979

techtracking@iop.org

Additional copies of this publication are available from:

Curran Associates, Inc.
57 Morehouse Lane
Red Hook, NY 12571 USA
Phone: 845-758-0400
Fax: 845-758-2633
Email: curran@proceedings.com
Web: www.proceedings.com

TABLE OF CONTENTS

IOP Conf. Series: Materials Science and Engineering **676** (2019) 011001 doi:10.1088/1757-899X/676/1/011001

• Preface

We are delighted that we can introduce the proceeding of the 2nd EPI International Conference on Science and Engineering (EICSE2018). The conference is an annual international conference held by Center of Technology (COT), Faculty of Engineering, Hasanuddin University. The conference is one of the activities conducted to commemorate the anniversary of Faculty of Engineering, Universitas Hasanuddin annually. The main aim of the conference is to strengthen research collabocations based on what so-called "triple helix" which consists of university, industry and government (UIG). Therefore, the adopted theme of the year 2018 conference is " Strengthening UIG-Based Collaborations in Anticipating the era of Industry Disruption 4.0."

The conference is also supported by Japan International Cooperation Agency (JICA) through C-BEST (Capacity Building in Engineering, Science and Technology) project. This year conference has also been certified by the Japanese embassy in Indonesia as part of activities commemorating 60 years of Indonesia-Japan relationship.

We would like to express our appreciation to all authors for their manuscript's contribution and also to other participants of the conference. We also thank all other parties including our university and our faculty leaders and staff for their support on conference organization and other matters.

On behalf of the EICSE2018 Committee,
Conference Chair

Dr. Faisal Mahmuddin

• Conference photographs

- **Editors**
  - Dr. Faisal Mahmuddin
  - Dr. Muhammad Anshar
  - Dr. Intan Areni

IOP Conf. Series: Materials Science and Engineering **676** (2019) 011002 doi:10.1088/1757-899X/676/1/011002

Conference Committee

➢ International Scientific Committee (ISC)

- Prof. Yoshihiro Narito, Hokkaido University (Japan)
- Prof. Shinya Honda, Hokkaido University (Japan)
- Prof. Katsuhiko Sasaki, Hokkaido University (Japan)
- Prof. Jun Ando, Kyushu University (Japan)
- Prof. Shigeru Kashihara, Nara Institute of Science and Technology (Japan)
- Prof. Akio Miyara, Saga University (Japan)
- Prof. Tsubasa Otake, Hokkaido University (Japan)
- Prof. Hideaki Yasuhara, Ehime University (Japan)
- Prof. Masao Furusho, Kobe University (Japan)
- Prof. Satrio Soemantri Brodjonegoro, Institut Teknologi Bandung (Indonesia)
- Prof. Herman Parung, Hasanuddin University (Indonesia)
- Prof. Dato Nor Aieni binti Haji Mokhtar, University Malaysia Terengganu (Malaysia)
- Prof. Wan Mohd Norsani Bin Wan Nik, University Malaysia Terengganu (Malaysia)
- Prof. Ahmad Fitriadhy, University Malaysia Terengganu (Malaysia)
- Prof. Nasaruddin Salam, Hasanuddin University (Indonesia)
- Prof. Imran Umar, Hasanuddin University (Indonesia)
- Prof. Ramli Rahim, Hasanuddin University (Indonesia)
- Prof. Slamet Trisutomo, Hasanuddin University (Indonesia)

➢ Organizing Committee (OC)

- Steering board :
  – Dr. Ir. Muhammad Arsyad Thaha, MT.
  – Prof. Baharuddin Hamzah, ST., M.Arch., Ph.D.
  – Dr. Eng. Muhammad Rusman, ST., MT.
  – Mukti Ali, ST., MT., Ph.D.
- Coordinator : Dr. Eng. Nasruddin ST., MT.
- Chair : Dr.Eng. Faisal Mahmuddin, ST., M.Inf.Tech., M.Eng
- Secretary : Muhammad Zubair Muis Alie ST., MT., Ph.D
- Treasurer : Nurfitriani Seilah, ST.
- Members :

– Dr. Eng. Ir. Muhammad Ramli, ST., MT.
– Ir. Syerly Klara, MT.
– Dr. Eng. Intan Sari Areni, ST., MT
– Dr. Indar Chaerah Gunadin, ST., MT.
– Dr. Eng. Purwanto, ST., MT.
– Dr. Eng. Muralia Hustim, ST., MT.
– Siti Hijraini Nur, ST., MT.
– Dr. Eng. Abd. Mufti Radja, ST., MT.
– Hamzah, ST., MT.
– Ir. Zaenab, MT.
– Dr. Eng. Ilham Alimuddin, ST., M.ArcGIS
– Dr. Eng. Amil Ahmad Ilham, ST., MT
– Dr. Ir. Zuryati Djafar, MT.
– Dr. Eng. Muh. Anshar, ST., MT.

Dr. Eng. Ulva Ria Irvan, MT.
– Dr. Ir. Rosmalina Hanafi, M.Eng.
– Sabaruddin Rahman, ST., MT, P.h.D
– Dr. Eng. Achmad Yasir Baeda, ST., MT.
– Amos Datu Bala, S.Kom.
– Jumadil, ST., M.M.
– Herawati, S.Sos
– Susanti Alwi, S.Sos
– Jayadi, S.Sos
– Syamsir Alam, SKM
– Hasrawati, SE.
– Supriani, ST.
– Baharuddin
– Kaharuddin
– Samsuddin
– Dirman Abdillah, S.Pd, BASc
– Juliana Palambean, A.Md

IOP Conf. Series: Materials Science and Engineering **676** (2019) 011003 doi:10.1088/1757-899X/676/1/011003

Peer review statement

All papers published in this volume of *IOP Conference Series: Materials Science and Engineering* have been peer reviewed through processes administered by the proceedings Editors. Reviews were conducted by expert referees to the professional and scientific standards expected of a proceedings journal published by IOP Publishing.

IOP Conf. Series: Materials Science and Engineering **676** (2019) 012001 doi:10.1088/1757-899X/676/1/012001

Stability Improvement by Reducing Voltage Fluctuation using SVC in Penetration Wind Power System

A Siswanto[1]*, I C Gunadin[2], S M Said[2] and A Suyuti[2]

[1]Program Study of Electrical Engineering, Universitas 17 Agustus 1945, Cirebon
[2]Department of Electrical Engineering, Engineering Faculty, Universitas Hasanuddin

*Email: asiswanto.untagcrb@gmail.com

Abstract. Currently, wind energy is a renewable energy that is growing because some of its advantages are connected to the grid system. This study is focus on presenting a probabilistic theory to design power system stability (PSS) and excitation of the FACTS system controller generator via Static Var Compensator (SVC) connected to wind turbines in Sidrap, region of South Sulawesi system. The purpose of the test system is on oscillation, profile voltage and influence stability of wind turbine penetration. An oscillating system requires system performance compensation from the side of the machine via PSS and transmission via SCV to increase the depth of the electro mechanical mode. So, the simulation results on 44 buses in the South Sulawesi system show the stability of the system and rapid voltage improvement and can reduce the oscillation that occurs. This system is tested using a time domain-based PSAT software simulation on the system.

1. Introduction

Modern power systems have come in the current era which has an impact on the limitation of variation and stress stability problems in the transmission network when operating. Historically, the South Sulawesi transmission system was more vulnerable to operating customer restrictions and uncertainty issues due to a network of reactive power requirements, deregulation of the electricity industry and the use of various renewable energy sources and different operations. At present the electric power system continues to evolve, load growth is always followed by transmission to connect power plants from new renewable energy such as wind turbines and PV. The load and growth are at the root of the load stability problem arising in accordance with the operating pattern so that the use of new technologies and controls is needed, to improve operation under oscillation conditions. Stability problems due to penetration of wind turbines have been carried out [1-3]. On the experience of steady state stability using Capital Analysis and stability, at [4, 5], and [6, 7]. Improper operating patterns can result in frequency stability, voltage stability and interred oscillation. However, currently the power stabilizer system (PSS) has been used to generate and control the voltage and frequency tuned using the metaheuristic method of neural network [8].

A lot of research has been successful on PSS presented in the literature [9]. On multi-machine systems, and the Coordinated SVC and AVR studies on voltage control on hybrid wind diesel systems have been carried out by [10, 11]. SVC using optimal load flow using SIMULINK has been done by [12]. Optimization using Eigen value analysis for transient stability of SVC has been carried out by [13]. In [14] has discussed the optimal sizing and location of SVC devices for improvement. Research has been conducted on the penetration of wind turbines [3] but the conventional SVC strategy still

IOP Conf. Series: Materials Science and Engineering **676** (2019) 012001 doi:10.1088/1757-899X/676/1/012001

needs to be developed. The proposed method is implemented on the South Sulawesi 44 system bus using PSAT software, because the wind direction characteristics in South Sulawesi are classified into two, namely west wind and east wind and it will affect the output of wind turbine so stability can be disturbed. The Sulawesi system displays using the PSAT bus voltage profile before being penetrated when the wind turbine is before and after it is installed. A voltage controller on the Sulawesi system is a function to fluctuate with oscillation due to variations in wind turbine output.

2. Fundamental Theory

The basic purpose of PSS installation is to widen the stability limit by modulation of the excitation generator to provide attenuation to synchronous when oscillations occur due to changes in load. Analytically, PSS can function as a transfer obtained by PSS from wash-out and the lead lag shown in Figure 1. The lead-lag aims to provide a suitable phase lead to compensate for the phase of the excitation lag and the torque generator in the system.

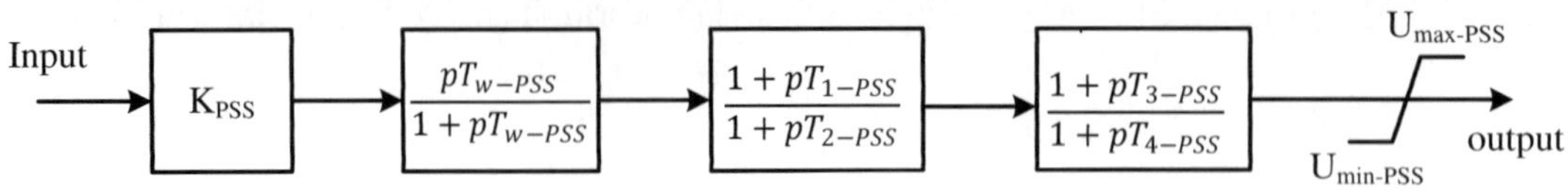

Figure 1. Typical controller with two lead lag stages

3. SVC Controller

Good damping to reduce the oscillation of the power system interconnected between the load and the filter. In addition to PSS, devices found on the transmission network side are FACTS devices to improve system stability. Static Var Compensator (SVC) is the most common FACTS device used in transmission networks. The main SVC application is maintaining a voltage bus bar at a predetermined value to supply the load. However, there has been a growing tendency to use SVC to help system stability when oscillations occur due to changes in load and the effect of penetration from wind turbines. An SVC equipped with a voltage regulator can provide synchronization torque but can be ignored by damping torque on a swinging system [15]. SVC can be operated in two different modes: voltage mode and VAR control mode. When SVC operates the voltage setting mode, implementing the following V-I characteristics is shown in Figure 2.

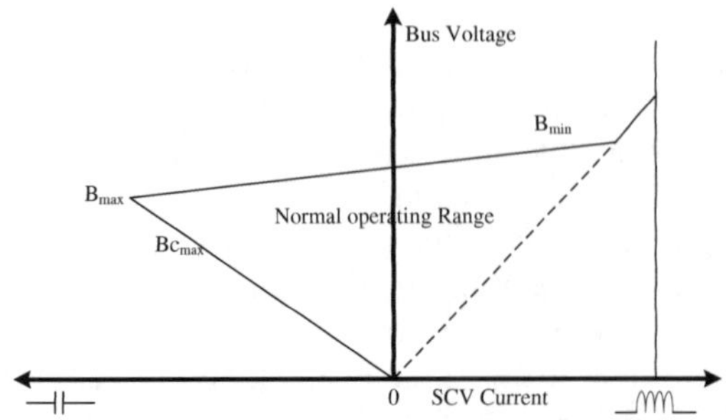

Figure 2. V-I characteristic of SVC

An additional damping controller is needed for extra attenuation when a voltage drop occurs due to additional load. SVC is more effective at controlling swinging power at higher power transfer rates; 150kV accuracy of SVC locations contributes greatly to compensation. A typical SVC configuration with inductors is controlled by thyristors and fixed capacitors, connected to the bus bar via step-down transformers, for BC and BL, each of which is an arrangement of capacitors and inductors. SVC behaviour can be represented by the following equation:

$$X_{SVC} = X_T - \frac{1}{B_C+B_L} = \frac{1}{Y_{SVC}} \quad (1)$$

$$\begin{bmatrix} V_R \\ V_j \end{bmatrix} = X_{SVC} \begin{bmatrix} -I_J \\ I_R \end{bmatrix} \quad (2)$$

$$\begin{bmatrix} \Delta I_R \\ \Delta I_J \end{bmatrix} = \frac{1}{X_{SVC}} \left\{ \begin{bmatrix} \Delta V_J \\ \Delta V_R \end{bmatrix} - \frac{1}{(B_C+B_L)^2} \begin{bmatrix} I_R \\ I_J \end{bmatrix} \Delta B_L \right\} \quad (3)$$

ISVC are the amount of voltage and current in the SVC terminal.ΔVSVC and ΔBL is the controller output. Δ in Equations (1) and (2) for use in minor interference techniques. For equation (3) is obtained so that the SVC module with 4-pin in Figure 3. is installed in the system network. Thyristor controller connection system for SVC and the system in Figure 4 with the provisions of ΔV_{SVC} dan ΔI_{SVC} can be derived from V_R and V_J (real and imaginary power parts of SVC voltage) then to signal $\Delta V_{SVC} = \{(V_R / V_{SVC})\, \Delta V_R + (V_J / V_{SVC})\, \Delta V_J\}$ The signal

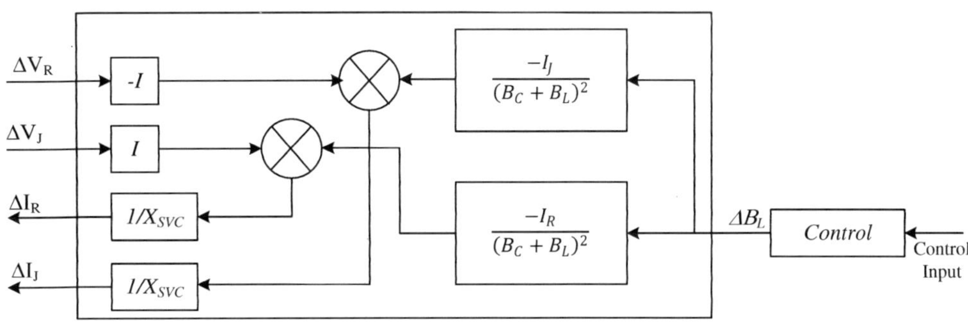

Figure 3. SVC Model with four pins

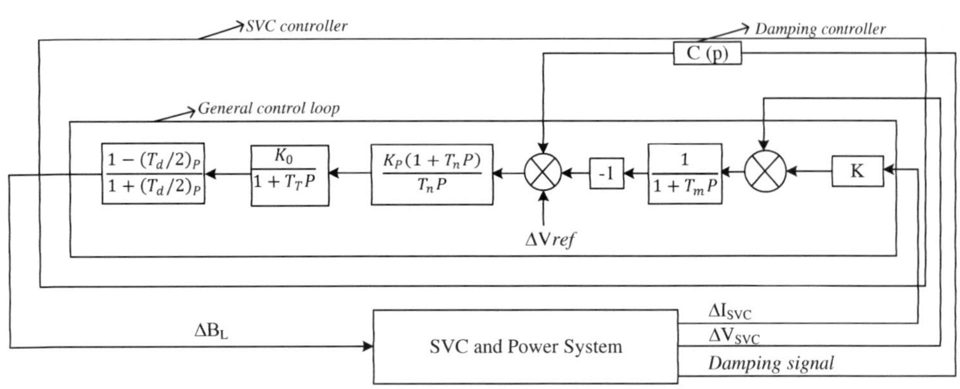

Figure 4. Signal and connection of SVC controller

To analyze the stability, a mathematical model of the differential algebraic equation (DAE) is needed. The set is used for Eigen value analysis in PSAT software. The form of DAE is

$$x = f^{\cdot}(x, y) \quad (4)$$

$$\mathbf{0} = \boldsymbol{g}(\boldsymbol{x}, \boldsymbol{y}) \tag{5}$$

In Equations (4) and (5) where x is a state variable and y is a vector of algebraic variables. The state matrix is calculated by Jacobean matrix. The state matrix is used to calculate Eigen values. There are two variables of positive Eigen value in a system using SVC shown in Figure 7, this shows the system is in stable condition. The objective function based on Eigen values is used to design SVC and PSS. Here, the scale function is used. complex pairs of Eigen values

$$\lambda = \boldsymbol{\sigma} + \boldsymbol{j\omega} \tag{6}$$

Note that σ is the real part of the Eigen value and jω is the imaginary part of the Eigen value, σ0 is the constant attenuation factor which is (-1). According to this objective function, relocate the oscillation mode from the right side to the left side image. so, the optimal value of the objective function is less than zero as shown in Figure. 5.

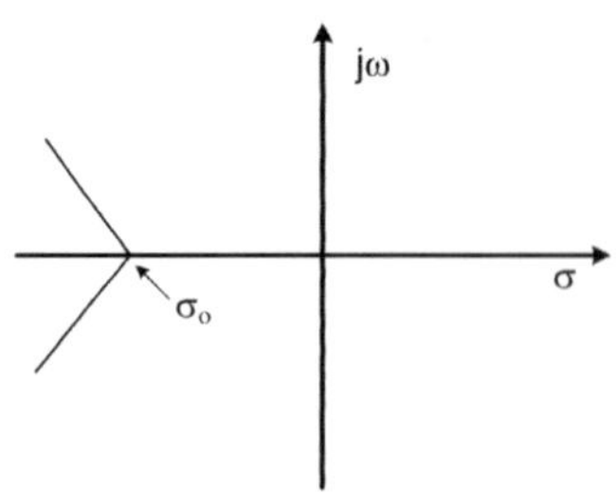

Figure 5. Objective function

4. Proposed Method

The testing of wind turbine penetration with SVC installation in the South Sulawesi network system is shown in Figure 6. The system consists of swing bus, fifteen bus generators, and 44 bus substations. This simulation uses the Matlab Toolbox to analyze and simulate electric power systems. Wind turbine used is 70MW installed on bus 2 in Sidrap.

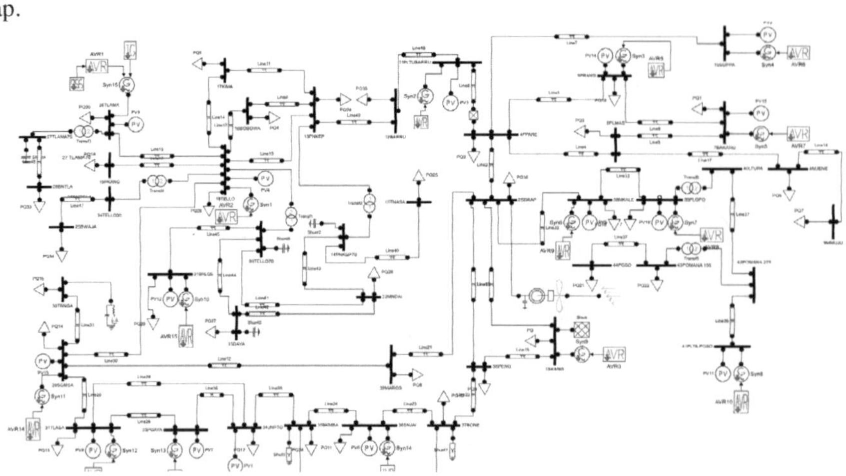

Figure 6. South Sulawesi 44-bus test system

5. Simulation and Results

Using PSAT software, the simulation was tested on the South Sulawesi 44 bus power system. This simulation aims to obtain the effect of the penetration of the wind Turbine on the bus 2 Sidrap on stability system using the SVSI indicator. Active and reactive power load fluctuations provide a different SVSI response due to the bus voltage deviation generated.

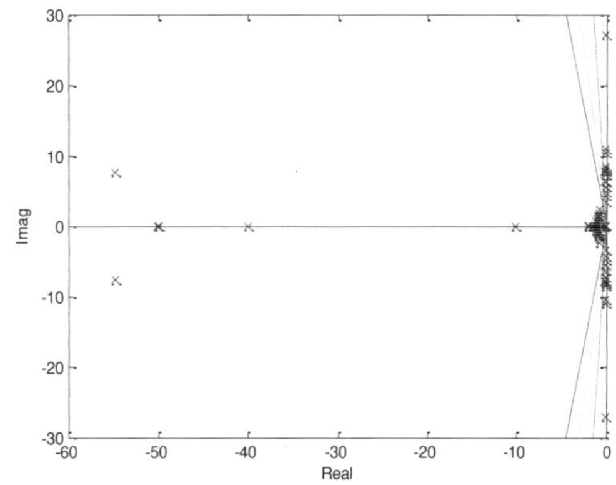

Figure 7. Eigen value analysis

Table 1. Eigen value analysis of system with SVC

	Without Wind, PSS and SVC	With Wind and PSS	with wind, PSS and SVC
Dinamic Order	111	117	119
Buses	44	44	44
Positive Eigen	0	0	0
Negative Eigen	110	116	117
Complex pairs	28	28	26
Zero Eigen	1	1	2

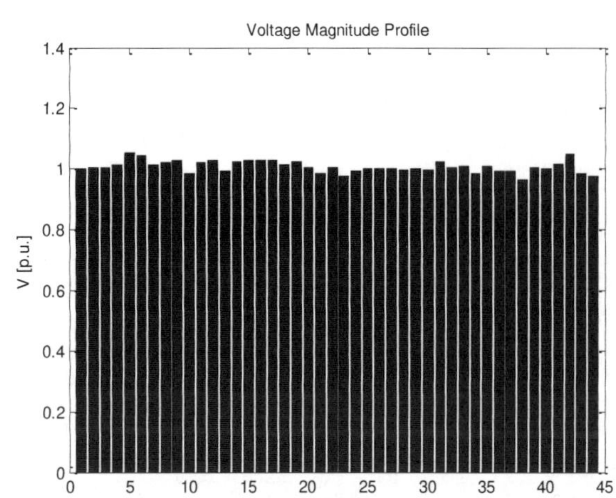

Figure 8. Voltage with wind, PSS and SVC

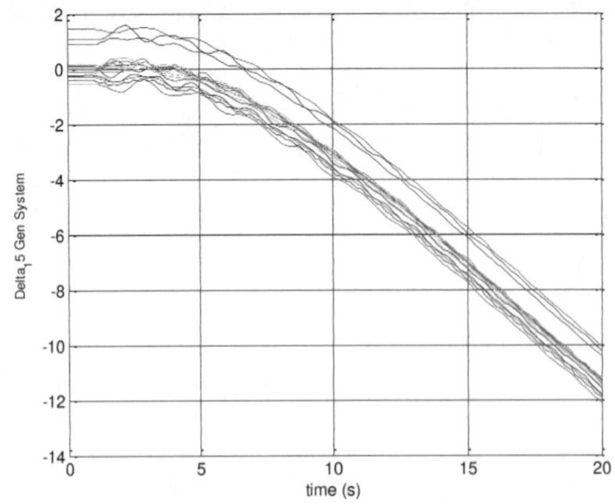

Figure 9. Delta 15 Gen system using wind, PSS and SVC

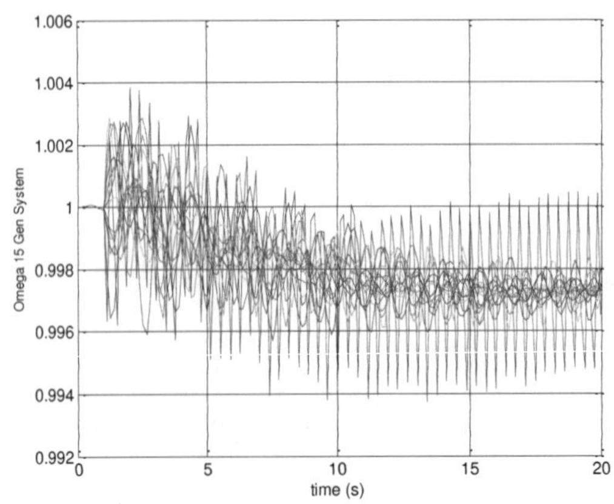

Figure 10. Omega 15 Gen system using wind, PSS dan SVC

Tabel 2. Test system parameter values

bus	Real power generation [pu] Without PSS, Wind and SVC	Reactive power generation [pu] Without PSS, Wind and SVC	Real power generation [pu] with PSS, Wind and SVC	Reactive power generation [pu] with PSS, wind, and SVC
Bus 1	1.2741	0.15092	2.2947	0.14243
Bus 2	-0.265	-0.103	-0.265	-0.103
Bus 3	-0.141	-0.034	-0.141	-0.034
Bus 4	-0.187	-0.047	-0.187	-0.047
Bus 5	0.099	-0.48739	0.099	-0.4501
Bus 6	-0.171	-0.041	-0.171	-0.041
Bus 7	0.013	0.86583	0.013	0.87468
Bus 8	-0.233	-0.037	-0.233	-0.037
Bus 9	-0.096	-0.048	-0.096	-0.048
Bus 10	0.311	-0.09708	0.311	-0.03173
Bus 11	0.604	-0.18632	0.604	-0.13574
Bus 12	-0.101	-0.024	-0.101	-0.024
Bus 13	-0.221	-0.08	-0.221	-0.08
Bus 14	0	0.13352	0	0.13335
Bus 15	-0.189	-0.0206	-0.189	-0.0206
Bus 16	0.331	-0.0154	0.331	-0.0154

Bus 17	-0.18	-0.058	-0.18	-0.058
Bus 18	-0.432	3.2572	-0.423	3.3367
Bus 19	-0.683	-0.177	-0.432	3.2572
Bus 20	0	0.21456	0	0.21441
Bus 21	0.062	-0.28016	0.062	-0.27682
Bus 22	-0.243	-0.026	-0.243	-0.026
Bus 23	-0.245	0.18617	-0.245	0.186
Bus 24	0	0	0	0
Bus 25	0	0	0	0
Bus 26	0.029	-0.7146	0.029	-0.7146
Bus 27	0	0.12435	0	0.12435
Bus 28	-0.265	-0.077	-0.265	-0.077
Bus 29	0.043	-1.1396	-0.457	-0.01295
Bus 30	-0.552	-0.167	-0.952	-0.999
Bus 31	0.584	0.45407	0.584	0.55431
Bus 32	-0.186	-0.005	-0.186	-0.005
Bus 33	1.961	-0.37138	1.961	-0.39327
Bus 34	0.451	-0.03284	0.451	-0.0265
Bus 35	-0.271	0.00095	-0.271	0.00107
Bus 36	0.031	0.0642	0.031	0.06249
Bus 37	-0.321	-0.01589	-0.321	-0.01605
Bus 38	-0.1108	0.25944	-0.1108	0.26881
Bus 39	-0.488	-0.02359	-0.1108	0.2359
Bus 40	0	0	0	0
Bus 41	195	0.25782	195	0.25782
Bus 42	0	0	0	0
Bus 43	-0.049	-0.005	-0.049	-0.005
Bus 44	-0.995	-0.018	-0.995	-0.018

6. Conclusion

In this paper, the problem of stability due to wind turbine penetration on the Sulawesi system on 44 buses is measured using SVSI. This study shows that wind turbine integration affects the value of voltage stability. Changes in load and the level of wind penetration installed on bus 2 in the sideway area determine the critical value of the load bus. The weakest voltage on bus 30 obtains with the active power value of -0.552 and the active power of -0.167, before setting up PSS wind and SVC. On the swing bus, the active power value from 1.2741 becomes 2.2947, while the reactive power from 0.15092 becomes 0.14243 when the power is increased but the reactive power decreases by 0.00849 pu.

References

[1] Rahman, M.M., M.A.H. Baky, and A.K.M.S. Islam, *Electricity from Wind for Off-Grid Applications in Bangladesh: A Techno-Economic Assessment.* 2017, 2017. 6 (1): p. 10.

[2] Sainz, L., et al., *Effect of wind turbine converter control on wind power plant harmonic response and resonances.* IET Electric Power Applications, 2017. 11(2): p. 157-168.

[3] Siswanto, A., et al., *Stability improvement of wind turbine penetrated using power system stabilizer (PSS) on South Sulawesi transmission system.* AIP Conference Proceedings, 2018. 1941(1): p. 020036.

[4] Gunadin, I.C., et al., *Steady-State Stability Assessment Using Neural Network Based on Network Equivalent.* TELKOMNIKA (Telecommunication Computing Electronics and Control), 2013. 9 (3): p. 411-422.

[5] Gunadin, I.C., *Determination of Steady State Stability Margin Using Extreme Learning Machine.* 2012.

[6] Gunadin, I.C., Z. Muslimin, and A. Siswanto. *Transient stability improvement using allocation power generation methods based on moment inertia*. in *2017 International Conference on Electrical Engineering and Informatics (ICELTICs)*. 2017.

[7] Rahman, Y.A., et al., *Evaluating the effect placement capacitor and distributed photovoltaic generation for power system losses minimization in radial distribution system.* AIP Conference Proceedings, 2018. 1941(1): p. 020027.

[8] Xu, D.-j., R.-m. He, and T. Xu. *ANN based multiple power system stabilizers adaptive and coordinated control*. in *Proceedings. International Conference on Power System Technology*. 2002.

[9] Esmaili, M.R., et al., *New Coordinated Design of SVC and PSS for Multi-machine Power System Using BF-PSO Algorithm.* Procedia Technology, 2013. 11: p. 65-74.

[10] Vachirasricirikul, S., I. Ngamroo, and S. Kaitwanidvilai, *Coordinated SVC and AVR for robust voltage control in a hybrid wind-diesel system.* Energy Conversion and Management, 2010. 51 (12): p. 2383-2393.

[11] Yuli Asmi Rahman, Agus Siswanto, and irwan Mahmudi, *Stability Issues in Presence Variable Distributed Generation Into Radial Distribution Network* International Conference on Industrial Electrical and Electronics (ICIEE) 2018.

[12] Singh, B. and G. Agrawal, *Enhancement of voltage profile by incorporation of SVC in power system networks by using optimal load flow method in MATLAB/Simulink environments.* Energy Reports, 2018. 4: p. 418-434.

[13] Karthikeyan, K. and P.K. Dhal, *A Hybrid BBO-DE optimization with Eigen value analysis based transient stability improvement by coordinated design of SVC.* Materials Today: Proceedings, 2018. 5(1, Part 1): p. 1239-1249.

[14] Savić, A. and Ž. Đurišić, *Optimal sizing and location of SVC devices for improvement of voltage profile in distribution network with dispersed photovoltaic and wind power plants.* Applied Energy, 2014. 134: p. 114-124.

[15] Bian, X.Y., et al., *Coordinated design of probabilistic PSS and SVC damping controllers.* International Journal of Electrical Power & Energy Systems, 2011. 33(3): p. 445-452.

IOP Conf. Series: Materials Science and Engineering **676** (2019) 012002 doi:10.1088/1757-899X/676/1/012002

Utilising the See-and-Follow Method for Enhancing Robot Learning Ability

Muh Anshar[1]*, D Halim[1] and C Yohannes[1]

[1]Research Lab Indonesia Australia Social Cognitive Robotics, Electrical Department, Faculty of Engineering, University of Hasanuddin

*Email: muh.anshar@gmail.com

Abstract. The rapid growth and development of robots today can be reflected from variations of research in the robotic field, and one of them that has been flourishing in the area is robot learning. Learning is one of the most important aspect needs to be implemented in a robot that allows the robot to improve its performance through experience-driven knowledge. There are several forms of learning methods and one of the most favourable approach is the see-and-follow method. This paper aims to develop robots that can learn to do the same movements with human movements through the see-and-follow method. Kinect sensor is used as a technology to support the process of recognizing the human body and its movements; which later to be transformed into the robot dimension. The results show that Kinect sensor can be used to measure the distance and recognize the joints of the human body along with the movements produced within a distance of 1 meter to 3.75 meters with an average reading error of 0.04 meter. The average delay time for the whole movement is 2.6 seconds. The average percentage of similarity between robot and human movements for the whole movement scenarios is 93.45%. Hence, high level of similarity in movement concludes that the method of see-and-follow successfully increases the learning skills of the robot.

1. Introduction

According to Mitchell [1], learning is a process to improve performance through experience. Learning is one of the important aspects that needs to be implemented in a robot, which allows the robot to improve various skills through experience. This experience can be obtained either from the robot mechanism itself or from its environment [2]. Robot can be said to learn if when doing a work, its ability increases over time correspond with its experience. One method that can be used in robot learning is to see-and-follow or typically known as mimicking method.

The method of see-and-follow models human as the object to be imitated and a robot moves according to what the human model is doing. Implementation of this method requires specific sensor that has the ability to recognize human body and human motions. These motions would be the input for the robot, which latter to be imitated during mimicking process without to manually move the robot hands to follow the human motions. Through this way, the see-and-follow method makes the Human-Robot Interaction become more user-friendly where controlling robots no longer requires a gamepad, joystick, keyboard, and mouse.

This paper aims to develop a robot framework that support learning process in order to acquire a certain skill such as motions which are performed by a human model through see-and-follow method.

IOP Conf. Series: Materials Science and Engineering **676** (2019) 012002 doi:10.1088/1757-899X/676/1/012002

Kinect sensor is used as a sensor technology to support the recognition process of the human body and its movements. Testing scenarios are designed to demonstrate the learning process by see-and-follow method and to access the performance of the learning.

2. Related Studies

The see-and-follow method has been widely implemented in the robot learning as a research area in robotics. Algan [3] in 2003 introduced a robotic arm that could mimic the movements of the human arm where the position of the arm was represented through camera images. In 2014, Jonas [4] introduced a humanoid robot that can mimic the movement of the entire human body in real time by utilizing the Xsens' MVN motion capture system. In the following year, 2015, Rajesh [5] also made a robotic arm that could mimic the movements of the human arm by utilising the Kinect sensor to determine the slope angle of the arm then transform it into joint positions of the robot.

3. Methods

In this study, the overall design of the framework robot learning system is shown in Figure 1 below. It consists of one Kinect sensor, 1 PC, 2 pieces of Arduino Mega 2560, a pair of 433 MHz RF transceiver-receiver communication module, IC 74LS241 driver, and a humanoid robot.

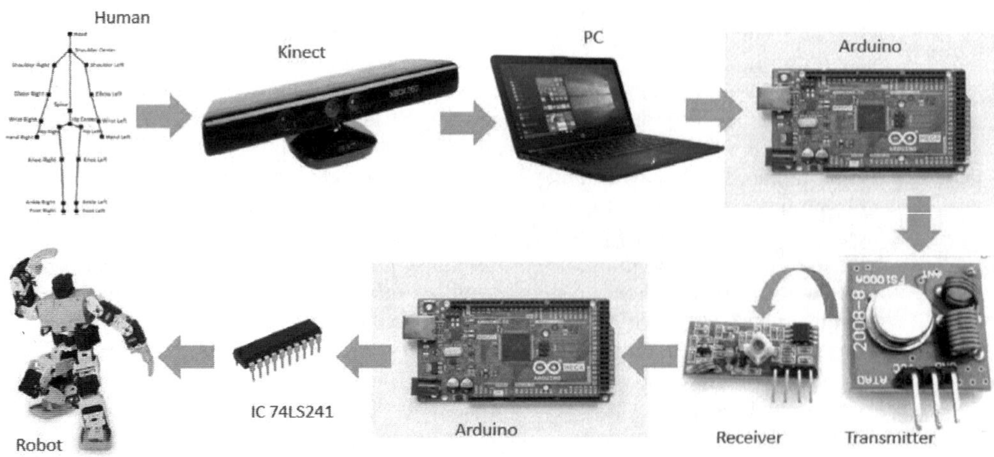

Figure 1. Overall design of robot learning system

The Kinect sensor is used for detecting the coordinates of the joint on the human body, i.e. the upper limb that constitutes to arm joints of the robot. The whole set of data is then processed by the PC for recognition phase and transformed into a sequence of joint data that associated with human arm motions. This signal is then forwarded to Arduino Mega 2560 as the local processing unit. The resulting data is sent via wireless to the robot a 433 MHz RF communication module. After the data is received, another Arduino Mega 2560 (local processor) will process the incoming data and convert the data into joint parameters, which will actuate the joint on the robot. Actuation of robot joints will produce the robot motions. The local processing at the robot side, requires an IC 74LS241, which serves as an interface that functions to sort the data whether it is addressed as Transmitter or Receiver. The scenario of the experiment that will be conducted is the specification of the working distance from the sensor system, the minimum and maximum coordinates of 3 dimensions (x, y, z) of the left hand joint, right hand joint, and head joint, the 3 dimensional coordinates (x, y, z) of human joints when doing movement, time delay between robot movements and human movements, and the degree of similarity between robot movements and human movements. The object being tested is a human operator with a height of 175 cm standing in front of the Kinect sensor where the sensor is placed at a height of 90 cm from the floor.

IOP Conf. Series: Materials Science and Engineering **676** (2019) 012002 doi:10.1088/1757-899X/676/1/012002

3.1. Testing of sensor working distance

This experiment aims to determine the minimum and maximum working distance from the Kinect sensor to be able to read the joints of the human body. To be able to measure the distance, two types of measurement methods are used. The first is using the Kinect sensor to read the z coordinates of the operator and the second is using the meter to measure the actual distance (real) between the Kinect sensor and humans. The working distance is denoted by d measured in meters and will be compared with the Kinect working distance based on theory.

The test results are shown in Table 1.

Table 1. Sensor working distance specifications

d (real)	d (Kinect)	Joint Status	Error
0	0	not detected	0
0.25	0.3	not detected	0.05
0.5	0.54	not detected	0.04
0.75	0.79	not detected	0.04
1	1.04	detected	0.04
1.25	1.29	detected	0.04
1.5	1.54	detected	0.04
1.75	1.8	detected	0.05
2	2.04	detected	0.04
2.25	2.29	detected	0.04
2.5	2.58	detected	0.08
2.75	2.8	detected	0.05
3	3.03	detected	0.03
3.25	3.29	detected	0.04
3.5	3.54	detected	0.04
3.75	3.78	detected	0.03
4	4.04	not detected	0.04

From Table 1, the Kinect sensor is able to recognize the joint of the human body only when the operator stands between a distance of 1 meter to 3.75 meters in front of the Kinect sensor based on the actual distance calculation (real). As for the calculation using Kinect sensors, joint operators can only be identified at a minimum distance of 1.04 meters to a maximum of 3.78 meters. Overall the error reading of distance in average is about 0.04 meters.

Table 2 below shows the comparison of the sensor Kinect working distance between the theory [6] and the test results.

Table 2. Comparison of working distance between theory and test result

d min theory (meters)	d min real (meters)	Error (meters)	d max theory (meters)	d min real (meters)	Error (meters)
0.8	1	0.2	4	3.75	0.25

Figure 2 depicts the comparison of joint detection range by Kinect sensor based on the theory and test result. Based on the graph, the minimum distance required by the Kinect sensor to detect the joint operator is greater than the minimum working distance by theory. Experiment show the maximum distance of Kinect sensor work only reaches 3.75 meters which does not reach 4 meters as stated in the theory.

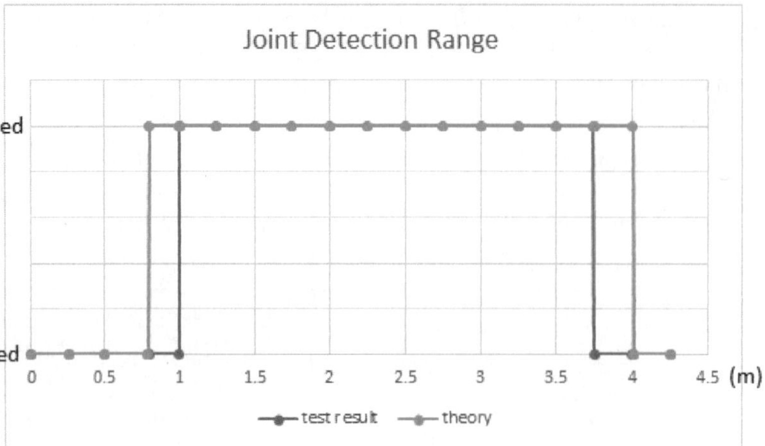

Figure 2. Joint detection range

3.2. Testing minimum and maximum 3-dimensional coordinate system of joint detection

This test aims to determine the minimum and maximum coordinates of the head joint, right hand joint, and left-hand joint of the human body that can be read by the Kinect sensor so that the operator can position themselves in front of the Kinect sensor. To be able to find out the coordinates, the measurement is done by two methods. The first is using measurements from the Kinect sensor and the second is using a meter. The Figure below shows an illustration of a 3-dimensional coordinate system (x, y, z) with detection by Kinect.

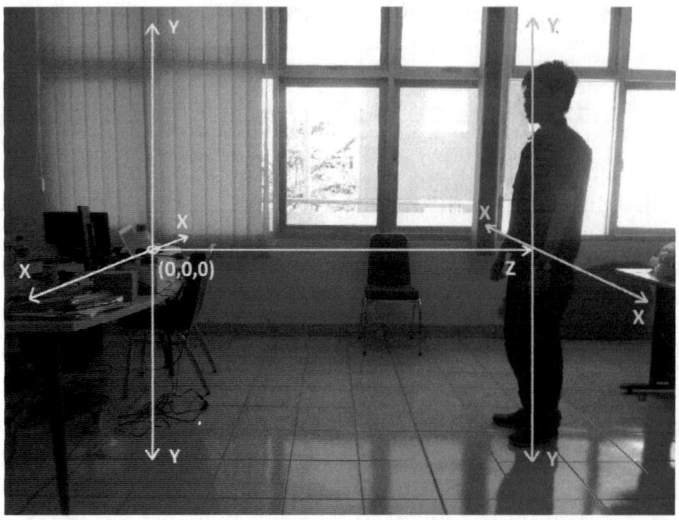

Figure 3. Illustration of 3-dimensional coordinate system

Table 4 shows that Kinect sensor can detect the right-hand joint with a maximum distance of up to 1.3 meters from the Kinect sensor (x coordinate) and left hand up to 1.32 meters from the Kinect sensor (x coordinate). The biggest error is 0.11 meters or in percentage is 9.1%.

Table 3. Minimum and maximum 3-dimensional coordinate system of joint detection (meters)

Joint	x max Kinect (meters)	x max real (meters)	Error (meters)	y max Kinect (meters)	y max real (meters)	Error (meters)
Head	0.84	0.8	0.04	0.81	0.85	0.04
Right Hand	1.3	1.36	0.06	1.1	1.2	0.1
Left Hand	1.32	1.36	0.04	1.09	1.2	0.11
Joint	z min Kinect (meters)	z min real (meters)	Error	z max Kinect (meters)	z max real (meters)	Error (meters)
Head	1.04	1	0.04	3.8	3.7	0.1
Right Hand	1.01	1	0.01	3.78	3.7	0.08
Left Hand	0.99	1	0.01	3.75	3.7	0.05

3.3. Testing coordinates of joint when doing the movement

This test aims to determine the coordinates of the head joint, right-hand joint, and left-hand joint of the human body when doing the movement. The movement includes straighten hands forward, raise hands up, lowering hands down, and stretching hands. Coordinate testing is done by Kinect sensor calculation where the results are shown in the Table 4 below.

Table 4. 3-Dimensional coordinates of joint when doing the movement

Movement	Head (meters)			Right-Hand (meters)			Left-Hand (meters)		
	x	y	z	x	y	z	x	y	z
Lowering hands down	-0.11	0.4	2.06	0.07	-0.42	1.89	-0.29	-0.41	1.91
Stretching hands	-0.32	0.4	2.14	0.4	0.32	2.1	-1.04	0.23	2.05
Raise hands up	-0.3	0.41	2.13	-0.13	0.79	2.05	-0.42	0.78	2.01
Straighten hand forward	-0.34	0.36	2.19	-0.18	0.32	1.6	-0.43	0.35	1.61

From Table 4, it can be concluded:

- When lowering hands down, the y coordinates of the right-hand joint and the y coordinates of the left-hand joint are smaller than zero
- When stretching hands, the x coordinates of righthand joints are greater than the x coordinates of head plus 0.45 meters and the x coordinates of the left-hand joint x are smaller than the x coordinates of head minus 0.45 meters
- When raising hands up, both y coordinates of right-hand and left-hand joint are bigger than the y coordinates of head
- When stretching hand forward, both z coordinates of right-hand and left-hand are bigger than the coordinates of head plus 0.45 meters

3.4. Testing time delay between human and robot movement

This test aims to find out how long it takes the robot to respond to the movements of the operator. From the Table 5 below, the smallest delay required by the robot in order to respond to the movement of the operator is 2 seconds and the biggest delay time is 4 seconds.

Table 5. Time delay between human and robot movement

Human movement	Robot movement	Time delay for experiment				
		1	2	3	4	5
Lowering hands down	Lowering hands down	2	3	2	3	2
Stretching hands	Stretching hands	2	3	3	3	2
Raising hands up	Raising hands up	3	3	2	2	2
Stretching hands forward	Stretching hands forward	2	3	4	2	4

Figure 4 below shows the average time delay between the motion of the robot and the movement of the operator for each movement.

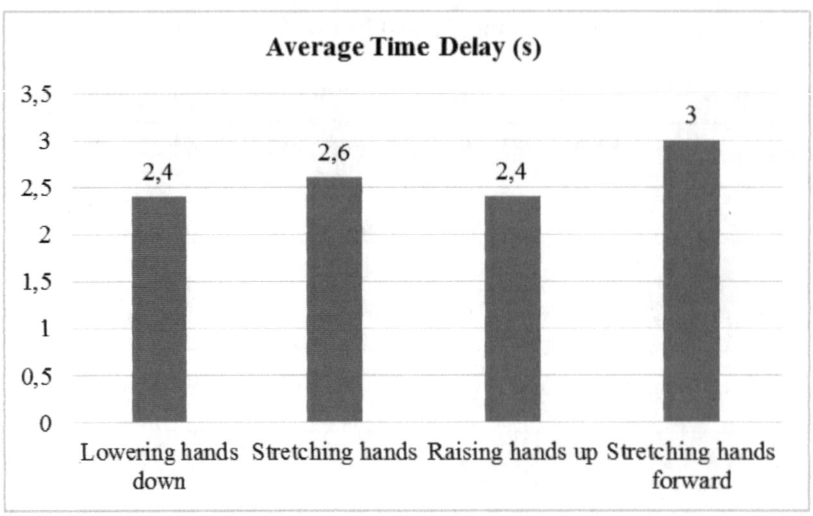

Figure 4. Average time delay for each movement

3.5. Testing the similarity between robot and human movement

The testing process is conducted by giving the opportunity to several respondents to test the robot that has been made. After testing, the questionnaire will be distributed to respondents to find out their opinions about how much the percentage of similarity of the movements produced by the robot to the movements they do.

Table 6. Percentage of similarity between human and robot movement

Respondent	Lowering hands down (%)	Stretching hands (%)	Raising hands up (%)	Stretching hands forward (%)
1	90	90	90	90
2	99	99	99	99
3	98	98	98	98
4	95	95	95	95
5	100	100	100	90
6	80	80	80	80
7	90	100	100	100
8	10	100	100	100
9	95	95	95	95
10	70	90	100	100
11	90	100	100	100
12	95	90	95	95
13	90	100	100	100
14	85	85	85	85
15	85	85	86	88

The average percentage of similarity between robot and human for each movement is above 90% where overall the average is 93.45%

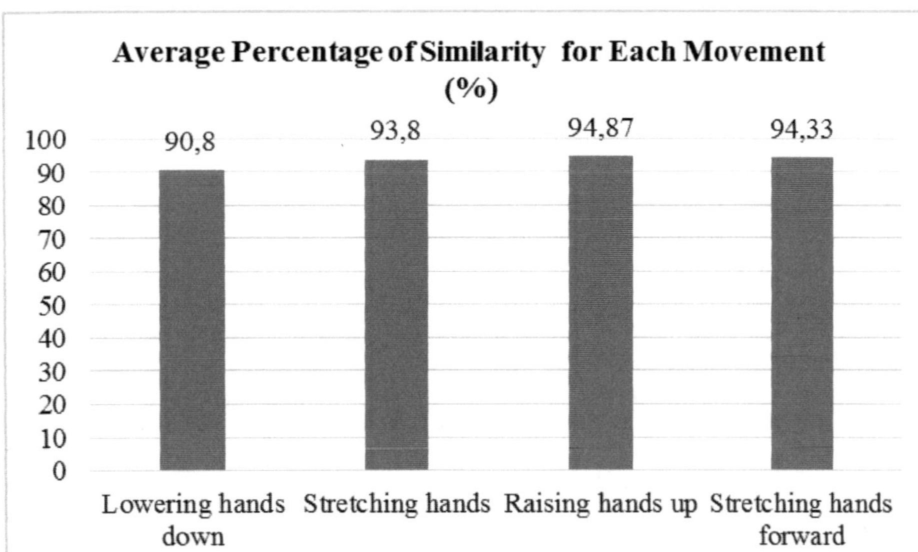

Figure 5. Average percentage of similarity between human and robot movement

4. Conclusions

The test results show that the Kinect sensor can be used to measure the distance and recognize the joints of the human body along with the movements produced within a distance of 1 meter to 3.75 meters with reading error in average is about 0.04 meters. While on the x-axis, the detection range of joint by Kinect sensor is 1.3 meters to the right and 1.32 meters to the left. The smallest delay required by a robot to respond to the movement of the operator is 2 seconds and the biggest delay is 4 seconds where the average delay time for the whole movement is 2.6 seconds. The average percentage of similarity between robot and human movements for the whole movement is 93.45%. Due to the high level of similarity in movement, it can be concluded that the method of see and follow successfully increases the ability of learning robots.

Acknowledgments

Our appreciation to all members of the Indonesia-Australia Cognitive and Social Robotics Research Lab for participation and support throughout the project. Authors also would like to thanks the Faculty of Engineering UNHAS for providing partial funds for the project under the LBE project funding scheme.

References

[1] Mitchell T 1997 *Machine Learning* (Singapore: McGraw-Hills International Editions).
[2] Anshar M 2003 *Directed Learning-based Evolutionary Approach for Legged Robot Motion* (Australia: Lambert Academic Publishing).
[3] Uskarci A, Alatan A, Dindaroglue S 2003 *Robot Mimicking: A Visual Approach for Human Machine Interaction Proc., Symp., Int., Conf., 2nd*.
[4] Koenemann J, Burget F, Bennewitz M 2014 *Real-time imitation of human whole-body motions by humanoids,* pp. 2806-2812.
[5] Rajesh K, Deepansh M, Nitin A, Nihil S 2015 *Implementation of Gesture Control in Robotic Arm using Kinect Module (Applied Mechanics and Materials* vol 786 p 378-382.
[6] Abhijit J 2012 *Kinect for Windows SDK Programming Guide* (UK: Packt Publishing).

Hardware and User Perspective Assessment on Application of Smart Door Access

Muh Anshar[1]* and Nasri Anas[1]

[1]Research Lab Cognitive Social Robotics and Advanced Artificial Intelligence Research Centre, Electrical Department, Faculty of Engineering, Hasanuddin University

*Email: muh.anshar@gmail.com

Abstract. Nowadays, applications of technology in various aspects of human life have been widely adopted. This increasing application relies on the degree of technology involvement, which supports an interconnected communication among resources available in the location. Smart room is a room equipped with technology where electronic devices contained in it can connect and communicate with each other so that it becomes a system that can be controlled thoroughly. One part of the smart room is the door. The door can be accessed using a variety of technologies. some of them are RFID tag technology, PIN code, Fingerprint technology, and Infrared sensor technology. This study discusses the satisfaction of smart room users about the time, distance and technology used. System 1 uses RFID tag technology, and Infrared sensor, and on system 2 uses RFID tag technology, PIN code, Fingerprint technology, and Infrared sensor technology. The time used is 10 seconds on system 1 and 3-6 seconds on system 2 is considered sufficient. The distance of around 3.5 -4.2 is also considered quite good and is also considered to be close. The use of RFID tag technology, PIN code, Fingerprint technology for incoming access and Infrared sensor technology for outgoing access is considered appropriate. Respondent satisfaction in system 1 was around 79% and in system 2 around 74%. System 1 is the most desirable technology, then system 2, then the door with conventional locking.

1. Introduction

Nowadays, applications of technology in various aspects of human life have been widely adopted. This increasing application relies on the degree of technology involvement, which supports an interconnected communication mechanism among available resources in the location. A smart room, for example, is a room that equipped with this interconnected mechanism, which allows communication among each other [1]. Hence, it becomes a unified system that can be controlled systematically and efficiently. One common resource to be implemented is door access. The door is an important part of a room or building. The door function is to connect one room to the other rooms and also to connect the outside of a building with the inside.

Applications of smart door, in larger scale, constitutes to smart room terminology and this is achieved typically by utilising electronic door locks. The door lock works by using electricity and electrical components to control the lock activation. This control can be done in various ways such as using RFID technology, fingerprint, PIN code, and so on. The technology has advantages and disadvantages of each, but whether the shortcomings and advantages of these technologies can be well received by users. This study will examine user satisfaction with the use of RFID technology, PIN codes, infrared sensors and fingerprint on electrical doors.

IOP Conf. Series: Materials Science and Engineering **676** (2019) 012003 doi:10.1088/1757-899X/676/1/012003

2. Related Studies

Electronic locking technology has been widely used and one of them is the one used by Anshar *et al.* [1] on smart room door access. Which on the smart room door uses RFID tag technology, Fingerprint technology, PIN code, and Infrared sensor technology. RFID tag technology is also used by Farooq [2] as a user identifier for access in and out of doors. Verma [3] also applies RFID tag technology to digital doors from a special room so that only certain RFID tags can access the room, namely RFID tags registered on the system. Padmajothi [4] uses fingerprint to access car vehicles, he judges that the use of fingerprints is safer because it cannot be used by others as well as conventional keys and PIN codes. Fair [5] also prefers to use fingerprint compared to RFID technology and password because according to him fingerprint is more efficient and reliable.

3. Method

This study will examine the use of satisfaction in two different systems. One system will use Arduino Mega 2560, magnetic door lock, UPS 12V / 5A, RDM6300 along with antennas as RFID Reader tags, relays, and Infrared sensor modules. The other system uses ZKTeco X6 with fingerprint, RFID Reader tag and PIN code, Infrared sensor module, relay, magnetic door lock, UPS 12V / 5A.

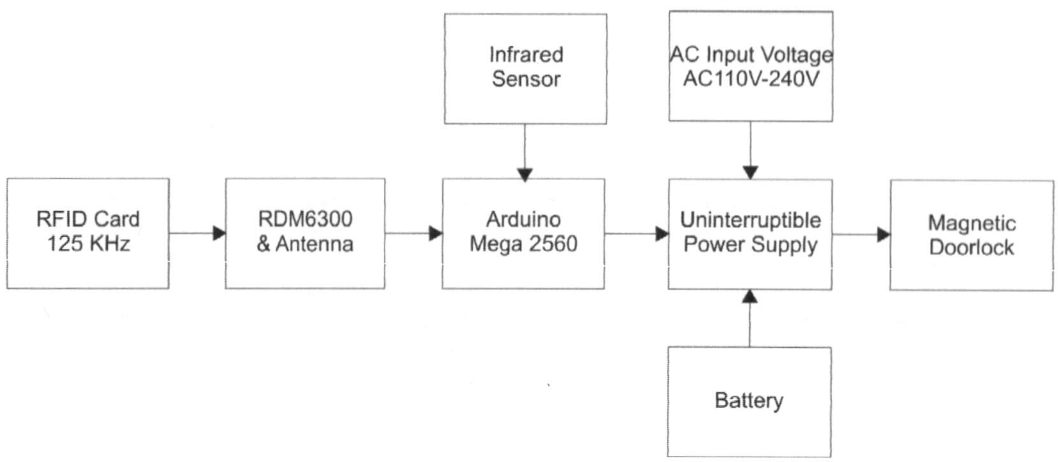

Figure 1. Block diagram of system 1

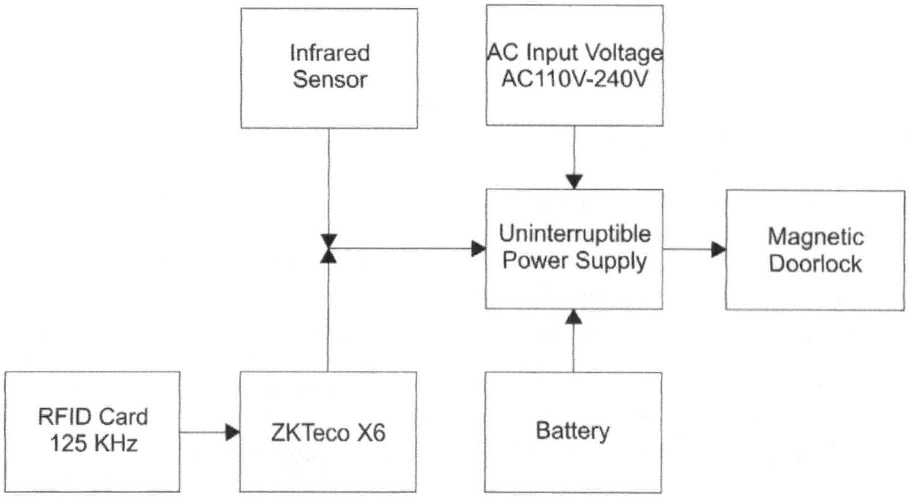

Figure 2. Block diagram of system 2

System 1 uses Arduino Mega 2560 as a controller. Each RFID card has its own unique code, the unique code is read on RDM6300 using an antenna on the outside of the door, then the data is read by Arduino as a controller. Arduino then determines whether to give access to the card owner or not, based on the suitability of the unique code registered with Arduino that was registered during programming. If the code reads on RDM6300 and is registered with Arduino accordingly, Arduino will send the control signal to the relay and forward it to the magnetic door lock via UPS.

To access the door, the user uses an infrared sensor installed in the room. Only by covering the sensor so that the infrared beam sent by the transmitter LED is received on the receiving LED, then the door can be opened. The process is almost the same as access. The difference lies in the suitability of the data. Exit access does not require data compatibility so that anyone can access the door to exit. Unlike the entry access, only certain people can use it.

System 2 uses ZKTeco X6 on access to the door. ZKTeco has its own controller so that reading RFID cards, fingerprints, PIN codes, and infrared sensors does not need to use Arduino. After reading in accordance with those registered with ZKTeco X6 then send a control signal to the UPS to deliver electricity or not to magnetic door lock. Likewise, with the infrared sensor installed next to the room. The control signal is then channeled to the UPS with the help of relay components.

To obtain the results of satisfaction from users, a survey was conducted in the form of filling out a questionnaire. The questionnaire consists of several things including satisfaction regarding the operating time on the system. Another parameter is about operating distance in the system. And also, satisfaction with the system as a whole. Questionnaires were also asked about satisfaction with the door access method used, namely the RFID tag technology, PIN code, fingerprint, and Infrared sensors.

4. Result

The number of respondents in this study is 28 people, all of them are students with 25 men and 3 women with an age range of 18-22 years. The questionnaire was distributed on 26-27 September 2018. Most respondents had the experience of using an electronic key on the door, which is about 19 out of 28 respondents as shown in the circle diagram in Figure 3. 10 of them had used it at the hotel, 7 had used it on campus, and 2 have used it elsewhere. 9 out of 28 respondents claimed they had never used it.

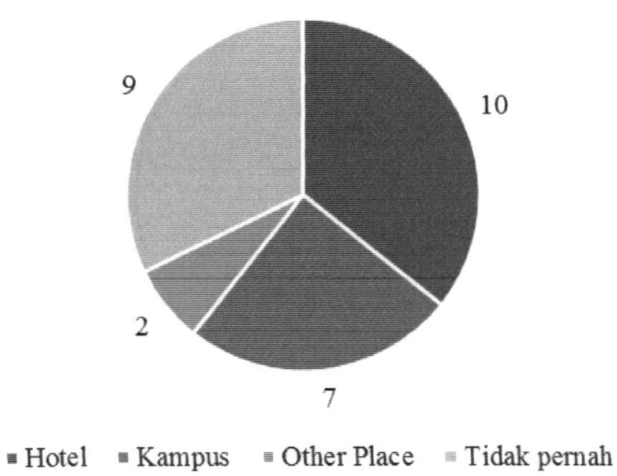

Figure 3. Circle diagram of the experience of using electronic lock

4.1. Satisfaction with operating time

The operating time on access in and out of system 1 is for 10 seconds the door is open, while for the operating time on the access to system 2 is 1.5 seconds the card reads and opens for 4.5 seconds. And the operating time on the outgoing access on system 2 is for 3 seconds. Most respondents considered the time provided by the two systems was enough as shown in the Table 1. 20 of the 28 respondents considered the time on system 1 enough, while 5 respondents considered it a bit long and 2 respondents rated it rather quickly and 14 of 28 respondents considered the time on system 2 enough. While 8 respondents considered it a bit long and 5 of them considered it rather fast. 1 person on system 1 and 2 chose not to answer. From these results, the time used in systems 1 and 2 is good enough and according to needs.

Table 1. Respondent's assessment of usage time

Type of System	Assessment (respondent)					
	too slow	rather slow	enough	rather quickly	too fast	no answer
System 1	0	5	20	2	0	1
System 2	0	8	14	5	0	1

4.2. Satisfaction with operating distance

The operating distance of the access exit is in system 1 which is around 4.2 cm, while the operating distance of access is on system 1 which is about 3.5 cm. And the operating distance of access out on system 2 is about 3.8 cm. while the operating distance of access in system 2 is about 3.6 cm. the questionnaire in this section is an open questionnaire. 12 respondents considered the distance that was quite good on system 1 and on the system 2. 4 respondents considered the distance close enough to system 1 and system 2 and 1 respondent rated the distance very close. While other respondents chose not to answer. From these results, the distance used in both systems is good enough and appropriate. Respondents who answered others considered the distance to be close.

4.3. Satisfaction with the technology used

Respondent's assessment of the use of RFID tag technology, PIN code, fingerprint technology, and Infrared sensor technology can be seen in Table 2. Based on the results, it can be seen that most respondents considered the technology used was appropriate.

Table 2. Respondent's assessment of the technology used

Technology used	Assessment (respondent)					
	very appropriate	appropriate	do not know	not appropriate	very inappropriate	no answer
RFID tag	7	15	5	1	0	0
Infrared	8	14	3	2	1	0
Fingerprint	4	14	2	7	1	0
PIN code	5	16	3	3	0	1

4.4. Satisfaction with the overall system

Respondent satisfaction with system 1 was about 79% and 74% in system 2. It can be seen in Figure 4 and Figure 5. So that it was assessed that respondents were quite satisfied with system 1 which used RFID tag technology and Infrared sensor technology. Respondents were also considered quite satisfied with system 2 which used RFID tag technology, PIN code, Fingerprint, and Infrared sensor technology. And Figure 6 describes which system is preferred by respondents to their homes. Respondents chose system 1, which was 11 respondents, then system 2 with 9 and then 7 respondents chose to keep using conventional keys. 1 respondent chose to use another system.

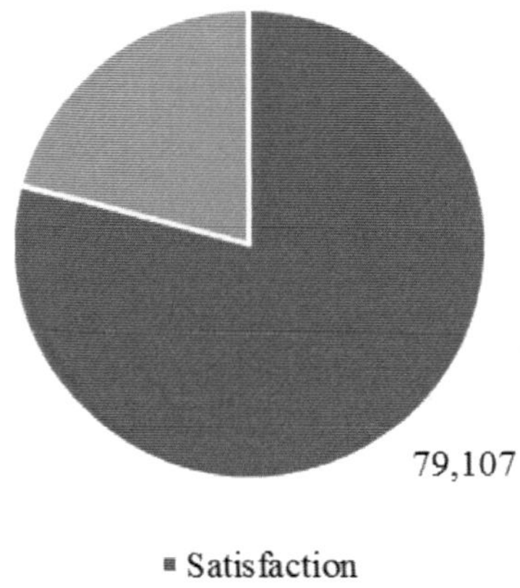

Figure 4. Respondent satisfaction with system 1

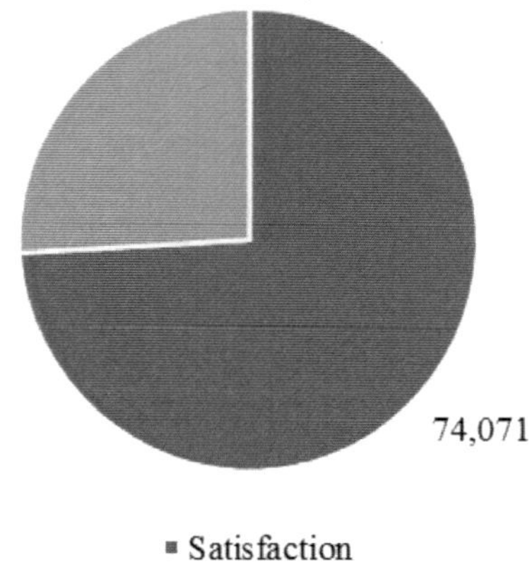

Figure 5. Respondent satisfaction with system 2

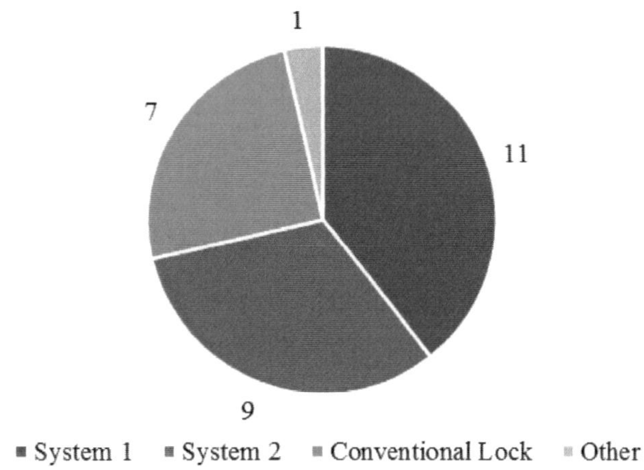

Figure 6. The type of system chosen for use in each home

5. Conclusions

Most respondents or about 20 of the 28 respondents considered the time used in system 1 was 10 seconds of sufficient operating time and 14 of 28 respondents rated the time in system 2 as 6 seconds for access and 3 seconds for outgoing access was considered sufficient, so that the time was assessed used in system 1 and system 2 are appropriate. Also, most of the respondents who answered questions about distance considered that the distance used was quite good at around 3.5 - 4.2 cm, although some of the other respondents tended to judge the distance to be rather close. The technology used was considered appropriate according to the majority of respondents, namely RFID tag technology, PIN code, and Fingerprint technology for access, as well as the use of Infrared sensor technology for outgoing access. Respondent's satisfaction with system 1 was around 79% while in system 2 it was around 74%. With system 1 that uses RFID tag technology for incoming access and Infrared sensors for access out is the most desirable system, after which is system 2 with the addition of fingerprint technology and PIN code.

Acknowledgments

A great appreciation to all members of the Cognitive Social Robotics and Advanced Artificial Intelligence Research Centre for participation and support throughout the project. Authors also would like to thanks the Faculty of Engineering UNHAS for the providing funds and support for the project under the LBE project funding scheme.

IOP Conf. Series: Materials Science and Engineering **676** (2019) 012003 doi:10.1088/1757-899X/676/1/012003

References

[1] Anshar M, Ejah A U, Zaenab, Rachmaniar I, Emir A, Alwi, Nurindasari, Dicky H, Anas N, Kurniawan A & Akkas M 2017 Smart Room Design: A Pilot Project *Proc. The 1st EPI Int. Conf. on Science and Engineering* no 4 pp 19-26.

[2] Farooq U, Hasan M, Amar M, Hanif A, and Usman M A 2014 RFID Based Security and Access Control System *IACSIT Int. Journal of Engineering and Technology* vol 6 no 4 pp 309-314

[3] Verma G K & Tripathi P 2010 A Digital Security System with Door Lock System Using RFID Technology *Int. Journal of Computer Applications* vol 5 no 11 pp 6-8.

[4] Padmajothi V, Harish R, Vishwanath M & Reddy T B 2017 Biometric Car Door Security System Technology *Int. Journal of Pure and Applied Mathematics* vol 115 no 6 pp 365-368.

[5] Adil S M, Umair A, Shaikh U, Kamble T & Sagvekar V 2018 Fingerprint Based Door Access *System Int. Journal of Engineering Research in Computer Science and Engineering (IJERCSE)* vol 5 no 3 pp 126-127.

Influence of Tropical Environment on Electrical Properties of Electrical Insulation Materials

S Manjang[1], I Kitta[1]*, Gassing[1], I R Sahali[1] and F Maricar[1]

[1]Department of Electrical Engineering, Hasanuddin University, Makassar, Indonesia

*E-mail: ikhlaskitta@gmail.com

Abstract. The development of optimal electrical insulation materials in the tropics continues. One of the development models is to look for fillers that allow it to be mixed with insulating materials (polymers, ceramics, and glass). This study aims to obtain the properties of electrical insulation materials in the form of polymeric materials (silicon rubber) which are filled with fly ash (coal-fired power plant waste) so that it can be used as a high-voltage electrical insulator. The activity that has been carried out was testing the performance of electrical insulating materials filled with fly ash under conditions under high electric fields and tropical environments. The research was conducted at the Electrical Engineering Building, Hasanuddin University, Makassar City, Indonesia. The performance evaluation of electrical insulation materials (silicon rubber plus fly ash) refers to the results of measurements of water diffusion characteristics and relative permittivity characteristics. The results of the study were in the form of information about the feasibility of coal-fired power plants to be used as fillers for electrical insulators in tropical environments. The results of this study are Fly ash added to silicone rubber to improve the electrical properties of the insulating material.

1. Introduction

Isolator is a material that is used with the aim to be able to separate the parts that are voltage with parts that are not voltage. Electrical insulators are widely used in almost all electrical equipment ranging from generators, transformers, electrical power distribution systems and even electronic devices. Generally, insulators that are often used in electricity networks in Indonesia are glass and porcelain ceramic insulators. This material has a disadvantage if it operates in humid conditions, because it has water absorbing properties, namely the contact angle of water is at a level less than 30°, this condition will get worse if it is in the field of work that is contaminated by natural conditions open as a seafront or industrial area, which results in a small amount of contact with the surface material and water droplets.

The phenomenon of the performance limitations of the insulating material referred to above, is very inviting the guts of high voltage equipment experts, so in the last few decades researchers have begun to develop types of insulators from polymer materials, such as Silicone Rubber (SIR). Silicone rubber material has the advantage of being able to resist water [1], besides that it can transform the hydrophobic nature of the material which is being affected by pollution due to the open natural conditions that it passes [2]. The ability of an insulator to function properly, depends greatly on the state of the surrounding environment (air pressure, humidity, temperature, rain and pollution) [3, 4].

Despite various advantages, until now the use of silicone rubber insulators in some countries is still limited, such as Indonesia because of the high production costs of silicone rubber insulators, so that many researchers have optimized materials that can be mixed with silicone rubber, especially in the case of filler material concentrated on silica (SiO_2) and alumina (Al_2O_3) [5].

Currently looking for a source of filler is not easy, so technology is needed to get the filler. So the research continues on the search for possible filer sources. One source of filler material that can be utilized as an alternative filler in silicone rubber insulators is coal fly ash, where coal fly ash contains chemical elements such as silica (SiO_2) and alumina (Al_2O_3).

Indonesia is a tropical region. The use of silicone rubber as an insulating material must be resistant to exposure to various factors of high-intensity tropical climate such as UV (ultra violet) radiation from the sun around 12 hours during the day, air temperature between 16-35 °C, relative humidity close to 100% between early morning until morning and high annual rainfall between 40-500 mm. These factors will simultaneously hit the silicone rubber insulator that is installed outdoors. So that this paper contains the results of research in the form of experimental studies of silicone rubber performance filled with fly ash conditioned in the tropics.

2. Research Method

2.1. Types of research

This paper is based on the results of experimental studies to test and analyze the feasibility of fly ash as a filler of silicone rubber insulating materials, especially high voltage electrical insulators. The location of the study was carried out at the Electrical Engineering Department of the University of Hasanuddin High Voltage Engineering.

2.2. Raw material and test material making proses

The materials used in this research are coal fly ash, Silicon rubber RTV 683. Fly ash or coal fly ash is used as filler for silicone rubber insulation material. Fly ash is meant here is the residue from burning coal in a steam power plant (coal-fired power plant waste) in the form of smoke or dust that comes out through a gas exhaust pipe and deposited by electrostatic precipitators. The coal fly ash used in this study was taken directly from the silo of Tonasa coal-fired power plant. The content of silica (SiO_2) and alumina (Al_2O_3) was 59.64%. Silicone rubber used in this study is polydimethylsiloxane with type RTV (Room Temperature vulcanization). Silicone rubber branded silicone rubber RTV 683.

The test material is made in two types, namely silicon rubber test material without filler material called MATERIAL-0, and a test material in the form of a mixture of silicone rubber with fly ash called MATERIAL-30, where this composition fly ash 30%. The process of making material test material that will be tested consists of 4 stages, namely: mixing, air bubble repeating, printing of test materials, removal of water content in the test material.

2.3. Outdoor testing

The results of the manufacture of test materials will then be tested for performance under exposure to the natural tropical climate. Analyze and characterize its properties using various measurement methods as described in the previous section. Parameters are measured every day after being burdened with a natural tropical climate, which are sample weight and relative permittivity. This percentage of water diffusion measurement is intended to see the state of water absorption in the material, where the change in weight as a function of aging is an indicator of the amount of water absorbed by the test material based on the percentage of water absorption. While the measurement of dielectric constants is intended to give an overview of the behavior of the silicone rubber dielectric properties to the environmental conditions [6].

3. Result and Discussion

3.1. Meteorological data

Tropical climate data in the form of relative humidity, air temperature and UV radiation on the Hasanuddin University campus and its surroundings are presented in Figures 1 and 2. From the beginning to the 40th day there was a fluctuation in daily average humidity from 60.2% to 93.0 % as shown in Figure 1. Variations in daily average air temperature changes are shown in Figure 1. Minimum daily average air temperature is 24.8 °C and maximum is 32.7 °C. This change in temperature is caused as a result of the initial change of the rainy season to the dry season where the occurrence of a torrential rain is quite heavy.

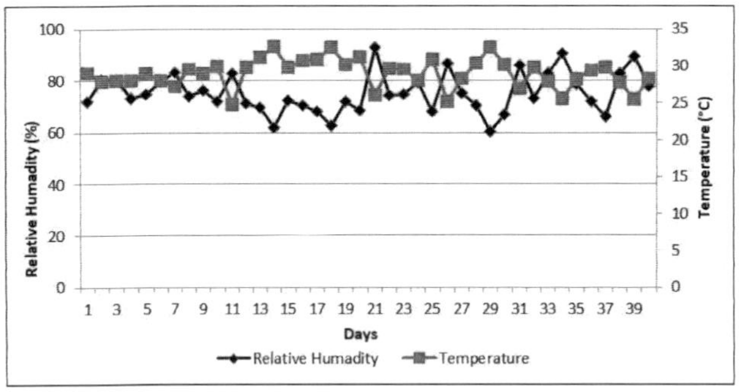

Figure 1. The relative humidity and The air temperature of research location

In Figure 2 shows the intensity fluorescence curve of UV radiation from the sun in Makassar. From the first day of the study to the 40th day of the study, the recorded average daily UV radiation intensity fluctuated from 0.02 watt/m^2 to 5.9 watt/m^2. This condition is the beginning of the dry season, where the density of clouds is lower than the rainy season.

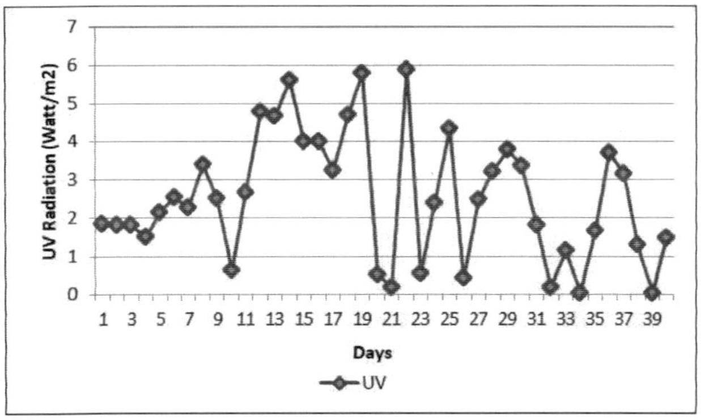

Figure 2. The UV radiation of research location

3.2. Water diffusion measurement result in test material

Figure 3 shows the change pattern curve for the percentage value of water diffusion of Material-0 and Material-30 test materials. The first day of aging, the percentage value of MATERIAL-0 water diffusion of 0.75% and MATERIAL-30 of 0.36% showed an increase in weight due to water uptake in the test material. On the first day of aging, the weight increase of the MATERIAL-0 test material was slightly higher than that of the MATERIAL-30 test material. This increase is due to absorption of water by the test material from the humidity of the surrounding air and rain. The process of diffusion of water into the MATERIAL-0 test material and MATERIAL-30 until close to saturation in the open air only lasts less than one day. The speed of diffusion of water into silicone rubber until saturation matches the results of the research by soaking the silicone rubber test material into 50 °C water [6] until the increase reaches saturation weight which only takes 2 days. This illustrates that the water molecule does not get much free space in the test material.

On the second day of aging, the weight of the MATERIAL-0 test material decreased slowly until the 40th day, where the water diffusion percentage value of 0.75% decreased to -0.47% (the test material became lighter). This weight reduction when associated with climate factor curves (Figures 1 and 2) shows that from the first day to the 40th day the climate shifts to the dry season where the average temperature in the environment is around an average of 28.9 °C. This gives an indication that the water absorbed by the MATERIAL-30 might evaporate or deabsorb during this time. The process of absorption and deabsorption of water in polymeric materials has been proposed by several researchers as a physical process. The MATERIAL-0 weight change pattern curve does not appear to follow the climate curve pattern (humidity and temperature) that occurs. Weight loss in MATERIAL-0 was affected by exposure to UV radiation from the sun.

On the contrary the MATERIAL-30 test material undergoes heavy changes following the pattern of daily climate change, where the change in weight starts from the first day until the 40th day is almost constant. On the first day the percentage of water diffusion value of MATERIAL-30 was 0.36% and on the 40th day the percentage of water diffusion value of MATERIAL-30 was 0.43%. A slight increase in weight occurs in MATERIAL-30, this indicates that the MATERIAL-30 test material is not affected by UV exposure from the sun.

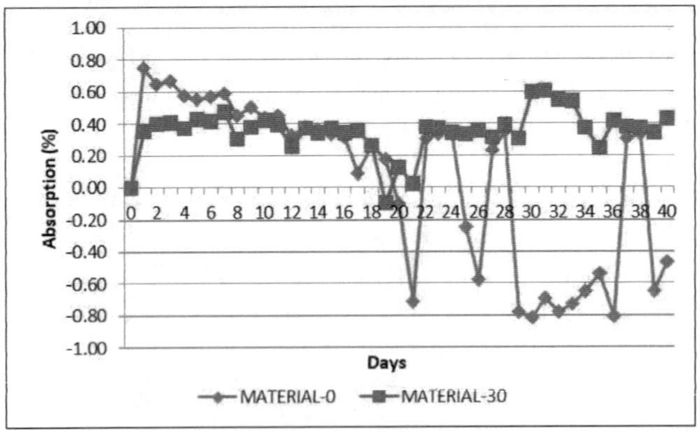

Figure 3. The results of water diffusion measurements for each sample

3.3. Relative permitivity measurement

The relative permittivity values of MATERIAL-0 and MATERIAL-30 test materials are presented in Figure 4. The relative permitivity of MATERIAL-0 and MATERIAL-30 after the first day (day 2) of aging showed a sharp increase (MATERIAL-0 = 1.81 and MATERIAL- 30 = 2.09) where previously in the initial conditions, MATERIAL-0 was 0.37 and MATERIAL-30 was 0.70. After that, in the following days the relative permittivity value decreased until the 10th day ie MATERIAL-0 of 1.06 and MATERIAL-30 of 1.23. Furthermore, on the 11th day the relative permittivity values of both test materials increased, namely MATERIAL-0 of 1.44 and MATERIAL-30 of 1.43. On the 12th day the relative permittivity value of the MATERIAL-0 test material decreased again until the 13th day, where the MATERIAL-0 value was 1.18. The relative permittivity value of MATERIAL is 0 until 40th day (MATERIAL-0 = 1.10). For MATERIAL-30 which on the 10th day has a relative permittivity value of 1.23, it rises to a value of 1.46 on the 13th day. Next down and saturation until the 40th day (MATERIAL-30 = 1.28).

The influence of climate on the relative permittivity of test materials exposed to this natural tropical climate causes the relative permittivity curve to tend to follow the daily climate pattern. This can be seen in the pattern of daily humidity and daily temperature which is almost the same as the daily pattern of relative permittivity of MATERIAL-0 test material and MATERIAL-30. And, the effect of daily UV radiation exposure also dominates because of a decrease in the relative permittivity value at the beginning of the measurement.

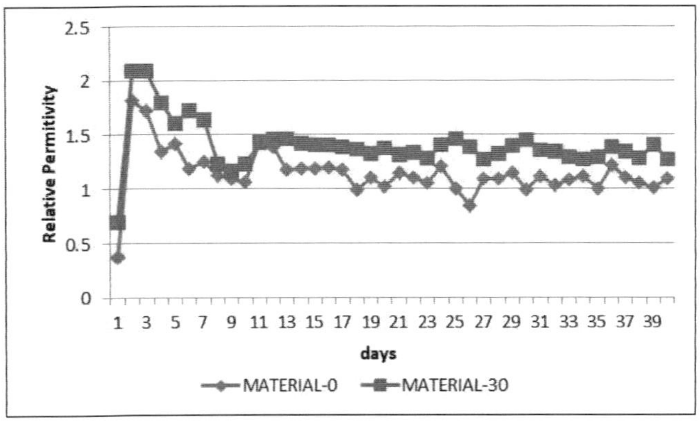

Figure 4. The measurement results of the relative permittivity of each test material

4. Conclusion

The filler added to silicone rubber is to improve the electrical properties of tropical climate environmental stresses. In the natural tropical climate, the impact of filler on silicone rubber for weight changes (percentage of diffusion of water) is to maintain the weight of silicone rubber as shown in MATERIAL-30 which is not like non-filler silicone rubber (MATERIAL-0), where the weight is getting decreased due to exposure to UV radiation from the sun. For the relative permittivity parameter, the role of coal fly ash filler on silicone rubber is to maintain the relative permittivity value, this is evidenced by the formation of the same pattern and the almost close value with non-filler silicone rubber (MATERIAL-0). Changes in the relative permittivity value of the test material are strongly influenced by climate and UV radiation from the sun.

Acknowledgments
The authors gratefully acknowledge Indonesia Government of Ministry of Research and Higher Education for financial support of this research.

References

[1] Hillborg H, Kornmann X, Krivda A, Meier P, Schmidt L E 2007 Hydrophobic Stability of Silicone Rubber after Water Immersion *2007 Annual Report Conference on Electrical Insulation and Dielectric Phenomena* IEEE.

[2] Naqvi S A R, Imdad S K 2013 Temperature and hydrophobicity of silicone rubber *Electrical And Electronics Engineering: An International Journal (ELELIJ)* **2** : 1.

[3] Kim J, Chaudury K, Owen M J 1999 Hydrophobicity Loss and Recovery of Silicone HV Insulation *IEEE Transaction on Dielectrics and Electrocal Insulation* **6** : 5.

[4] Arshad, Nekahi A, McMeekin S G 2018 Measurement of surface resistance of silicone rubber sheets under polluted and dry band conditions *Electrical Engineering* Springer Berlin Heidelberg **100** : 3.

[5] Thong-om S, Payakcho W, Grasasom J, Marungsri B 2011 Study of Ageing Deterioration of Silicone Rubber Housing Material for Outdoor Polymer Insulators *International Journal of Chemical, Molecular, Nuclear, Materials and Metallurgical Engineering* Waset **5** : 8.

[6] Manjang S, Kitta I, Muhammadia K B, Mulya R N I 2015 Effect of water diffusion on dielectric behavior of polymer insulators, *Proceedings - 5th International Conference on Electrical Engineering and Informatics: Bridging the Knowledge between Academic, Industry, and Community*, ICEEI.

IOP Conf. Series: Materials Science and Engineering **676** (2019) 012005 doi:10.1088/1757-899X/676/1/012005

Temperature and Salinity Gradients Analysis for a Solar Pond Prototype

Z Muslimin[1]*, I C Gunadin[1], Muh Anshar[1] and A Siswanto[2]

[1]Department of Electrical Engineering, Engineering Faculty, Hasanuddin University
[2]Electrical Engineering Study Program, 17 Agustus 1945 Cirebon University

*E-mail: zaenab@unhas.ac.id

Abstract. In this study, a solar pond prototype is used to analyze the temperature and salinity gradients of alternative energy source processes. The study used two types of solar pond to investigate the temperature and salinity of gradients from each solar pond. The first solar pond utilizes sunlight as a heat source that emitted directly on the solar pond and the second solar pond utilizes the spotlights as a direct heat source emitted on the solar pond. The first solar pond is made of a cylindrical plastic tank of 1.05 m tall, 0.8 m in diameter placed outdoors in direct sunlight and the second solar pond model is made of a trapezium plastic tank with a height of 0.9m, the upper diameter 0.54m and a bottom diameter of 0.49m placed indoors with floodlights. The gradient temperature salinity and temperature measurement in the fest solar pond are smaller than temperature in the second solar pond however, the T_3 temperature in the outdoor solar pond is larger than the indoor solar pond. The observation results show that the first solar pond generates the mean value of temperature and salinity of successive gradients 32 (°C); 30 % and the second solar pond has the average value of temperature and salinity of successive gradients 31 (°C); 28 %.

1. Introduction

Renewable energy is an important source of global energy because of rising crude oil prices due to the burning of fossil fuels. Any electricity generated from fossil and any other types of non-renewable fuels have the potential of polluting the environment. Besides, with limited fossil fuel sources and an increased demand for power to accommodate the increase in human population, power generation from renewable energy is the most promising solution to maintain sustainable electricity supply for the future [1]. The national energy policy aims to reduce fossil fuels-driven electricity, especially petroleum. Utilization of renewable energy is required for the sustainability of supply and national energy reserves. Indonesia as a maritime country has the longest coastline in the world, that is ± 80.791,42 Km and Indonesia is an archipelagic country as it has more than 5000 islands. This natural state is a great potential for the development of renewable energy, such as geothermal, micro hydro, wind, solar, or energy sourced from the sea that spread wide and very abundant [2]. One source of renewable alternative energy sourced from the sea that is currently undeveloped in Indonesia is solar pond. Indonesia geographically has enormous potential in developing solar ponds because Indonesia has a tropical climate that has considerable high solar intensity [3] and, as a maritime country which surrounded by sea, the availability of raw materials of salt water is huge [4]. Regardless of its simple technology, solar pond requires high complexity of equipment to be used [5, 6]. Development of renewable energy in this research used solar pond prototype to observe the temperature and salinity of sea water gradient. The first solar pond model made of cylindrical plastic tank with height 1.05 m, diameter 0.8 m which is placed outdoors directly get the rays solar and the second solar pond model is made of a trapezium-shaped plastic tank with a height of 0.9 m, an upper diameter of 0.54m and a bottom diameter of 0.49m placed indoors gets rays through a spotlight.

2. Literature Review

Solar pond is a salt water pond, which is capable of collecting solar thermal energy, and the resulting heat energy will be converted into electrical energy [7, 8]. This technology is a fundamental technology and very easy to use with enough land and proper design. The solar pond technology is aptly used in areas that have plenty of sunlight and solar ponds that are environmentally friendly and pollution-free [9]. Seawater so it can absorb the sun's heat more optimally. The sea water is accommodated in a pond called the solar pond. The sea water at the bottom of the solar pond that gets sunlight can absorb the heat of the sun naturally which resulting the rising of the temperature[10]. Sea water contains salt which would make the density increased when it is exposed to the sunlight. Hot ocean water will descend to the bottom of the solar pond, while the cold sea water will rise to the surface. This causes a temperature difference between the bottom of the solar pond and the surface of the solar pond, this temperature difference causes a different heat . This heat difference will be used as a source of electrical energy.

The solar pond is divided into three layers as shown in Figure 1 which consists of:

a. Upper Convective Zone; This zone is called the surface layer. This surface layer has the lowest gradient salinity level, and the temperature is close to ambient temperature. This coating is important to prevent the lower layers so that the temperature is not reduced due to the evaporation process and the influence of the wind.

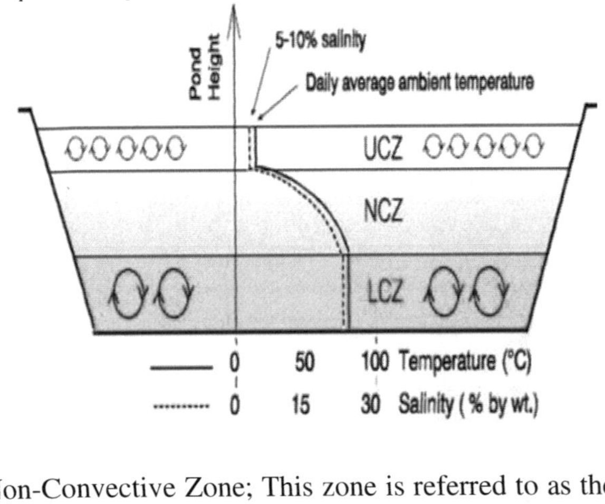

Figure 1. The division of zones in the solar pond is based on the salt content [1]

b. Non-Convective Zone; This zone is referred to as the gradient layer or middle layer. This layer lies between the upper zone and the lower zone of the solar pond, at this layer the temperature and salinity of the gradient increases with the depth of the solar pond. This layer is not homogeneous and if the salinity of the gradient is large enough it will prevent a convective phenomenon.

c. The lower Convective zone; this zone is called a homogeneous layer, this homogeneous layer has a high temperature and salinity to produce high heat.

3. Research Methods

To analyze the temperature and salinity of the gradients of both solar pond models shown in Figure 2 and Figure 3. Gradient salinity data retrieval is conducted on the saltwater surface of both solar pond while the measurement of temperature data is performed at several points from both solar pond.

1. The first solar pond is made of a cylindrical plastic tank with a height of 1.05 m, a diameter of 0.8 m placed outside the room that directly receives sunlight. The thermometer is installed at 4 points:

IOP Conf. Series: Materials Science and Engineering **676** (2019) 012005 doi:10.1088/1757-899X/676/1/012005

T_0 (measuring ambient temperature), T_1 (measuring salt water temperature at a distance of 0.35 m from the surface of the solar pond), T_2 (measuring salt water temperature of 0.7 m from the surface of the solar pond), T_3 (measuring water temperature salt at a distance of 1.05 m from the surface of the solar pond) and gradient salinity measurements are performed on the surface of the solar pond. The measurement results are shown in Table 1.

2. The second solar pond is made of a trapezoidal plastic tank with a height of 0.9 m, an upper diameter of 0.54 m and a bottom diameter of 0.49 m placed in a room with an area of 12 m2. Measurements are made using spotlights as a heat source to obtain the temperature and salinity of the gradients in the solar pond.

Figure 2. The solar pond model is cylindrical shape

The thermometer is installed at 4 points: T_0 (measuring ambient temperature), T_1 (measuring the temperature of brine at a distance of 0.3 m from the surface of the solar pond), T_2 (measuring salt water temperature at a distance of 60cm from the surface of the solar pond), T_3 (measure the temperature of salt water at a distance of 90cm from the surface of the solar pond). The measurement results are shown in Table 2.

Figure 3. The solar pool model is a trapezoidal form

Table 1. Measuring ambient temperature

Time	Date / T_0 (°C)											
	1-8-2017	2-8-2017	3-8-2017	4-8-2017	5-8-2017	6-8-2017	7-8-2017	8-8-2017	9-8-2017	10-8-2017	11-8-2017	12-8-2017
06.00	24	24	24	24	24	25	25	25	26	28	27	25
09.00	28	29	28	28	29	31	30	31	30	30	29	27
12.00	33	31	32	31	30	31	31	33	32	33	31	30
15.00	32	33	32	30	31	34	33	31	31	32	30	29
18.00	29	28	28	29	29	30	31	31	31	31	30	29

Table 2. The measurement of salt water temperature at a distance of 0.35 m from the surface of the solar pond

Time	Date / T_1 (°C)											
	1-8-2017	2-8-2017	3-8-2017	4-8-2017	5-8-2017	6-8-2017	7-8-2017	8-8-2017	9-8-2017	10-8-2017	11-8-2017	12-8-2017
06.00	32	32	29	29	29	29	29	29	29	30	29	28
09.00	29	30	29	29	29	29	30	30	30	30	29	28
12.00	32	32	31	31	32	32	31	31	33	31	31	30
15.00	33	33	33	33	33	33	33	32	33	32	31	30
18.00	32	32	31	32	32	32	32	33	32	31	31	30

Table 3. The measurements of water temperature salt at a distance of 1. 05 m from the surface of the solar pond

Time	Date / T_2 (°C)											
	1-8-2017	2-8-2017	3-8-2017	4-8-2017	5-8-2017	6-8-2017	7-8-2017	8-8-2017	9-8-2017	10-8-2017	11-8-2017	12-8-2017
06.00	31	31	30	30	30	30	30	30	30	31	29	29
09.00	30	31	30	30	30	30	31	31	31	31	30	29
12.00	33	33	32	32	33	33	32	32	33	32	32	30
15.00	34	34	34	34	34	34	33	34	35	34	32	31
18.00	33	33	33	33	33	33	33	32	34	32	31	30

Table 4. The measurements of water temperature salt at a distance of 1.05 m from the surface of the solar pond

Time	Date / T_3 (°C)											
	1-8-2017	2-8-2017	3-8-2017	4-8-2017	5-8-2017	6-8-2017	7-8-2017	8-8-2017	9-8-2017	10-8-2017	11-8-2017	12-8-2017
06.00	30	30	31	31	29	29	30	30	30	30	30	29
09.00	30	31	30	30	31	29	30	32	31	31	30	29
12.00	33	33	32	33	33	32	32	32	33	32	31	30
15.00	34	34	33	34	34	34	34	34	34	33	32	31
18.00	33	32	32	33	33	33	33	33	33	31	30	31

Table 5. The graph of salinity and temperature at some point versus time in the second solar pond

Time	Date / S (‰)											
	1-8-2017	2-8-2017	3-8-2017	4-8-2017	5-8-2017	6-8-2017	7-8-2017	8-8-2017	9-8-2017	10-8-2017	11-8-2017	12-8-2017
06.00	30	30	32	31	32	33	32	32	32	31	30	33
09.00	30	31	31	32	32	32	32	32	32	31	31	33
12.00	30	30	31	31	31	31	31	31	31	32	32	32
15.00	30	30	30	30	30	30	30	30	30	32	32	32
18.00	30	30	30	30	31	31	30	30	30	32	33	32

Table 1 is the first solar cell temperature measurement result. The temperature measurement results show that at 06.00 the average environmental temperature is 25 °C; 09.00 time average environment temperature is 29 ºC; at 12.00 the average environmental temperature is 32ºC; at 15.00 the average temperature of the environment is 32 °C and the time 18.00 the average environmental temperature is 30 °C.

Table 2 is the result of temperature measurement at the first solar pond at a distance of 0.35m from the water surface. The temperature measurement results show that at 06.00 the average temperature is 30 °C; at 09.00 the average temperature is 29 ºC; 12.00 the average temperature is 31 °C; at 15.00 the average temperature is 32 °C and the average temperature 18.00 is 32 °C.

Table 3 is the result of temperature measurement at the first solar pond at a distance of 0.7m from the water surface. The temperature measurement results show that at 06.00 the average temperature is 30 °C; at 09.00 the average temperature is 30 ºC; 12.00 the average temperature is 32 °C; at 15.00 the average temperature is 34 °C and the average temperature 18.00 is 33 °C.

Table 4 is the result of temperature measurement at the first solar pond at a distance of 1.05m from the water surface. The temperature measurement results show that at 06.00 the average temperature is 30 °C; at 09.00 the average temperature is 30 ºC; 12.00 the average temperature is 32 °C; at 15.00 the average temperature is 33 °C and the average temperature 18.00 is 32 °C.

Table 5 is the result of salinity measurements in the first solar pool on the water surface. Salinity results show that at 6:00 the average salinity is 32 (‰); at 09.00 the average salinity is 32 (‰); at 12.00 the average salinity is 31 (‰); at 15.00 the average salinity is 31 (‰) and at 18.00 the average salinity is 31 (‰).

Table 6. The measurements of ambient temperature

Time	Date / T_0 (ºC)											
	1-8-2017	2-8-2017	3-8-2017	4-8-2017	5-8-2017	6-8-2017	7-8-2017	8-8-2017	9-8-2017	10-8-2017	11-8-2017	12-8-2017
06.00	26	26	25	25	26	26	28	30	32	31	28	26
09.00	28	27	29	30	29	29	30	30	30	30	30	28
12.00	30	29	33	33	31	30	32	32	32	32	32	30
15.00	31	30	32	32	29	33	33	33	33	33	31	32
18.00	30	30	30	28	30	30	31	32	31	32	31	29

Table 7. The measurements of the temperature of brine at a distance of 0.3 m from the surface of the solar pond

Time	Date / T_1 (°C)											
	1-8-2017	2-8-2017	3-8-2017	4-8-2017	5-8-2017	6-8-2017	7-8-2017	8-8-2017	9-8-2017	10-8-2017	11-8-2017	12-8-2017
06.00	29	29	31	31	30	32	33	39	32	31	30	32
09.00	29	33	33	33	32	33	34	41	34	40	32	34
12.00	33	33	35	35	34	34	36	42	35	41	33	36
15.00	35	37	36	36	36	35	38	42	37	42	36	37
18.00	36	38	38	37	38	38	39	42	39	43	39	38

Table 8. The measurement of the salt water temperature at a distance of 60cm from the surface of the solar pond

Time	Date / T_2 (°C)											
	1-8-2017	2-8-2017	3-8-2017	4-8-2017	5-8-2017	6-8-2017	7-8-2017	8-8-2017	9-8-2017	10-8-2017	11-8-2017	12-8-2017
06.00	28	28	31	30	30	32	32	36	32	31	29	31
09.00	31	32	32	32	31	33	33	37	33	34	30	31
12.00	32	32	33	33	32	33	34	38	34	35	33	32
15.00	33	34	34	34	33	33	35	38	34	36	33	32
18.00	34	34	35	35	35	34	36	41	36	39	36	33

Table 9. The measurement of the temperature of salt water at a distance of 90cm from the surface of the solar pond

Time	Date / T_3 (°C)											
	1-8-2017	2-8-2017	3-8-2017	4-8-2017	5-8-2017	6-8-2017	7-8-2017	8-8-2017	9-8-2017	10-8-2017	11-8-2017	12-8-2017
06.00	28	28	30	29	29	31	31	34	31	30	29	31
09.00	31	31	31	30	30	32	31	35	32	30	29	31
12.00	30	31	32	30	31	33	32	35	32	32	30	31
15.00	31	33	31	31	31	32	33	35	33	34	32	32
18.00	33	33	32	32	34	33	34	37	34	37	36	32

Table 10. The graph of salinity and temperature at some point versus time in the second solar pond

Time	Date / S (°/oo)											
	1-8-2017	2-8-2017	3-8-2017	4-8-2017	5-8-2017	6-8-2017	7-8-2017	8-8-2017	9-8-2017	10-8-2017	11-8-2017	12-8-2017
06.00	30	30	32	32	33	35	34	30	35	31	30	38
09.00	31	30	31	30	32	32	32	30	34	32	31	37
12.00	32	31	32	32	31	30	31	30	33	32	32	36
15.00	29	29	29	29	30	31	30	30	33	33	35	35
18.00	29	28	29	29	31	28	31	31	31	33	35	33

Table 6 is the first solar cell temperature measurement result. The temperature measurement results show that at 06.00 the average environmental temperature is 25 °C; 09.00 time average environment temperature is 29 °C; at 12.00 the average environmental temperature is 32 °C; at 15.00 the average temperature of the environment is 32 °C and the time 18.00 the average environmental temperature is 30 °C.

Table 7 is the result of temperature measurement at the first solar pond at a distance of 0.35m from the water surface. The temperature measurement results show that at 06.00 the average temperature is 30 °C; at 09.00 the average temperature is 29 °C; 12.00 the average temperature is 31 °C; at 15.00 the average temperature is 32 °C and the average temperature 18.00 is 32 °C.

Table 8 is the result of temperature measurement at the first solar pond at a distance of 0.7m from the water surface. The temperature measurement results show that at 06.00 the average temperature is 30 °C; at 09.00 the average temperature is 30 °C; 12.00 the average temperature is 32 °C; at 15.00 the average temperature is 34 °C and the average temperature 18.00 is 33 °C.

Table 9 is the result of temperature measurement at the first solar pond at a distance of 1.05m from the water surface. The temperature measurement results show that at 06.00 the average temperature is 30 °C; at 09.00 the average temperature is 30 °C; 12.00 the average temperature is 32 °C; at 15.00 the average temperature is 33 °C and the average temperature 18.00 is 32 °C.

Table 10 is the result of salinity measurements in the first solar pool on the water surface. Salinity results show that at 6:00 the average salinity is 32 (‰); at 09.00 the average salinity is 32 (‰); at 12.00 the average salinity is 31 (‰); at 15.00 the average salinity is 31 (‰) and at 18.00 the average salinity is 31 (‰).

4. Results and Discussion

The results of gradient temperature and salinity measurements for 12 days have been presented in tables 1-5 for the first solar pool (outdoor) and in tables 6-10 for the indoor solar pool. Measurements were made five times at 06.00 hours; 09.00; 12.00; 15.00 and 18.00 hours are categorized as 1, 2, 3, 4 and 5 in the picture. The temperature and gradient salinity measurements expressed in Table 1-10 are represented as graphs shown in Figure. 4-10. Figure 4-7 shows the results of temperature measurements at each point versus time on outdoor solar pond and indoor solar pond. Figure 8 is a graph of salinity gradient versus time in the first solar pond (outdoor) and solar pond (indoor). Figure 9-10 is a temperature graph at some points and salinity gradient versus time in outdoor pond and indoor pond.

The temperature versus time graph shown in Figure 4-6 shows the average temperature of an indoor pool greater than the average temperature in an outdoor pool, this is because the ambient temperature of indoor pond is cloistered while for outdoor pond is strongly influenced by weather and wind. However, the temperature graph (T3 = °C) in Figure 7 averages outdoor pool temperature is larger than average indoor pool temperature, this is due to the effects of ultraviolet rays of sunlight. For salinity gradient versus time value varies between outdoor solar pool and indoor solar pool, but in the outdoor solar pool the salinity value increases with time.

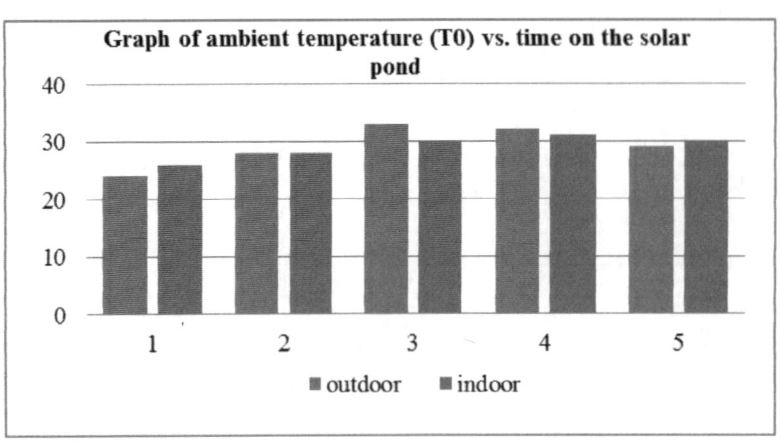

Figure 4. Graph between ambient temperature (T_0 = °C) versus time in the first solar pond and second solar pond

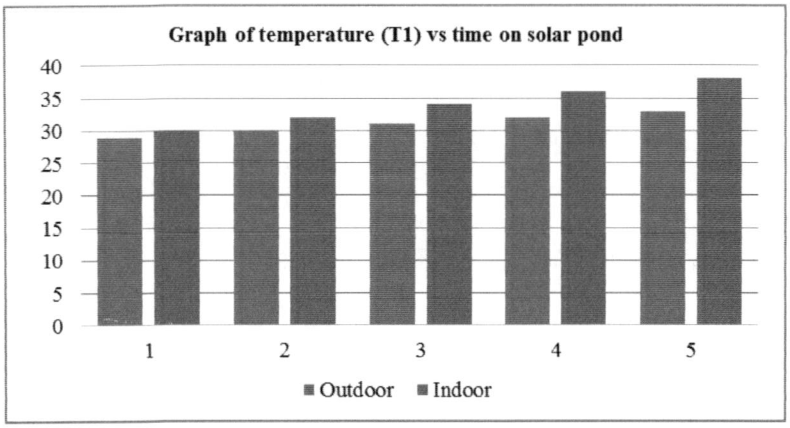

Figure 5. Graph between temperature (T_1 = °C) versus time in the first solar pond and second solar pond.

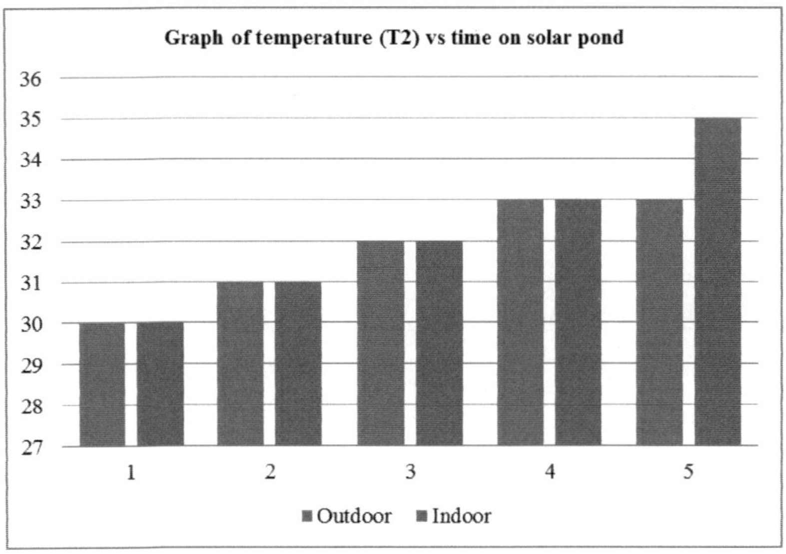

Figure 6. Graph between temperature (T_2 = °C) versus time in the first solar pond and second solar pond

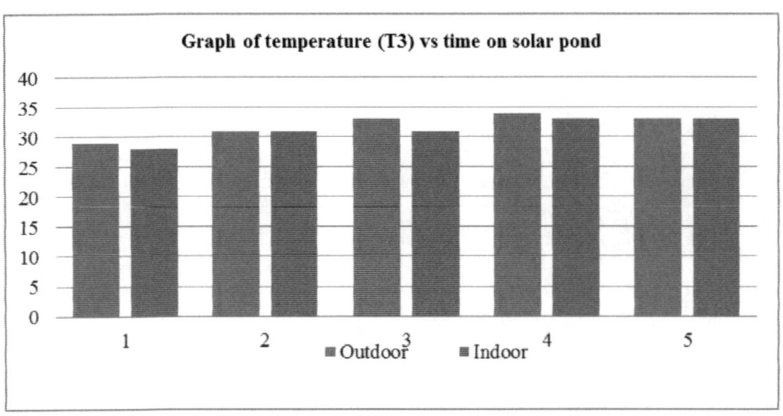

Figure 7. Graph between temperature (T_3 = °C) versus time in the first solar pond and second solar pond

IOP Conf. Series: Materials Science and Engineering **676** (2019) 012005 doi:10.1088/1757-899X/676/1/012005

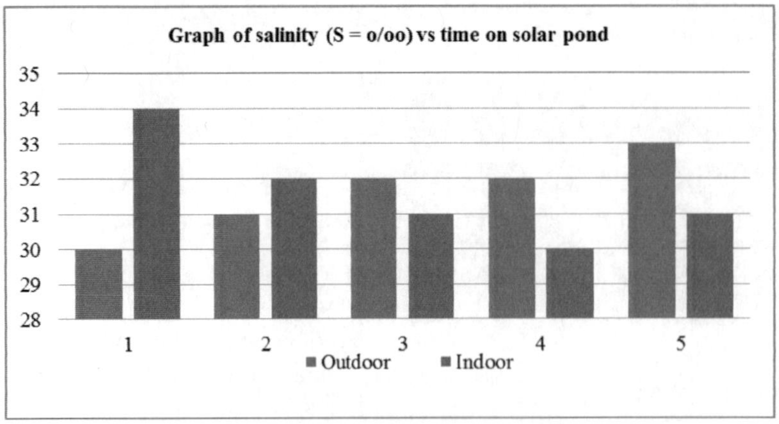

Figure 8. Graph between salinity (S = o/oo) versus time in the first solar pond and the second solar pond

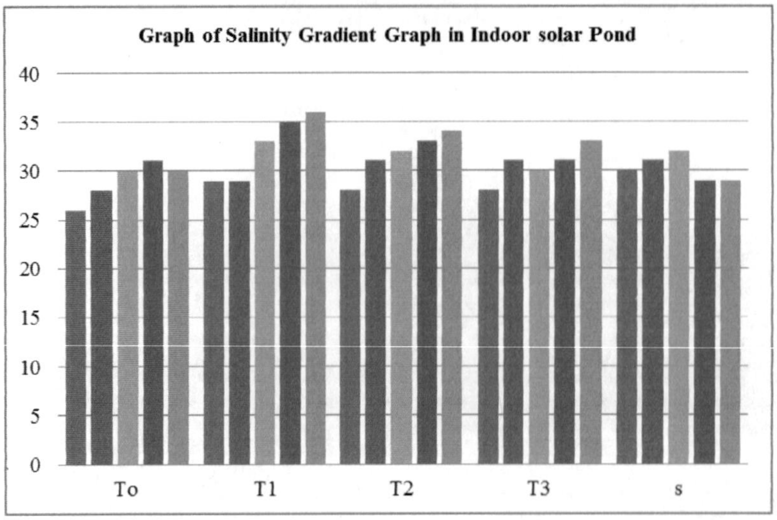

Figure 9. Graph of salinity and temperature at some point versus time in the second solar pond

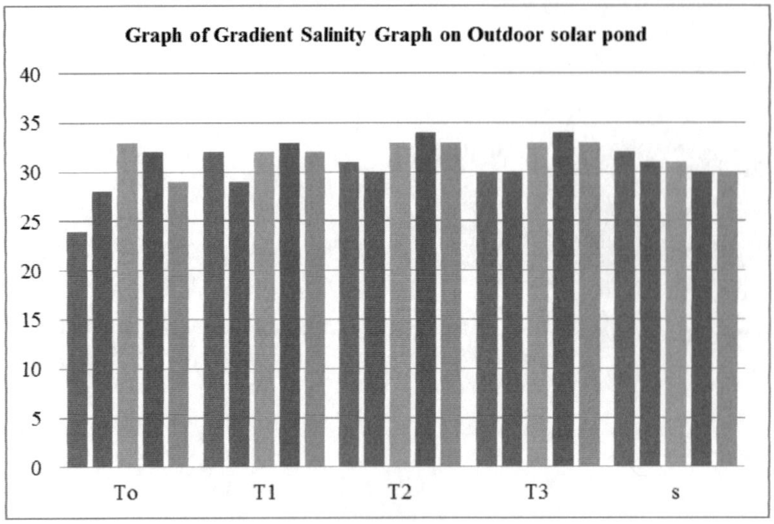

Figure 10. Graph of salinity and temperature at some point versus time in the first solar pond

5. Conclusion

The gradient temperature salinity and temperature measurement in the fest solar pond are smaller than temperature in the second solar pond, however the T_3 temperature in the outdoor solar pond is larger than the temperature in indoor solar pond. The observation results show that the first solar pond generates the mean value of temperature and salinity of successive gradients 32 (°C); 30 % and from the second solar pond the average value of temperature and salinity of successive gradients 31 (°C); 28 %.

Acknowlegment
This research is funded by Institue of Research and Community Service of Hasanuddin University internal grant to Benua Maritim Indonesia Spesifik (BMIS) research.

References

[1] Singh, B., et al., *Small scale power generation using low grade heat from solar pond* Procedia Engineering 49 2012: p. (2012) 50 – 56
[2] Sugiawan, Y. and S. Managi, *The environmental Kuznets curve in Indonesia: Exploring the potential of renewable energy.* Energy Policy, 2016. 98(Supplement C): p. 187-198.
[3] Zikra, M., Suntoyo, and Lukijanto, *Climate Change Impacts on Indonesian Coastal Areas.* Procedia Earth and Planetary Science, 2015. 14(Supplement C): p. 57-63.
[4] Nugroho, D., et al., *Modelling explicit tides in the Indonesian seas: An important process for surface sea water properties.* Marine Pollution Bulletin, 2017.
[5] Simic, M. and J. George, *Design of a System to Monitor and Control Solar Pond: A Review.* Energy Procedia, 2017. 110(Supplement C): p. 322-327.
[6] Amouei Torkmahalleh, M., et al., *Key factors impacting performance of a salinity gradient solar pond exposed to Mediterranean climate.* Solar Energy, 2017. 142(Supplement C): p. 321-329.
[7] Assari, M.R., H. Basirat Tabrizi, and A. Jafar Gholi Beik, *Experimental studies on the effect of using phase change material in salinity-gradient solar pond.* Solar Energy, 2015. 122(Supplement C): p. 204-214.
[8] Bernad, F., et al., *Salinity gradient solar pond: Validation and simulation model.* Solar Energy, 2013. 98(Part C): p. 366-374.
[9] Khalilian, M., *Assessment of the overall energy and exergy efficiencies of the salinity gradient solar pond with shading effect.* Solar Energy, 2017. 158(Supplement C): p. 311-320.
[10] Sayer, A.H., H. Al-Hussaini, and A.N. Campbell, *Experimental analysis of the temperature and concentration profiles in a salinity gradient solar pond with, and without a liquid cover to suppress evaporation.* Solar Energy, 2017. 155(Supplement C): p. 1354-1365.

IOP Conf. Series: Materials Science and Engineering **676** (2019) 012006 doi:10.1088/1757-899X/676/1/012006

Insertion of 275 kV Transmission Line for Improving the Voltage Profile and Efficiency of Electrical Power System

I Kitta[1]*, S Manjang[1], I R Sahali[1] and F Maricar[1]

[1]Department of Electrical Engineering, Hasanuddin University, Makassar, Indonesia

*E-mail: ikhlaskitta@gmail.com

Abstract. The location of electricity generation generally lies in the centers of renewable energy sources which are generally far from urban centers, such as in the province of South Sulawesi, Indonesia. This condition becomes a separate issue related to the quality of the voltage and the efficiency of the electrical system that uses 150 kV transmission lines to transmitted electrical energy from the power plants. Therefore, to avoid waste and lack of quality electricity, it is necessary to insert a new transmission line from these power plants to the city of Makassar. Where the transmission line built is 275 kV transmission line because it increases the voltage profile and is efficient compared to 150 kV transmission line even though the network is added to the bank's capacitor. Therefore, this paper shows the reasons why a 275 kV transmission line was chosen for the improvement of voltage and reduction of power losses in South Sulawesi system.

1. Introduction

At present the use of electricity has become a very important need for humans. The need for electricity is increasing for urban areas. This condition causes the distribution of electrical energy to these areas. An electric power system is said to have a high level of reliability if the system is able to provide a continuous supply of electrical energy needed by consumers [1]. There are several problems faced by an electric power system to be able to supply electrical energy continuously. One of them is related to the problem of quality of electricity voltage and small power losses [2]. Because the load center is generally located in a location far from the center of electricity generation, the distribution of electrical power is needed to reach consumers.

Constant voltage values will maximize the performance of electrical equipment used by consumers. Whereas with small power losses will maintain the electricity supply to consumers as desired and can reduce financial losses during the electricity transmission process.

The process of repairing the voltage on the network can be done using the voltage regulation method which is by using a bank capacitor and a transformer tap [3]. The installation of capacitors is expected to reduce power losses and improve the voltage profile of transmission line systems [4]. With minimization of losses on the network, the bus voltage profile can be maintained at allowable values so that the continuity and quality of the operation of the electric power system can always be maintained.

Transmission line losses are directly proportional to the resistance of the conductor and inversely proportional to the transmission line voltage square [5], so the reduction in power losses obtained due to the transmission line voltage increase is far more effective than the reduction in power losses by reducing the conductor resistance value. Consideration of increasing transmission line voltage is what tends to be done to reduce power losses in the transmission line.

Therefore, one of the efforts to reduce power losses and improve the quality of voltage in the South Sulawesi System is the insertion of 275 kV transmission line. In addition to these reasons, the development of 275 kV transmission line in the South Sulawesi system is to connect power plants in the area of renewable electricity to the load center in South Sulawesi, namely the City of Makassar. South Sulawesi has many primary energy sources, especially in the form of hydropower which can be developed into Hydro Power Plant. The potential of hydropower that can be developed into a Hydro Power Plant around 1,996 MW. In addition, there is also the potential for natural gas for Natural Gas Power Plant and Combined Cycle Gas Turbine Power Plant with measured reserves of 470 BSCF, and there is potential for coal for Coal-fired Steam Power Plant of 37.3 million tons[3] [6].

To see how much influence 275 kV transmission line has on the South Sulawesi system, the research is carried out, the results of which are outlined in this paper.

2. Research Method

2.1. Test system description

The case analyzed in this study is the South Sulawesi electricity system. The data used in this study are secondary data obtained from the interconnection system of South Sulawesi. The data includes the layout of the South Sulawesi system illustrated in the one-line diagram, covering the capacity of the generators, transformers and transmission lines, and the working voltages. Equipped with data on system loading and the current operating system patterns.

The electricity load of the province of South Sulawesi is mostly in the southern part of the city of Makassar, while the renewable energy plant is located in the northern and central parts of South Sulawesi. South Sulawesi's electricity system currently consists of 70 kV and 150 kV interconnection systems shown in Figure 1. The South Sulawesi system is supplied from hydro power plant, coal-fired steam power plant, gas turbine power plant, combined cycle gas turbine power plant, diesel engine power plant, and wind turbine power plant. The total installed capacity of the power plant in South Sulawesi is 1367 MW. The power of the generator is around 1102 MW, while the peak load is 950 MW. The number of existing substations in South Sulawesi is 33 with a total capacity of 1888 MVA. The ratio of the number of household customers with electricity is 84.46%.

2.2. Simulation procedure

The reason for adding a network with an 275 kV transmission line is based on the following power losses equation (1) :

$$\Delta P_t = \frac{p^2 \cdot R}{V_r^2 \cos^2 \varphi} \tag{1}$$

with, P is load power at the receiving end of transmission line (watt), V_r is voltage at the end of the transmission receiver (volt), R is Resistant to transmission wire (ohm), and Cos φ is Load power factor.

It can be seen that the transmission line power losses can be reduced in several ways, including by increasing the transmission line voltage, reducing the conductor resistance, and increasing the load power factor.

In the research that has been carried out, simulations are carried out with the use of bank capacitors as an attempt to raise the bus voltage level and optimize the distribution of power to the system. Usage shunt capacitors are useful as additional reactive resources to compensate for inductive power due to loading by performing reactive power compensation, which aims to keep the voltage profile always at allowable limits. The installation of a shunt capacitor is also expected to also reduce losses which means saving electricity energy which means a decrease in the electrical current flowing in the load.

IOP Conf. Series: Materials Science and Engineering **676** (2019) 012006 doi:10.1088/1757-899X/676/1/012006

Data processing is carried out using Load Flow simulation software [7]. There are several stages through which data is processed in five calculation scenarios. The scenario is :

1. Scenario 1 (basic scenario)
   This scenario is a basic or existing scenario, where the system is conditioned on conditions without any addition of transmission lines. This basic condition will be used as a basis to see the effect of 275 kV transmission line insertion, 150 kV transmission line, and the addition of bank capacitors.
2. Scenario 2, add a 275 kV transmission line in the South Sulawesi system.
   Scenario 2 is a system condition that has been added with 275 kV transmission line with ACSR type, 2 x 430 mm^2 type, along 195 kmc. The location of the 275 kV transmission line addition is shown in Figure 1, where the base of the transmission line is in the Palopo bus, and the base of the network receiver is in the Kima bus which is a bus representing the load center (Makassar city).

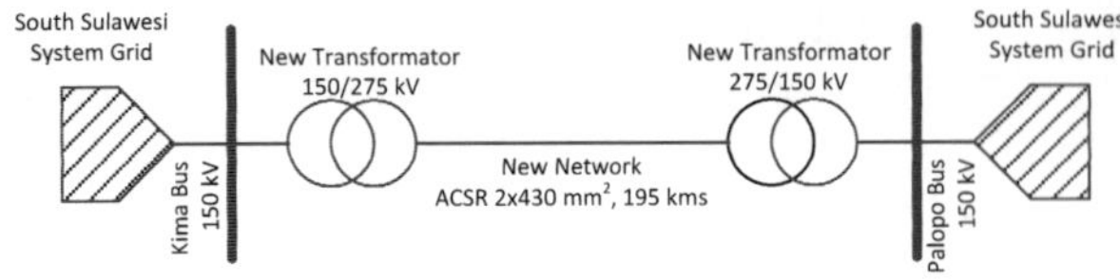

Figure 1. The model of Scenario 2

3. Scenario 3, add a 150 kV transmission line
   In this condition a 150 kV transmission network is inserted in the South Sulawesi system, where the sending bus is the Palopo bus and the receiving bus is the Kima bus. This 150 kV transmission line has the same network parameters as the 275 kV transmission line parameter in scenario 2, conductor type is ACSR 2 x 430 mm^2 for 195 kmc. As shown in Figure 2.

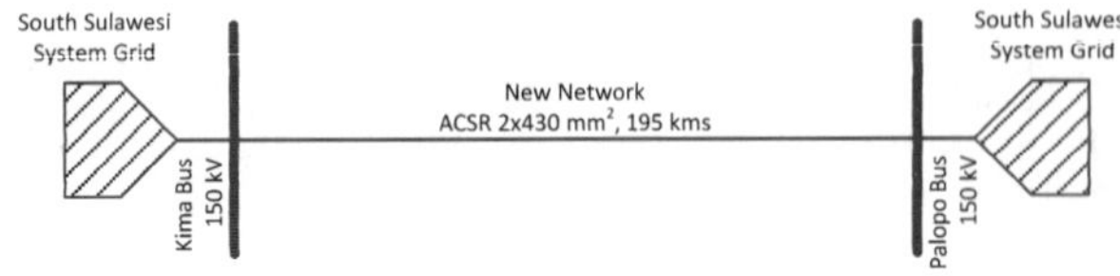

Figure 2. The model of Scenario 3

4. Scenario 4, 5, dan 6, add a 150 kV transmission line with conductor type of ACSR 2 x 430 mm^2 along 195 kmc, and add bank capacitors.
   - Scenario 4 is insertion scenario a 150 kV transmission line and installation of 20 MVar bank capacitors on sending bus side (Palopo bus), As show in Figure 3.

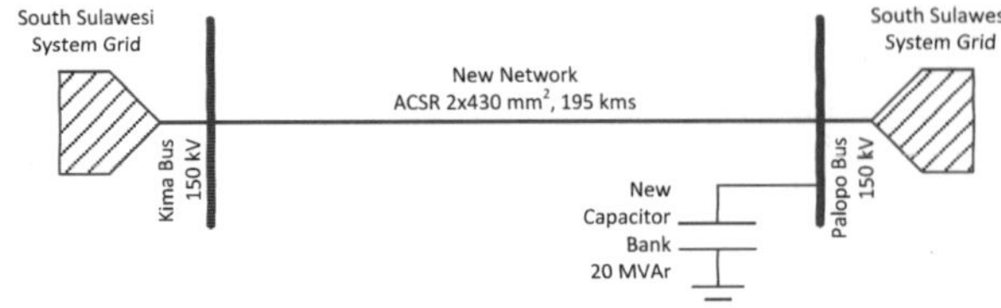

Figure 3. The model of Scenario 4

   - Scenario 5 is the condition of the system with the installation of 20 MVar bank capacitor on the Kima bus and the 150 kV transmission line insertion. As show in Figure 4.

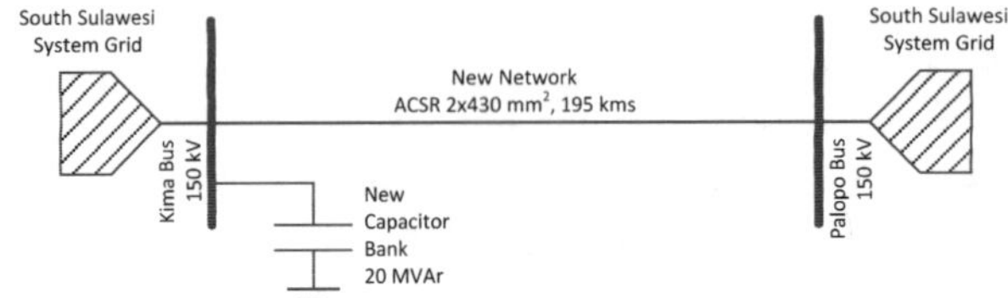

Figure 4. The model of Scenario 5

- Scenario 6 is the condition of installing a bank capacitor on the side of sending bus and the receiving bus which is 10 MVar each. And also inserting 150 kV transmission line to connect the two bus. As show in Figure 5.

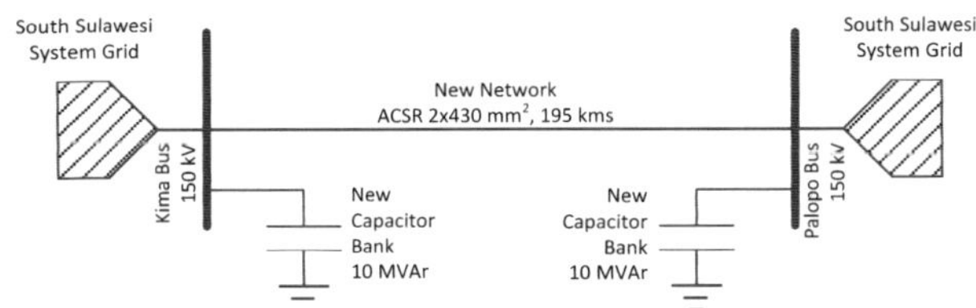

Figure 5. The model of Scenario 6

Based on the scenarios and data of the South Sulawesi system, power flow calculations are performed under normal conditions or conditions before insertion of new transmission lines and addition of bank capacitors (scenario 1). From the acquisition, it will be seen how the condition of the network on the system interconnection as well as how much power losses caused by the South Sulawesi network.

In multi-bus systems, it is necessary to complete the power flow to obtain voltage profiles and network power losses using Newton Raphson method. The Method solves the problem of power flow by using a set of non-linear equations to calculate the voltage and voltage phase of each bus [7].

3. Result and Discussion

3.1. Power flow analysis results

Based on the calculation results of power flow analysis for South Sulawesi system, the results are obtained as in Table 1.

Table 1. The calculation results of power flow analysis for South Sulawesi system

		Scenario 1	Scenario 2	Scenario 3	Scenario 4	Scenario 5	Scenario 6
Power Gen.	MW	1058.53	1045.08	1048.65	1048.47	1048.55	1048.5
	MVar	541.98	474.09	496.74	472.65	471.06	471.93
Load	MW	1024.66	1024.66	1024.66	1024.66	1024.66	1024.66
	MVar	284.47	284.47	284.47	284.47	284.47	284.47
Losses	MW	32.87	20.42	23.99	23.81	23.89	23.84
	MVar	353.28	288.52	310.09	306.45	305.53	305.94
Bus Volt. Max.	%	100.13	100.22	100.07	100.57	100.30	100.39
Bus Volt. Min.	%	94.59	96.11	95.65	95.89	96.05	95.97

The power flow at any time in the system is always changing, this is due to changes in load that occur at any time. In this study, the analysis of power flow was carried out in the condition of the last highest peak load that had occurred in the South Sulawesi electricity system until the time of this research. The highest peak load at that time was 1024.66 MW and the total generation was 1057.53 MW. Jeneponto Coal-fired steam power plant bus as a slack bus, namely a bus that must bear all the power shortages in the system after all other power plants operate, generate power of 37.8 MW. This power of power plant still meets the capacity of the Jeneponto bus at 114 MW.

3.2. Effect of bus voltage profile

The results of the calculation of the bus voltage profile for each scenario are shown in Figure 6, where when the initial conditions (scenario 1) based on the study of power flow obtained the voltage profile of each bus ranged from 94.59% to 100.13%. The minimum value of the working voltage of the South

Sulawesi system is below the minimum limit of the SNI standard voltage and IEEE Standard, which is the lower limit of -5% and the limit of + 5% of the nominal voltage (100%). Bus number 31 namely Sinjai bus which has the lowest voltage compared to other buses, which is 94.59%.

In scenarios 2, 3, 4, 5, and 6, the voltage on all buses has improved. Scenario 2 which inserts 275 kV transmission line into the South Sulawesi system gets results that have an impact on bus voltage, which can increase the voltage profile on all buses in the system, which is changed to 96.11% until 100.22%. Likewise in scenario 3, the bus voltage profile becomes 95.65% - 100.07%. For scenario 4 the bus voltage profile is between 95.89% - 100.57%. Also for scenario 5, the results of the bus voltage profile between 96.05% - 100.30% are shown. And the scenario 6 also produces a voltage profile that is better than scenario 1, which is between 95.97% - 100.39%.

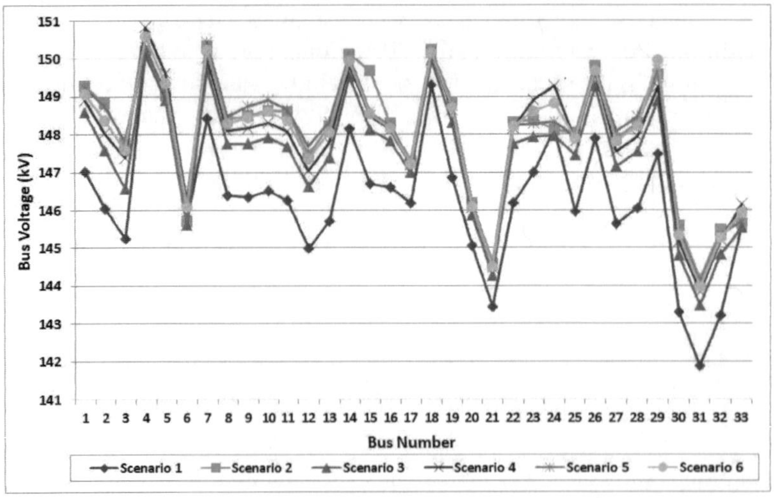

Figure 6. The bus voltage profile for each scenario

3.3. Effect of system power losses

From the Table 1 it can be seen that the initial conditions are conditions that exist in the system without transmission line insertion or addition of bank capacitors, active power and reactive power generated by all plants in the South Sulawesi system of 1057.53 MW and 1045.08 MVar with power losses in the system of 32.87 MW and 353.28 MVar. When the South Sulawesi system was inserted a 275 kV transmission line, the total power generation was 1045.08 MW and 474.09 MVar, while the power losses dropped to 20.42 MW and 288.52 MVar.

In the comparative scenario, scenarios 3, 4, 5, and 6 were obtained for scenario 4 that inserting a 150 kV transmission line on the system obtained total generation power of 1048.65 MW and 496.74 MVar, power losses of 23.99 MW and 310.09 MVar. Table 1, it show about the power generation and power losses in scenario 4, scenario 5, and scenario 6, differing slightly from scenario 3.

If Figure 6 is analyzed, the voltage profile value when inserted is 275 kV transmission line (scenario 2) and 150 kV transmission line (scenario 3, 4, 5, and 6) are almost the same. However, scenario 2 allows more optimal reduction in power losses compared to other scenarios. Scenario 2 active power losses is 288.52 MW or down 12.45 MW, while scenario 3 is 310.09 MW, as well as scenario 4, 5 and 6 which differ only slightly compared to scenario 3. So scenario 2 has a better impact on the South Sulawesi system.

Likewise in the reduction of the reactive power of the South Sulawesi system, seen in Table 1, the application of 275 kV transmission line inserts will reduce reactive power losses by 64.76 MVar which previously amounted to 353.28 MVar.

4. Conclusion

Insertion of 275 kV transmission line in the South Sulawesi system can increase the bus voltage profile and reduce power losses. The scenario shows that the busbar voltage on all busbars is included in the permissible voltage range. This means that the average voltage level of this scenario is greater than the voltage level of the basic scenario. Scenario 2 also shows that after insertion of 275 kV transmission line, system power losses is less than initial conditions (scenario 1). Scenario 2 also shows the insertion of 275 kV transmission line has more optimal impact in terms of voltage quality and power transmission efficiency compared to the 150 kV transmission line insertion scenario and the addition of bank capacitors (scenarios 3, 4, 5, and 6). This can be seen in the average voltage level of scenario 2 is greater than the voltage level of scenarios 3, 4, 5 and 6. Similarly, the reduction in power losses, where the application of 275 kV transmission line in the South Sulawesi system reduces losses more system power compared to 150 kV transmission line (scenarios 3, 4, 5, and 6).

Acknowledgments

The authors gratefully acknowledge Indonesia Government of Ministry of Research and Higher Education for financial support of this research.

References

[1] Tonmitr K, Ratanabuntha T, Tonmitr N and Kaneko E 2016 Reduction of power loss from corona phenomena in high voltage transmission line 115 and 230 kV *Procedia Computer Science* **86** 381-384.
[2] Johnson D O and Hassan K A 2016 Issues of power quality in electrical systems *International Journal of Energy and Power Engineering* **5** 4 148-154.
[3] Onyemaechi A and Isaac O O 2014 Minimization of power losses in transmission lines *IOSR Journal of Electrical and Electronics Engineering (IOSR-JEEE)* **9** 3 23-26.
[4] Kumar S, Rehman C K F and Shaikh S A 2018 Voltage improvement and power loss reduction through capacitors in utility network *International Conference on Computing, Mathematics and Engineering Technologies (Sukkur, Pakistan, 3-4 March 2018) IEEE.*
[5] Pande S and Ghodekar J G 2012 Computation of technical power loss of feeders and transformers in distribution system using load factor and load loss factor *International Journal of Multidisciplinary Sciences and Engineering* **3** 6 22-26.
[6] ESDM Ministry 2016 *PLN Electric Power Supply Business Plan for 2016 – 2025* (Jakarta: Ministry of Energy and Mineral Resources of the Republic of Indonesia).
[7] Saadat H 1999 *Power System Analysis* (New-York: McGraw-Hill).

IOP Conf. Series: Materials Science and Engineering **676** (2019) 012007 doi:10.1088/1757-899X/676/1/012007

Impact Optimal DG Placement Against Harmonic Distribution on Reconfiguration Distribution Network on Microgrid System

M D Faraby[1]* and O Penangsang[1]

[1]Department of Electrical Engineering, Institut Teknologi Sepuluh Nopember, Indonesia

*E-mail: muhira16@mhs.ee.its.ac.id

Abstract. Increases in harmonic distortion in Radial Distribution System (RDS) will result in additional shorter insulation lifetime, higher temperature and also malfunction. These are undesirable as they cause power losses and also affect the voltage profile. In this paper the harmonic analysis is carried effect and performance placement Distributed Generation (DG) after reconfiguration distribution network in active radial distribution system or microgrid concept. The optimal size of DG is determining using K-Means Clustering method and Load Curtailment method to designing distribution network reconfiguration after restoration. The load flow analysis and harmonic flow analysis is carried out with direct approach using BIBC and BCBV matrices. The proposed method is carried out on IEEE 33 bus system having the harmonic sources injected as current sources using MATLAB and ETAP software. The result show DG placement on 4 bus points using K-Means Clustering can improve voltage levels and reduce harmonic dispersion. After restoration and reconfiguration of the system, the harmonic spread is strongly influenced by the current flowing on the bus and the location of the source harmonics on the system.

1. Introduction

Radial distribution system is the simplest type of network that is commonly used. Along with the development of the radial distribution system which is getting bigger and wider it is certainly followed by several major problems in the form of greater power losses and deteriorating voltage profiles [1]. With increasing interest on power generation in an environmental friendly way, the development of renewable energy based distributed generation (DG) has received great attention in recent years as renewable energy resource are pollution free and abundant in nature [2].

Irrespective of the advantages of radial distribution, this system is very span to the spread of harmonics. The development of the age followed by technological developments will surely lead to the use of non-linear loads such as semiconductor devices [3]. The source of harmonics produces a non-sinusoidal current which is flawed where each periodic wave is not a sinus form containing harmonics. Non-linear loads are the cause of high current harmonics on the system and can cause losses in the distribution of electrical energy and the destruction of existing equipment [4].

So, to reach to these targets, loss reduction and voltage profile improvement, together with THD reduction planning of the electric system with the presence of DG requires the definition of several factors such as, the best technology to be used, the number and the capacity of the units, the best location, the type of network connection and etc. the problem of DG allocation and sizing is of great importance. The installation of DG units at non-optimal places and with non-optimal sizing can result in an increase in system losses, damaging voltage state, voltage flicker, protection harmonic, stability and implying in an increase in costs and, therefore, having an effect opposite to the desired [5].

Reconfiguring a distribution network is one of the techniques of resetting network configuration by operating a switch tie or in other words changing the status of an open or closed switch on a distribution network [6, 7]. Different method have been developed to solve reconfiguration problem in distribution network. Used of reconfiguration technique for service restoration and power loss reduction is proposed in [8].

The development of research on improving power quality especially harmonic distortion in radial distribution systems related to placement and sizing DG in the concept of microgrid by using various methods based on numerical methods, artificial intelligent, etc. Harmonic, power loss, and voltage deviation are main issue in power system quality. The installation of DG at non-optimal location and size can lead the increasing of power loss, power quality decline, protection coordination and power stability problems [9]. But for the side of the reliability of the system, when there is a disturbance where one of the buses in the system is required to undergo restoration, the action is carried out by the operator, configuring the network distribution system. In previous studies [8], research on the spread of harmonics on a passive radial distribution system after reconfiguration due to the restoration was influenced by the distance of the harmonic source to the swing bus. This paper will discuss the continuation of the previous research where the renewal of this paper is in the condition of active radial distribution system or microgrid. The effect of placement and sizing DG on harmonic distribution after restoration will be examined in this paper.

2. Basic Theory

2.1. Microgrid and type of distributed generation

Microgrid is a small-scale power system network. The generator ranges from 50kW-1kW using available energy sources such as microhydro, solar, wind, and others.. By utilizing local renewable energy sources that are available in each region, microgrids can be used as a solution to realize energy independent areas for areas that are not affordable by electricity [10]. Different types of DGs include Synchronous generators, Photovoltaic generators and Asynchronous generators [11].

2.2. Harmonic, harmonic index and standard harmonic

In harmonic and interharmonic distortion, the alternating current voltage wave is not pure sinusoidal in other words distorted. Based on the fourier wave can be decomposed into several sinusoidal waves with a higher frequency of fundamental frequencies. In interharmonic distortion, wave distortion is formed from sinusodial waves with a frequency of multiples of the fundamental wave frequency, with multiples in the form of fractions. Whereas in harmonic distortion, waveform distortion is formed from sinusoidal waves with frequency multiples of the fundamental frequency. Harmonic frequency equation can be written as follows [4, 12].

$$f_h = h \, x \, f_d \tag{1}$$

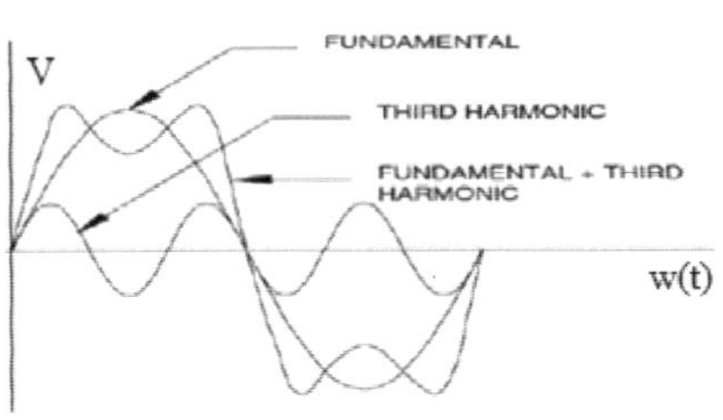

Figure 1. Non-linear waves by adding harmonics of three fundamental frequencies [4]

According to the IEEE, IHD is the ratio between the rms value of each harmonic current compared to the RMS value of the fundamental flow or the first harmonic. The first harmonic or fundamental flow is always worth 100%. Where the IHD equation is:

$$IHD_n = \frac{I_n}{I_1} \times 100\% \tag{2}$$

THD is the ratio between the rms value of all the harmonic components and the rms value of the fundamental expressed in percent (%) :

$$I_H = \sqrt{\left(I_2^2 + I_3^2 + I_4^2 + I_5^2 + I_6^2 + I_7^2 ...\right)} \tag{3}$$

$$THD = \frac{I_H}{I_1} \times 100\% \tag{4}$$

The commonly used THD index is :

$$THD = \sqrt{\frac{\sum_{n=2}^{N} V^2 n}{V_1}} \tag{5}$$

Defining the ratio of the rms value of the harmonic component to the rms value of the base component is usually expressed in percent form. This index is used to measure the deviation of periodic waveforms containing harmonics of perfect sine waves. The value of THD is zero in the event of a perfect sine wave.

% VTHD is the percentage of total voltage distorted by harmonics to fundamental frequency and% ITHD is the percentage of total current distorted by harmonics to fundamental frequency. Standard voltage harmonics used according to IEEE Std. 519-1992.

2.3. K-means clustering

K-means Clustering is a method of grouping a group of data objects with a certain number based on the equation of a particular characteristic or attribute [1]. This method has proven to be reliable in solving several distribution system problems such as determining the optimal location of the Phasor Measuring Unit (PMU) and determining the optimal location of DG.

Grouping data objects by the K-means clustering method is done by calculating the closest distance of each data object to the center point of the cluster (centroid). In the initial stage of this method, the number of clusters is determined first based on the number of data objects. In addition, in the initial stage the centroids of each cluster are randomly selected from the available data objects. Then this method will test each data object and group it into one cluster based on the closest distance to each centroid. The value of the centroid will be updated repeatedly by calculating the average value of the data objects in each cluster. Then testing and grouping data objects into clusters with the new centroid is done again. The iteration will continue to be repeated until the value of the new centroid does not change from the previous centroid (converging)[1].

2.4. Reconfiguration system

Load Curtailment must have three conditions met for selecting the appropriate switch tie. This requirement is used to narrow down the optimal tie switch search solution and tie switch candidates will be selected in the use of this method. It aims to maximize the number of restored loads. This requirement is required with respect to the amount of load between sectionalizing switches [6].

- I_{LC}^{avail} (available current) is the maximum amount of load that may be subtracted from the upstream bus to increase the current spare capacity. The reduced load is between the branch and I_{LC}^{avail} . It can provide power to each candidate tie switch.
- I_M is the current spare capacity. Spare capacity is the capacity of the reserve flow that can meet the load requirements I_M obtained from :

$$I_M = I_{rating\ cb} - I_{bus\ current} \tag{6}$$

IOP Conf. Series: Materials Science and Engineering **676** (2019) 012007 doi:10.1088/1757-899X/676/1/012007

I_M the selected value is the largest candidate of the tie switches to be used,

- Z_{path}

Z_{path} is the impedance distance between buses. Z_{path} to increase the bus voltage by shortening the distance between buses in a way:

$$Z_{path} = \frac{V_{ts}}{\sum I_{L,k}} \tag{7}$$

Z_{path} is the smallest or the shortest distance is to be selected among the candidate tie switches.

3. Methodology

3.1. FBS load flow analysis

The Forward-Backward Sweep (FBS) method is used for the power flow equation based on two matrices namely BIBC (Bus Injection to Branch Current) and BCBV (Branch Current to Bus Voltage). [BIBC] is the relation matrix between the injection current on the bus with the current at the branch, whereas [BCBV] is the relation matrix between the branch current and the bus voltage [13].

$$[B] = [BIBC] * [I] \tag{8}$$

$$[\Delta V] = [BCBV] * [B] \tag{9}$$

where [BCBV] is the relation matrix between the bus voltage and the branch current. Substituting equation (8) with (9) is obtained by:

$$[\Delta V] = [BIBC] * [BCBV] * [I] \tag{10}$$

$$[\Delta V] = [DLV] * [I] \tag{11}$$

where [DLV] is the relation matrix between the voltage drop and the injection current on the bus.

The power flow solution is obtained by means of an iteration process so that it can be written as follows:

$$I_i^k = \left(\frac{P^{sch} - jQ^{sch}}{v_i^k} \right) \tag{12}$$

$$\left[\Delta V^{k+1}\right] = [DLV] * \left[I^k\right] \tag{13}$$

$$\left[V^{k+1}\right] = [V_0] - \left[\Delta V^{k+1}\right] \tag{14}$$

The iteration process stops when the absolute difference between the previous iterated bus current injection and the current iteration bus current injection is less than or equal to the specified tolerance ε.

$$\left| I_i^{k+1} - I_i^k \leq \varepsilon \right| \tag{15}$$

Total loss of power can be written

$$P_{loss} = [R]^T * [BIBC] * [I]^2 \tag{16}$$

3.2. Main step K-means clustering

K-means Clustering consists of several types such as Partition, Hierarchical, Density based, Model based, and Grid based. In this study K-means clustering of Partition type is used. Grouping data objects by K-means clustering is done by calculating the closest distance of each data object to the center point of the cluster (centroid). Centroid will be updated every time iteration until convergent [14].

The stages of the K-means clustering algorithm in more detail :

- Determine the number of clusters according to the number of data objects according to the following equation.

$$K_{group} \approx \sqrt{\frac{n}{2}} \tag{17}$$

  where : K_{group} is number of cluster; n is number of bus
- Initiate the centroid by randomly choosing between available data objects
- Calculate the distance of each data object to the centroid using the Euclian distance equation as follows

$$d\left(x_j, C_i\right) = \sqrt{\sum_{j=1}^{n} \sum_{i=1}^{m} \left(x_j - C_i\right)^2} \tag{18}$$

  where : d is Euclidean distance; x_j is data object; C_i is centroid; n is number of data object; m is number of cluster.
- Group each data object into one cluster with the closest distance
- Update the centroid value of each cluster by calculating the average data object in each cluster
- Check the centroid value. If the new centroid changes from the previous centroid then the algorithm returns in step 3. But if the new centroid does not change from the previous centroid then the solution is considered convergent so the K-means clustering algorithm is complete

3.3. Modeling of DG

Grouping data objects using clustering techniques in the previous step produces candidate DG mounting buses on each cluster [1]. After getting the grouping results for each bus in the system, each data object will be sorted based on the average of the LSF normalization value and the voltage of each bus. The average normalization value of LSF and bus voltage is called the bus index. The bus index value will not exceed 1 and less than 0 because the data object is a normalization value on a scale (0.1).

Each bus will be grouped according to the value of its data object and the bus with the highest bus index value of each cluster will be the candidate for the DG installation location on the system [1]. Then each bus index for each cluster will be calculated on average to get the cluster index. Candidates for DG installation locations are on buses belonging to clusters that have high LSF values and low bus voltage [15]. Therefore, candidates for the DG installation location are on the bus with the highest bus index value in the cluster with the highest cluster index value. If bus A is a bus with the highest bus index value of a cluster X and cluster X is a cluster with the largest cluster index value, then bus A is determined to be a candidate for the DG location.

Active and reactive power capacity DG can be calculated using the following equation

$$P_{DG} = \frac{P_{lkeff} + \alpha Q_{lkeff}}{1 + \alpha^2} \tag{19}$$

$$Q_{DG} = \frac{P_{lkeff} + \alpha Q_{lkeff}}{1 + \beta} \tag{20}$$

In the equation it is assumed that θ is the DG power factor angle that will be installed on the system. To calculate DG capacity considering optimal power factors, it can be stated to be [16]:

$$\alpha = \tan\theta = \frac{Q_{lkeff}}{P_{lkeff}} \tag{21}$$

$$\beta = \cot\theta = \frac{P_{lkeff}}{Q_{lkeff}} \tag{22}$$

with power factors obtained using the following equation:

$$PF_{DG} = \frac{P_{lkeff}}{\sqrt{P_{lkeff}^2 + Q_{lkeff}^2}} \tag{23}$$

3.4. Harmonic load flow

The backward sweep algorithm of HLF is used to obtain the matrix [A] representing the relationship between the branch current and the bus injection current for the h harmonic sequence. The forward sweep produces a [HA] matrix representing the relationship between harmonic bus voltage and harmonic bus injection current. For more details, the harmonic power flow algorithm, RDS n-bus is shown on the single-line diagram as follows [17]:

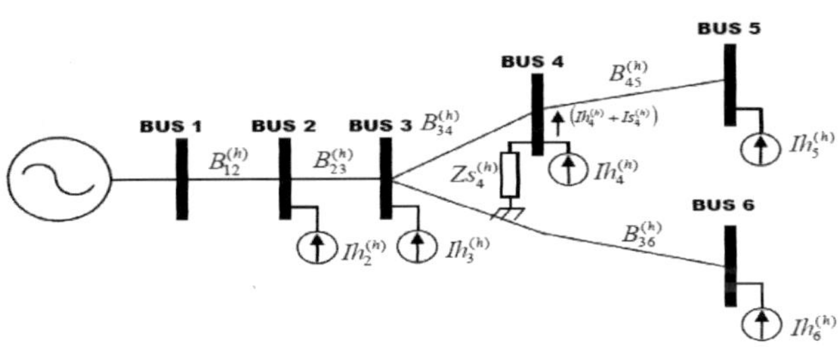

Figure 2. Single line diagram RDS 6-bus [13]

It should be noted that the harmonic current vector of the filter contains one element ie the harmonic current absorbed by the filter on bus 4. If the number of filters connected to RDS is equal to nl then the vector harmonic vector dimension of the filter is *(nl x 1)*. Thus the vector of the h harmonic bus current is:

$$\left[I^{(H)}\right] = \begin{bmatrix} Ih^h \\ \cdot \\ \cdot \\ \cdot \\ Is^h \end{bmatrix} \tag{24}$$

To find the current flowing from the i-j branch the equation is used;

$$\left[B_{ij}^{(h)}\right] = \left[A_{ij}^{(h)}\right]\left[I^{(h)}\right] \tag{25}$$

$$\left[A_{ij}^{(h)}\right]=\begin{bmatrix} Ah_{ij}^{(h)} \\ \cdot \\ \cdot \\ \cdot \\ Ah_{ij}^{(h)} \end{bmatrix} \tag{26}$$

Total loss of power can be written with vector form as follows:

$$Ploss(h)=\left[R(h)\right]T\left[A(h)\right]\left[I(h)\right]T \tag{27}$$

The voltage values of rms bus i (V_{rms_i}) and THD can be calculated as follows:

$$Vrms_i=\sqrt{\left|V_i^{(1)}\right|^2+\sum_{h=h_0}^{h_{max}}\left|V_i^{(h)}\right|^2} \tag{28}$$

Total Harmonic Distortion (THD) can be calculated as follows :

$$THD_i(\%)=\frac{\sqrt{\sum_{h=h_o}^{h_{max}}\left|V_i^{(1)}\right|^2}}{\left|V_i^{(1)}\right|^2} \tag{29}$$

3.5. Main step load curtailment

Main step from load curtailment algorithm :

- Determine candidate tie switch (TS) and sectionalizing switch (SS).
- Determine candidate load bus to be restored.
- Select and operate one TS candidate and implement the load curtailment algorithm in an out-of-service restoration area.
- Select and operate the TS and SS that have been chosen and implemented load curtailment algorithm.
- Determine load with priority scale to be restored by load curtailment and open SS.
- Run load flow distribution system

3.6. Constraints

The constraints are defined in terms of bus voltage limit and the total harmonic distortion limit. Bus voltage limit : The magnitude of bus voltages should be within limit defined as :

$$V_{\min}\le\left|V_i\right|\le V_{\max} \tag{30}$$

V_{min} and V_{max} is the lower and upper limit of bus voltage, V_i is the *rms* value of *ith* bus voltage

$$\left|V_i\right|=\sqrt{\left|V_i\right|^2+\sum_{h=h_o}^{h_{max}}\left|V_i^{(h)}\right|^2} \quad I = 2, 3, \ldots n \tag{31}$$

Total harmonics Distortion Limit : The THD of the system should be less than equal to maximum allowable limit

$$THD_i(\%)\le THD_{max} \tag{32}$$

The maximum allowable limit for THD is 5% [18].

3.7. IEEE 33 bus and study case
In this research, the system to be tested IEEE 33 Bus Radial Distribution System consisting of 32 SS (normally close) and 5 TS (normally open).

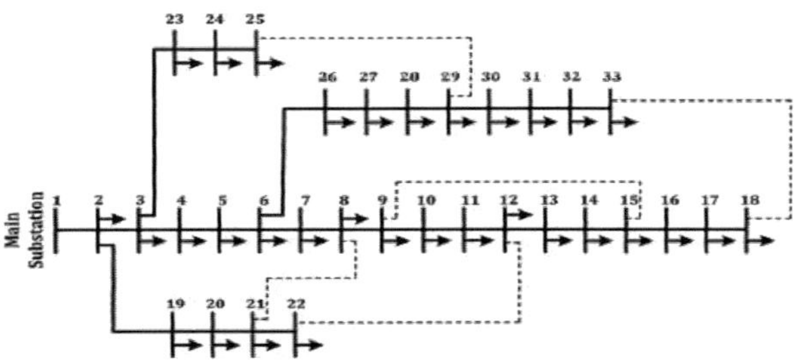

Figure 3. Single line diagram distorted IEEE 33 bus [8]

In this research, we give harmonic source on bus 4, 21 and 29 which then in select some bus for restoration there are bus 6, 19 and 26 and placement 4 unit DG in system. In general, the flowchart of this study can be seen in the flowchart below :

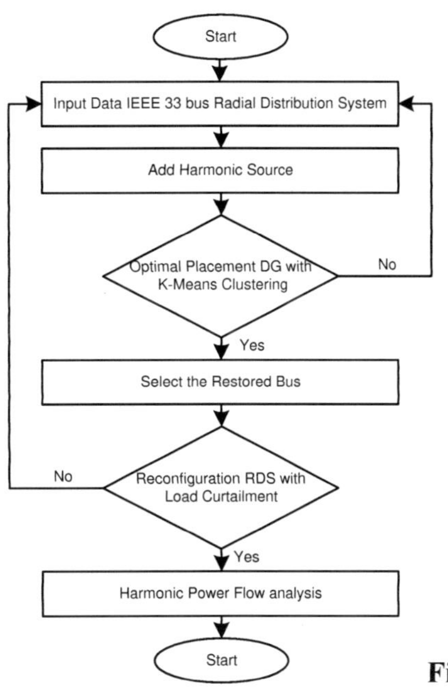

Figure 4. Flowchart of research

4. Result and Discussion

The system IEEE 33-bus radial distribution system with 32 bus sectional switching and 5 tie switch. The base power and the base voltage taken for system are 100 MVA and 12.6 kV. Total load in terms of active power and reactive power connected is 3.72 MW and 2.30 MWAR. The total active power loss is 210 kW. After injecting harmonic source on bus 4, 21 and 29, the system will generate harmonics on the order 11, 13, 23, 25, 35, 37, 47 and 49. The result of DG placement optimization with K-Means Clustering Method can be seen in Table 1 and Figures 5 and 6.

Table 1. DG placement optimization results using K-Means Clustering

Priority	DG in Bus	Group Cluster bus	P (kW)	Q (kVAR)	PF
1	6	3, 6	2107	1522	0.810
2	13	7, 8, 9, 10, 11, 12, 13, 14, 15, 16, 17, 18, 26, 27, 28, 29, 30, 31, 32, 33	451.7	206.9	0.909
3	5	4, 5, 23, 24, 25	2205	1585	0.813
4	20	1, 2, 19, 20, 21, 22	270.1	120.2	0.913
Losses Power (kW)			74.4		

In Table 2 DG placement optimization results using K-Means Clustering with a total active power capacity of 5033.7 kW. The optimization results produce 4 clusters with a priority scale. Each cluster is placed in a DG on one of the buses that can cover the other buses in each cluster. In addition there is a decrease in losses of 135.6 kW or 64.57 %.

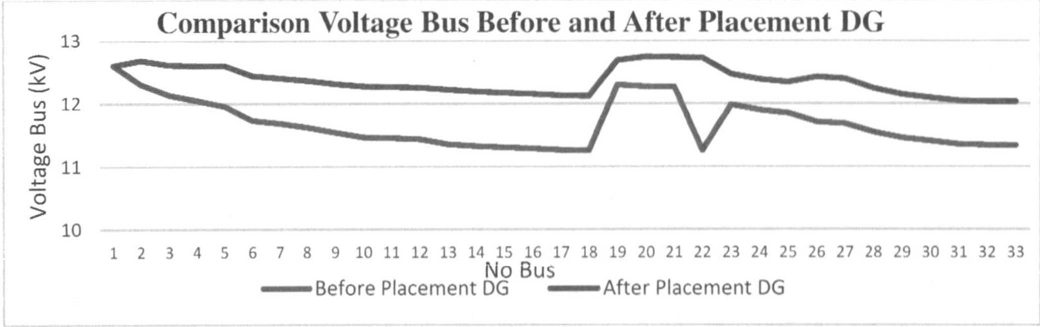

Figure 5. Comparison voltage bus before and after placement DG

In Figure 5 Placement of 4 DG units on buses 5, 6, 13 and 20 results in improved voltage levels at all bus points. It was also seen that there were several bus points under voltage conditions that could be repaired so that all bus points were in normal conditions. In this case the biggest voltage deviation was located on bus 18 at 7.69%.

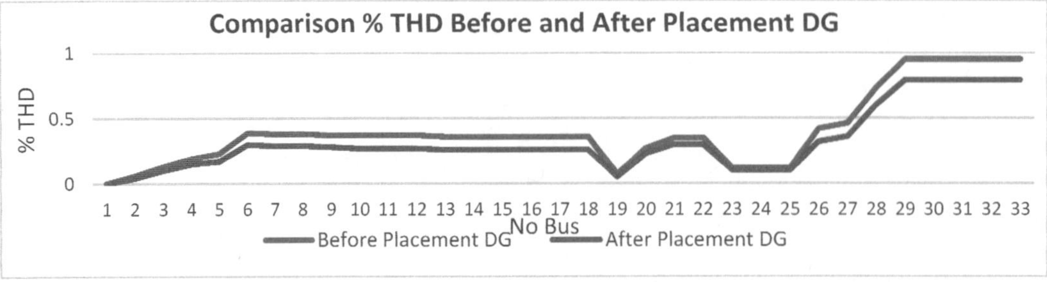

Figure 6. Comparison % THD before and after placement DG

In Figure 6, the % THD value is decrease in all bus points. This is due to the DG placement affecting the spread of harmonics which has an effect on the decline of fundamental currents in the system.

Table 2. Comparison voltage bus, % THD, active tie and losses before and after reconfiguration

Bus	NORMAL CONDITION AFTER DG PLACEMENT		RESTORATION BUS 6		RESTORATION BUS 19		RESTORATION BUS 26	
			Active Tie	12-22 & 25-29	Active tie	8-21	Active tie	18-33
			Losses	21.469 kW	Losses	21.957 kW	Losses	13.562 kW
	%THD	V (kV)	%THD	V (kV)	%THD	V (kV)	%THD	V (kV)
1	0	12.600	0.00	12.600	0.00	12.600	0.00	12.600
2	0.04	12.681	0.05	12.470	0.02	12.480	0.02	12.554
3	0.1	12.614	0.12	12.403	0.10	12.417	0.10	12.548
4	0.15	12.610	0.14	12.396	0.14	12.409	0.13	12.581
5	0.17	12.606	0.14	12.394	0.16	12.404	0.13	12.621
6	0.3	12.447	0.00	0.000	0.27	12.404	0.10	12.737
7	0.29	12.408	0.23	12.586	0.27	12.376	0.09	12.710
8	0.29	12.368	0.23	12.577	0.28	12.341	0.10	12.678
9	0.28	12.317	0.23	12.593	0.27	12.316	0.10	12.646
10	0.27	12.278	0.23	12.599	0.27	12.297	0.10	12.620
11	0.27	12.265	0.23	12.598	0.27	12.295	0.10	12.617
12	0.27	12.255	0.23	12.414	0.27	12.293	0.10	12.612
13	0.26	12.227	0.22	12.600	0.26	12.293	0.10	12.602
14	0.26	12.193	0.22	12.573	0.26	12.266	0.10	12.570
15	0.26	12.178	0.22	12.557	0.26	12.249	0.10	12.549
16	0.26	12.157	0.22	12.541	0.26	12.232	0.10	12.527
17	0.26	12.134	0.22	12.517	0.26	12.208	0.10	12.492
18	0.26	12.127	0.22	12.510	0.26	12.200	0.10	12.479
19	0.05	12.689	0.06	12.467	0.00	0.000	0.05	12.552
20	0.23	12.740	0.21	12.456	0.84	12.608	0.22	12.552
21	0.3	12.731	0.27	12.443	0.86	12.320	0.30	12.543
22	0.3	12.725	0.25	12.429	0.86	12.592	0.30	12.535
23	0.1	12.464	0.32	12.352	0.09	12.372	0.31	12.503
24	0.1	12.387	0.49	12.254	0.09	12.288	0.48	12.420
25	0.1	12.345	0.65	12.199	0.09	12.245	0.66	12.378
26	0.32	12.428	1.11	12.600	0.30	12.380	0.00	0.000
27	0.36	12.394	1.11	12.572	0.33	12.349	1.09	12.593
28	0.6	12.244	1.11	12.451	0.57	12.209	1.09	12.595
29	0.79	12.142	1.11	12.191	0.75	12.108	1.09	12.593
30	0.79	12.092	1.11	12.321	0.75	12.064	1.09	12.538
31	0.79	12.049	1.11	12.271	0.75	12.014	1.09	12.594
32	0.79	12.038	1.11	12.260	0.75	12.002	1.09	12.590
33	0.79	12.031	1.11	12.257	0.75	11.999	1.09	12.463

As seen in Table 2, the % THD and voltage values of buses 6, 19 and 26 are zero (in yellow highlighted cell). This is because the bus is in a condition of inactivity due to restoration action on the bus after reconfiguration using load curtailment. Reduced losses after reconfiguration is caused by the effect of restoration or release of load on the system so that the power supplied by the swing bus decreases.

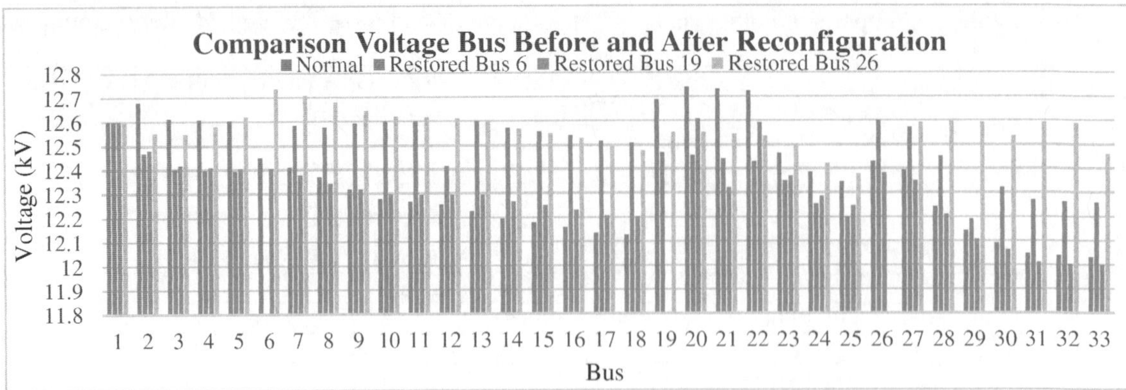

Figure 7. Comparison voltage bus before and after reconfiguration.

The result shows that load curtailment is able to determine the correct tie switch and closed and open CB in IEEE 33-bus Radial distribution system after placement DG with K-Means Clustering Method where all voltage bus can still keep the voltage value according to standard allowed. The voltage level change after reconfiguration and varies with a minimum voltage deviation value compare the previous research [8]. In figure 6, the existence of DG placement is able to improve the voltage level in all buses on the system despite the restoration that occurs on one bus. The voltage level for each bus as a whole is close to 1 pu.

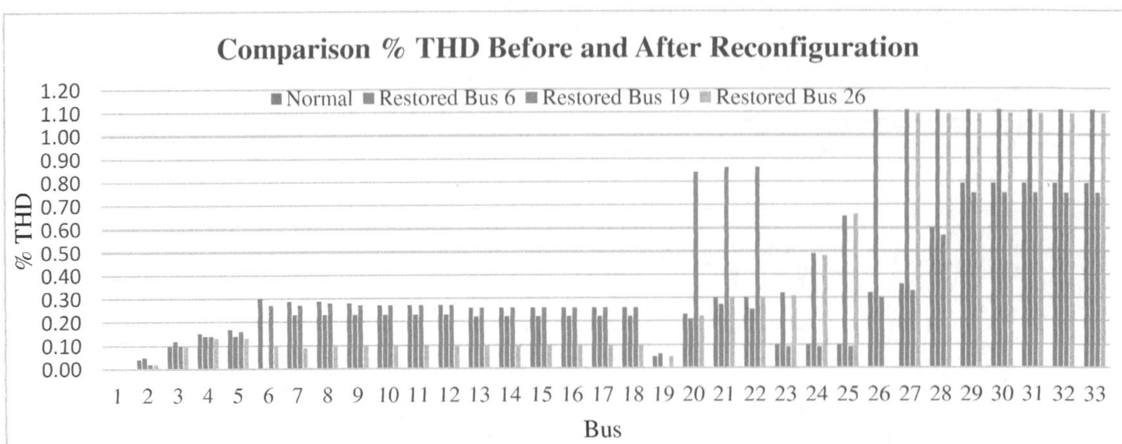

Figure 8. Comparison % THD before and after reconfiguration

As seen in Figure 7, the % THD value varies after reconfiguration. There is a reduction of % THD on buses that are placed DG as well as buses that are around the swing bus but different from other load buses. The occurrence of an increase in the value of% THD at some load bus points around the harmonic source due to the greater current flowing on the load bus and the greater impedance value due to the distance to the swing bus or PV bus getting farther away.

5. Conclusion

DG placement is a hot issue about developing power quality on radial distribution systems. The use of K-Means Clustering is considered to be able to improve the voltage level and reduce the harmonic spread in the Radial Distribution System. The reconfiguration action due to the restoration of one of the buses affects the DG that has been placed. The Load Curtailment method can determine the status of the tie switch that will be used. Maintain voltage levels on permitted conditions and decrease power losses due to DG placement after reconfiguration but not with harmonic spreads. Harmonic injection is the source of a given bus and the amount of current flowing on the bus.

References

[1] T Kurniawan, D. F. U. Putra & O. Penangsang, 2017, "Study of 3 Phase Active Load Flow on Radial Distribution System with Optimal DG Placement and Sizing using K-Means Clustering Method", Thesis Department of Electrical Engineering ITS Surabaya, Indonesia.

[2] X. Liang & A. Chowdhury, 2017, "Harmonic Mitigation through Advanced Control Method for Grid-Connected Renewable Energy Source", *2017-PSEC-0801*, **pp. 1-12**, Canada.

[3] A.I. Maswood, M.H. Haque, 2002, "Harmonics Source Effect and Mitigation Techniques", *Second International Conference on Electrical and Computer Engineering (ICECE)* **pp. 26-28**, Dhaka, Bangladesh.

[4] Aswin Rosyida, 2017, "Optimal Placement and Sizing Passive Filter in Radial Distribution System Using Whale Optimization Algorithm (WOA)", Thesis Electrical Engineering ITS Surabaya, Indonesia.

[5] O. Amanifar, 2013, "Optimal Distributed Generation Placement and Sizing for Loss and THD Reduction and Voltage Profile Improvement in Distribution System Using Particle Swarm Optimization and Sensitivity Analysis", Department of Electrical Engineering Isfahan University of Technology, Iran.

[6] A. Asri, O. Penangsang, R.S. Wibowo, 2015, "Optimizing Distribution Network Reconfiguration Using Load Curtailment to Isolate Interference in Distribution Automation System in Microgrid System", Thesis Electrical Engineering ITS Surabaya, Indonesia.

[7] Z.O. Prabandaru, O. Penangsang, & R.S. Wibowo., "Optimizing Location of Capacitor and Reconfiguration for Minimal Power Losses in Radial Distribution System Using Genetic Algorithm", Thesis Electrical Engineering ITS Surabaya, Indonesia.

[8] M.D. Faraby, & O. Penangsang., 2018, "A Study of Harmonic Spreading Against Distribution Network Reconfiguration in Passive Radial Distribution System", *4th International Conference Science and Technology 2018* **pp. 154**, Yogyakarta, Indonesia.

[9] Umar, Firdaus, M. Ashari & O. Penangsang, 2016, Optimal Location, Size and Type of DGs To Reduce Power Losses and Voltage Deviation Considering THD in Radial Unbalanced Distribution System", *2016 International Seminar on Intelligent Technology and Its Application*, **pp. 577-582.** Surabaya, Indonesia.

[10] Reza Rahmani, S. S. Aghaee, S. H. Hosseinian & S. H. H. Sadeghi, 2017, "Determining Maximum Penetration Level of Distributed Generation Source in Distribution Network Considering Harmonic Limit and Maintain Protection Coordination Scheme", *22nd Electrical Power Distribution Conference*, **pp. 196-199**, Semnan, Iran.

[11] Abraham Olatoke & Mohamed Darwish, 2012, "Relay Coordination and Harmonic Analysis in a Distribution Network with over 20% Renewable Source", Brunel University, UK.

[12] M.D. Faraby, M. Ashari., & O. Penangsang, 2017, "Optimal Placement and Sizing Single Tuned Filter With Inertia Weight Particle Swarm Optimization Method", *National Seminar Microwave, Antenna and Propagation 2017,* Universitas Jendral Achmad Yani Cimahi, Indonesia.

[13] Teng, J. dan Chang, C., 2007, "Backward-Forward Sweep-Based Harmonic Analysis Method for Distribution System", *IEEE Transaction on Power Delivery*, Vol. 22, No. 3, **pp. 1665-1672.**

[14] S. Florina, G. Gheorghe, C. Gianfranco and C. Gheorghe, 2012, "Using K-Means Clustering Method in Determiination of The Optimal Placement of Distributed Generation Sources in Electrical Distribution Systems," *IEEE.*

[15] R. Florina, C. Gianfranco, G. Gheorghe and C. Gheorghe, 2012, "Two-Stage Distributed Generation Optimal Sizing with Clustering-Based Node Selection," *Electrical Power and Energy System, ELSEVIER,* vol. 40, **pp. 120-129**.

[16] R. N. Rao, 2013, "Harmonic Analysis of Small Scale Industrial Loads and Harmonic Mitigation Techniques in Industrial Distribution System", *International Journal of Engineering Research and Applications*, Vol. 3, No. 4, **pp. 1511-1540**.

[17] A. K. Tagore & A. R. Gupta, 2017, "Harmonic Load Flow Analysis of Radial Distribution System in Presence of Distributed Generation", *2017 International Conference on Power and Embedded Drive Control (ICPEDC)*, **pp. 147-151**, India.

[18] A. K. Tagore & A. R. Gupta, 2017, "Impact of DG and D-STATCOM Allocation in Radial Distribution System for Reducing Harmonics", *8th ICCCNT 2017*, India.

IOP Conf. Series: Materials Science and Engineering **676** (2019) 012008 doi:10.1088/1757-899X/676/1/012008

Classification of News on "Radar" Tarakan Online Using K-Nearest Neighbor Method with N-Gram Features

E D Bintari[1]*, Gunawan[2] and A Indriani[1]

[1] STMIK PPKIA Tarakanita Rahmawati, Tarakan, Kalimantan Utara, Indonesia
[2] Sekolah Tinggi Teknik Surabaya, Surabaya

*Email: evidianti@ppkia.ac.id

Abstract. Classification or text categorization is one of the most common themes in analysing complex data. The classification in this research aims to define the class of an object that has not been known to the class. The classification process will be performed in the process of learning and testing against a known class dataset object. As an online news information, Radar Tarakan classifies news to ease readers to find the desired news and to ease the work of the administrators to do the automatic classification. The website of Radar Tarakan daily news (www.radartarakan.co.id) publishes approximately 480 local news everyday. Local news is the news in the surrounding areas of Tarakan in North Borneo. The website contains the classification of eight local news including the dimensions of the area surrounding the town of Tarakan. This research uses 720 news to build the system of news classification automatically. A text classification method used in this research is the K-Nearest Neighbor (KNN) with N-grams. The accuracy of matching classes generated by the K-Nearest Neighbor with N-Gram is 85.65% (out of 180 test data) and that of the sub class is 70.35 %. The classification time is around 2.7 second per news.

1. Introduction

The reliability of Internet technologies was instrumental in fulfilling the information needs. The people currently getting smarter to choose online media to fulfill information needs every time, because it can be accessed anytime and anywhere. Information/news can be seen through the website of the newspaper online. The daily newspaper Radar Tarakan have a Website (www.radartarakan.co.id.).

The website contains of news that have a several classifications, such as the economic news headlines, entertainment, sports and news regarding areas around Tarakan. Before the news was publish on its website, the first such news is classified in accordance with its classification. The process of the news is still done with the selection process of classification in the form of news, so the chances of selection news category error may occur. From the above condition that occurs, then it takes a discipline of text mining to overcome it.

The algorithms that are used in this classification is the K-Nearest Neighbor. Steps to be taken in the research starting from later entering information in the form of news from the website (www.radartarakan.co.id) and then do the preprocessing [1] with the groups of news, after that the preprocessing performed then an N-gram document of the next news feature n-gram into training and materials will eventually do the calculation with K-Nearest Neighbor method for the determination of the proper classification with the news that will be published on the website www.radartarakan.co.id. The end result of this research is the classification of local news and news that has the exact number of which are classified.

IOP Conf. Series: Materials Science and Engineering **676** (2019) 012008 doi:10.1088/1757-899X/676/1/012008

2. K-Nearest Neighbor

K-Nearest Neighbor algorithm (K-NN) classify the document test based on k nearest Neighbors [2]. Examples of training can be considered as vectors in a multidimensional feature space. This space is partitioned into a region with locations and labels of the training samples. Usually Euclidean distance or cosine similarity is used. During this stage, the data classification training (classes that need to be identified) is also represented as a vector in space features. Distance or similarity of test vectors for all vector training calculated and k nearest training samples are selected. There are several ways to classify a test vector for certain classes. Classic K-NN algorithm determines the class with a majority of the electors of the K-Nearest Neighbor. This research proposes to use a simple algorithm KNN without weighting features for text categorization. This paper proposes to use the method of selection of the features so finding the relevant features in the task of learning interactions (based on interdependence). KNN algorithm can be reduce a lot of number of features that was not be uses, so it is important for the size of a very large document such as the world wide web [3]. The ease of this algorithm makes more efficient computation time but can also be easily used by users instead of the expert. Conceptually, the example of document x called instance that was represented as a vector with a length of | F | so the size of the vocabulary are symbolised as follows:

$$< w1, w2(x)(x), w3(x), \dots \dots\, wF(x) >$$

Where the Wj(x) is the weight of the term j. Weights calculated from the obtained frequencies. KNN algorithm, distance was uses for the basis weighting for the contribution of each neighbor k in the process of determining the class. Next defined confidence document d to class c as:

$$\text{Confidence (c, d)} = \frac{\sum_{ki' \in k \mid (Class\,(ki')=c)} Sim\,(ki',d)}{\sum_{ki \in k} Sim\,(ki,d)} \tag{1}$$

Where the sim is the value returned by the functions in common used to compare with its Neighbors. that is, for each neighbor in the set of neighbors K (size k) belongs to a certain class c, then summed up the similarities to and document d divided by the sum of all the similarities of neighbors k relating to the document d. To compare the document d with instance i for example, selected functions very simple CosSim using term frequency weighting approach:

$$\text{CosSim (i, d)} = CosSim\,(i,d) = \frac{C}{\sqrt{A*B}} \tag{2}$$

Where C is the number of terms is the same between the document d with instance i, A is the number of term frequent on instance i (test data) and B is the number of frequent terms in document d (training data). So the highest score got the ranking for the determination of class.

3. N-Gram

An N-gram is a sequence of sub item N that already given from sequence. N-gram can be understood as long as window that moves as much as N of the text, the contents of this window is N-grams. N-gram level can be either character or word level, level statement. N-gram method is very widely used in natural language processing statistics. It depends on the language and works well in the case of the text that has the noise [4]. N-gram can be described in two differences. The first is to take a token by each character in a sentence. The second is from every word. For example, it is assumed that there is a sentence "A special utility is available to import images and insert them us links in the questions," then it can be extracted two different logic for N-Grams as follows:

- Logic first:

  -Trigram : {asp, spe, pec, eci, cia, ial,…}
  -4gram : {aspe, spec, peci, ecia, cial,ialu, ….}
  -5gram : {aspec, speci, pecia, ecial, cialu, ialut, …}

- The second Logic

  -Trigram : { A special utility, special utility is, …}
  -4gram : { A special utility is, special utility is available, …}
  -5gram : {{ A special utility, special utility is available, …}

In Figure 1 system architecture, the main process is described in detail in research on radar online local news classification method using the K-Nearest Neighbor (K-NN) with N-grams, system architecture consists of 3 stages, namely the process of data acquisition, data praproses and stage classification with detail calculation KNN with N-grams.

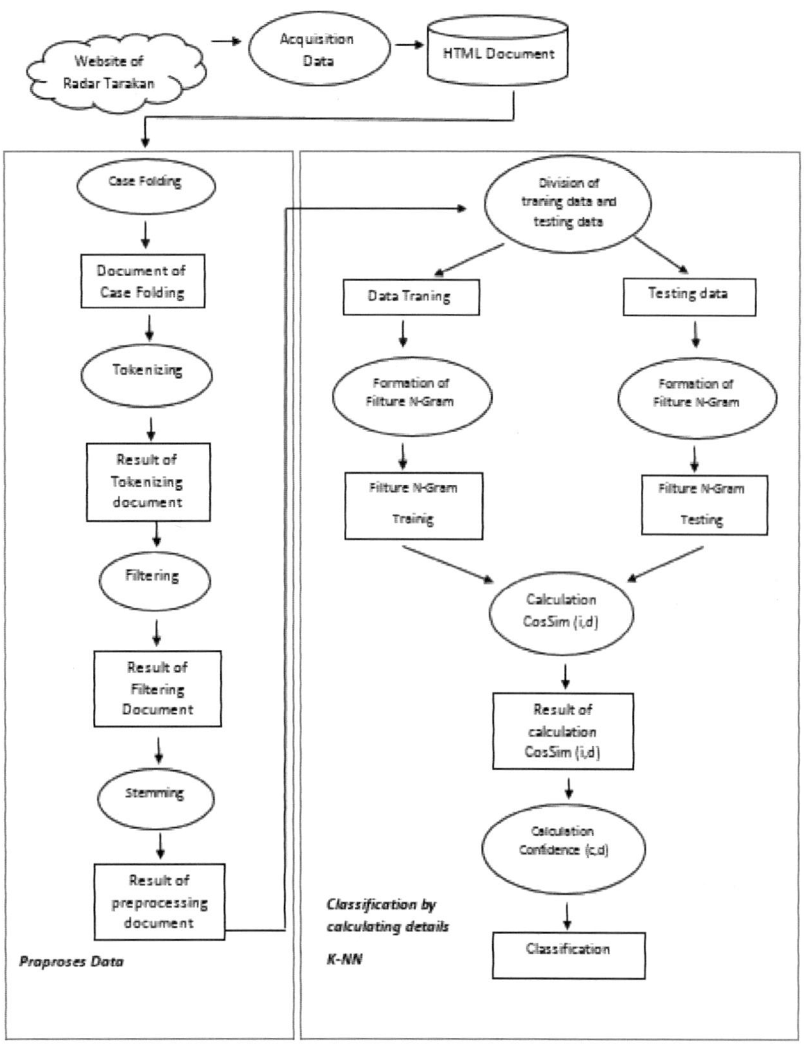

Figure 1. System architecture

In the early stages of the classification of local news radar Tarakan, data acquisition is performed by using a crawler to get documents from the news website radar Tarakan, after it's done pre processing the data consists of stage tokenizing, Filtering/stop word removal and stemming then conducted the process of division of training data and test data for N-gram feature formed, further from the N-gram feature done by the method of classification of K-Nearest Neighbor. Training data used to get classified documents while the test data used to test in the determination of the classification with the K-Nearest Neighbor method and achieved the expected accuracy.

IOP Conf. Series: Materials Science and Engineering **676** (2019) 012008 doi:10.1088/1757-899X/676/1/012008

Before doing the experiment, prepared 720 local news documents comprising the class dimension of the Tarakan to Kaltara with all the sub class of economy, politics, sports, education, development, crime and sub classes etc. In Table 1 and 2 can be seen the level of accuracy of the classification of local news with Unigram and N-Gram

Table 1. Unigram features accuracy results.

Data Training	Data Testing	Accuracy of classes	Accuration of sub classes
540	180	86.27%	64.71%
264	74	84.93%	64.38%
168	178	83.89%	65.10%
117	54	77.78%	72.22%
	Average	83.22%	67.34%

Table 2. Bigram features accuracy results.

Data Training	Data Testing	Accuracy of classes	Accuration of sub classes
540	180	88.24%	70.63%
264	74	86.30%	71.49%
168	178	86.58%	68.44%
117	54	81.48%	72.22%
	Average	85.65%	70.35%

After a trial of all features with a wide variety of comparative testing and the training dataset a different conclusion that the classification of local news with N-Gram increase accuracy of 2.43% and 3.01% for the class to the accuracy of the sub class with an average processing time of classification features of unigram 1 304.36 seconds, while the time it takes to process a classification feature N-Grams on average by 2, 371.58 seconds. Analysis of the writer can convey is can the occurrence classification accuracy trials have been conducted that training composition data distribution comparisons that balanced against a trainer all data classes and sub classes join in influencing whether or not high accuracy of classification of local news Radar online Tarakan.

In previous research, namely research that presents the results of classification of text documents of Arabic-speaking produces the value 87.62% precision and recall 63.41% experienced an increased precision of 10% of the unigram feature use. On research the author did an increase in accuracy precision value of 11%. The author can convey analysis is the use of the term N-Gram can provide increased precision and accuracy values because the system gets a choice selection of features against the dataset news, further distribution of the amount of data a balanced composition of the training classes and sub classes are very helpful to the process of classification has high accuracy.

The trial has been conducted on the decline the value of accuracy and precision and recall the sub class. Note on process tests the value of accuracy class can reach 84.84% while the value of accuracy class sub of 71,87%. Analysis of the author's response to the significant difference between the accuracy of the classes and sub classes is due at the sub class there is a sub class of others. Sub class of others that many hold datasets with different keywords.

4. Conclusions

From the discussion that already studied in this experiment, the conclusion can be drawn as follows:

1. Using text classification methods, namely the method of K-Nearest Neighbor (KNN) with N-Gram, in terms of the suitability of different types of classes and sub classes that are generated by the KNN against the determination of the types of classes that are conducted by the information (IT) Radar Tarakan belongs either. Percentage the accuracy of the match type of class against class 180 test data of 83.22% and sub class of 67.34%.

2. Features of the N-Gram has accuracy better than/bag unigram features of words/single term, with increased accuracy of ± 3%. Test data for as many as 180 features unigram/single term class type of match accuracy 85.65% and sub class of 70.35%.
3. KNN Method has high accuracy if the existing system has many categories (classes), because in the process of testing the results of prediction classes that do not fit with the manual class average is a sub class of others. The author's analysis is because a sub class of others hold many data sets that are diverse, making it difficult to find a suitable pattern for test data (testing).

5. Acknowledgment

This research was supported by STMIK PPKIA Tarakanita Rahmawati Tarakan in North Borneo, Radar Tarakan newspaper for our research object and Sekolah Tinggi Teknik Surabaya (STTS). We are thankful to our colleagues who provided expertise that greatly assisted the research.

6. References

[1] Byeongkyu Ko, Chang Choi, Pankoo Kim, Dongjin Choi, Junho Choi, 2012, Document Classification Through Building Specified N-Gram, Dept. of computer engineering Chosun University Gwangju, South Korea.

[2] Jiawei Han and Micheline Kamber, Data Mining : Concepts and Techniques, 2nd edition, Morgan Kaufmann, 2006.

[3] Raymond J. Mooney. 2006, CS, Machine Learning Text Categorization, University Of Texas at Austin.

[4] Riyad Al-Shalabi dan Rasha Obeidat, 2008, Improving KNN Arabic Text Classification with N-Grams Based Document Indexing, Jordan University For Science and Technology.

IOP Conf. Series: Materials Science and Engineering **676** (2019) 012009 doi:10.1088/1757-899X/676/1/012009

Monthly Rainfall Prediction Using Statistical Downscaling with Combination of Grid Boxes and Adaptive Neuro Fuzzy and Inference System in Lombok

A Safril[1]* and A Ulfiana[2]

[1]Sekolah Tinggi Meteorologi Klimatologi dan Geofisika, Tangerang Selatan, Banten, Indonesia
[2]Balai Besar Meteorologi Klimatologi dan Geofisika Wilayah IV, Makassar, Indonesia 90231

*E-mail: agus.safril@stmkg.ac.id

Abstract. Lombok is an island in Indonesia that has more people live in this area. Flood is a natural disaster that cause severe impact on society. To manage the risk of natural disaster, seasonal prediction of rainfall for early warning system are needed. However, prediction is difficult to predict due to El Niño and La Niña. Seasonal prediction of rainfall using Global Circulation Model (GCM) is useful to capture rainfall variability, but has a lack information because it has coarse of grid resolution (more than 200 kilometers). This model cannot provide in the local area so statistical downscaling model is used to get detail information. To optimize the result of prediction from global model, a method to capture the extreme condition is needed. Taking a best predictor that will be used to predict the rainfall is done to get the high performance of model used Singular Value Decomposition (SVD). ANFIS (Adaptive Neuro Fuzzy and Inference System) is a model prediction to capture the rainfall variability. Predictor is chosen based on the physical condition of atmosphere and ocean that have impact on Lombok Region. Selected area of predictor (atmospheric variable) are grid boxes in the Maritime Continent (80°E - 150°E and 12.5°N – 12.5°S) and Nusa Tenggara (105°E - 120°E, 10°S - 0). Eight variable predictors are used for prediction, they are air temperature at 2 m (T2M), geopotential height at 500 mb (Z500), zonal and meridional wind at 850 mb (U850 and V850) and at 200 mb (U200 and V200), sea level pressure (SLP), and air temperature at 850 millibar (T850). Both predictor variables (grid box combination) are used to predict for lead time one month. The result of research shows that rainfall prediction can capture the rainfall variability (correlation 0.70) with RMSE (Root Mean Square Error) is 69.

1. Introduction

Lombok is a small island with a mountain range in it. Moreover, Lombok also has rivers whose upstream is in mountainous area and downstream is in lowlands area. These conditions cause various natural disasters such as droughts, floods, or landslides. In West Lombok, floods occurred several times during the rainy season and experienced drought during the dry season [1].

Besides having the potential hazards of flooding and drought, Lombok is also known as earthquake prone area. This has an effect on potency of landslide. Earthquake and high rainfall area, high potential of the landslide, areas with ground cracks due to the earthquake with steep slope can result in floods and landslides during the rainy season. One of the causes of soil movement or landslides is high rainfall as a trigger [2]. Monthly rainfall predictions are needed to obtain information of rainfall condition in the future so that they can be used for mitigation (actions to reduce the impact of disaster both life and property). In addition of disaster mitigation needed, information of rainfall prediction also useful for decision making in agriculture caused by climate variability due to ENSO (El Niño Southern Oscillation) [3]. It is because Lombok is also known as the center of paddy, some secondary crops, and vegetables farming [4]. Rain prediction information is used to determine the cropping pattern when planting will begin and the suitable crops.

Monthly rainfall predictions using Global Circulation Model (GCM) can capture rainfall patterns in the next few months, but information obtained in low resolution (more than 200 kilometers). Therefore, a statistical downscaling method is needed to get rainfall prediction in local scale [5]. Predictors that used in this study are various atmospheric variables (air temperature, air pressure, air humidity, etc.), while predictand (that which is predicted) is rainfall.

In previous studies related to statistical downscaling using grid box (single) and global predictors for predictions in Indonesia region with grid boxes in the study area (Sumatra) [6]. Predictors characteristics of local grid box area around the study area are needed to obtain local patterns. In that context, research is carried out by combining grid boxes with global and local regions in the grid of study area.

2. Data and Method

2.1. Data

Predictor variables data that will be used for prediction are hindcast data (predictions of the past period) output of the Global Circulation Model (GCM) from National Centers for Environmental Prediction (NCEP) in monthly period with the output of 15 ensemble members. This result is averaged so that the average ensemble is obtained and used for prediction one month ahead (one-month lead time). Data reanalysis can be downloaded at NOAA (https://www.esrl.noaa.gov/) to perform a physical analysis of the selection of predictor boxes. Data resolution for analytical purposes is 2.5° (250 kilometers).

Data of predictor variables used for this study are eight variables consisting of zonal and meridional wind at 850 millibars (U850 and V850) and at 200 millibars (U200 and V200), surface air temperature (T2M), air temperature at 850 millibars (T850), air pressure at sea level (SLP), and geopotential height at 500 millibars (Z500) as in Table 1 offered at 2.5° (250 kilometers) resolution GCM. The 25-year data period is hindcast data (predictions that were run in the past between 1981-2005. Predictor data is used for prediction time period for the next month

Table 1. GCM atmospheric independent variables (wind, air pressure, air temperature, geopotential height) and symbols used (U850, V850, U200, V200, T200, T850, and Z500)

No	Independent Variables	Symbols
1	Zonal wind at 850 millibars	U850
2	Meridional wind at 850 millibars	V850
3	Zonal wind at 200 millibars	U200
4	Meridional wind at 200 millibars	V200
5	Surface air temperature (at 2 meters)	T2M
6	Air temperature at 850 millibars	T850
7	Air pressure at sea surface	SLP
8	Geopotential height at 500 millibars	Z500

The rainfall data as predictor variable is monthly rainfall data sourced from Indonesia Agency for Meteorology, Climatology and Geophysics (BMKG) from 1981-2005 namely Dasan Tereng (latitude: -8.584052, longitude: 116.18441 E, Elevation: 156 m), Narmada District, West Lombok Regency, West Nusa Tenggara (Figure 1).

IOP Conf. Series: Materials Science and Engineering **676** (2019) 012009 doi:10.1088/1757-899X/676/1/012009

Figure 1. Study area (rainfall station) in Narmada District, Lombok

The selected predictor areas (grid boxes) are used for predictors (Figure 2), which include:

1. Regional area (Maritime Continent) in the 80°E-150°E, 12.5°N-12.5°S
2. Java Sea (around Lombok) in the 105°E-120°E, 0-10°S.

To obtain the best predictor variable, correlation analysis was performed on all 8 predictor variables (U850, U200, U850, V850, T850, T2M, SLP, and Z500) from the two grid boxes. The result of a predictor window combination of two predictor variables from these two predictor boxes is used as the input for prediction using the statistical downscaling method.

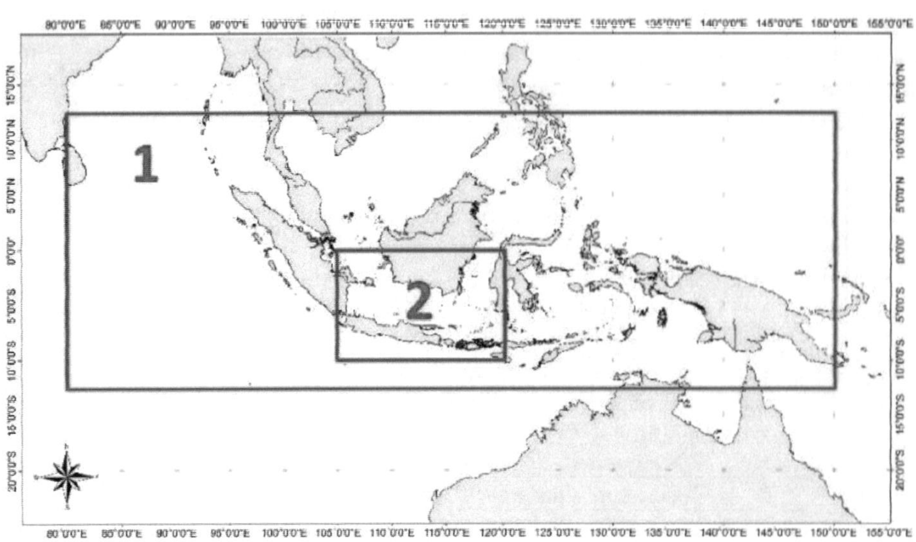

Figure 2. Grid boxes of predictors on Maritime Continent and Java Sea. Grid box 1 represents the global region and grid box 2 for the local area

2.2. Method

2.2.1. Grid boxes selection based on atmospheric physical characteristics. Selection of the grid boxes for the predictor area is based on physical characteristics that affect rainfall in the Maritime Continent (Figure 3). The Maritime Continent region has high solar radiation so that it becomes a source of water vapor for Indonesia region [7]. In addition, the islands in Indonesia are influenced by several global factors such as monsoon and El Nino Southern Oscillation (ENSO) so that the variability of rainfall becomes high [8]. The monsoon pattern is also dominant in the territory of Indonesia. For the Nusa Tenggara region, rainfall pattern is Monsoon A with a peak rainfall [9]. In December-January-February, the wind pattern is westerly (the west coast region of Sumatra, the Western Pacific, and the South China Sea) which brings rainfall in the Nusa Tenggara Islands region as in Figure 3 (big box) (80°E-150°E, 12.5°N -12.5°S), while the small box for predictors is around the Nusa Tenggara region (local characters) (105°E-120°E, 0-10°S)

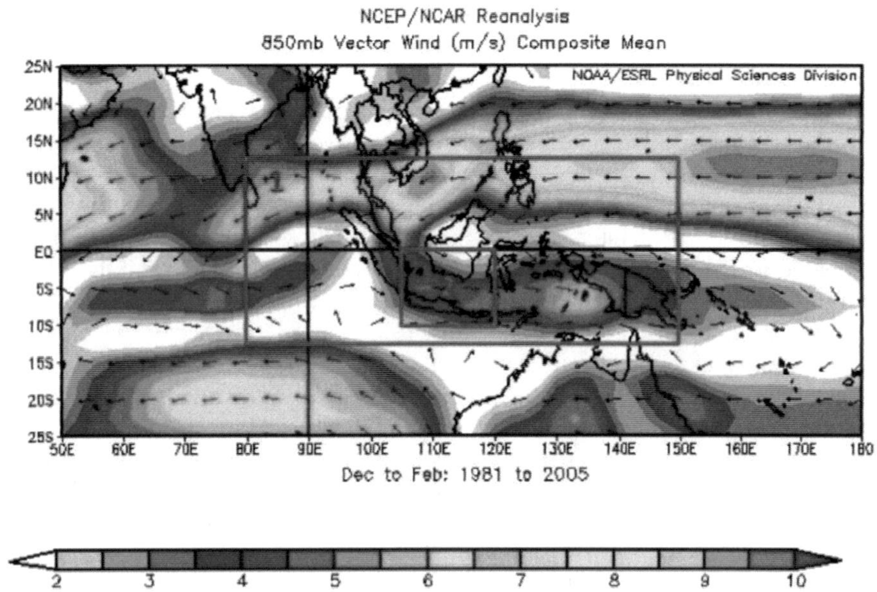

Figure 3. Wind pattern average in December-January-February. Big box (grid box of Maritime Continent area) and small box (Nusa Tenggara region)

2.2.2. SVD correlation by reducing multicollinearity and dominant mode. Singular Value Decomposition (SVD) or Principal Component Analysis (PCA) is a method to see dominant factors that influence the characteristics of rainfall in the study area and eliminate multicollinearity. The SVD method provides a physical interpretation of the results of the relationship between predictor variables and rainfall statistically [10]. The dominant pattern used is mainly in capturing monsoon variability. The Nusa Tenggara region is an area with an annual rainfall pattern that is strongly influenced by the monsoon. This method identifies pairs of fields (matrices) to patterns of space and time. Each pair of fields describes a covariance fraction. In the dominant mode, the strength of the relationship between predictor variables with rainfall is seen with the correlation method. Relationship between variables is strong when approaching 1 and weak when approaching 0 (zero).

The independent variable **X** is expressed in a matrix with two predictor variables from the GCM model in the grid box 1 (Maritime Continent) and grid box 2 (Nusa Tenggara). The correlation between each predictor variable and Lombok observation rainfall (matrix **Y**) is computed to find the highest correlation. The results of the variables that have the highest correlation are used as predictor variables for each grid box. The best variables in the first and second grid boxes are then used as independent variables.

Next the covariance matrix (**XT** can be calculated if it has the same number of rows as **Y** as the time series).

$$C = XT\ Y \tag{1}$$

IOP Conf. Series: Materials Science and Engineering **676** (2019) 012009 doi:10.1088/1757-899X/676/1/012009

After the covariance matrix **C** is formed, then the Singular Value Decomposition (SVD) is calculated and matrices **U, V** and diagonal matrix **L** is obtained, so that:

$$C = ULV\ T \tag{2}$$

where the singular vector **X** is the columns in **U** and the singular vector of **Y** is the columns of **V**. Then from this matrices mode (coefficient) expansion is obtained in the time series between time series (predictor variables) and precipitation series (predictand) and correlated. Correlation values ranged from 0 to 1. Correlation approaching 1 shows the strong relationship between predictand and predictor variables, when approaching 0 indicates a weak relationship between the two variables.

2.2.3. Regression model to capture extreme rainfall pattern.
Indonesia region is influenced by various climate phenomena so a model is needed to capture the variability. The Maritime Continent region is influenced by several global factors so that the variability of rainfall becomes high. This is caused by the Indonesian Maritime Continent which is predominantly influenced by seasonal variability due to the Asian monsoon [11]. In addition, it is also influenced by the phenomenon of inter-annual variability such as El Niño-La Niña and IOD (Indian Ocean Dipole) alone or simultaneously [12]. This condition causes high rainfall variability so that extreme rainfall (high or low) and irregular often occur.

Based on these conditions, a prediction model that can capture these extreme patterns is needed. ANFIS (Adaptive Neuro Fuzzy Inference System) is a regression method that can capture nonlinear patterns and chaos. This regression prediction model based on artificial intelligence used to capture the behavior of past rain anomalies that are used for the prediction model in the future. For the purposes of predictive capability analysis, 22 years of training data and 3-year testing data are used with the *gbellmf* membership function and *Sugeno* type. Prediction with ANFIS uses the *Sugeno* type, with membership numbers (*numMFs*) are two, the membership function pattern input is '*gbellmf*' and the membership function output is '*linear*'. The ANFIS structure is presented in Figure 4. The figure shows two data inputs (x, y).

In this ANFIS structure, EBP (Error Back Propagation) method is carried out in the first layer and LSE (Least Square Estimator) [13]. In the ANFIS structure the process for updating the fuzzy membership parameters is done at the first layer by EBP and for updating the linear parameters of the system output at layer four by LSE. ANFIS has the ability to recognize past rainfall anomalies and reduce prediction errors.

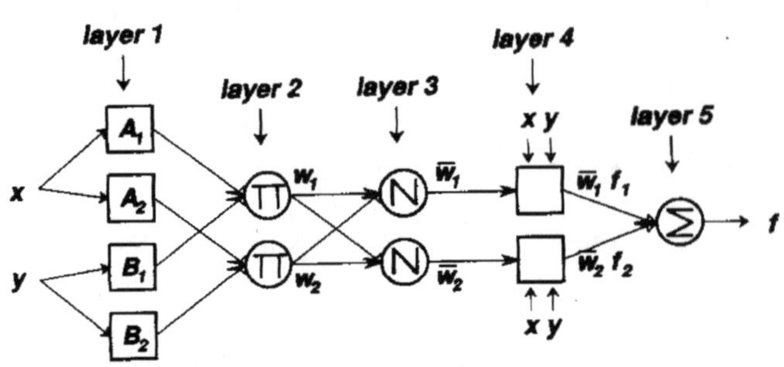

Figure 4. ANFIS multilayer structure [14]

2.2.4. Prediction results verification.
After prediction, to see the accuracy of the model, a model verification analysis is performed. The first analysis uses qualitative descriptive analysis using graphs by looking at the comparison between the rain signal patterns of prediction results and observation. Quantitatively, the reliability of the prediction model is calculated by the regression variance or RMSE (root mean square error) which is obtained from the square of the difference between the predictions y_k and observations o_k rainfall [15]. RMSE values are getting smaller (approaching zero) indicates better model. In addition, a correlation

is also made between the results of prediction and observation. High correlation value shows the model is getting better.

$$MSE = \frac{1}{n-u}\sum_{k=1}^{n}(y_k - o_k) \quad (3)$$

$$RMSE = \sqrt{MSE} \quad (4)$$

3. Results and Discussion

3.1. Correlation results between predictors and predictand

Correlation results between predictor variables and rainfall using Singular Value Decomposition (SVD) show strong correlation. The highest correlation is indicated by surface air temperature (2 meters) variable with a correlation on the Maritime Continental grid box (0.72) and local (0.71) as in Table 2. Predictions are made with multiple independent variables with predictors from the Maritime Continent and local grid boxes. This best variable becomes the input regression model using ANFIS.

Table 2. Correlation between predictor variables and rainfall

No	Variables	Grid Box	
		Maritime Continent	Local
1	U850	0.70	0.42
2	V850	0.68	0.57
3	U200	0.52	0.54
4	V200	0.67	0.60
5	T2M	**0.72**	**0.71**
6	T850	0.70	0.65
7	SLP	0.69	0.55
8	Z500	0.30	0.29
Best	T2M	**0.72**	**0.71**

3.2. Rainfall pattern in Lombok Island and model verification

Scatter analysis of rainfall from 25 years observations between monthly rainfall. Monthly rainfall in Lombok shows a monsoonal pattern (one peak of rainfall). Rainfall with value is greater than 150 millimeters per month occur in November to April with the peak rainfall in December. Dry conditions begin to occur in April and the driest in July continued to August (Figure 4). This dry soil condition and cracking due to the earthquake and when filled with water will cause a subsequent landslide.

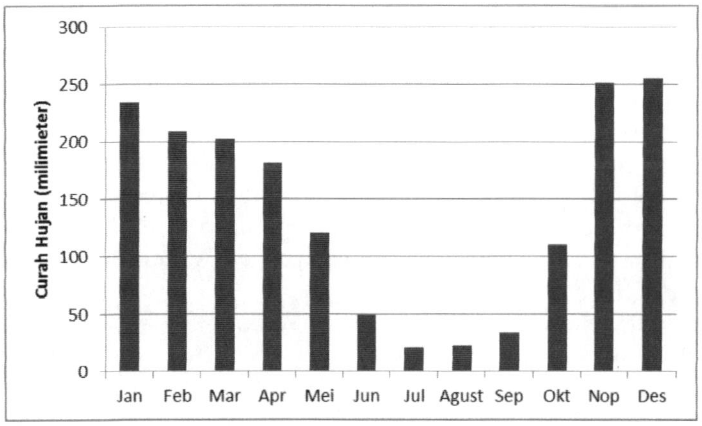

Figure 5. Monsoonal rainfall pattern in Lombok

The prediction result shows the rainfall signal between the prediction (thick line) and observation (dashed line) can capture the observation rainfall pattern (monsoon) as in Figure 6. In a few years the high variability between November and July has not fully captured the pattern local rainfall, but can quantitatively capture extreme rainfall that exceeds 250 millimeters per month.

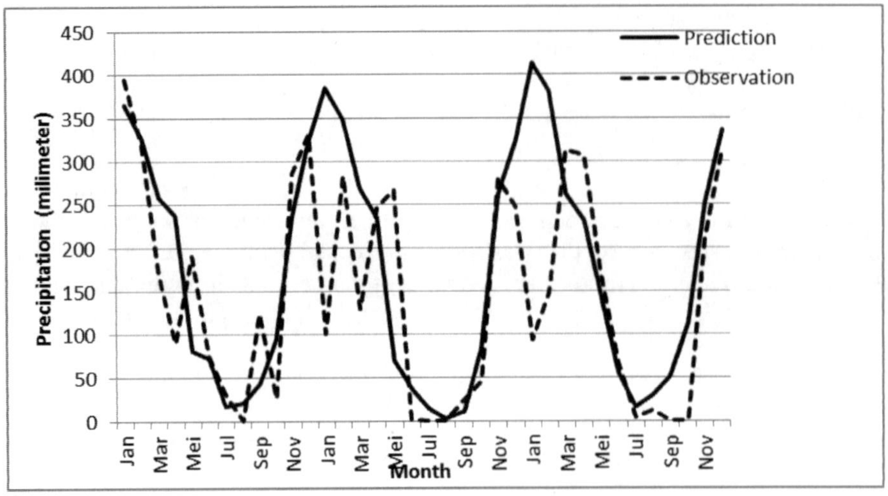

Figure 6. Comparison between prediction result and observation rainfall pattern

Furthermore, based on quantitative analysis calculated the correlation value and RMSE value. The result of the correlation between prediction and observation by simulating the prediction of 2003-2005 shows that rainfall prediction can capture high rainfall patterns (Figure 6). Correlation values between prediction and observation is strongly related (0.70) and RMSE (Root Mean Square Error) is 69 millimeters.

4. Conclusions

Correlation analysis between predictors and predictand shows the best variable for being grid box predictor to be used for rainfall prediction in Lombok is surface air temperature (T2M) with relatively strong correlation (0.72). From the best predictor variable, it is obtained a regression model with predictor combination based on grid boxes in the Maritime Continent and local around West Nusa Tenggara which is able to capture extreme rainfall patterns of monsoon and inter-annual variability. Predictive model capability shows a strong relationship between prediction and observation results, shows high correlation values (0.70) and RMSE 69 millimetres of monthly rainfall which generally ranges from 100 millimetres per month.

References

[1] Marchiavelly M I, Narieswari L, Munajati S L, Sumaryono, Santoso W E and Martha S 2012 Hazard risk mapping on tourism region of West Lombok Regency, West Nusa Tenggara *Majalah Ilmiah Globe* **14** 187-99.

[2] Sipayung S B, Cholianawati N, Susanti I, Aulia A and Maryadi E 2014 Development of empirical equation model in predicting the occurrence of landslide at watershed of Citarum (West Java) based on the TRMM satellite data *Jurnal Sains Dirgantara* **12** 12-21.

[3] Utami A W, Jamhari and Hardyastuti S 2017 El Nino, La Nina, dan penawaran pangan di Jawa, Indonesia *Jurnal Ekonomi Pembangunan* **12** 257-71.

[4] BPS-Statistics of Lombok Barat Regency 2016 *Lombok Barat Regency in Figures* (Gerung: BPS-Statistics of Lombok Barat Regency) p 239.

[5] Juneng L, Tangang F T, Kang H, Lee W J and Seng Y K 2010 Statistical downscaling forecasts for winter monsoon precipitation in Malaysia, using multimodel output variables *J. Clim.* **23** 17-27.

[6] Safril A, Hadi T W, Hadi S and Tjasjono B H K 2013 Prediksi curah hujan bulanan menggunakan Adaptive Statistical Downscaling *Jurnal Meteorologi dan Geofisika* **14** 25-31

[7] Ramage 1968 Role of tropical maritim continent in the atmospheric circulation *Mon. Weather Rev.* **96** 6.

[8] Chang C P, Wang Z, Ju J and Li T 2004 On the relationship between Western Maritim Continent monsoon rainfall and ENSO during northern winter *J. Clim.* **17** 665-72.

[9] Aldrian E and Susanto R D 2003 Identification of three dominant rainfall regions within Indonesia and their relationship to sea surface temperature *Int. J. Climatol.* **23** 1435-52.

[10] Björnsson H and Venegas S A 1997 *A Manual for EOF and SVD Analyses of Climatic Data* (Montreal: Department of Atmospheric and Oceanic Sciences and Centre for Climate and Global Change Research, Mc Gill University) p 23.

[11] Harijono S W B 2008 *Interaksi Fenomena El Nino dan Dipole Mode Secara Simultan serta Monsun Musim Panas India Terhadap Variabilitas Curah Hujan di Sumatera Bagian Utara* (Bandung: Institute Technology Bandung)

[12] Tjasyono B H K, Lubis A, Junaeni I., Ruminta and Harijono S W B 2008 Dampak variasi temperatur Samudera Pasifik dan Hindia Ekuatorial terhadap curah hujan di Indonesia *Jurnal Sains Dirgantara* **5** 83-95.

[13] Fariza A, Helen A and Rasyid A 2007 *Proc. Seminar Nasional Aplikasi Teknologi Informasi (SNATI)* (Yogyakarta: Jurusan Teknik Informatika, Fakultas Teknologi Industri, Universitas Islam Indonesia) pp 77-82.

[14] Jang J S R 1993 ANFIS: Adaptive-Network-Basec Fuzzy Inference System *IEEE Trans Syst Man Cybern* **23** 665-685.

[15] Wilks D S 1995 *Statistical Methods in the Atmospheric Sciences* (New York: Cornel University) p 279.

IOP Conf. Series: Materials Science and Engineering **676** (2019) 012010 doi:10.1088/1757-899X/676/1/012010

Integration of LoRa-Cellular: Design and Implementation of Data Communication in Vehicle Tracking Systems

A A Ilham[1]*, Adnan[1] and R Angriawan[2]

[1]Department of Informatics, Engineering Faculty, Hasanuddin University, Indonesia
[2]Department of Electrical Engineering, Engineering Faculty, Hasanuddin University, Indonesia

*E-mail: amil@unhas.ac.id

Abstract. The development of communication technology has improved efficient communication in tracking mobile objects. Several types of communication links are available for the tracking systems. Among them, cellular is the most widely used communication network for tracking inter-city areas. However, the implementation of IoT using cellular network requires the availability of communication network from network service providers. The communication will be lost for the area which is not covered by the cellular network. To overcome this problem, in this paper we proposed the Integration of LoRa-Cellular (ILC) using hybrid communications: LoRa network and the cellular network to trace a vehicle position. The decision which network used for communication is decided based on the level of RSSI. The energy consumption is also a parameter in the communication link selection. The transition process uses soft-handover to maintain the availability of communication links and reduce data loss. Our experiments show that the threshold of LoRa RSSI level uses -120 dBm and the threshold of cellular RSSI level uses -85 dBm. Energy consumption of Lora and cellular while sending data has a significant difference. Lora consumption is 292 mW and cellular consumption is 497 mW.

1. Introduction

The mobile object tracking system is a system designed for remote observation, security, and the location of objects while cruising away. It involves tracking humans [1], animals [2], and vehicles [3] as shown in a recent study. Location information, transmission media, and storage media are the main parameters for location accuracy and real-time monitoring [4].

The mobile object tracking system with various transmission media and communication network was developed by researchers. This communication network is cellular [5], Zigbee [6], and LoRa [7]. Several aspects considered in recent studies for the development of tracking systems. These aspects are the area of coverage, power consumption, and implementation costs. The focus in developing tracking systems is to create powerful tracking with low cost [8].

Every network in the communication system has their own drawback. As with the cellular network, the tracking area relies heavily on the network infrastructure of the service provider. A problem arises when a region is not covered by network service in some tracking areas (blank spot area) such as in suburban areas and mountain areas. Another example such as LoRa network, The Lora network has limited area coverage, expansion of coverage requires additional infrastructure and more costs. The implementation of the Lora network on vehicle tracking has constraints on unlimited mobility and large tracking areas. These problems need costs to expand the coverage area. Example case, the application of the Lora network consists of node and gateway. It provides long-distance communication, around 10 km [9]. If the tracking area is more than 10 km, a new gateway is needed.

Muhammad et al in his research made a Vehicle Tracking System and history recorder of vehicle journey using GSM (Cellular) as a communication network, but the tracking area depends on the availability of the GSM network around the tracking area [10]. Another study by Kim et al. Used the LoRa as a communication link on vehicle tracking systems. The proposed system does not depend on the availability of the network service provider [7]. Next research by Team et al developed an energy-efficient and low-cost vehicle tracking system using LoRa as a communication system [9]. Next research by ham develops IoT system to track and monitor patients with mental disorders using Lora as a communication link. This research evaluates the performance of Lora in the conditions of Line of Sight (LoS), Non-Line of Sight (N-LoS), and the "crowded area" [11].

A good object tracking system is a tracking system with communication networks available anytime and anywhere, but every communication system has their own limitations in providing network service. Therefore, in this research, we developed a tracking system that provides an alternative to track vehicles in areas not covered by cellular networks. The alternative is the Lora network, our system is Integration of LoRa-Cellular (ILC). ILC design has two communication network services. Lora and Seluler are integrated to expand the coverage area and always provide network services during the monitoring process. Not only that, the system that we propose can also be implemented in other cases that require a backup of communication.

2. Architecture and Design of System

2.1. Tracking architecture

Figure 1 shows an overview of the vehicle tracking architecture. The main parts of the development are Node-ILC and Lora-Gateway, Node-ILC is the transmitter of the location obtained from GPS via one of the Lora networks or cellular networks, and Lora-Gateway is the recipient of data from Nodes-ILCs and bridges to connect to the internet.

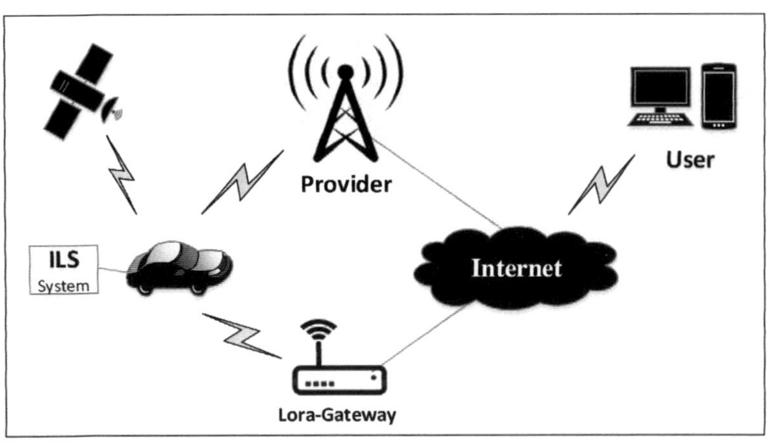

Figure 1. Tracking Architecture on ILC

2.2. Data delivery scheme

The proposed system has two different communication links with the same destination from Node-ILC to the internet. First destination: Node-ILC → LoRa-Gateway → Cellular BTS → internet. Second destination: Node-ILC → Cellular BTS → internet. The process of sending location data from the

IOP Conf. Series: Materials Science and Engineering **676** (2019) 012010 doi:10.1088/1757-899X/676/1/012010

Node-ILC to Users starts from the device selects communication link to be used according to the available network, if the device chooses LoRa, the location data will be sent to the LoRa-Gateway then forwarded to the Internet, if the Node-ILC selects Cellular, the location data will be sent to the BTS then forwarded to the internet.

3. Suggestion Method

This chapter discusses case studies and methodologies for the proposed system. Figure 2 shows the transition scenario of the communication link between LoRa network and cellular network. Area A and area B are cellular network coverage and area C is blank spot area. Assume that Lora-Gateway will be installed on Area C as an alternative data transmission. If a vehicle in either area A or area B, Node-ILC will use cellular services. But when a vehicle crosses an area where there is no cellular network available, the Node-ILC will use Lora to send data. In another case, when a vehicle crosses an area covered by both the cellular network and the Lora network (A-C / C-B), ILC Node measures RSSI level. ILC will use RSSI level measurements as a reference for the choice of the communication link in accordance with RSSI level measurements in table V. Transitions Service are carried out when the vehicle is between the area A-C / C-B.

The type of handover on ILC is soft-handover, the soft-handover steps on ILC are as follows:

- Measurement step, the first step is to measure the signal strength of the cellular network and LoRa network.
- Decision step, the measurement results are compared with the RSSI Level threshold in table V as a reference for the handover process.
- Execution step, execution of the handover will take place at this step

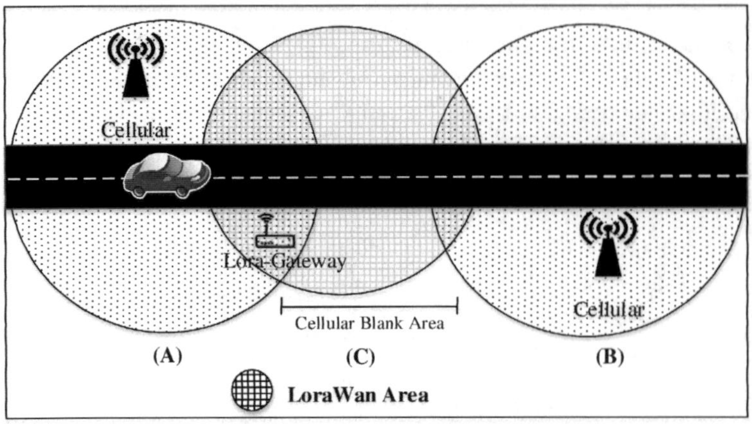

Figure 2. Transition Scenario

4. Hardware and Software Implementation

4.1. Hardware

In the design of hardware, the hardware part is divided into two functional parts, namely Node-ILC and LoRa-Gateway. Components of the Node-ILC are Arduino Mega, GPS Module NEO-6M, 915Mhz Dragino LoRa Shield, external antenna 3 dBi and SIM900 Mini V4.0. Node-ILC voltage source is a power bank. For components of the LoRa-Gateway are Arduino Uno, Dragino LoRa Shield 915Mhz, Ground Plane Antenna FPV Telemetry, SIM900 Mini V4.0 and a voltage source using a power bank.

IOP Conf. Series: Materials Science and Engineering **676** (2019) 012010 doi:10.1088/1757-899X/676/1/012010

4.2. Software implementation
Figure 3 shows the flowchart of the proposed system. Part A is Node-ILC software and part B is LoRa-Gateway software. The Node-ILC software is designed to send the location to LoRa-Gateway or cellular BTS. The LoRa-Gateway software is designed to forward the packet from the Node-ILC to the internet.

We use the Arduino Integrated Development Environment (Arduino-IDE) to develop the proposed tracking system. We use RadioHead library and TinyGPS++ library.

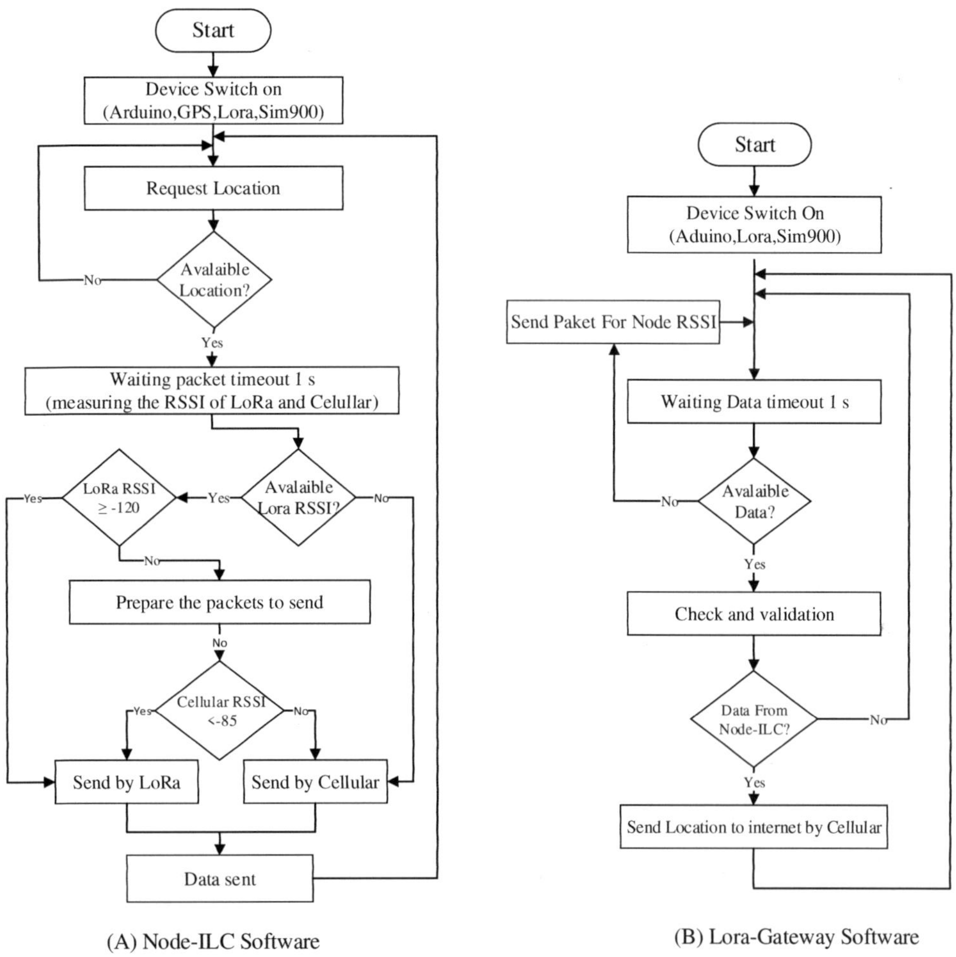

Figure 3. Flowchart of the proposed system

5. Experiment and Result

5.1. Experiment scenario
We conducted some experiments to obtain the RSSI level, packet delivery ratio, and energy consumption. The following are the experimental step of RSSI level and packet delivery ratio:

- The Node-ILC is installed on the vehicle.
- Vehicles will move away from LoRa-Gateway (if using LoRa) or cellular base transceiver station (if using cellular)
- The Node-ILC measures RSSI levels and acquires the current location from the GPS module

- Next process, the Node-ILC combines several data such as current location, RSSI levels and the current time.
- The Node-ILC sends packet periodically, each packet has a serial number for the packet identifier.
- We get the packet delivery ratio from the number of packets received

Details of the packets sent are the location of the vehicle, the current time and RSSI level (LoRa and cellular) with a total size of 50 bytes. The location experiment of LoRa network was on Malino street, Borongloe, Bontomarannu, Gowa Regency, South Sulawesi (Faculty of Engineering, Hasanuddin University). The location experiment of the cellular network was on Inspeksi PAM street, Manggala, Makassar City, South Sulawesi. Table 1 shows the LoRa configuration.

The measurement of energy consumption in LoRa and cellular (Sim900A) uses Adafruit INA219 Current Sensor Breakout in two conditions. The conditions are transmission and standby.

Table 1. Lora configuration

Lora Configuration	Setting Value
Bandwidth	125 kHz
Preamble Length	8
Coding Rate	4/5
Spreading Factor	7
TX Power	13 dBm

5.2. Analyzing performance

This chapter discusses transmission performance on LoRa network and cellular network based on RSSI Level. The results of the experiments are in table 2 and table 3. A good RSSI level is RSSI level near 0. For example, the signal strength of -120 dBm is better than the signal strength of -100 dBm. The lowest RSSI level of LoRa is between -120 dBm to -130 dBm with packet delivery ratio of 67%. The lowest RSSI level of Cellular is between -80 dBm and -90 dBm with packet delivery ratio of 76%. The both lowest RSSI level is the threshold for reference in the handover process.

Table 2. Lora Performance

Packet Detail	$-100 \leq X < -90$	$-110 \leq X < -100$	$-120 \leq X < -110$	$-120 \leq X < -130$
Send Packet	100	100	100	100
Received Packet	98	93	80	67
Lost Packet	2	7	20	37
Received Rate	98%	93%	80%	67%

Table 3. Cellular (Sim900A) Performance

Packet Detail	$-60 \leq X \leq -50$	$-70 \leq X < -60$	$-80 \leq X < -70$	$-90 \leq X < -80$
Send Packet	50	50	50	50
Received Packet	49	46	43	38
Lost Packet	1	4	7	12
Received Rate	98%	92%	86%	76%

IOP Conf. Series: Materials Science and Engineering **676** (2019) 012010 doi:10.1088/1757-899X/676/1/012010

5.3. Comparison energy consumption
Energy consumption is an additional parameter for choosing a communication link. Table 4 shows energy consumption on Sim900 (cellular) and LoRa.

Table 4. Energi consumption

Device		Current (mA)	Power (mW)
Cellular	Standby	55	247
	Send Packet	107	497
LoRa	Standby	6	21
	Send Packet	85	292

5.4. Communication link selection
The results of the experiment provide information about the proper condition of the RSSI level to send data. Table 5 shows the threshold of RSSI level. It is the reference for the communication link selection. When the LoRa RSSI level is greater than -120 dBm, the LoRa network will become the communication link. While the cellular network becomes a communication link when the RSSI level is greater than -85 dBm and the RSSI LoRa is smaller at -120 dBm.

Table 5. Threshold of communication link selection

RSSI		Seluler (dBm)	
		≥ -85	**≤ -85**
LoRa (dBm)	**≥ -120**	LoRa	LoRa
	≤ -120	Seluler	LoRa

5.5. Implementation results of ILC
This section explains the results of implementing the ILC. Experiment results can be seen in table 5. The experimental results show the service transition process. The first transition occurs on line 10. The LoRa network and cellular network do not reach the threshold so that the communication link uses the default option (LoRa network). Next transitions occur on line 12. Cellular networks reach the threshold requirements and LoRa network does not reach the threshold requirement so communication links use the cellular network. Next, on line 15, RSSI level of Lora network reach the threshold so that the communication link uses the LoRa network.

Table 6. Experiment results of ILC

No	Longitude	Latitude	Cellular RSSI	Lora RSSI	Time Node-ILC	Description
1	-5.124391	119.54945	-90	0	06/30/2018-08:09:32	Cellular
2	-5.124383	119.54945	-90	0	06/30/2018-08:09:36	Cellular
3	-5.124378	119.54945	-90	0	06/30/2018-08:09:42	Cellular
4	-5.124376	119.54945	-90	0	06/30/2018-08:09:45	Cellular
5	-5.124369	119.54944	-90	0	06/30/2018-08:09:52	Cellular
6	-5.12437	119.54945	-90	0	06/30/2018-08:09:57	Cellular

7	-5.124376	119.54945	-90	0	06/30/2018-08:10:00	Cellular
8	-5.124373	119.54943	-90	0	06/30/2018-08:10:02	Cellular
9	-5.124383	119.54942	-90	0	06/30/2018-08:10:06	Cellular
10	-5.124383	119.54941	-86	-131	06/30/2018-08:10:09	Lora
11	-5.124382	119.54941	-86	-128	06/30/2018-08:10:11	Lora
12	-5.12439	119.54941	-84	-128	06/30/2018-08:10:15	Cellular
13	-5.124392	119.54942	-84	-124	06/30/2018-08:10:18	Cellular
14	-5.124372	119.54942	-84	-124	06/30/2018-08:10:22	Cellular
15	-5.12428	119.54947	-86	-119	06/30/2018-08:10:25	Lora
16	-5.12377	119.54962	-84	-118	06/30/2018-08:10:34	Lora
17	-5.123311	119.54976	-82	-118	06/30/2018-08:10:40	Lora
18	-5.123063	119.54989	-86	-115	06/30/2018-08:10:43	Lora
19	-5.122812	119.55009	-88	-116	06/30/2018-08:10:46	Lora
20	-5.12233	119.55125	-88	-116	06/30/2018-08:10:56	Lora

6. Conclusion

The availability of communication networks is an important part of the object tracking system. Integration of LoRa-Cellular is one of the studies to keep the availability of communication network. The focus on LoRa-Cellular Integration is to overcome the problem in the tracking system, like is a blank spot area and data loss.

The communication link selections use two parameters. The parameters are RSSI level and energy consumption. The threshold of LoRa RSSI level uses -120 dBm and the threshold of Cellular RSSI level uses -85 dBm. RSSI levels of both communication link use a threshold of -120 dBm and -85 dBm to form an intersection between LoRa coverage and cellular coverage. Table 4 shows energy consumption. LoRa uses low power than Cellular, so the communication link accentuates the LoRa network.

References

[1] D. Sunehra, P. L. Priya, and A. Bano, "Children Location Monitoring on Google Maps Using GPS and GSM Technologies," Proc. - 6th Int. Adv. Comput. Conf. IACC 2016, pp. 711–715, 2016.

[2] R. Fablet et al., "application to wild animals electronic tracking and satellite ocean observation data To cite this version : Multiscale analysis of geometric planar deformations : application to wild animals electronic tracking and satellite ocean," vol. 52, no. 6, pp. 1–15, 2014.

[3] A. Anusha and S. M. Ahmed, "Vehicle Tracking and Monitoring System to Enhance the Safety and Security Driving Using IoT," Proc. - 2017 Int. Conf. Recent Trends Electr. Electron. Comput. Technol. ICRTEECT 2017, vol. 2017–Decem, pp. 49–53, 2017.

[4] A. Muzaffar and W. Yan, "Long-Range Wireless Sensor Networks for Geo-location Tracking : Design and Evaluation," 18th Int. Electron. Symp. (IES), Bali, Indones. 29-30th Sept. 2016,

IOP Conf. Series: Materials Science and Engineering **676** (2019) 012010 doi:10.1088/1757-899X/676/1/012010

pp. 76–80, 2016.
[5] M. Rahman, J. R. Mou, K. Tara, and I. Sarkar, "Real Time Google Map and Arduino Based Vehicle Tracking System," no. December, pp. 8–10, 2016.
[6] K. S. Ravitej, M. V. M. Sukumar, and A. Y. Felix, "Multimode children tracking - Using ZigBee device," 6th Int. Conf. Comput. Power, Energy, Inf. Commun. ICCPEIC 2017, vol. 2018–Janua, pp. 691–694, 2018.
[7] D. H. Kim, J. Bin Park, J. H. Shin, and J. D. Kim, "Design and implementation of object tracking system based on LoRa," 2017 Int. Conf. Inf. Netw., pp. 463–467, 2017.
[8] Anupriya K, Jubin Sebastian E, J. Yomas, Hari Babu P, and T. Dwarakanath, "Integrating ZigBee and Sub GHz devices for long range networks," 2016 Online Int. Conf. Green Eng. Technol., pp. 1–5, 2016.
[9] T. Hadwen, V. Smallbon, Q. Zhang, and M. D'Souza, "Energy efficient LoRa GPS tracker for dementia patients," Proc. Annu. Int. Conf. IEEE Eng. Med. Biol. Soc. EMBS, pp. 771–774, 2017.
[10] O. A. Mohamad, R. T. Hameed, and N. Ţăpuş, "Design and Implementation of Real Time Vehicle Tracking System Based on Arduino Intel Galileo," Architecture, 2016.
[11] N. Hayati and M. Suryanegara, "The IoT LoRa system design for tracking and monitoring patient with mental disorder," 2017 IEEE Int. Conf. Commun. Networks Satell. COMNETSAT 2017 - Proc., vol. 2018–Janua, pp. 135–139, 2018.

IOP Conf. Series: Materials Science and Engineering **676** (2019) 012011 doi:10.1088/1757-899X/676/1/012011

Environmental Sensor Design Prototype for Genset Backups Power at Ground Station and Data Centers Based on Internet of Things Devices

A Hidayat[1]*, P R Ramadhan[1], Zainuddin[1] and H Zainuddin[2]

[1]Stasiun Bumi Penginderaan Jauh Parepare Sulawesi Selatan
[2]Balai Monitor Spektrum Frekuensi Radio Kelas II Makassar

*E-mail: arif.hidayat@lapan.go.id

Abstract. Earth station and data center devices require stable backup power. The remote sensing earth station has 2 generators that can operate simultaneously or alternately. This Generator Set has optimal ability to work at room temperature up to 50 degrees Celsius. This environmental sensor prototype consists of three main parts: temperature sensor, ESP wifi module and website to display updated information at any time. The generator set system is only able to work under room temperature 50 degrees Celsius, if the room temperature exceeds 50 degrees Celsius the system will turn off the generator. This sensor system is used to monitor the room temperature so that it is not more than 50 degrees Celsius. All of the above systems can be accessed through the Local Area Network within the Parepare Remote Sensing Earth Station. With this system is expected to natural disturbances that occur in the antenna can be minimized so that the antenna ability to track and data acquisition can be maximized. The end of this prototype sensor is expected to help extend the life of antennas operating in the remote earth station LAPAN Parepare.

1. Introduction

The remote sensing earth station of the national aviation and space agency (SBPJP LAPAN) is one of the parts that is sheltered by the remote sensing deputy LAPAN, maintaining the continuity of the availability of low, medium resolution remote sensing data [1]. And the high and international standard remote sensing remote satellite station is the vision and mission of the remote satellite sensing system. To support the ability to achieve the vision and mission, the remote sensing earth station is equipped with remote sensing satellite data reception equipment that is in accordance with the established standards [2]. One of the main equipment used is the antenna to receive signals emitted by remote sensing satellites that pass in Indonesian territory [3, 4].

Earth station operations and data centers have backup power. This backup power system Electric backup system in the form of a generator set is 2 units [5]. This system is able to work together or alternately in supporting the operation of earth stations and data centers in parepare. The ability of the generator when operating is to reach room temperature below 50 degrees [6]. The heat of the ambient temperature in South Sulawesi's Parepare is approximately 33 degree. The Generator set also generated the heat and its multiply with the environments room, conditions like this can cause the air temperature in the generator room to be out of control. To determine the increase in room temperature, a system that is able to monitor the temperature remotely is needed. By knowing the actual temperature conditions in the field, further action can be taken by opening the main door and

IOP Conf. Series: Materials Science and Engineering **676** (2019) 012011 doi:10.1088/1757-899X/676/1/012011

automatically turning on the additional fan, if the temperature approaches the maximum value of 50 degrees. Thus, the system can run optimally without experiencing power disruption.

2. Design System

This system uses several devices namely temperature and humidity sensor DHT 11, NodeMCU ESP8266 contains an Arduino module and wireless device, wifi router and PC monitoring. This system can be seen in the picture below.

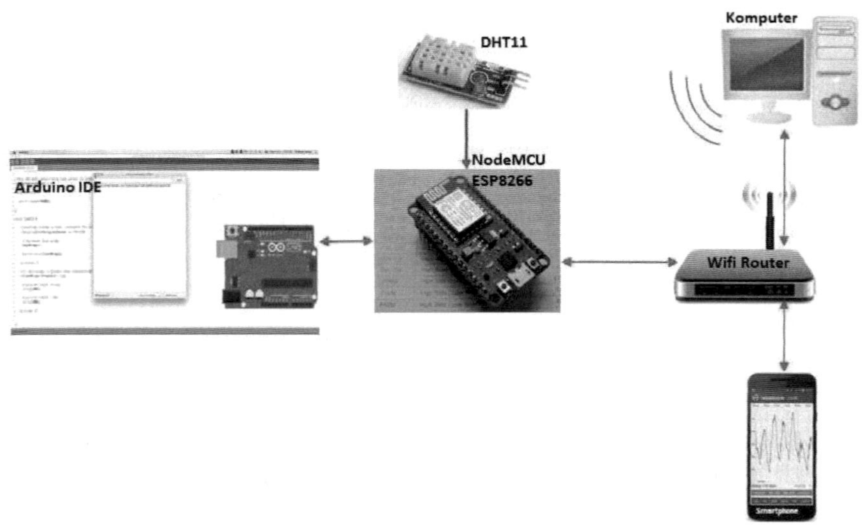

Figure 1. Design systems

2.1. Arduino module and system

The Arduino module in this series is intended to facilitate each component to be interconnected and to give orders to each component or as a terminal to process data sent by the sensor before being displayed in the user display [3].Here is a picture of Arduino and its datasheet. The Arduino module already inserted to RF Part in this application.

2.2. DHT sensor module 11

DHT11 is a digital sensor that can measure the temperature and humidity of the surrounding air. Has a good level of stability and calibration features. The calibration coefficient is stored in the OTP program memory, so that when the internal sensor detects something, this module includes the coefficient in its calculation, this DHT11 includes sensors that have the best quality, judged by the response, fast data reading, and anti-interference capability. Small size, and with signal transmission up to 20 meters, with specifications: Supply Voltage: + 5V, Temperature range: 0-50 °C error of ± 2 °C, Humidity: 20-90% RH ± 5% RH error, with digital specifications interfacing system. make this product suitable for many temperature and humidity measurement applications.

Table 1. Characteristics of air humidity / Humidity sensors

Model	DHT11
Power supply	3-5.5V DC
Output signal	digital signal via single-bus
Measuring range	humidity 20-90% RH ± 5% RH error temperature 0-50 °C error of ± 2 °C
Accuracy	humidity +-4%RH (Max +-5%RH); temperature +-2.0Celsius

Resolution or Sensitivity	humidity 1%RH; temperature 0.1Celsius
Repeatability	humidity +-1%RH; temperature +-1Celsius
Humidity hysteresis	+-1%RH
Long-term Stability	+-0.5%RH/year
Sensing period	Average: 2s
Interchangeability	fully interchangeable
Dimensions size	12*15.5*5.5mm

2.3. NodeMCU ESP8266

To be able to read all sensor devices and display them into an open source web site, a device that can function as a wifi modem is needed. The NodeMCU device can function as a modem or sensor device control unit (H Ardian, 2017). NodeMCU is an opensource IoT platform. Consists of hardware in the form of ESP8266 System On Chip from ESP8266 made by Espressif System, also the firmware used, which uses scripting programming language.

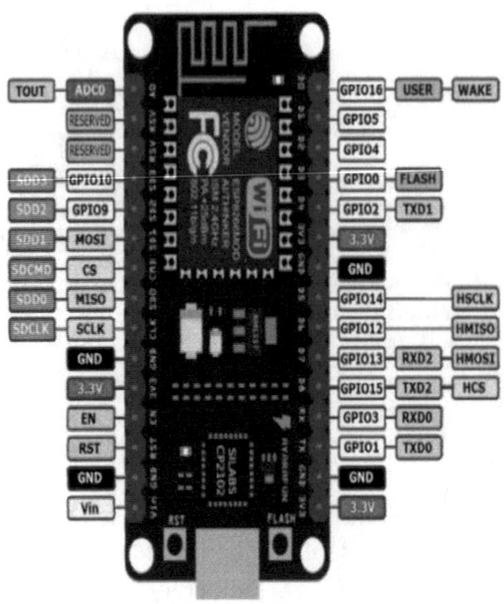

Figure 2. Layout module nodeMCU ESP8266

The system process starts from the device connection to Arduino, Arduino then performs sensor readings through the sensor port. If the port sensor is read successfully, the sensor readings are sent via the wireless network. Whereas if the reading process fails, the reading port process is repeated. The results of the sensor reading are then displayed on the display.

IOP Conf. Series: Materials Science and Engineering **676** (2019) 012011 doi:10.1088/1757-899X/676/1/012011

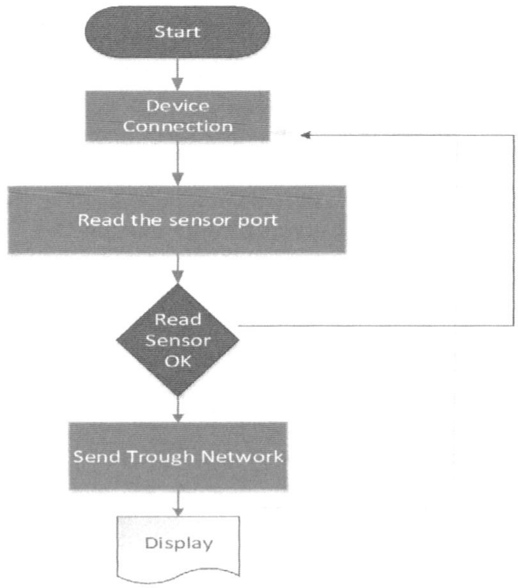

Figure 3. Flowchart system

3. Implementation and Analysis

The system design is then implemented in the generator system then the results are compared with thermometers that have been installed on the generator set.

Figure 4. Generator set position

Figure 5. Test system sensors

3.1. Implementation

After assembling and programming on the arduino IDE, a device is produced that is able to measure temperature and humidity in realtime. Testing is done in the room to find out the sensor function that has been programmed with the esf8266 module, the DHT11 temperature sensor has sensitivity in reading the room temperature conditions which are given several conditions in the testing stage. Sensor readings can be monitored through iOT Thingspeak connected to the esf8266 module, besides the iOT Thinspeak results are also found in the arduino IDE monitor series.

This tool rank consists of Module Nodemcu ESP8266, temperature sensor DHT11 and project board, after building the rank then connected to the PC to upload the program into the esp8266 module. Connections between ranks are:

Table 2. Pin position

Pin Nodemcu ESP8266	Pin DHT11
3 V	VCC
G	GND
D7	DATA

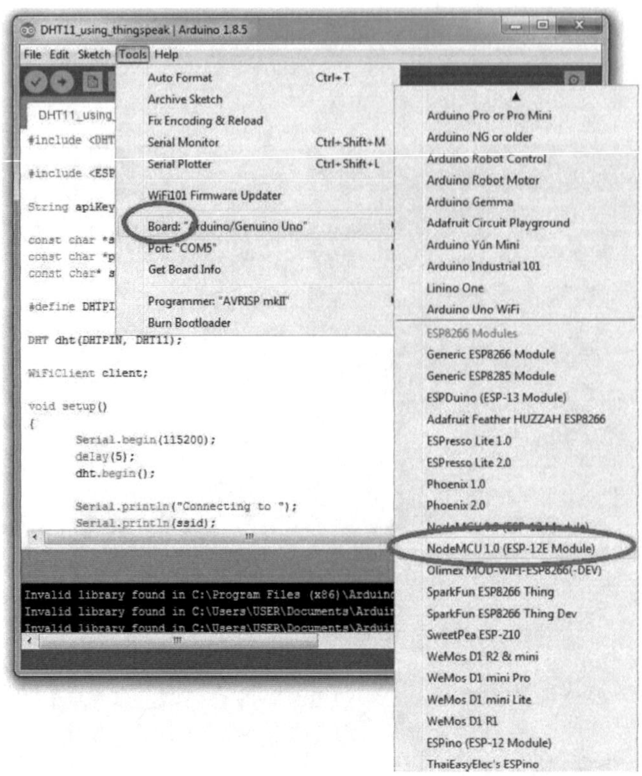

Figure 6. Configuration

IOP Conf. Series: Materials Science and Engineering **676** (2019) 012011 doi:10.1088/1757-899X/676/1/012011

3.2. Measurement result

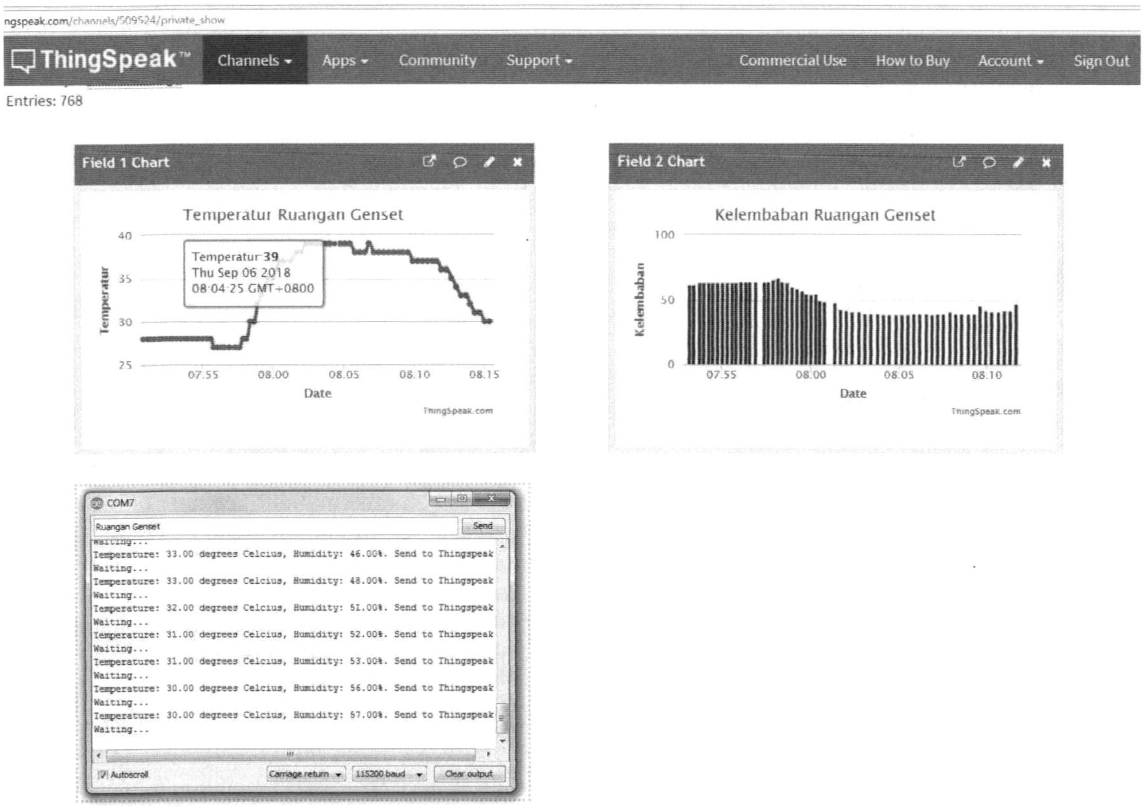

Figure 7. Test systems

Temperature can be viewed using the web by entering the URL or IP address of the device, thus making it easier to monitor using a web browser application. The test was carried out by bringing the genset engine closer to a higher temperature, it can be seen that the graph rises according to the measured temperature value.

Figure 8. Existing termometer

3.3. Analysis

The value of the measurement results with the value of the digital display thermometer in the generator has a difference of 1 degree centigrade. This system can also be monitored through a web browser or using an android application. This makes it easy in daily observations, the system can also be combined with automatic switch to turn on the exhaust fan and additional cooling fans.

4. Conclusions

Measurements using a DHT11 sensor with measurements using a digital sensor get a data difference of 1 degree Celsius. Temperature changes in the generator room can be monitored directly through a web based application. Thus, this system can be implemented as a system that is used to monitor indoor temperature based online.

References

[1] Hidayat., W. Hasbi, E. Palantei dan S. Syarif (2017). Kendali Sakelar Empat antena Bersusun Untuk Penjejakan Interferensi Frekuensi TTC Satelit LAPAN". Jurnal Teknologi Dirgantara, Vol 15 No 1.

[2] A. Hidayat, A (2017). Doppler Method Using 6 Antenna Array Switch In Real Time Remote Monitoring Direction Finder. (Magister Teknik Elektro Master Thesis), UNHAS (Universitas Hasanuddin), Makassar.

[3] E. Palantei, S. Syarif, A. Hidayat dan S.T.A Munawar (2017) "Low-cost Switched Array-wide Band Antenna for Search and Rescue Disaster Management", 2017 3rd International Conferences on Science and Technology-Computer (ICST) IEEE pp:131-134, 11-12 July, 2017.

[4] A. Hidayat, Ardiansyah, P. R. Ramadhan dan S.T.A. Munawar. Design and Implementation Web Based Expert System For Analizing Performance Of antenna Seaspace 5.1, Desain dan Implementasi Sistem Pakar Analisis Performansi Antena Seaspace Axyom 5.1 Berbasis Web". Jurnal Teknologi Dirgantara:, Vol : 12 (2), pp.154-162. , 2 Desember 2014.

[5] A. Hidayat, S.T.A. Munawar, S. Syarif dan A. Achmad, Leo Groud Station Analysis Using Fast Fourier Transform", 7 th International Anual engineering Seminar (IEEE: UGM), 1-2 Agustus, Yogyakarta, Indonesia.

[6] H. Ardian, Dedi Triyanto, dan Tedy Rismawan (2017), Sistem Kendali Lampu Dan Steker Terintegrasi Menggunakan Mikrontroller Berbasis Web Service. Jurnal Coding Sistem Komputer Untan, Volume 05, No. 3 (2017), hal 40-47.

IOP Conf. Series: Materials Science and Engineering **676** (2019) 012012 doi:10.1088/1757-899X/676/1/012012

Indonesian Sign Language Letter Interpreter Application Using Leap Motion Control based on Naïve Bayes Classifier

Ridwang[1]*, Syafaruddin[2], A A Ilham[1] and I Nurtanio[1]

[1]Department of Informatics, Hasanuddin University, Makassar, Indonesia
[2]Department of Electrical Engineering, Hasanuddin University Makassar, Indonesia

*E-mail: ridwan@uin-alauddin.ac.id,

Abstract - Most deaf children are born into hearing families and the majority do not begin their Sign Language learning before age 3-6 in school. As children learning Sign Language from infancy show higher language proficiency than those who do not, it underlines the importance of hearing parents learning how to sign. The recent innovation in the technology and introduction of new gesture control based devices, like leap motion controller. Technology that can track finger and hand gestures accurately at very low cost. The research method in Naïve Bayes algorithm, in order to classify 26 alphabet letter of Indonesian sign language including letter J and Z which using gesture for describe it. The Leap Motion controller is a consumer gesture sensor aimed to augment a user's interactive experience with their computer and it was utilized as an interface for hand motion tracking without the need of wearing any external instruments. It use for derive feature data from the deaf children gesture recording to set data in application. The accuracy level of the test result reached 96%.

1. Introduction

Persons with disabilities are groups of limitation persons that can hinder their participation in community. Communities that live with limitations are divided into three, namely physical, mental, and both (physical and mental). Examples of physical limitations include blindness, hearing impairment, speech impairment, disability, and other physical limitations.

At present there are a lot of deaf people. In Indonesia, from the Central Bureau of Statistics (BPS) in 2010, there are 3,024,271 million people (the total population in 2010 was 191,709,144 million) experiencing hearing and speech impairment [1].

Based on the results of interviews with several members of the Indonesian Deaf Welfare Movement DPC GERKATIN Makassar, one of the most important things to help with the daily activities of deaf-speech people is to communicate with normal people. Difficulty in communicating will affect the lives and interpersonal relationships in the deaf-mute community. Difficulty in communicating between deaf-mute sufferers and normal people can bring problems in the process of integrating deaf-speech sufferers into the wider community [2].

The sign language used for each country may be different. In Indonesia admit two sign language, the official sign language system used in Indonesia is the Sistem Isyarat Bahasa Indonesia (SIBI) where the system is similar to the sigh language applied in the United States, American Sign Language (ASL) and BISINDO used more sign language community in Indonesia. In this study using the Sistem Isyarat Bahasa Indonesia (SIBI).

The previous study by Wibowo et al [3] recognize 24 alphabet Indonesian letters in Indonesian sign language which resulted in average classification accuracy for Naïve Bayes is 95%. The introduction of J and Z letters cannot record from object because they using gesture for describe letter in Indonesian sign language (SIBI). Previous research using the Multilayer Perception (MLP) Neural Network (NN) method for the introduction of Arabic sign language resulted in an average accuracy of 88%. Then another research using Naive Bayes and Linear Discriminant Analysis (LDA) algorithm resulting in the accuracy of Naive Bayes that is better than LDA, but LDA shows faster time [4]. Marin et al. [5] investigated the performance of Leap Sensors Control by training SVM Classifier to recognize 10 different static mark with 1,400 sample counts capable of achieving an average accuracy of 80% as well as research focusing on challenges that include features for accurate signaling cues, and argues on how normalization to accommodate a reliable system for users with different hand sizes. Chen et al. [6] use the algorithm Hidden Markov Model (HMM) and Support Vector Machine (SVM) for dynamic hand movements. Accuracy using the HMM algorithm is 90.56% and SVM is 98.24%. This study detected numbers and alphabets with a total of 36 movements captured by utilizing Leap Motion.

Based on those issue above, an automated system that is useful as a translation from sign language into written language is needed. The research is based on Classifier Algorithm and sensor Leap Motion Control (LMC), focused on the dynamic alphabet letters and words of Indonesian sign language (SIBI).

2. Material and Method

2.1. Leap motion control

Leap Motion is a small device basis of a USB device that can allow a computer user to control or play a computer using motion. In Figure 1, there are two monochromatic cameras with three infrared LEDs in Leap motion. Leap motion captures independent hand movements and finger movements, as well as objects such as pens. In fact, 200x Leap motion is more sensitive than free touch technology in existing products and technologies in 2014. Based on Leap motion has a high detection accuracy so it is widely developed as a controller and also motion recognition.

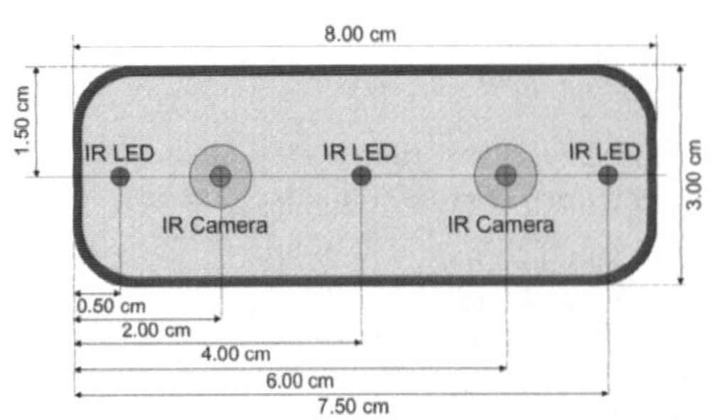

Figure 1. Leap motion control schematic

Leap Motion was developed by David Holz and Michael Buckwald. They developed a device similar to Microsoft Kinect but claimed to have a higher level of accuracy. The purposes of David Holz and Michael Buckwald are to replace the functions of the keyboard and mouse and allow users to explore the computer with just finger movements. The Leap Motion works by creating an interactive 4 cubic feet room that is able to detect fingers, hands and arm movements [7]. Data retrieval using Leap motion utilizes API (Application Programing Interface) designed specifically for developers who want to utilize Leap motion [3].

The Leap Motion Sensor controller device accompanied by API support from its provider. Through API support, the hand and finger data can be sent to the person who created the program to use the sensor as an alternative computer interface tool. Table 1 lists the feature of the hand and fingers obtained from the APIs in this study. Feature of a normal palm (vector uni direction), and speed (in millimeters per second). The LMS API also handles behind-the-scenes image analysis, reducing complex extracting features task of developer from raw input data. This allows research to focus more on feature selection, not the extraction process. Below are accuracy result of applying machine learning to different signification classification [3].

Table 1. Features obtained from LMC API

	PALM		FINGERS	
	Name	Type	Name	Fingers
Hand Features*	Normal	Vector	Direction	Vector
	Position	Vector	Length	In mm
	Velocity	Vector in mm/sec	Tip Position	Vector
	Confidence	a float in [0,1]	Tip Velocity	Mm/sec
	Pinch Strength	a float in [0,1]	Dip Position	Vector
	Grab Strength	a float in [0,1]	Pip Position	Vector
	Sphere Center	Vector	Mcp Position	Vector
	Sphere radius	In mm		

2.2. *Naïve Bayes classifier*

The Bayesian theorem provides a way to calculate the posterior probability based on assumed prior probabilities and the probability of different data object under given assumptions. Bayesian algorithm based on Bayesian theorem is a simple and commonly used classification algorithm which has rigorous mathematical theories as a basis. Assuming each attribute of the items to be classified is independent of each other, the constructed classification algorithm is Naïve Bayes. The basic idea of Naïve Bayes is: for a given item to be classified, calculate the probability of each category which the item belongs to, and the item will belong to the category with the largest probability [8].

The Bayesian classification approach is based on quantifying the trade-offs between various classification decisions using probability and the costs that accompany such decisions [9]. The classifier was developed based on Bayes probability rule which states:

$$P(Y|X_1,...,X_n)=\frac{P(X_1,...,X_n|Y)P(Y)}{P(X_1,...,X_n)} \quad (1)$$

This is the probability of obtaining Y given conditions X1toXn or the posterior probability of Y given a prior probability of Y and X1toXn likelihood. In our case, using the letters, this is interpreted as posterior probability of any letter given the likelihood parameter and a prior probability of that letter. The decision is taken based on the greatest of the calculated probabilities to achieve minimum error classification [9]. From the Bayes rule, the NBC is mathematically stated as:

$$P(X_1,...,X_n|Y)=\prod_{i=1}^{n}P(X_i|Y) \quad (2)$$

2.3. *Euclidean distance*

Euclidean Distance is the classifier that is most often used to calculate the similarity of 2 vectors. If the distance between 2 images is large, then it can be concluded that the two images are not similar, whereas if the distance between two images is small, then the image is similar [10]. The equation of the Euclidean distance is.

$$d_{ij}=\sqrt{\sum_{n-1}^{k}\left(x_{ik}-x_{jk}\right)^2} \quad (3)$$

2.4. Feature extraction

Leap motion produces sensor data in the form of the position of each hand and finger joint when there is a hand moving in the sensor capture room. These data are formed into a vector model which is then stored in the database as reference data. Each movement will be compared with this reference data to detect what the movement means.

In an initial attempt to create sign recognition software, using video analysis, the main challenge was not a feature that had to be put forward, but extraction itself. However, in the case of Leap Motion, the API provides a framework with a skeleton model for each hand, and reduces individual developers from heavy duty to analyze IR images to get relevant information [3].

Each hand object is built from the finger and the object of the palm, and each finger is formed by a set of bones. Each object has an appropriate 3-dimensional vector, referring to its direction and position in the Euclidian space in the sensor display.

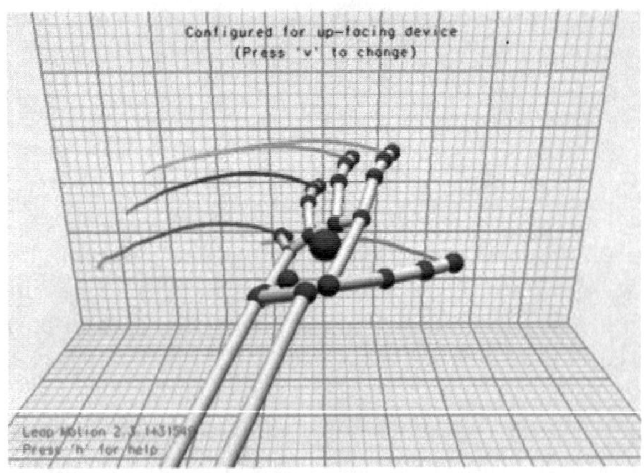

Figure 2. LMC Framework model

Dynamic features are dynamic motion features or features obtained from LMC detection of the hands [11]. The good feature formation will increase accuracy in sign language recognition. The features used are Distal Tips Fingers, Palm Direction and Palm Position.

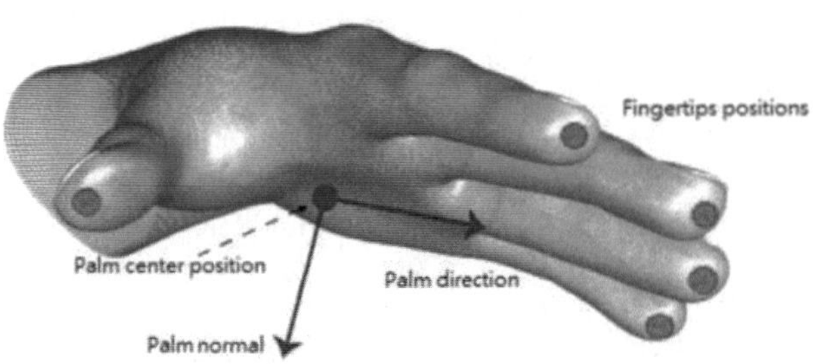

Figure 3. Hands features

2.5. Data acquisition

There are 26 letters of Indonesian sign language Alphabet (SIBI) used in this case static and dynamic as show in Figure 4. The total number of training use totals 1300 samples derived from 50 samples for each letter class of 26 alphabet sourced letter from 5 different hand respondents where each respondent made 10 training data.

Figure 4. Sign language Indonesia (SIBI) alphabet

In this paper the initiate time of capture is base on leap motion first time frame capture and while leap motion capture hand gesture time. The tools used for this data capture was written with referencing to controller SDK using Python.

3. Design System

System design consists of two main parts, namely the application of data training or builder and application of data testing. The training data application is used to record the user's hand movements without having to learn about programming languages first. Besides, it can be used to store data into a dataset and can extract data from datasets into vector shapes that will be used for classification processes. The interpreter application is used as a user interface to learn sign language. The interpreter application is connected to Leap Motion Service running with the Leap Motion API.

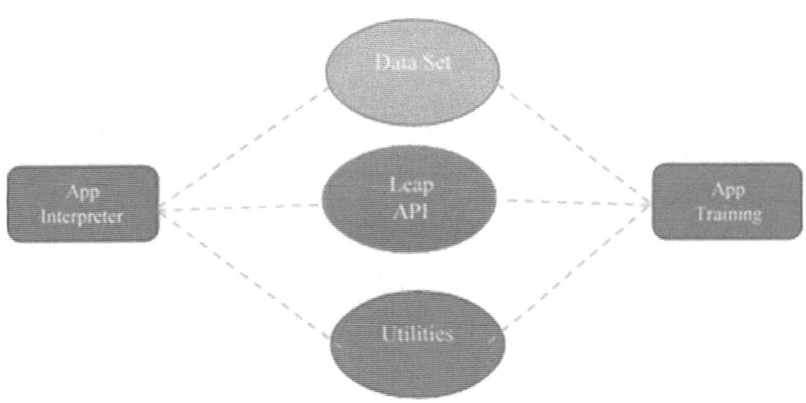

Figure 5. System Design

In the data training process, the vector data was obtained from the results of feature extraction by SDK Leap Motion and then data normalization was carried out on the three data items obtained from the reading of leap motion control namely Vector fingers, hand patterns or rolls and palms. Normalization of data on the fingers with the following formula:

$$Normalisasi = F_i - C \tag{4}$$

F_i is the coordinates of the fingers of each hand,
C is the coordinates of the position of the palm.

In the extraction stage, the hand direction feature is only made to change the value to degrees because the data obtained from the API Leap Motion is a Radian value. The formulas used at this stage are:

$$\deg = rad * 180/\pi \tag{5}$$

Whereas the process of test data is almost the same as the training data process, the difference lies in the last process where the training data process saves data to the dataset, while the test data process converts data from continuous data into discrete data using the Naïve Bayes Gaussian approach.

4. Classifier

This research used Naive Bayes for its basic classifier. This classifier was chosen due to its simplicity in technique and implementation. The basic principal of Naïve Bayes is assuming every feature in the feature-sets is not dependent each other. Its efficiency in using the data training, by using a small amount of data-set without reducing the accuracy of classification performance makes this classifier widely used on many gesture recognition researches [12].

Basically, Naive Bayes is a statistical approach with the probability formula as follows:

$$P((C_k|x)) = \frac{P(C_k)P((x|C_k))}{P(x)} \tag{6}$$

In order to Naïve Bayes the classifier can work at discrete values, one of which can assume a Gaussian distribution for each feature above all samples in the feature vector. So the probability of the feature x entering a class is

$$P(x|c) = \frac{1}{\sigma c\sqrt{2\pi}} e^{-\frac{(x-\mu c)^2}{2\sigma_c^2}} \tag{7}$$

Probability calculations often get a small number, and multiplying them together will risk losing precision. To overcome this, the logarithm of all three parts of the equation is taken, resulting in a formula:

$$\log(P(c)) + \log\left(\frac{1}{\sigma\sqrt{2\pi}}\right) - \frac{(x-\mu)^2}{2\sigma^2} \tag{8}$$

Calculation of Euclidean distance is used in samples of average values using formulas:

$$d(p,q) = \sqrt{(q_1 - p_1)^2 + (q_2 - p_2)^2} \tag{9}$$

Simple classification of vectors based on the highest probability of all class probabilities :

$$\max_i \left(P(c_i)\sum_j^k P(x_j|c_i)\right) \tag{10}$$

5. Experiment and Result

Base on the stage shown in the system design, the system development as a result of this writing can be seen in Figure 6 as app data training and Figure 7 AppInterpreter, for the recognition of Indonesian sign language pattern using leap Motion Control and Naïve Bayes.

IOP Conf. Series: Materials Science and Engineering **676** (2019) 012012 doi:10.1088/1757-899X/676/1/012012

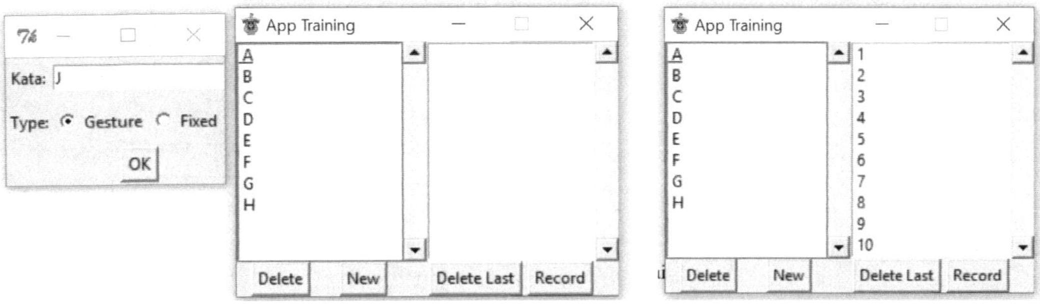

Figure 6. Result App Data Training

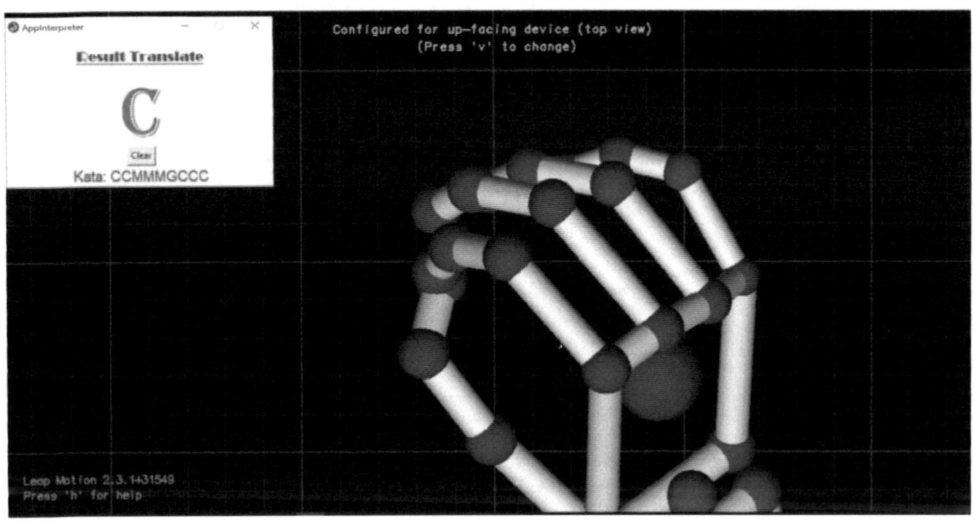

Figure 7. App Interpreter

5.1. Experiment

Using a tool titled as Tkinter (GUI of Python) for presenting and evaluating the classifier method base on model generated from feature-sets which is formatted on P file. Python already the default Naïve Bayes classifier as well, so that we just need import library from scikit.

The research tried to classify the gesture using gesture classification for leap motion, we classifying the 26 letters of Indonesian sign language (SIBI) using the hand of the respondents. Through the testing process with the loop count 6 times to test the 26 letters of SIBI alphabet consist of A, B, C, D, D, F, G, H, I, J, K, L, M, N, O, P, Q, R, S, T, U, P, W, X, Y, Z. Total data set amounted 156 samples and the respondent's distance hand to the leap motion sensor device are 12 cm. this data classify using naïve Bayes algorithm.

5.2. Result

Classification result showed on the Table 2, where are letters known as another letter because pattern or gesture same other letter. For example, N recognized as M or N.

Table 2. Classification Result

Alphabet	Experiment		Known As other Letter
	True	False	
A	6	0	
B	6	0	
C	6	0	
D	6	0	
E	6	0	
F	6	0	
G	6	0	
H	6	0	
I	6	0	
J	4	2	I
K	6	0	
L	6	0	
M	6	0	
N	4	2	M,O
O	6	0	
P	6	0	
Q	6	0	
R	6	0	
S	6	0	
T	6	0	
U	6	0	
V	6	0	
W	6	0	
X	6	0	
Y	6	0	
Z	4	2	U,L

Table 3. Leap Motion Classification

Result Of	Samples	Percent
Correctly Classified Instance	150	96,16 %
Incorrect Classified Instance	6	3,84 %
Total Instance	156	

Feature extraction and data normalization for data testing from Leap Motion API same with training data processing. Testing data take from 17 frame with contain XYZ finger distal, pitch and roll coordinate. Classification result shown on the Table 4 which data shorted from the highest probability to low probability. Beside determined from probability level, so determine from Euclidean distance. If there are alphabet had highest probability value and the closest distance between training data with testing data then found the valid data.

Table 4. Probability count result of each class

Class	Probability	Euclidance
A	0.49829063002951625	0.12427167771365978
T	0.4964730754720572	0.6618753674332744
M	0.4964615506098093	0.9584814326190543
S	0.4964276936817576	0.4256425352416988
N	0.49626077181041944	1.0204044337833829
P	0.4946699394927318	1.4140232621574138
X	0.49443004115831723	0.597754565193763
Q	0.4938637754273052	1.5787214251343185
Y	0.49381156705336626	1.261124779865341
J	0.49377800601617323	1.461864793829387
O	0.4934157826912663	0.913767149405667
I	0.49287800601917325	1.0242695823042858
E	0.49177978614464	1.278849553767374
H	0.4911575131250296	1.350758627322214
Z	0.4910607418904194	1.121487434075456
C	0.49095721853010466	1.2303544651418814
G	0.4899545038668552	1.333429433914674
V	0.48961781454498543	0.8201949635276955
D	0.4890006075788114	0.5960133709134533
L	0.4885066079953518	1.094546589476957
R	0.48437733856275167	0.5770244486141277
W	0.4803115273695701	1.4605308147142804
U	0.47989742792023293	0.8555228250937538
K	0.47855049630856755	1.208564425582325
F	0.47749830469325505	1.1384021651862253
B	0.22240801991810477	1.4040570409006827

Based on Table 4, probability values from A letter is great, and also have closest distance so that classification result of A letter is valid.

5.3. Discussion

As shown on the table classification result above (Table 2) there are letters recognized as another letter. For instance, N recognized as M and O, Z recognized as U and L, J recognized I. the main problems from their shapes resembles the position coordinate and finger figure one to another on this study. For get training data, please do be carefully because can influence result of the data classification. As shown on the table Leap Motion Classification most of gesture and fixed classification generate above 96 %. During leap motion usage, it is frequently failed to recognize the hidden fingers considering its hardware limitation from sensor view. The mistake could occur because no match data from respondent with data from leap motion API. The real finger shown versus the finger are in the model API skeleton are different on the screen.

IOP Conf. Series: Materials Science and Engineering **676** (2019) 012012 doi:10.1088/1757-899X/676/1/012012

6. Conclusion

In this paper, Dynamic Hand Recognition for Indonesian Sign Language (SIBI) using Leap Motion Control approaches has been proposed Euclidean Distance and Naïve Bayes as a method for get the high accuracy. The experiment result, show increase accuracy until 96% with gesture and fixed hand. As the future work, it can combine two sensors for get more accuracy and combine more methods or algorithms.

References

[1] Sensus Penduduk, "Penduduk Menurut Wilayah dan Tingkat Kesulitan Mendengar," 2010. [Online]. Available: http://sp2010.bps.go.id/index.php/site/tabel?tid=276&wid=0.

[2] L. Potter, J. Araullo, and L. Carter, "The Leap Motion controller: a view on sign language," *Proc. 25th Aust. Comput. Interact. Conf. Augment. Appl. Innov. Collab.*, pp. 175–178, 2013.

[3] M. D. Wibowo, I. Nurtanio, and A. A. Ilham, "Indonesian sign language recognition using leap motion controller," *2017 Int. Conf. Inf. Commun. Technol. Syst.*, vol. 17, pp. 17–22, 2017.

[4] L. Alexander, "Sign Language Recognition using Leap Motion," no. March, 2016.

[5] P. Marin, G. ;Dominio, F.;Zanuttigh, "Hand Gesture Recognition With Leap Motion And Kinect Devices Department of Information Engineering , University of Padova," *Image Process. (ICIP), 2014 IEEE Int. Conf.*, pp. 1565–1569, 2014.

[6] P. Kumar, R. Saini, S. Kumar Behera, D. Prosad Dogra, and P. Pratim Roy, "Real-Time Recognition of Sign Language Gestures and Air-Writing using Leap Motion," vol. 1, pp. 1–4, 2017.

[7] J. Shin and C. M. Kim, "Non-Touch Character Input System Based on Hand Tapping Gestures Using Kinect Sensor," *IEEE Access*, vol. 5, no. c, pp. 10496–10505, 2017.

[8] A. Haghpanah, "A non-parametric mixture of Gaussian naive Bayes classifiers based on local independent features," no. 1, pp. 209–212, 2017.

[9] M. Mohandes, S. Aliyu, and M. Deriche, "Arabic sign language recognition using the leap motion controller," *2014 IEEE 23rd Int. Symp. Ind. Electron.*, pp. 960–965, 2014.

[10] N. Fitriyah, "Analisis dan simulasi sistem pengenalan wajah dengan metode fisherface berbasis outdoorvideo."

[11] A. Z. Shukor, M. F. Miskon, M. H. Jamaluddin, F. Bin Ali Ibrahim, M. F. Asyraf, and M. B. Bin Bahar, "A New Data Glove Approach for Malaysian Sign Language Detection," *Procedia Comput. Sci.*, vol. 76, no. Iris, pp. 60–67, 2015.

[12] Khamid, A. D. Wibawa, and S. Sumpeno, "Gesture Recognition for Indonesian Sign Language Systems (ISLS) Using Multimodal Sensor Leap Motion and Myo Armband Controllers Based-on Naïve Bayes Classifier," *2017 Int. Conf. Soft Comput. Intell. Syst. Inf. Technol.*, pp. 1–6, 2017.

IOP Conf. Series: Materials Science and Engineering **676** (2019) 012013 doi:10.1088/1757-899X/676/1/012013

Designing Board Games to Foster Tolerance

M H Suprapto[1]*, L P S Hartanti[2] and R Latumahina[3]

[1] Faculty of Psychology, Pelita Harapan University Surabaya, Indonesia
[2]Faculty of Engineering Pelita Harapan University Surabaya, Indonesia
[3]Faculty of Law, Pelita Harapan University Surabaya, Indonesia

*Email: maria.suprapto@uph.edu

Abstract. Today, tolerance for diversity in society begins to erode. The low tolerance can threaten the integrity and unity of the Unitary Republic of Indonesia (NKRI). Tolerance towards diversity should be inculcated since childhood. One way to develop tolerance in children is to use the playing method. This study aimed to develop children's play equipment that can increase children's tolerance for diversity in society. One type of children's play equipment is board games. This board game is expected to educate children to be more tolerant of various differences in society such as: ethnicity, race, religion, and class. This increase in tolerance will support the unity and unity of the Indonesian people. The method used to design this board game was a creative method. Data collection was done by conducting surveys, observations, and literature searches. Audience target characteristics are children aged 6-10 years and live in big cities in Java such as: Jakarta, Bandung, Surabaya, and Semarang.

1. Introduction

Indonesia has a very diverse population in terms of ethnicity, religion, race and class. Existing diversity is the wealth and characteristics of the Indonesian Nation which is famous for its motto "Bhinneka Tunggal Ika". Awareness and tolerance of diversity in society is a key factor for the establishment of harmony and unity in Indonesia.

Nowadays, tolerance for diversity begins to erode. One example of a trigger erosion of tolerance was a conflict election in Jakarta in April 2017 [1]. Children who did not have the right to vote in political activities could be affected by the attitude of their parents. Understanding of children who had not been well formed makes children unable to filter good things and bad things from political activities. The negative impact was that children could lack tolerance. Children tended to choose friends based on similarities possessed such as: tribe, social class, religion, or race. The younger generation no longer understood the importance of maintaining tolerance with a diverse and diverse society. Lack of tolerance can be a serious problem and a threat to the integrity of the Republic of Indonesia.

Childhood is a golden age to form children who have noble character and personality. Young children who are tolerant are expected to be a strong and peaceful young generation so that they are able to maintain the unity and integrity of the nation. Tolerance towards diversity should be inculcated since childhood. Children have an ethnic identity at the age of 3 years [2], even children who are 5 years old have been able to express their racial attitudes and bias in assessing people who are different from them (race, ethnicity, religion, or social class) [3-5].

One way to develop tolerance in children is to use the playing method. Children love games. Games are a powerful way to develop children's social and emotional abilities [6]. The game can provide stimulation to develop language skills and problem-solving skills [7]. The game has been widely used in various contexts to develop insight, empathy, prosocial skills and improve children's positive behavior [8]. This game can be a means to overcome prejudice and increase tolerance among diverse communities. This study aims to develop a game prototype that can increase children's tolerance for the diversity of the communities around them.

2. Methods

2.1. Design methods
The research method used in designing this game was a creative method. Data collection techniques carried out was observation, interviews, and literature search.

2.1.1 Observation.
Observation aimed to find an overview of board games suitable for children aged 6-10 years. Observations were carried out in the board game café and or shop, namely at Toys Toys, Surabaya; Games on Café, Jakarta; and Dakon Library, Jogjakarta. Boardgame cafe rented a variety of board games and provides a variety of food and drinks. Observations are carried out by observing existing games. Dakon (Dolanan Karo Konco) Library is a boardgame library founded by a community of board game players and board game makers called Dakon.

2.1.2 Interview.
The interview aimed to explore the method of making board games, and tips on making board games. The interview subjects were two game designers, who had made board games.

2.1.3 Literature search.
Literature search was used to gather information about children's cognitive development, ways to develop tolerance, and board game design methods. Literature search was done by reading and composing theories from the materials that have been collected.

2.2 Audience target characteristics
The specifications of the target audience segmentation were as follows:

2.2.1 Demographic. primary target:
- Children: 6-10 years old
- Gender: unisex (female and male)
- Education: Elementary school
- Social class: middle-up
- Various ethnic, religious and racial backgrounds

2.2.2 Psychographic
- Audience has high curiosity.
- Audience likes challenging games.
- Audience likes to play in groups'.

2.2.3 Geographical. Audience lives in big cities in Java such as: Jakarta, Bandung, Surabaya and Semarang.

2.2.4 Secondary target. The secondary target of this game was parents of children aged 6-10 years and teachers who taught in class I-IV elementary school, as well as gamers.

2.3. Unique selling point
Board games can be used to educate children about how to show tolerance to others. This game had content in the form of everyday situations, namely situations in the family, at school, and in the neighborhood around the house, which invited children to be tolerant of others. The game was equipped with questions about Citizenship Education material, especially the material on Pancasila, unity in diversity, tolerance and mutual cooperation. These questions were packaged in the form of challenges so that they can help children to play while learning Civics Education subject matter.

3. Design Concepts
The main concept in the design of this board game was to design game products to increase children's tolerance for diversity. This board game can be used as an interactive learning medium that involves interaction between players, where children will learn while playing. This game can be played 2-4 people and has certain rules that players must obey. This board game was dominated by event cards (situation cards) equipped with a response card.

3.1. Format / Media form
This board game is called I am An Archipelago Kids (Aku Anak Nusantara) with the slogan "Smart and Tolerant" which means that this game teaches children in Indonesia to be able to show tolerance to others (family and friends). This board game is a race game genre. Players will be required to have a strategy to win the game. This game is categorized as a roll and move game, because it uses dice.

3.2. Content menu
In this game consists of game board, character cards, situation cards, response cards, challenge cards, dice and player's pawn. Character cards are 6 cards. Each character has special abilities that can benefit the player. Special ability can only be used twice in the game. This character card is taken randomly by the player by shaking it first.

A number 50 event card will describe events or problems that children often encounter in the family, at school, or in the neighbourhood. The response card is the response of the player to the event or problem. The response card is 65 cards. Response cards can be divided into five types of cards, namely: "Happy" Card, "Sad" Card, "Happy or Sad" Card, "Take Two Response Cards", and "Throw Questions to Other Players" Cards.

Challenge cards number 40 cards. This challenge card contains questions about Civics Education subject matter, namely: Pancasila, unity in diversity, tolerance and mutual cooperation. Each card is equipped with an answer

3.3. Game flow
The game flow is describes as follows
- The player shuffles the cards first.
- The player determines the turn to play by performing a hompimpa or rock-paper-scissors.
- The player picks up a character card in the stack of characters that have been shaken.
- The player chooses the pawn and places the pawn in number 1.
- The player picks up a Response card in the shuffled response card.
- The game goes clockwise.
- Situation cards are taken and read by the next player, for example the situation card for the first player, read by the second player, and so on.
- The first player takes one response card, and the second player picks and reads the situation card for the first player. If the symbol shown on the response card matches the one on the incident card, the player has the opportunity to throw the dice then move the pawn on the game board as much as the number that comes out is combined. However, if the symbol shown on the response card does not match the one on the incident card, the player will lose the opportunity to move the pawn on the game board.

- If the dice shows the number six, the player has the right to throw the dice once more. If a player gets six points for three times in a row, then the player gets a Challenge Card.
- If the player arrives in the red box, the player gets a Challenge Card. The next player will take and read the Challenge Card. If the player can answer the questions on the Challenge Card, the player has the right to advance according to the number of points on the Challenge Card. If the player can't or wrongly starts the question on the Challenge Card, then the player must go back according to the number of points on the Challenge Card.
- If the player arrives in a green box, the player can get a bonus to step according to the number shown on the game board; or get a trap so he must retreat according to the number shown on the game board.
- If there are two players who occupy the same box, then the player who occupies the box first.
- Players can throw questions to other players using the Response Card "Throwing Questions to Other Players." If another player succeeds in answering correctly, the player who throws the question does not get a chance to throw the dice. If another player is wrong or can't answer correctly, the player who throws the question will get the chance to throw the dice.
- The game will end if one of the players has reached the finish line first or the player who is at the front (after shuffling the response card twice).

3.4. Communication strategy

This game was designed for children aged 6-10 years. The game is one of the fun learning media for children. The communication used in this game is the introduction of various situations or events that may be experienced by children in daily life in the family, at home, and in the neighborhood. The introduction of this situation is stated in the Situation Card. The response or attitude of the child to the incident is stated through the Response Card. If the pawn occupies a red box, the player gets a Challenge Card. This Challenge Card contains questions about Citizenship Education material, especially the material on Pancasila, unity in diversity, tolerance and mutual cooperation. This question indirectly invites children to relearn the Citizenship Education subject matter and invites them to love Indonesia more.

3.5. Visual strategy

The theme of the game raised was how to be tolerant in everyday life. The cards were drawn with cartoon illustrations so that the images seem funny and humorous. This funny impression can attract children's attention to play longer.

All cards in this game are: Character Cards, Situation Cards, Response Cards, and Challenge Cards, equipped with pictures that clarify the writing. Game characters were named according to the typical name of the region's character, such as: Gomgom from North Sumatra (Batak), Selvi from East Nusa Tenggara (Kupang). Each character was given a background that is characteristic of the origin of the character, for example: Gomgom was depicted in the background of a traditional North Sumatra house, Selvi wore traditional East Nusa Tenggara (Kupang) clothes and next to her was a picture of Sasando musical instruments. Board games were designed with cultural themes (dances) and typical Indonesian animals (such as orangutans, and rhinos).

Color selection is important for a design. Game boards and game cards were designed using bright colors. Bright colors can create a cheerful impression and can attract interest in playing with children.

3.6. Media strategy

Media is a means or means of connecting to convey messages to the target audience.

3.6.1 Main media. The main media for this game are game boards, Character Cards, Situation Cards, Response Cards, Challenge Cards, and Dice.

3.6.2 Supporting media. This game uses supporting media so that this game can be understood and played by children. Supporting media used in this game are:

- User Guide
  This game guide book is used to help children to learn how to play. This guide can also be used by parents or teachers to teach children how to play this game.
- Pawn
  Pawn is a game marker for children to distinguish the first player, second player, and so on.
- Game Box
  The Game Box is used to provide a glimpse of information about the game and become a game card storage area so that it is not easily scattered.

4. Conclusion

The Game I am An Archipelago Kids (Aku Anak Nusantara (AAN)): Smart and Tolerant was designed to foster tolerance in children aged 6-10 years. This attitude of tolerance is sharpened through learning about the right response or attitude in dealing with situations or events in the family, school, and in the surrounding environment. Tolerance was also strengthened by an understanding of Civic Education lessons, especially Pancasila, unity, tolerance and mutual cooperation. This game was expected to be used at home and at school in shaping the personality of children who love homeland and are tolerant.

Acknowledgements

This research was supported and funded by the Ministry of Research and Higher Education (*Kemenristekdikti)* under Program of Excellent Applied Research 2018 (*Penelitian Unggulan Terapan Perguruan Tinggi*).

References

[1] Nurfitri W n. d. *Politik pilkada mengancam rasa toleransi anak*. Retrieved from http://www.parenting.co.id/usia-sekolah/politik-pilkada-mengancam-rasa-toleransi-anak.
[2] Varadi L 2014 *Youth trapped in prejudice: Hungarian adolescents' attitudes toward the Roma*. (Berlin: Springer Fachmedien Wiesbaden).
[3] Milner D 1987 Children and racism In K. McLeod (Ed.) *Multicultural education: a partnership* (Toronto: Canadian Council for Multicultural and International Education).
[4] Ramsey P G 1987 *Teaching and learning in a diverse world: Multicultural education for young children* (New York: Teachers College Press).
[5] Roe M 1982 *Multiculturalism, racism and the classroom: A CEA report* (Toronto: Canadian Education Association).
[6] Hromek R 2005 *Games to promote social and emotional intelligence for children aged 4 to 14*. (London: Paul Chapman publishing).
[7] Smilansky S and Shefatya L 1990 *Facilitating play: A medium for promoting cognitive, socio-emotional and academic development in young children* (Gaithersburg, MD: Psychosocial and Educational Publishers).
[8] Tingstrom D H Sterling-Turner H E and Wilczynski S M 2006 The good behavior game: 1969 2002 *Behavior Modification* 30 225.

IOP Conf. Series: Materials Science and Engineering **676** (2019) 012014 doi:10.1088/1757-899X/676/1/012014

Comparison of Accuracy in Extreme Learning Machine Based on Hidden Node Structure Variation for Lung Cancer Classification

S Tandungan*[1], Indrabayu[2] and I Nurtanio[2]

[1]Department of Electrical Engineering, Engineering Faculty, Hasanuddin University
[2]Departement of Informatics, Engineering Faculty, Hasanuddin University

*E-mail: sofyantandungan@gmail.com

Abstract. This paper present Extreme Learning Machine to classify lung cancer nodules. Lung cancer is a type of lung disease that requires fast and specified treatment. Skills, facilities and multidisciplinary approach are required for diagnosing lung cancer. The use of Computed Tomography (CT) to detect lung cancer can reduce the number of deaths from lung cancer, but it increases the workload of the radiologist because CT screening process produces many medical images. Computer systems become one of the potential solutions to help radiologists solve the problem. Extreme Learning Machine is an algorithm that able to provide good generalization at fast learning time which is essential to help radiologists in analyzing lung cancer nodules images. In this paper, there were 877 nodules extracted from LIDC-IDRI dataset. All nodules used in this experiment consist of lung cancer nodules that diagnosed to four different level of malignancy and annotated by up-to four different radiologists. The result shows Extreme Learning Machine achieve 85.17%, 85.58% and 84.87% in accuracy and Matthew Correlation Coefficient 0.755, 0.762 and 0.749 using Hardlimit, Radial basis Function and Triangular Basis function, respectively.

1. Introduction

Cancer is one of the leading causes of death in the world. In 2012 the number of cancer deaths is 8.2 million. The biggest causes of death caused by cancer include lung cancer, liver, stomach, colorectal, and breast cancer. Lung cancer is a cancer with the highest number of deaths as many as 1.59 million people [1]. Lung cancer is a type of lung disease that requires fast and specified treatment. Skills, facilities and multidisciplinary approach are required for diagnosing lung cancer. Radiological examination is one of the tests performed to detect and diagnose lung cancer. Radiological examination to see the condition of the lungs can be done using X-rays and Computed Tomography (CT). Currently, CT is an excellent imaging technique for early detection of lung cancer [2, 3]. Many kinds of research have harness computer vision technology to develop system to assist medical practitioner in cancer diagnosis [4-7].

Computer-aided Detection and Diagnosis (CAD) is a computer system that aims to improve radiologist accuracy and reduce the time needed to analyze images. CAD system is divided into two namely Computer-aided Detection (CADe) and Computer-aided Diagnosis (CADx). CADe was built to locate pulmonary nodules in medical images. Several studies have shown that the use of CAD systems can improve radiologist performance in detecting pulmonary nodules in CT images [3]. Computer-aided Diagnosis System (CADx) takes the characteristic values contained in the image and then uses the classification method to determine benign or malignant tumors. The use of CT to detect lung cancer can reduce the number of deaths from lung cancer. But it increases the workload of the radiologist because CT screening process produces many medical images. Computer systems become one of the potential solutions to help radiologists solve the problem [8].

In the lung cancer screening process using CT scan, radiologist should check or analyze the data in large quantities. In this paper we present Extreme Learning Machine performance as CADx to predict malignancy of lung cancer nodule from CT scan images. Several methods have been introduced whether to detect or classify lung cancer in CT images using different approaches. Firmino et al. developed a system to detect and diagnose pulmonary nodules in CT images. The system used Watershed and HOG to detect nodules and features that presented by McNitt-Gray et.al which is calcification, internal margin, shape, texture, lobulation, spiculation [9]. Zayed and Elnemr present analysis of Haralick feature to classify abnormal lung to cancerous or edema [10]. El-Baz et al. introduced the nodule diagnose method using three-dimension feature from spherical harmonic that represent the surface of nodule in three dimensional [11]. J. Yuan et al. use the combination of Fisher Vector based on the SIFT algorithm to classify six types of nodules based on the intensity distribution and position [12]. Zhu et al. classified the malignancy of lung cancer using CT image data using GBM and 3D Dual Path Network which achieve a comparable result with human level [13]. Lee et al. [14] use two step approaches by combining genetic algorithm with Random Subspace Method for feature selection and LDA as the classifier and show possibility to predict the malignancy for solitary pulmonary nodules. Han et al. [15] compare 2D and 3D texture feature of Haralick, Gabor and LBP on CADx system for lung nodule using SVM to classify benign and malignant nodules.

The result showed a 3D extension of Haralik feature show potential gain in accuracy if optimal direction can be found. Madero Orozo et al. [16] use Daubechies and Wavelet Sub-band to extract feature from lung images and classify to contain cancer or not using Support Vector Machine. The system able to reduce the complexity of CADx by not using the segmentation process like others CADx system. Kumar et al. [17] diagnose lung cancer using autoencoder as the feature extraction algorithm and L-BFG optimization. In this paper, Extreme Learning Machine (ELM) performance for classification of lung nodules level of malignancy was presented. The experiment conducted with different hidden node number and activation function and measured using Accuracy, Matthew Correlation Coefficient, Sensitivity and Specificity.

2. Materials and Methods

2.1. Datasets

The Dataset used in this paper obtained from Lung Image Database Consortium (LIDC) consist of 1010 patients with one or more lung cancer nodules that have been annotated and diagnosed by up-to four radiologists [18]. For malignancy diagnosis from radiologists, the nodules are labelled with different level of malignancy, which is "1", "2", "3", "4", and "5". Figure 1 shows sample images of patient lung cancer. There were 877 cases extracted for this study which consist of 439 cases labelled as "1" or highly unlikely malignant appearance, 98 cases labelled as "2" or moderately unlikely malignant appearance, 99 cases labelled as "4" or moderately suspicious malignant appearance and 241 cases labelled as "5" or highly suspicious malignant appearance. Nodules with label "3" was not included because it's diagnosed as uncertain malignancy by the radiologists. Each nodule is segmented using annotation from radiologists. Figure 2 shows segmented nodules with different malignancy levels. Each nodules images size is set to 64 x 64 and then will be used as inputs for Extreme Learning Machine (ELM).

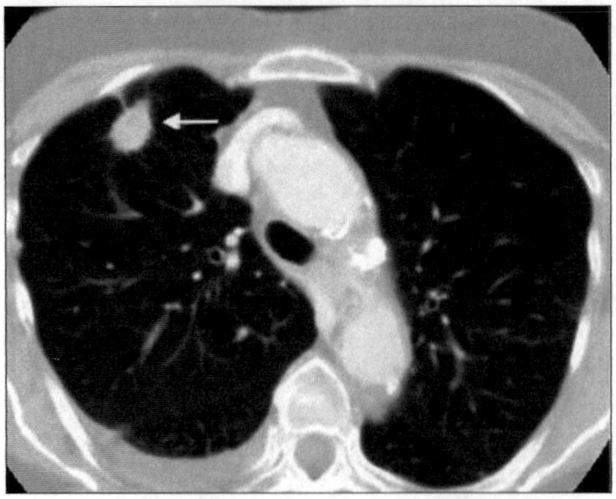

Figure 1. CT Lung image that diagnosed have cancer
Source: The Lung Image Database Consortium

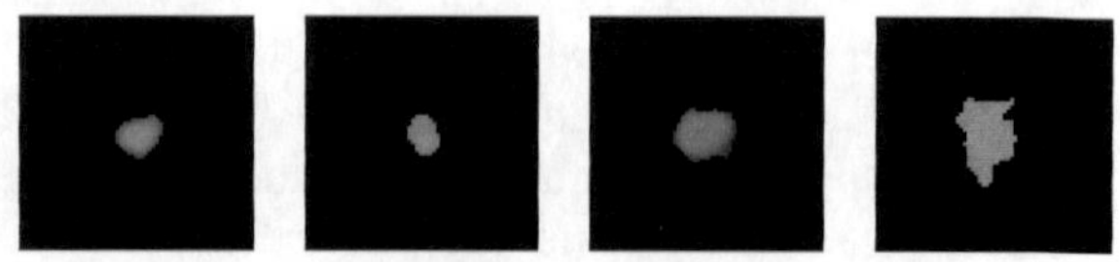

Figure 2. Extracted lung cancer nodules based on annotation from radiologists. From left to right is nodule with malignancy level 1, 2, 4, and 5

2.2. Classification
This study employs Extreme Learning Machine (ELM) as lung cancer nodule classification method. ELM was introduced by Huang et al. [14] based on single hidden layer feedforward neural networks (SLFNs). ELM algorithm tends to provide performance with good generalization and fast learning time. There are two important steps in ELM, which are calculate the output weights and to determine the number of nodes to optimize the networks. The process of ELM algorithm for lung cancer nodule classification is shown in Figure 3.

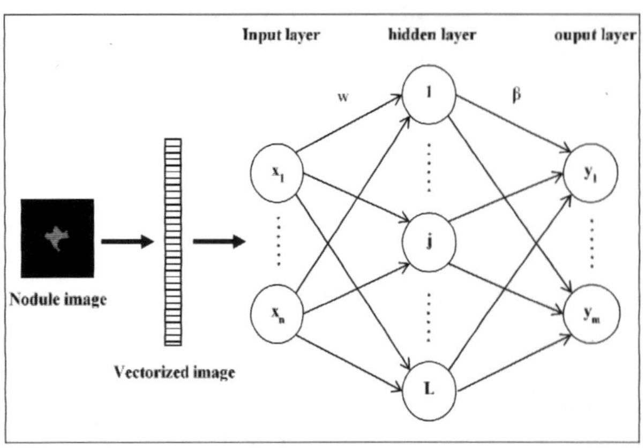

Figure 3. Lung Cancer Nodules classification with ELM

Given N training data (x_i, y_i), for i = 1, ……, N, where x_i consist of input data with n dimensions, and y_i is target data with m dimensions. The number of hidden nodes in hidden layer is denoted as L. Training process in Extreme Learning Machine Algorithm is conducted as follows:

1. Randomly generate hidden neuron parameter (e.g. input weight vector w and bias b) based on uniform distribution.
2. Calculate matrix $g(w, x, b) = w*x + b$.
3. Calculate $\beta = H^{\dagger} * Y$.
4. Calculate the result decision function, $Y = H * \beta$.
   Where H, β, and Y are defined as follows.

$$H = \begin{bmatrix} g(w_1, x_1, b_1) & \dots & g(w_L, x_1, b_L) \\ \dots & \dots & \dots \\ g(w_1, x_N, b_1) & \dots & g(w_L, x_N, b_L) \end{bmatrix}_{N \times L} \tag{1}$$

$$\beta = \begin{bmatrix} H_1^\dagger * y_1 \\ \dots \\ H_L^\dagger * y_m \end{bmatrix}_{L \times m} \tag{2}$$

$$Y = \begin{bmatrix} y_1 \\ \dots \\ y_m \end{bmatrix}_{N \times m} \tag{3}$$

$H^\dagger$ is the Moore-Penrose inverse matrix of H, β is the output weight vector, and g is activation function. This study applied Sigmoid, Hardlimit, Radial basis function, Triangular Basis function and tanh as activation function and various hidden node number to find the best configuration. All the activation function used in this experiment shown in equation (4)-(8).

$$g(n) = \frac{1}{1 + e^{-n}} \tag{4}$$

$$g(n) = \begin{cases} 1, & if\ n \geq 0 \\ 0, & if\ n < 0 \end{cases} \tag{5}$$

$$g(n) = \exp(-n^2) \tag{6}$$

$$g(n) = \begin{cases} 1 - abs(n), if\ -1 \leq n \leq 1 \\ 0, otherwise \end{cases} \tag{7}$$

$$\tanh(x) = \frac{2}{1 + e^{-2x}} - 1 \tag{8}$$

2.3. Evaluation

ELM performance on lung cancer nodule classification is measured using confusion matrix and time needed for training and testing process. Accuracy, sensitivity, specificity, F1 score and Matthew Correlation Coefficient (MCC) are calculated from confusion matrix to see ELM performance., which defined in equation (9)-(13).

$$Accuracy = \frac{TP + TN}{TP + TN + FP + FN} \tag{9}$$

$$Sensitivity = \frac{TP}{TP + FN} \tag{10}$$

$$Specificity = \frac{TN}{TN + FP} \tag{11}$$

$$F_1 = \frac{2TP}{2TP + FP + FN} \tag{12}$$

$$MCC = \frac{\#TP \times \#TN - \#FP \times \#FN}{\sqrt{(\#TP + \#FP)(\#TP + \#FN)(\#TN + \#FP)(\#TN + \#FN)}} \quad (13)$$

where True Positive (TP) is the number of outcomes where the model predicts positive class correctly. False Positive (FP) is the number of outcomes where the model predicts negative class as positive class. False Negative (FN) is the number of outcomes where the model predicts positive class as negative class. True Negative (TN) is the number of outcomes where the model predicts negative class correctly.

3. Result and Discussion

Experiment on lung cancer classification was performed in Matlab Ver 9.3 that running on PC Intel Core i7 with RAM 8 GB. There were 877 data use in this experiment that consist of 614 training data and 263 testing data. ELM was trained and tested for 10 times and the result presented as the highest averages performances at different configurations on activation function and number of hidden neurons. The average accuracy and computation time from each activation function are shown in Table 1.

Table 1. ELM classification average accuracy on different activation function

Activation Function	Number of Hidden Nodes	Training		Testing	
		Acc (%)	Time(sec)	Acc (%)	Time(sec)
Hardlim	4900	100.00	3.42	85.17	0.59
Radbas	4900	100.00	3.80	85.59	0.67
Sigmoid	100	74.41	0.08	77.07	0.03
Tanh	4900	100.00	3.45	79.96	0.65
Tribas	5000	100.00	3.75	84.87	0.61

The best results for training accuracy in Table 1 are for all the activation function except sigmoid function. In testing process, Radial basis Function achieved highest accuracy but compared with other activation function, Hardlimit (Hardlim) showed a less computational complexity. Table 2 shows ELM performance in terms of Matthew Correlation Coefficient (MCC). For multiclass case, generalized MCC presented in [19] was used. MCC values is in range [-1, 1], with 1 demonstrate perfect classification. MCC considered more useful for evaluating classifier, because it's the only measurement to consider the unbalanced condition occurs in dataset [20] which is also occurs in this experiment. The result of average MCC values was presented in Table 2. The average MCC values were similar among Radial basis, Hardlim and Triangular basis function.

Table 2. MCC measurement for training and testing

Activation Function	Number of Hidden Nodes	Training	Testing
Hardlim	4900	1	0.755
Radbas	4900	1	0.762
Sigmoid	100	0.53	0.541
Tanh	4900	1	0.675
Tribas	5000	1	0.749

For more in-depth result, the average sensitivity and specificity of Hardlimit, Radial basis function and Triangular basis function for each class used in the testing process are presented in Table 2. Hardlimit, Radial basis and Triangular basis function have good ability to correctly identified nodule that have level 1 malignancy or highly unlikely malignant characteristic which proven with high

Sensitivity values. High Specificity values at level 2 and 4 means that Hardlimit, Radial basis and Triangular basis function have good ability to ignore lung nodule without the characteristic of nodule level 2 and 4 malignancy. For level 5 malignancy, Hardlimit, Radial basis and Triangular function have high Sensitivity and Specificity which means that the activation functions have good ability to recognize lung nodules with level 5 malignancy characteristic and ignore lung nodules without the characteristic of nodules with level 5 malignancy.

Table 3. Sensitiviy and specificity measurement for each level of malignancy

Level of Malignancy	Sensitivity			Specificity		
	Hardlim	Radbas	Tribas	Hardlim	Radbas	Tribas
1	0.951	0.949	0.947	0.820	0.818	0.817
2	0.538	0.517	0.490	0.983	0.987	0.991
4	0.514	0.576	0.467	0.978	0.979	0.983
5	0.882	0.892	0.913	0.959	0.961	0.941

4. Conclusion

The ELM performance in classification of lung cancer nodules has been presented. ELM algorithm was implemented with various activation function and number of hidden nodes to find the best configuration to predict malignancy level of lung cancer nodules. The ELM algorithm was trained and tested using LIDC-IDRI dataset. All the activation function used in this study achieved highest average accuracy for training process except Sigmoid function. Meanwhile, Hardlimit, Radial basis function and Triangular basis function show good testing performance in lung nodule malignancy classification but Hardlimit function show less computational complexity compared with other Radial basis and Triangular function. ELM with Hardlimit, Radial basis and Triangular basis function have good ability to recognize lung nodules with the characteristic of level 1 malignancy or highly unlikely malignant and level 5 malignancy or highly likely malignant. The algorithm also shows good ability to ignore lung nodules without the characteristic of level, 2, 4 and 5 malignancy. However, there was still improvement needed in the ability to ignore lung nodules without the characteristic of level 1 malignancy and to recognize lung nodules that classify as level 2 and 4 malignancy.

References

[1] Stewart B W and Wild C P 2014 *World Cancer Report 2014* (Lyon: International Agency for Research on Cancer/World Health Organization).

[2] Firmino M, Angelo G, Morais H, Dantas M R and Valentim R 2016 Computer-aided detection (CADe) and diagnosis (CADx) system for lung cancer with likelihood of malignancy *Biomed Eng Online* **15** 2.

[3] Rubin G D, Lyo J K, Paik D S, Sherbondy A J, Chow L C, Leung A N, Mindelzun R, Schraedley-Desmond P K, Zinck S E, Naidich D P and Napel S 2005 Pulmonary Nodules on Multi–Detector Row CT Scans: Performance Comparison of Radiologists and Computer-aided Detection *Radiology* **234** 274–83.

[4] Indrabayu, Fatmasari A R and Nurtanio I 2017 A Colour Space Based Detection for Cervical Cancer Using Fuzzy C-Means Clustering *Proceedings of the 6th International Conference on Bioinformatics and Biomedical Science - ICBBS '17* the 6th International Conference (Singapore, Singapore: ACM Press) pp 137–41.

[5] Lee H and Chen Y-P P 2015 Image based computer aided diagnosis system for cancer detection *Expert Systems with Applications* **42** 5356–65.

[6] Mosquera-Lopez C, Agaian S, Velez-Hoyos A and Thompson I 2015 Computer-Aided Prostate Cancer Diagnosis from Digitized Histopathology: A Review on Texture-Based Systems *IEEE Reviews in Biomedical Engineering* **8** 98–113.

IOP Conf. Series: Materials Science and Engineering **676** (2019) 012014 doi:10.1088/1757-899X/676/1/012014

[7] Yang Q, Li L, Zhang J, Shao G and Zheng B 2014 A new quantitative image analysis method for improving breast cancer diagnosis using DCE-MRI examinations: Quantitative image analysis of breast DCE-MRI *Medical Physics* **42** 103–9.

[8] Team T N L S T R 2011 Reduced Lung-Cancer Mortality with Low-Dose Computed Tomographic Screening *New England Journal of Medicine* **365** 395–409.

[9] Firmino M, Angelo G, Morais H, Dantas M R and Valentim R 2016 Computer-aided detection (CADe) and diagnosis (CADx) system for lung cancer with likelihood of malignancy *BioMedical Engineering OnLine* **15.**

[10] Zayed N and Elnemr H A 2015 Statistical Analysis of Haralick Texture Features to Discriminate Lung Abnormalities *International Journal of Biomedical Imaging* **2015** 1–7.

[11] El-Baz A, Nitzken M, Khalifa F, Elnakib A, Gimel'farb G, Falk R and El-Ghar M A 2011 3D Shape Analysis for Early Diagnosis of Malignant Lung Nodules *Information Processing in Medical Imaging* vol 6801, ed G Székely and H K Hahn (Berlin, Heidelberg: Springer Berlin Heidelberg) pp 772–83.

[12] Yuan J, Liu X, Hou F, Qin H and Hao A 2018 Hybrid-feature-guided lung nodule type classification on CT images *Computers & Graphics* **70** 288–99.

[13] Zhu W, Liu C, Fan W and Xie X 2017 DeepLung: 3D Deep Convolutional Nets for Automated Pulmonary Nodule Detection and Classification.

[14] Lee M C, Boroczky L, Sungur-Stasik K, Cann A D, Borczuk A C, Kawut S M and Powell C A 2010 Computer-aided diagnosis of pulmonary nodules using a two-step approach for feature selection and classifier ensemble construction *Artificial Intelligence in Medicine* **50** 43–53.

[15] Han F, Wang H, Zhang G, Han H, Song B, Li L, Moore W, Lu H, Zhao H and Liang Z 2015 Texture Feature Analysis for Computer-Aided Diagnosis on Pulmonary Nodules *Journal of Digital Imaging* **28** 99–115.

[16] Madero Orozco H, Vergara Villegas O O, Cruz Sánchez V G, Ochoa Domínguez H de J and Nandayapa Alfaro M de J 2015 Automated system for lung nodules classification based on wavelet feature descriptor and support vector machine *BioMedical Engineering OnLine* **14.**

[17] Kumar D, Wong A and Clausi D A 2015 Lung Nodule Classification Using Deep Features in CT Images (IEEE) pp 133–8.

[18] Armato S G, McLennan G, Bidaut L, McNitt-Gray M F, Meyer C R, Reeves A P, Zhao B, Aberle D R, Henschke C I, Hoffman E A, Kazerooni E A, MacMahon H, van Beek E J R, Yankelevitz D, Biancardi A M, Bland P H, Brown M S, Engelmann R M, Laderach G E, Max D, Pais R C, Qing D P-Y, Roberts R Y, Smith A R, Starkey A, Batra P, Caligiuri P, Farooqi A, Gladish G W, Jude C M, Munden R F, Petkovska I, Quint L E, Schwartz L H, Sundaram B, Dodd L E, Fenimore C, Gur D, Petrick N, Freymann J, Kirby J, Hughes B, Vande Casteele A, Gupte S, Sallam M, Heath M D, Kuhn M H, Dharaiya E, Burns R, Fryd D S, Salganicoff M, Anand V, Shreter U, Vastagh S, Croft B Y and Clarke L P 2011 The Lung Image Database Consortium (LIDC) and Image Database Resource Initiative (IDRI): A Completed Reference Database of Lung Nodules on CT Scans: The LIDC/IDRI thoracic CT database of lung nodules *Medical Physics* **38** 915–31.

[19] Gorodkin J 2004 Comparing two K-category assignments by a K-category correlation coefficient *Computational Biology and Chemistry* **28** 367–74.

[20] Chicco D 2017 Ten quick tips for machine learning in computational biology *BioData Mining* **10.**

Cloud Classification Based on Images Texture Features

I Nurtanio[1]*, Z Zainuddin[1] and B H Setiadi[2]

[1]Department of Informatics, Engineering Faculty, Hasanuddin University, Indonesia
[2]Department of Electrical Engineering, Engineering Faculty, Hasanuddin University, Indonesia

*Email: ingrid@unhas.ac.id

Abstract. An identification of cloud imagery is part of the cloud observation process which is very important to know the potential for weather changes, especially in the Sultan Hasanuddin airport area. The purpose of this research is to build an artificial intelligence model to identify and classify texture patterns of cloud images. The research used 80 clouds images data contained in the Sultan Hasanuddin Airport area. The data consist of four types of clouds, Altocumulus, Cirrus, Cumulonimbus and Cumulus. In this research, a feature extraction process using Gray Level Co-occurrence Matrix (GLCM) algorithm and Support Vector Machine (SVM) is used for the classification process. We used a set of 4 GLCM features. The 4 selected features are contrast, correlation, energy and homogeneity. Training and testing data using cross validation method with three stages validation. The highest level of accuracy is found in the third stage validation with an accuracy value of 85%.

1. Introduction

In aviation, the cloud is one of the things that determines flight fluency and safety, especially the type of Cumulonimbus (CB) cloud. This cloud is much feared in flight because it can cause updraft, downdraft and wind shear (unexpected wind speed changes). If the plane is in or under this cloud at the time after takeoff, before landing, or during flight will cause dangerous aircraft position instability [1].

Cloud identification model is one of the things needed in the field of Meteorology and in the field of Air Traffic Regulators. This research creates a model to determine the shape and type of clouds using artificial intelligence which is expected to help Observer in the field of Meteorology or Air Traffic Regulator.

Previous research has discussed identifying cloud patterns and textures. and the effect of cloud cover on solar cell voltage (photovoltaic) [2]. The study calculated the effect of cloud cover on sunlight which is the input of solar cell voltage which is divided into four parts of the cloud conditions, clear skies, thin clouds, thick clouds and dark clouds. Zhao Zhen used the Gray Level Co-occurrence Matrix (GLCM) for extracting cloud features and SVM for the classification process. The success rate of accuracy using this method is 96.44%. Other research discussed using the GLCM and SVM methods for classification of acne based on texture [3]. In this research, texture analysis uses values from feature extraction in the GLCM method. Feature extraction is obtained by calculating the neighboring relationship between two pixels at an orientation of 0°, 45°, 90° and 135°. The five statistical characteristics used in the GLCM method are *contrast, correlations, dissimilarity, energy, entropy*. Accuracy generated in the study reached 72%.

Based on previous research, the GLCM and SVM method can analyze texture patterns well. In this research, identification of cloud images has some fundamental problems. That is the extraction of

IOP Conf. Series: Materials Science and Engineering **676** (2019) 012015 doi:10.1088/1757-899X/676/1/012015

cloud image features and building cloud image classification models. There are 4 types of clouds in the area of Sultan Hasanuddin Airport which are the basis of this research, i.e. *Altocumulus, Cirrus, Cumulus* and *Cumulonimbus*. The extraction process of cloud image features uses the GLCM method and the cloud classification process uses the SVM method.

2. Materials and Method

2.1. Materials
This research uses primary image data taken using a Canon EOS 750D DSLR camera with 24 MP sharpness and Canon Power Shoot SX600 HS with 16 MP sharpness and a resolution of 6000 x 4000 pixels. In this research only took four types of cloud images found around the area of Sultan Hasanuddin Airport in Makassar. The types of cloud are *Altocumulus, Cirrus, Cumulonimbus* and *Cumulus*. Each type of cloud has 20 images, consisting of 15 training images and 5 test images. The total data is 80 cloud images. We used Matlab as software for processing the data.

2.2. Method
In this research, the steps of the identification method of the cloud images are as follows:
1. Preprocessing: Preprocessing stages are data preparation before entering the feature extraction process. At this stage the original cloud image is 6000 x 4000 pixels (Figure 1) has gone through the process of setting parameters so that the original size changes to 200 x 200 pixels (Figure 2). Furthermore, the cloud image is changed to grayscale (Figure 3).

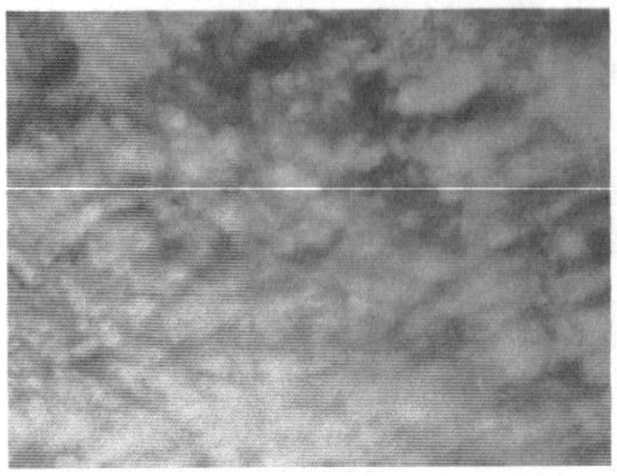

Figure 1. Real 6000 x 4000 pixel cloud image

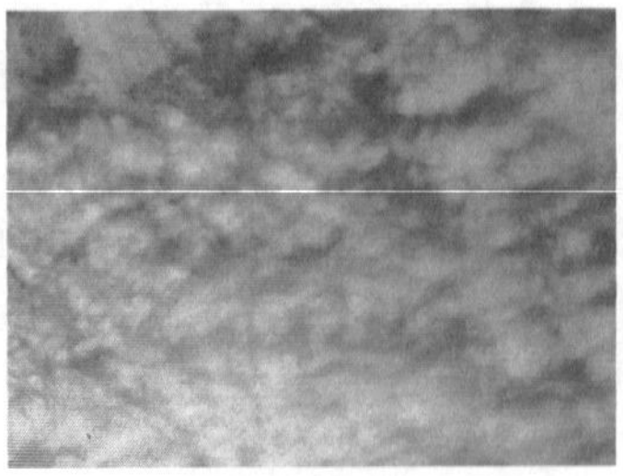

Figure 2. Cloud image 200 x 200 pixel

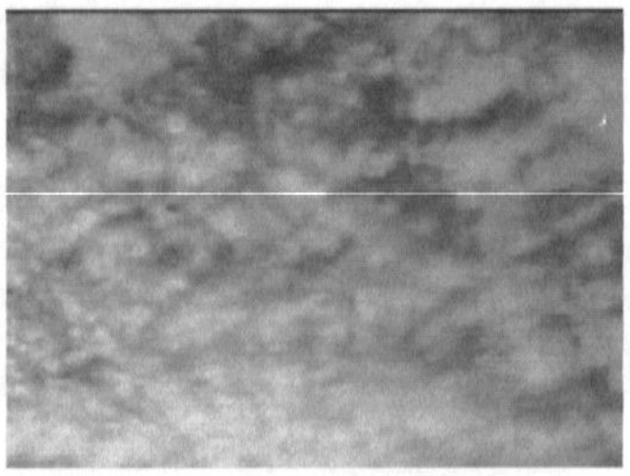

Figure 3. Grayscale cloud image

2. Feature Extraction: this step is to find a feature set of cloud images using GLCM and that can accurately to be identified.
3. Classification: this section discusses about training and testing data using the SVM method for cloud images identification.
4. Validation: using cross validation method to look for accuracy values.

2.3. Gray level co-occurrence matrix (GLCM)
The feature extraction based on gray level co-occurrence matrix (GLCM) is the second order statistics that can be used to analyze images as a texture [4]. In addition to the horizontal direction (0°), GLCM can also be formed for the direction of 45°, 90° and 135° as shown in Figure 4 below.

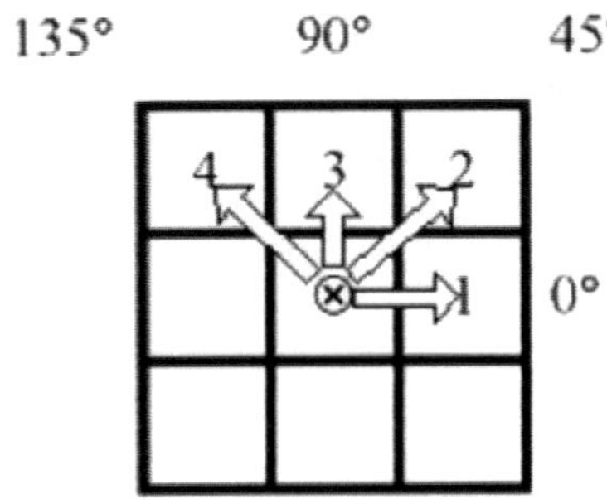

Figure 4: Direction of GLCM generation. From the center (⊗) to the pixel 1 representing direction = 0° with distance d =1, to the pixel 2 direction = 45° with distance d = 1, to the pixel 3 direction = 90° with distance d = 1, and to the pixel 4 direction = 135° with distance d = 1.

Haralick and his colleagues [5] extracted 14 features from the co-occurrence matrix, but in this research we used only 4 features, i.e. contrast, correlation, energy, and homogeneity with 4 direction and distanced d =1. The following below is the equation of the GLCM feature used in this research.

$$Contrast = \sum_{i,j=0}^{N-1} (i-j)^2 \, p(i,j) \tag{1}$$

This statistic measures the spatial frequency of an image and is difference moment of GLCM. It is the difference between the highest and the lowest values of a contiguous set of pixels. It measures the number of local variations present in the image. A low contrast image presents GLCM concentration term around the principal diagonal and features low spatial frequencies [6].

$$Correlation = \sum_{i,j=0}^{N-1} \frac{(i-\mu_i)(j-\mu_j)\,p(i,j)}{\sigma_i \sigma_j} \tag{2}$$

The correlation feature is a measure of gray tone linear dependencies in the image [6].

$$Energy = \sum_{i,j=0}^{N-1} p(i,j)^2 \tag{3}$$

It measures the textural uniformity that is pixel pair repetitions. It detects disorders in textures. Energy reaches a maximum value equal to one. High energy values occur when the gray level distribution has a constant or periodic form [6].

$$Homogeneity = \sum_{i,j=0}^{N-1} \frac{p(i,j)}{1+(i-j)^2} \tag{4}$$

This statistic is also called as Inverse Difference Moment. It measures image homogeneity as it assumes larger values for smaller gray tone differences in pair elements. It is more sensitive to the presence of near diagonal elements in the GLCM. It has maximum value when all elements in the image are same [6].

2.4. Support Vector Machine (SVM)

Support vector machine, is a supervised learning technique that seeks an optimal hyperplane to separate two classes of samples. Kernel functions are used to map the input data into a higher dimensional space where the data are supposed to have a better distribution, and then an optimal separating hyperplane in the high-dimensional feature space is chosen. The idea behind SVMs is to map the original data points from the input space to a high dimensional, or even infinite-dimensional, feature space such that the classification problem becomes simpler in the feature space. The mapping is done by a suitable choice of a kernel function (Figure 5).

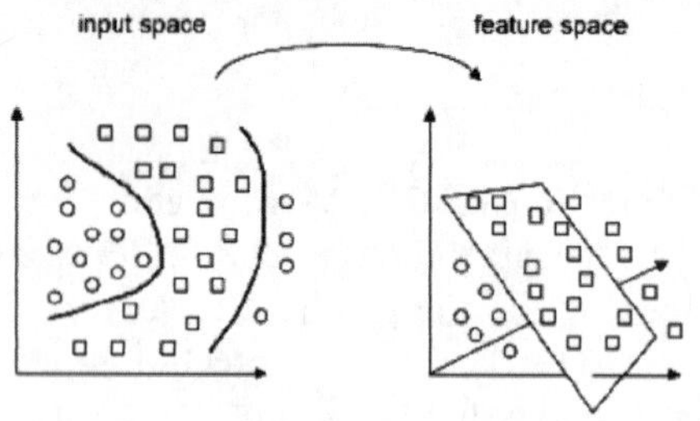

Figure 5. SVMs allow mapping of the data from the input space to a high-dimensional feature space where a linear separation is obtained [7]

Clouds data that have trained and achieved the desired results need to be tested to determine the ability when studying the training data given. Testing is conducted to observe the performance of the system that has been created by looking at the value of the minimum error. Training and testing results can be analyzed by observing the accuracy of output target. The training and testing process of clouds classification is shown in Figure 6.

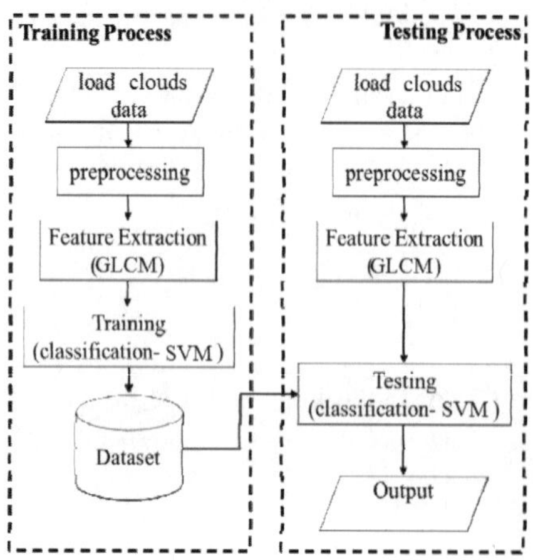

Figure 6. Flowchart training and testing process in clouds classification

2.5. *Cross validation*

In this research, process validation using cross validation method with three stages of validation. Clouds data used consist of 60 training data and 20 testing data for four types of clouds. Cross validation method for each cloud is shown in Table 1.

Table1. Cross validation method

Clouds Data	First Validation	Second Validation	Third Validation
1	Training Data	Training Data	Training Data
2	Training Data	Training Data	Training Data
3	Training Data	Training Data	Training Data
4	Training Data	Training Data	Training Data
5	Training Data	Training Data	Test Data
6	Training Data	Test Data	Test Data
7	Training Data	Test Data	Test Data
8	Training Data	Test Data	Test Data
9	Training Data	Test Data	Training Data
10	Training Data	Test Data	Training Data
11	Training Data	Training Data	Training Data
12	Training Data	Training Data	Training Data

Clouds Data	First Validation	Second Validation	Third Validation
13			
14			Test Data
15			
16			
17			
18	Test Data	Test Data	Training Data
19			
20			

3. Result and Discussion

All the experiments were conducted in Matlab version 9.0 running on a PC Intel-Pentium Core i5 with 4 GB of RAM. A total of 80 clouds data measuring 200x200 pixel are processed using the GLCM-SVM method. Table 2 below shows the results of GLCM feature extraction on all four types of clouds.

Table 2. The result of GLCM feature extraction

No	Real Clouds Data	*Grayscale*	GLCM Output					
				0	45	90	135	average
1	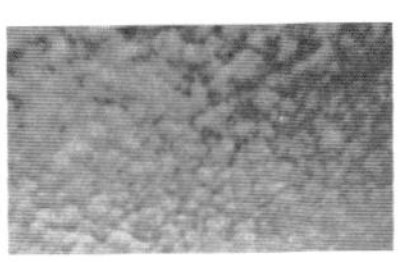		Contrast	0.16515	0.23645	0.22256	0.27025	0.2236
			Correlation	0.90333	0.86126	0.86945	0.84138	0.86885
			Energy	0.22221	0.19656	0.20051	0.18533	0.20115
			Homogeneity	0.91814	0.88601	0.89205	0.87116	0.89184
				0	45	90	135	average
2	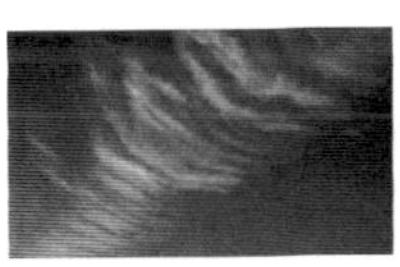	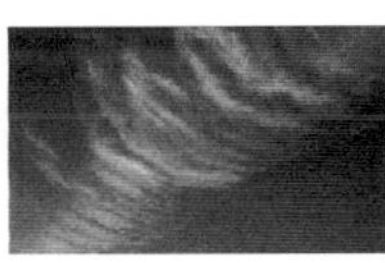	Contrast	0.077206	0.15437	0.11991	0.10855	0.11501
			Correlation	0.9657	0.93143	0.94674	0.95177	0.94891
			Energy	0.24072	0.21487	0.22582	0.22768	0.22727
			Homogeneity	0.96143	0.92444	0.9405	0.94581	0.94305
				0	45	90	135	average
3	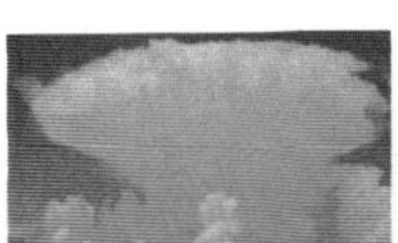		Contrast	0.041867	0.064118	0.048455	0.0634	0.05446
			Correlation	0.97381	0.95989	0.96969	0.96029	0.96592
			Energy	0.2787	0.26797	0.27494	0.26801	0.27241
			Homogeneity	0.97928	0.96881	0.97604	0.96885	0.97325
				0	45	90	135	average
4			Contrast	0.04995	0.10361	0.09495	0.10431	0.088205
			Correlation	0.9785	0.95548	0.95918	0.95519	0.96209
			Energy	0.26658	0.25118	0.25297	0.25074	0.25537
			Homogeneity	0.97572	0.95371	0.95695	0.95336	0.95993

The output extraction feature in this GLCM method is a matrix. This matrix is the characteristic of each cloud image. Table 2 showed that no specific value about the four types of clouds. That's why SVM as classification method was chosen to classify the type of clouds. To choose the best training data for high accuracy, we used cross validation method with three stages of validation. The results of validation shown in Figure 7.

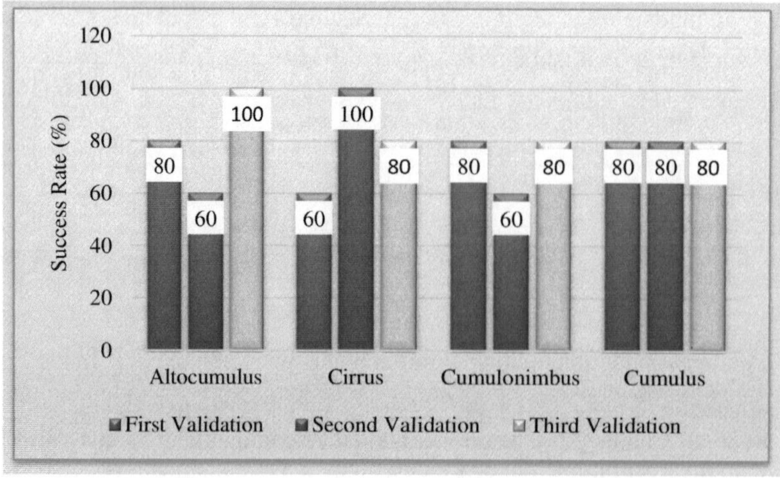

Figure 7. The graphic result of cloud classification

Based on the graph in Figure 7, it could be seen that the validation of stage three Altocumulus cloud types reached 100% success rate. Cirrus, cumulonimbus, and cumulus cloud types reached 80% success rate. Likewise with the Cirrus cloud type on second stage of validation the success rate reached 100%, but altocumulus and cumulonimbus cloud types just reached 60% success rate. At the first stage, there is 60% success rate for cirrus cloud type. In Cumulonimbus clouds the success rate of testing reached 80% in the first and third stages of validations. While in Cumulus's cloud the test success rate reached 80% in all validations. Thus the third stage of validation was chosen, because in the four types of clouds, the lowest success rate reached 80% while in the first and second stage of validations, there was still a success rate of 60%. From the average of three stage validation, the third stage also occupied the highest value, which is 85% success rate. The results of the average value of three stages validation shown in Figure 8.

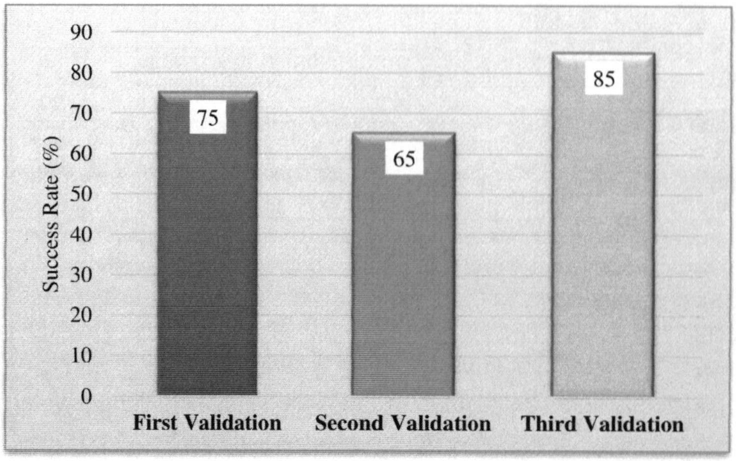

Figure 8. The average values for three stages of validation

Now we can see in Figure 8, in the first stage of validation had accuracy level of 75%, while in the second stage of validation had accuracy level of 65%, then in the third stage of validation had accuracy level of 85%.

IOP Conf. Series: Materials Science and Engineering **676** (2019) 012015 doi:10.1088/1757-899X/676/1/012015

4. Conclusion

Based on experimental results, it concludes that Gray Level Co-occurrence Matrix (GLCM) with Support Vector Machine (SVM) method, can be used for cloud identification. Clouds data used consist of 60 training data and 20 testing data for four types of clouds. Process validation using cross validation method with three stages of validation the highest accuracy value in this research achieved in third stages of validation and reach 85% accuracy.

References

[1] Anonymous, Meteorological, Climatological, and Geophysical Agency: *Comulonimbus cloud alert until January 2017*: Okezone News," *https://news.okezone.com/.* [Online]. Available: https://news.okezone.com/read/2016/12/05/340/1559035/ [Accessed: 03-Sep-2018].

[2] Z. Zhen *et al.* 2015 SVM Based Cloud Classification Model Using Total Sky Images for PV Power Forecasting in *IEEE Power & Energy Society Innovative Smart Grid Technologies Conf. (ISGT)*, Washington, DC, USA, pp. 1–5.

[3] M. Ramadhani and D. Suprayogi 2018 Acne Classification Based on Texture Using GLCM Method in *e-Proc. of Engineering* : vol 5 No. 1 pp. 870-876.

[4] I. Nurtanio *et al* 2013 Classifying Cyst and Tumor Lesion Using Support Vector Machine Based on Dental Panoramic Images Texture Features in *IAENG Int. Journal of Computer Science* vol 40 (1) pp 29-37

[5] R. M. Haralick *et al* 1973 Textural Features for Image Classification in *IEEE Transactions on Systems, Man, and Cybernetics* vol SMC-3 no 6 pp. 610–621.

[6] D. Gadkari 2000 *Image Quality Analysis Using GLCM* thesis in University of Pune, Florida.

[7] J. Luts *et al* 2010 A tutorial on support vector machine-based methods for classification problems in chemometrics in *Analytica Chimica Acta* vol 665 no 2 pp. 129–145.

IOP Conf. Series: Materials Science and Engineering **676** (2019) 012016 doi:10.1088/1757-899X/676/1/012016

A Study for the Application of Steel Frames on a Traditional Wooden Fishing Boat

L Bochary[1], M R Firmansyah[1]*, G Sitepu[1] and S Asri[1]

[1]Departement of Naval Architecture, Engineering Faculty, University of Hasanuddin, Indonesia

*E-mail: mr.firmansyah@unhas.ac.id

Abstract. As its main material, specific timbers for supporting boat structure are required in the construction of traditional wooden fishing boat in Indonesia. The requirements for the specific timbers are vary such as it must be water resistant for ship skin planking and must be in the specific form for the boat frames in which the curvature will be different according to the specific curvature of each transversal boat section along the boat length. Currently, the required specific form of timbers for the boat frames are becoming difficult to be acquired especially in the midship area of the ship which requires extreme curvature. The difficulty to acquire the specific curvature form of frames as required have threaten the sustainability of wooden boat construction process in Indonesia. Hence, an alternative material for replacing the wooden material for boat frames is needed. The requirements for the alternative material are as follows; it must be easy to be formed according to the boat transversal section form along the boat length and must be strong enough to support the strength of the wooden boat hull. This research is conducted in order to analyse the feasibility of steel frame application for replacing wooden frame in the construction of traditional wooden fishing boat. Some alternatives of different steel frame dimensions with specific profile forms are analysed to determine the appropriate steel frame construction for certain type of wooden fishing boat. In the end, a recommendation of the steel frame dimensions for replacing wooden frame on a certain traditional wooden fishing boat is proposed.

1. Introduction

Wooden boat is a boat which almost all of its constructions are made from wood. The construction of wooden boat in Indonesia have been conducted since hundred of years ago. The ability to build the wooden boat have been inherited by the craftsmen ancestor generation by generation. Up to several years ago, wooden material for the construction of wooden boat is relatively easy to obtained as plenty of forests in Indonesia. But as the deforestation in some forest in Indonesia, wooden material as the main material for wooden boat is becoming difficult to obtain. Hence, the price for the material has become expensive [1] and there is no alternative material which been studied yet for replacing the wooden material in the wooden boat construction. If the condition continuously happens, some day in the future, the wooden boat can not be constructed in Indonesia anymore.

In order to support the sustainability for wooden boat construction in Indonesia, some efforts need to be done to find an alternative material for replacing wooden material in the construction of wooden boat. The alternative material must be on certain part of the wooden boat construction to maintain the specific form of the wooden boat.

In this research, wooden frame in the wooden boat construction have been selected to be analysed. Steel material have been chosen as an alternative material for frame construction to replace wooden frame in wooden boat [2]. The selection of frame construction in this research as it consumes significant amount of cubic volume of wooden material. Besides, this form of construction is the most difficult form of wooden boat construction to obtain.

2. Current Traditional Wooden Fishing Boat Frame Construction

Frame in any boat construction is the most important part of the construction since it being the skeleton to support the boat body and at the same time support the boat longitudinal strength. Frame in a wooden boat construction is installed on the boat transversal side along the boat length from keel to up to the side of the deck beam (Figure 1) [3]. In wooden boat construction, the shell plank as well as the inner shell plank are attached to the wooden frame [4].

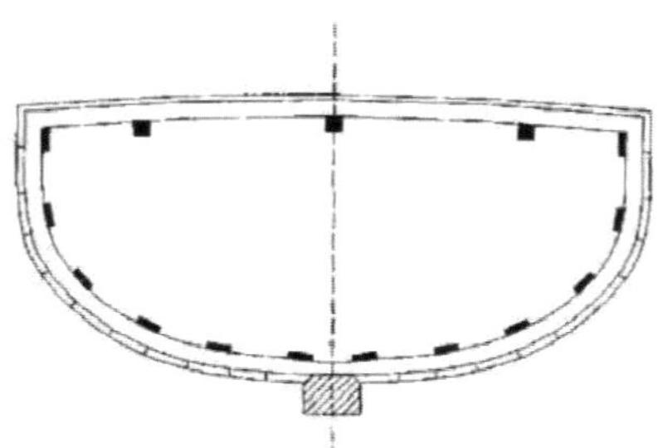

Figure 1. The construction of wooden boat

Unlike frames on the construction of steel ship, frame on a wooden boat construction is made from single frame and it is continuously constructed from starboard to portside according to the boat body curvature but they are attached and connected through wooden keel. The distance between frames in the wooden boat construction is measured between wooden axes and it is determined based on the shipping area of the respective wooden boat [5].

Wooden frame is generally made from naturally curved wooden as it will support the whole boat strength [6]. But sometimes, the wooden frame must be connected in order to get the required wooden frame for the boat construction [7]. Though, priority is given to the continuous wooden frame construction. Hence, as the tight requirements, the wooden frame is relatively difficult to obtained as required. On the other side, the need for the specific wooden frames is a must [8].

In order to solve this problem, an alternative material for wooden frame is required. The selection of steel frame for replacing the wooden frame become the most rational selection for some reasons such as it easy to get as steel frame can be found everywhere, it is stronger then wooden material, it is easy to be formed according to the curvature of the wooden boat body and it is more cheaper to compare with wooden frame price [9].

3. Feasibility of Steel Frame Application for Replacing Wooden Frames

Object for this research is traditional wooden fishing boat with 30 GT of its capacity. Data of the boat to be collected for this analysis consist of wooden boat main dimension as well as its construction dimension such as wooden frame, shell plank, inner shell plank and boat keel. The location for data collection is in a traditional wooden boatyard in Galesong Kota village, Takalar regency. The boat at the time of data collection is still under-construction as can be seen in Figure 2.

Figure 2. The construction of 30 GT traditional wooden fishing boat in Takalar regency

The main dimension of the 30 GT of wooden boat can be seen in Table 1 below.

Table 1. The main dimension of the 30 GT of traditional wooden fishing boat

Ship tonnage	:	30 GT
Length overall (LOA)	:	2280 cm
Length between perpendicular (LBP)	:	1850 cm
Breadth	:	510 cm
Height	:	180 cm
Draught	:	100 cm

3.1. The boat construction dimension

In this research, the main boat construction data to be collected is its keel, frame, shell plank, inner shell plank, distance between frame as well as distance between inner shell plank.

3.1.1. Frame dimension.

Wooden boat frame was measured for 10 samples of wooden frame dimension. As traditionally formed for the wooden frame construction, there is an inconsistency of wooden frame dimension along the frame. Hence, in the wooden frame measurement process, the frame was divided into 4 zones of measurement as can be seen in Figure 3. Zone A is the farthest from the boat keel while zone D is the closest from the boat keel. The average value of the wooden frame dimension will be the dimension of the wooden frame.

Figure 3. The measurement of wooden frame dimension

The result of the measurement can be seen in Table 2 below.

Table 2. The wooden frame dimension on the four different zones.

Zone	No	Wooden boat 30 GT			
		Width (cm)	x	Height (cm)	Distance between axis
A	**1**	**9.7**	x	**11.8**	**39.9**
B	**2**	**9.4**	x	**12**	**39.9**
C	**3**	**10.6**	x	**13**	**39.5**
D	**4**	**10.6**	x	**13.1**	**40**

3.1.2. Wooden inner planking dimension.

The dimension of the wooden inner planking for 30 GT of traditional wooden fishing boat is 20 cm wide and 5 cm for its thickness (Figure 4). The distance between inner planking is about 50 cm average (Table 3).

Figure 4. The wooden inner planking measurement

Table 3. The dimension of the inner planking and the distance between

No	Wooden inner planking dimension 25 x 5 cm			Distance between inner wooden plank
	Width (cm)	x	Thickness (cm)	Distance between axis
1	**25.018**	x	**4.85**	**50.05**

4. Wooden Boat Keel Dimension

Keel for the boat is made from the big and long beam. The boat keel measurement is conducted in some parts of the boat keel as the inconsistency of the boat keel dimension. The measurement process of the boat keel dimension can be seen in the Figure 5 and the result of the measurement can be seen in Table 4.

Figure 5. The measurement process of the boat keel

Table 4. The dimension of the 30 GT traditional wooden fishing boat

Boat	**Average height (cm)**	**Average width (cm)**
30 GT	20 ± 0.3	17 ± 0.5

4.1. Shell plank dimension

The process of the wooden boat shell plank measurement process can be seen in the Figure 6 while the result of the measurement can be seen in Table 5.

Figure 6. The measurement process of the boat shell plank

Table 5. The dimension of the wooden boat shell plank.

Boat	**Average thickness (cm)**	**Average width (cm)**
30 GT	**5 ± 0.3**	**20 ± 0.5**

5. Modelling Wooden Frame Structure

Boat construction data to be used in the modelling process is the data from the boat construction measurements in the midship part of the boat. The selection of this construction part is because midship part of the boat is bear the biggest load in the transversal construction of the boat [10]. Data obtained from the measurement process then translated into two and three dimensional of drawing as can be seen in Figure 7 below. Modelling uses x axis which refer to ship breadth, y axis which refer to ship height and axis z which refer to ship length [11].

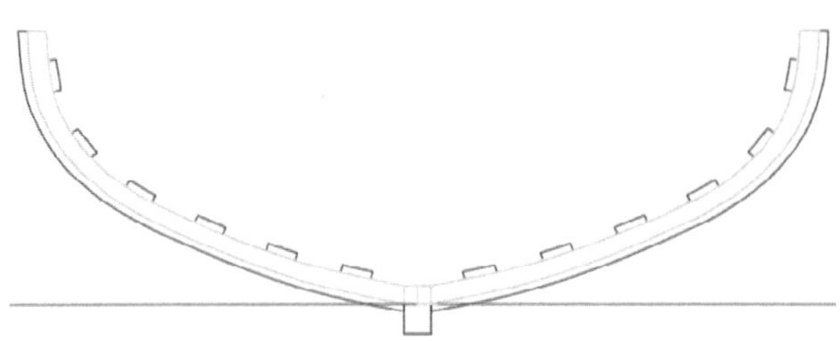

Figure 7. The transversal midship construction of the fishing boat

5.1. Load calculation on the boat

The load to be applied on the boat construction in this model is calculated based on the Indonesian Bureau Classification (BKI) rules Vol. II 2016 [12] for load calculation on a ship. It consists of two types of load which works on ship which are boat side and boat bottom load.

5.2. Boat side load

Side load according to BKI [12] rules consist of two directions of side load which are sea wave from the front and sea wave from the side of the boat. The result of the calculation based on the formula in this rules can be seen in Table 6 below.

Table 6. The result of side load on the boat for each height

Boat 30 GT		
Height	Sea wave from the front (N/mm^2)	Sea wave from the boat side (N/mm^2)
0	1.276 E-02	1.455 E-03
0.5	9.134 E-03	1.296 E-03
1	5.512 E-03	9.099 E-03
1.5	5.249 E-03	8.271 E-03
1.8	5.103 E-03	7.844 E-03

Based on the Table 6, the biggest side load being applied to the boat body is both when the sea wave is coming from the front and from the boat side at the height of 0 which is on the lowest part of the boat body.

5.3. Boat bottom load

Bottom load according to BKI rules consist of two directions of bottom load which are sea wave from the front and sea wave from the side of the boat. The result of the calculation based on the formula in this rules can be seen in Table 7 below.

Table 7. The result of bottom load on the boat for each height

Boat 30 GT		
Height	Sea wave from the front (N/mm2)	Sea wave from the boat side (N/mm2)
0	2.755 E-03	4.549 E-03
0.25	5.255 E-03	7.049 E-03
0.5	7.755 E-03	9.549 E-03
0.75	1.025 E-02	1.204 E-02
1	1.275 E-02	1.454 E-02

Similar to the side load, the biggest bottom load being applied to the boat body is both when the sea wave is coming from the front and from the boat side at the lowest part of the boat body.

5.4. Condition of the loading and restraint

The loading pressure is then being applied to the boat body which is based on the side and bottom load as previously calculated. The pressure on the boat body as the result of the calculated load can be seen in Figure 8 below.

Figure 8. Distribution of pressure load on the boat wooden plank

The load to be applied is the resultant load of the boat side and bottom load. The distribution load only be applied in the direction of the plane normal angle and hence, the load resultant can be found on the normal axis of the plane. The result of the load model can be seen in the following figure (Figure 9).

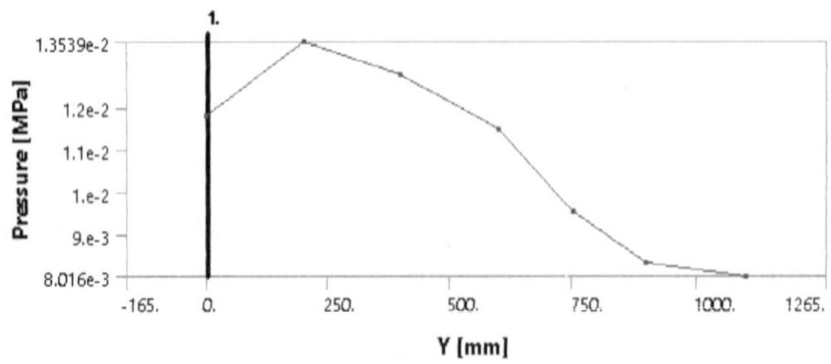

Figure 9. The result of the load on normal direction of the area

6. Recommendation of Suitable Steel Frame Dimensions for Replacing Wooden Frame

6.1. The selection of steel frame dimension

The form of the steel frame to be applied to replace the wooden frame in the construction of the wooden fishing boat is based on the Larengi [13] research which recommend the T form of the steel frame. The selection of the T from for the steel frame is based on the consideration of some aspects such as its form, easy to be constructed, ease for binding the frame to the boat shell plank and some other criteria. The T form of the steel frame construction can be applied on the boat shell plank by following the curvature of the boat body while the binding process using some nuts and bolts which is bound with the shell plank (Figure 10) [2]. For inner shell plank stand, some attachments according to the distance of the inner shell plank have been applied.

Figure 10. Model of the steel frame to be used [13]

The form of the steel frame to be used to replace the wooden frame in the construction of the wooden boat is according to frame form in Figure 10. The dimension of the steel frame is conducted by calculating its stiffness which is equal to the wooden frame or EI_{steel} is less then or equal to EI_{wood}. The profil dimension is calculated on trial and error method with the constraint of widht and the height of the steel frame is not more than the width and the height of the wooden frame.

6.2. Deflection

Respon structure analysis of the frame was conducted in order to identify its deflection. Deflection can be obtained by simulating the load on the structure using some alternatives of the steel frame. The deflection to be analysed is the deflection on the direction of x axis and y axis. Deflection from z axis can be ignored as the value of the deflection is very small and its not directly affect the structure [14]. The total deflection on the frame can be seen in Figure 11. Frame to be observed is frame located in the middle of the construction model as the result of the side load and bottom load. The observation is including the inner shell plank, shell plank as well as two or three frames next to the main observed frame.

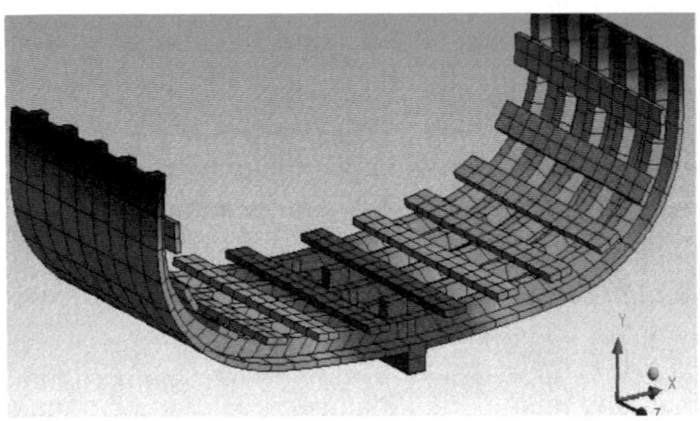

Figure 11. Total deflection on the model

In Figure 11, the colour determined the value of the deflection. The red colour shows the biggest deflection occurred to the construction model. The maximum value of the deflection for 30 GT of wooden fishing boat can be seen in Table 8. The result of the deflection then can be compared between wooden frame and steel frame and it can be seen in Table 9.

Table 8. Maximum deflection value in x axis direction in mm

Boat	Frame	
	δ	**Distance from CL**
30 GT	**0.979**	**2325.1**

Table 9. Result of the comparison of steel frame and wooden frame dimension

Boat	30 GT
Wooden dimension	121 X 100 mm
E_{wood}	11.866 Mpa
I_{wood}	14.763.008 mm^4
E_{steel} [15]	206.000 Mpa
$Isteel = \frac{AIwooden}{Esteel}$	0.65 X 8.50 E+05 mm^4
I_{steel} ; *trial and error*	5.73 E+05 mm^4
Steel frame dimension	100X80X8 mm
$\frac{EIsteel}{AIwooden}$	0.67

IOP Conf. Series: Materials Science and Engineering **676** (2019) 012016 doi:10.1088/1757-899X/676/1/012016

Based on the analysis, the steel frame dimension to replace the wooden frame in the construction of the wooden 30 GT fishing boat is 100 mm x 80 mm x 8 mm.

7. Conclusion

The use of steel frame for replacing wooden frame on a wooden boat construction shows promising result. In this study, steel frame was analyzed based on the characteristics of wooden frame in the construction of wooden boat. Some steel frame dimension was simulated in order to find its behaviour which is closest to the behaviour of wooden frame. The result shows that, when using steel frame dimension which have the same value of EIsteel/EIwood, deflection on steel frame is 35% smaller for 30 GT Wooden boat. It means that steel frame is strong enough to carry any load to compare with wooden frame. The dimension of steel frame which appropiate for 30 GT wooden boat is 100 x 80 x 8 mm.

References

[1] Ayuningsih, Dewi. 2007. *Techno economic for the wooden vessel construction on a traditional wooden shipyard in Gebang village, Cirebon, West Java.* Institut Pertanian Bogor. Bogor [in Bahasa].

[2] Bochary, L. dan Larengi, F, 2012, *An alternative for the use of steel frame on 30 GT wooden vessel construction.* Journal Riset Teknologi Kelautan. Vol 10. No. 2, pp 121-272. Makassar [in Bahasa].

[3] Biro Klasifikasi Indonesia. 1992. *Guidance for building motorized sailing ship.* Indonesian Classification Bureau, Jakarta [in Bahasa].

[4] BSN. 2013. *Design specification for wooden construction.* Badan Standar Nasional. Jakarta [in Bahasa].

[5] Rauf, A. M., dan Chaerunnisa, A. 2003. *The development of a method for traditional wooden boat construction in Bulukumba regency.* Lembaga Penelitian Pengembangan Masyarakat Universitas Hasanuddin, Makassar [in Bahasa].

[6] Hunggurami, Elia, Sudiyo Utomo, Beddy Y. Messakh. 2016. *Identification of strength reference of commercial wooden type in Kupang city based on SNI 7973:2013.* Undana. Kupang [in Bahasa].

[7] Karim, Abdul Azis , Mabsyur H., A. Haris Muhammad. 2011. *Study for the dimension and model connection of wooden boat construction in a traditional wooden boat shipyard in Bulukumba regency.* UNHAS. Makassar [in Bahasa].

[8] BKI. 1996. *Rules for wooden boat. Biro Klasifikasi Indonesia.* Jakarta [in Bahasa].

[9] BKI. 2014. Vol V Rules for Materials. Biro Klasifikasi Indonesia. Jakarta [in Bahasa].

[10] Gerr,dave. 2000. *Boat Strenght for Builders, Designers, and Owners. International Marine.* Camden.

[11] Pinem, Daud. 2013. *ANSYS Mechanical system analysis.* Wahana Ilmu Kita. Jakarta [in Bahasa].

[12] BKI. 2016. *Vol II Rules for domectic hull.* Biro Klasifikasi Indonesia. Jakarta [in Bahasa].

[13] Larengi, Farid. 2012. *An alternative for the use of steel frame on a 30 GT wooden boat construction.* Laporan Penelitian. Fakultas Teknik UNHAS. Makassar [in Bahasa].

[14] Infometrik. 2009. *Basic concept of Finite Elemen Method.* Online. Accessed: http://www.infometrik.com/2009/07/konsep-dasar-finite-element-method/ . 11 Maret 2017 [in Bahasa].

Simulating Torque Converter on a 3 GTs Fishing Vessel using Computational Fluid Dynamics

Zulkifly*[1], Baharuddin[1], H Rivai[1], M R Firmansyah[2] and T M Santoso[1]

[1]Marine Engineering Department, Engineering Faculty, University of Hasanuddin
[2]Naval Architecture Department, Engineering Faculty, University of Hasanuddin

*E-mail: navalarchitecture78@gmail.com

Abstract. Torque converter is needed to reduce the difference distance on an engine power curve and propeller operating load curve. Torque converter works by removing and cutting the flow of power from engine to transmission shaft automatically in order to adjust with fluctuative propeller load. Torque converter works on a fluid clutch principal hence it has ability to absorb and reduce excessive torque moments. This research aims to determine the effect of impeller distance against dynamic characteristics of torque converter on a 3 GT fishing vessel. Based on the designed torque converter dimension, a simulation using computational fluid dynamics (CFD) in an ANSYS 16.0 simulation software was conducted. The simulation was done on three distances between impellers i.e. 2.5 mm, 5 mm, and 7.5 mm and on five rotations i.e. 600 rpm, 1000 rpm, 1400 rpm, 1800 rpm, and 2200 rpm. Information obtained from the CFD simulation includes: pressure and shear stress distribution on impeller, flow velocity on the torque converter shell, torque value, as well as maximum transmission power of torque converter. The result shows that the ideal impeller gap for the 3 GTs fishing vessel from the dynamic characteristics parameter of torque converter is 7.5 mm.

1. Introduction

Motorization policy for fishing vessel in South Sulawesi by equipping the vessel with motor and modern technology fishing gear has been applied since the beginning of 1970. The aim of the policy package is to optimize the utilization of sea resources. This policy is in turn reduce significantly the use of sail as the vessel prime mover. Further, the motor in the vessel is supported by single reduction gearbox system. However, the weakness of this system is the existence of differences on the distance in an engine power curve and propeller operating load curve. The latter is constantly fluctuated according to the condition change during ship operation such as change on weather, wave, ship load or change of sea current.

In many cases, there are some damage reports on the shafting system of fishing vessel as the consequences of the condition [1-2]. Among them are misalignment of the engine shaft, drastically damage/worn out on the shaft and its bearing, broken unbalancing propeller as well as early overhaul on the ship main engine. An automatic torque converter for muffling and absorbing the load is needed to minimize the difference on the distance. The load to be reduced in this case is the propeller shaft load which exceed the power availability in the ship engine.

The working procedure for the torque converter (TC) is similar with an automatic transmission system on an automatic motor car. The main difference is on the system and working mechanism and on the working principle of fluid. The advantage of using torque converter on similar system is on its ability to absorb big torque moment. This ability makes it to feasible to be used on a ship to bear power transmission load. On the other side, the maintenance for this system is easy and comparatively cheap, the construction of the system is simpler and it has spontaneously output converter to compare with conventional continuously variable transmission (CVT).

2. Transmission System on a Ship

Transmission system on a ship proceeds power from the main engine to propeller. This system moves the ship at the required speed. As can be seen from figure 1, transmission system consists of main engine, fly wheel, thrust block, intermediate shaft, propeller shaft and propeller. Thrust from propeller is used to move the ship through transmission system. Shaft was equipped with thrust block, intermediate bearing and stern tube bearing [3].

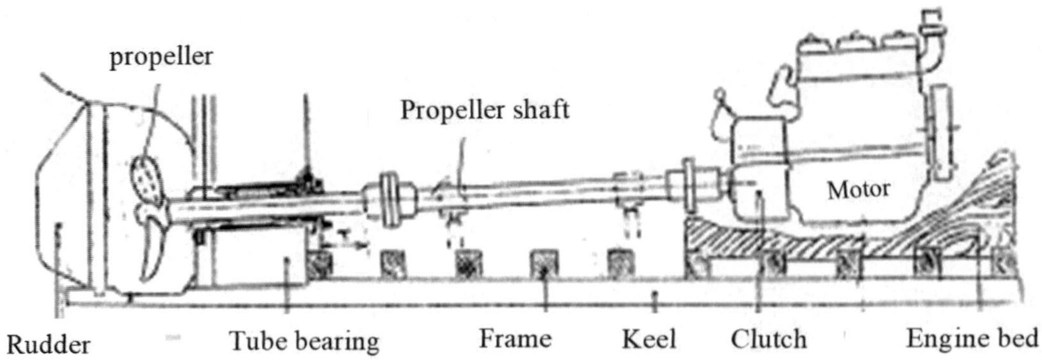

Figure 1. Transmission system on a motorized fishing vessel

Power resulted from a diesel engine is in the form of torsional moment or torsion (T, N.m) and rotation speed (n, r/min) through crank shaft which can be used for many needs such as source of energy for rotating dynamo shaft or can be used for moving both sea and land transportation vehicles. Output power from main engine is resulted from crank shaft as a shaft driver which connected with shaft which is moved by clutch connection.

Clutch is an engine element which connecting and stopping torsion from shaft driver to the driven shaft. On a diesel engine as a ship driver, the shaft rotation speed is bigger to compare with the need of propeller shaft rotation speed (ship propeller). To adapt the differences in speed, gearbox (transmission box) is needed before proceed to the propeller shaft [4]. By adopting an advanced automatic transmission system technology, the same thing can be done on a ship propeller system by using torque converter.

Torque converter is a coupling element which is installed between engine and shaft system. The advantage of the system is coming from power transfer of hydrodynamic principle from engine to transmission system which at the same time works as load diminisher in order to keep other mover component functions as well as to fix economic fuel of the ship fuel (Figure 2).

The development of the concept in the beginning is for hydraulic engines for heavy work (crane and heavy equipment) which have/experience failure because of the damage to the power of transmission gearbox [5].

This research is regarding the effect of the impeller angle speed on the magnitude of centrifugal force on hydraulic liquid of torque converter. The result of the research has provide some recommendation on some optimal angle of the impeller blades including the number of impeller blades. If impeller is rotated by the main engine, the hydraulic fluids inside the impeller shell will be rotate as well and bounce to the torque converter shell wall in the direction of the engine rotation.

The number of area and impeller blades is significantly having effect on torque output, area criteria on torque converter case in the determination of area ratio where the comparison between blade area with the object to be designed [6].

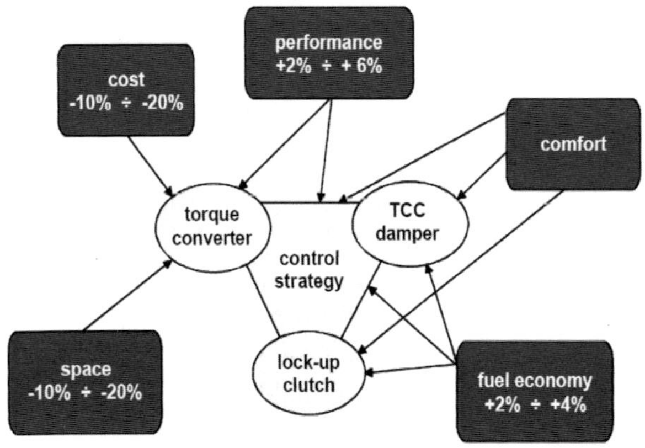

Figure 2. Torque converter system

In a modern automatic transmission, lock-up clutch has been implemented in a torque converter to lock engine and shaft transmission in a higher gear ratio. The main advantage of using the lock-up clutch is to increase the torque converter efficiency. The speed ratio and torque ratio without using lock-up clutch will be around 0.9 and 1 respectively. Hence the torque converter efficiency which can be defined as a product from speed and torque ratio can reach 90%. Mechanism in a lock-up clutch increases the efficiency value to be 100% as the rigid connection between pump torque and turbine shaft (Figure 3).

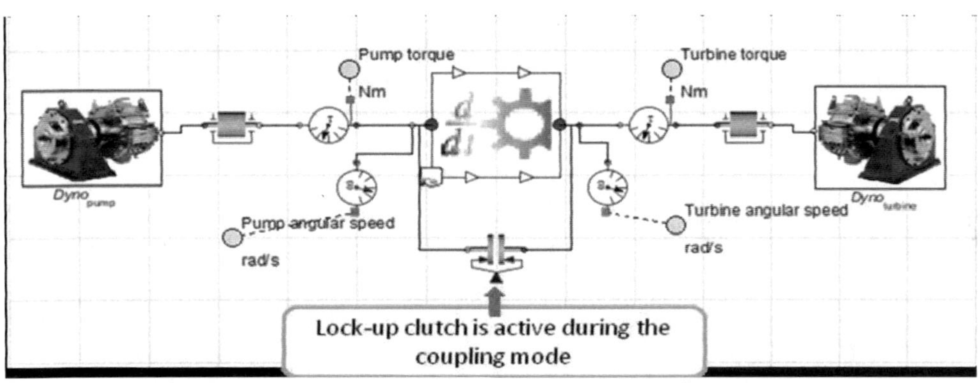

Figure 3. Scheme of torque converter modelling system

In a manual lock-up clutch, there are some unwanted torque values during the using of clutch which affect the driving process. Dynamic torque converter model can be represented by three equations of angle momentum for pump, turbine and stator as well as energy conversion equation. Dynamic model capable to capture transient response of a torque converter during automatic gear shifting as well as transition flow which is caused by engine brake phenomenon. Hence, dynamic torque converter model is being used to evaluate flow characteristic during gear shifting process and engine braking phenomenon. Torque converter model to be proposed is to be integrated with average value engine, transmission and dynamic vehicle longitudinal to evaluate torque converter characteristic in a powertrain model [7]. Based on this model, torque impeller and forces on impeller blades can be identified.

Based on the equation for torque calculation in Guan, Dong and Wu [8], the equation for forces working on torque converter impeller can be seen in the following:

$$TP = \rho Q\left(v\ uB_2\ rB_2 - v\ uB_1\ rB_1\right) \tag{1}$$

IOP Conf. Series: Materials Science and Engineering **676** (2019) 012017 doi:10.1088/1757-899X/676/1/012017

where:

TP	= Impeller torque (Nm)
ρ	= Fluid density (kg/m^3)
Q	= Flow rate (m^3/s)
$v\ uB_1$	= Inlet tangential velocity (m/s)
$v\ uB_2$	= Outlet tangential velocity (m/s)
rB_1	= Inlet blade radius (m)
rB_2	= Outlet blade radius (m)

Impeller rotation will produce kinetic energy on oil fluids which will be received by permanent turbine on output shaft. In this process, turbine blades are received kinetic energy (oil flow) from impeller which then transform it into mechanical energy. Hence, torque output from impeller can be described in the following equation [7]:

$$T_{out}(t) = T_{in}(t)\,GR\ Eff \tag{2}$$

where:

$T_{in}(t)$	= Impeller torque (Nm)
GR	= Ratio
Eff	= Efficiency

From equation (2), output power can be calculated using the following equation:

$$Power(kW) = \frac{torque(Nm) \times 2\pi \times rotation\ speed(RPM)}{60000} \tag{3}$$

3. Research Data

Data of ship to be reference in the torque converter planning is fishing vessel which operated in the area of South Sulawesi is KM Ilda Yanti ship (Figure 4) with ship length is 13 m, breadth 1.5 m dan height 1 m. this ship used main engine with the capacity of 23.1 kW and maximum rotation is 2200 rpm.

Figure 4. 3 GTs Fishing vessel KM Ilda Yanti

The location of the torque converter on the ship in this research is between main engine and the propeller boss with the geometry dimension of the torque converter is following the available room in that place.

IOP Conf. Series: Materials Science and Engineering **676** (2019) 012017 doi:10.1088/1757-899X/676/1/012017

4. Torque Converter Geometry Design

The design calculation and impeller area refer to the Zarotti and Nervegna [6] method (Hydraulic torque converter design) where to give comparison on impeller area with case with the value of 0.31. The value with the comparison of impeller area is obtained the impeller radius si 13 cm, the area of impeller blade is 0.01 m2 while the area of casing is 0.05 m2. Torque converter design is made in 3 difference variations of distances between impeller which are 2.5 mm, 5.00 mm, and 7.5 mm. This distance then will be simulated to analyzed variable which involve with the performance and characteristic of torque converter. Simulation of computational fluids dynamics using software ANSYS 16.0 with the solution fluent and meshing in curvature form with the minimum size is 0.27 mm and maximum size is 10 mm.

5. Simulation Result

The selection of condition as part of the simulation setup by using transient mode and based on the oil fluid density which enter torque converter. Condition variable on running setup is the speed fluid on inlet area of 2 m/s with the steady condition, fluid oil density is 845 kg/m3, and impeller rotation speed with variation 600 rpm, 1000 rpm, 1400 rpm, 1800 rpm, and 2200 rpm. The following arrangement is time step with 0.001 s, the number of each time step is twice and the number of iterations per time step is 20 time of repetition. The variables values for analysis are taken from the post process in simulation step using calculator function. They are consist of pressure value which is taken from the face and back side of the impeller while area value is taken from the face and back area of the impeller.

5.1. Pressure Profile on Impeller Blade and Fluid Flow Speed

The result of simulation with ANSYS on 5 different variation speeds of impeller rotation for pressure and flow speed shows that the impeller average pressure with the average fluid speed on torque converter shell is in line on the rotation up to 1000 rpm and begin to inversely proportional on rotation 1000 rpm until 2200 rpm (Figure 5).

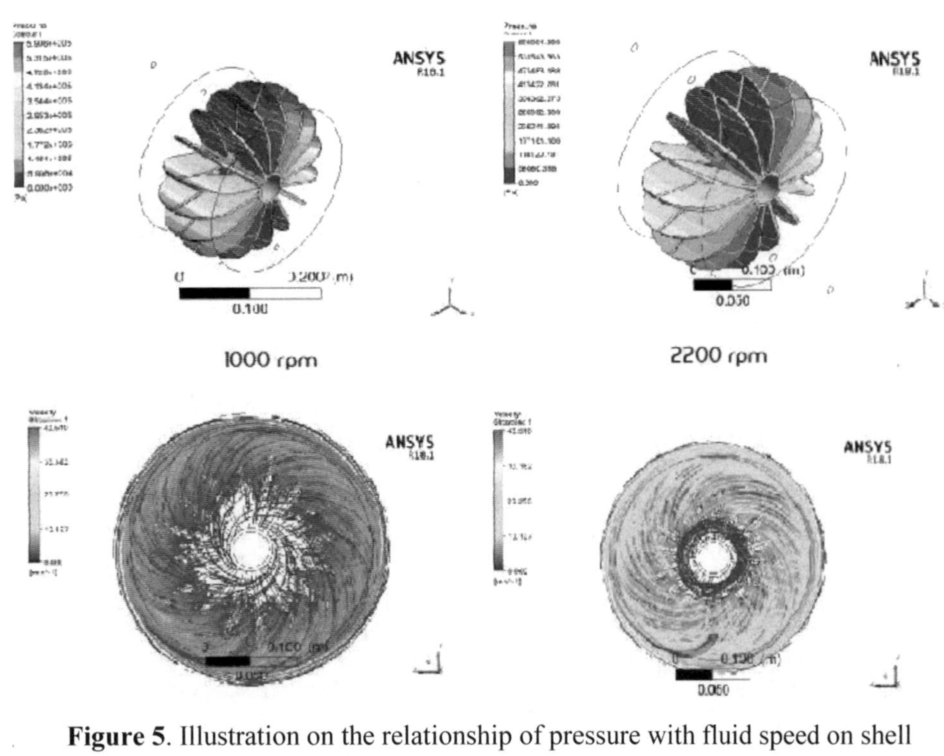

Figure 5. Illustration on the relationship of pressure with fluid speed on shell

IOP Conf. Series: Materials Science and Engineering **676** (2019) 012017 doi:10.1088/1757-899X/676/1/012017

The result of the simulation shows that the pressure distribution and the speed of fluid flow are occurring on the same area in impeller surface. On a 7.5 mm impeller gap, the maximum pressure is occurred. Besides, the minimum and maximum flow speed are occurred as well. Based on the visual observation on the impeller gap, the fluids flow turbulence is reduced as the impeller gap becoming wider.

5.2. Shear Stress

Characteristics of average shear stress distribution on impeller surface on rotation speed 1000, 1400, 1800 and 2200 rpm can be obtained from the following illustration in Figure 6.

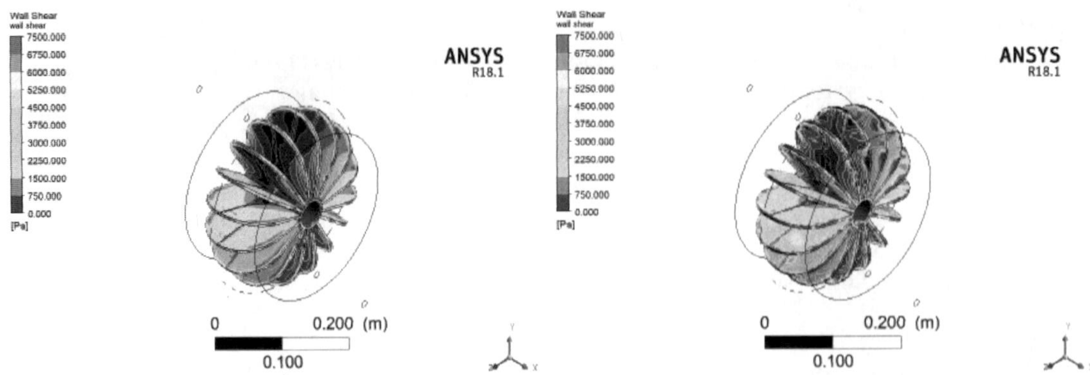

Figure 6. Shear stress distribution on rotation 1000 rpm (left) and 1400 rpm (right)

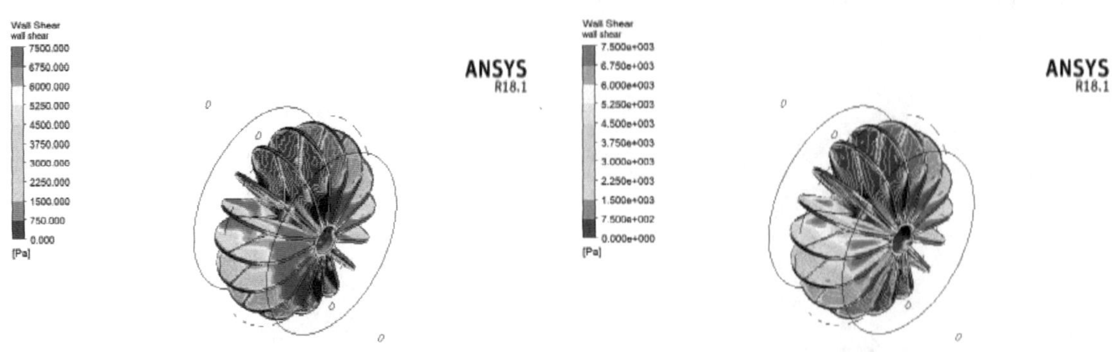

Figure 7. Distribution of shear stress on 2200 rpm (left) and 1800 rpm (right)

The result of simulation shows that the contour pattern of maximum shear stress occurred on impeller in 2200 rpm. The ability of impeller to hold drag and strain stresses can be identified from yield stress value. On a 2200 rpm for manganese bronze impeller, the stress value is above the permitted yield and tensile stress value. This is happened possibly because of fatigue on the material. Hence, both side of the impeller blade as well as on bosching part need to be thickened.

5.3. Torque and Transmission Power

The value of torque data is obtained from calculator function on menu result ANSYS and place the centre of torque calculation on impeller output with the assumption that the impeller output which become output torque and output power from torque converter. Hence, maximum power output can be obtained from the Eq. (3).

Correlation on the relationship of rotation on torque and impeller transmission power can be seen in Figure 8:

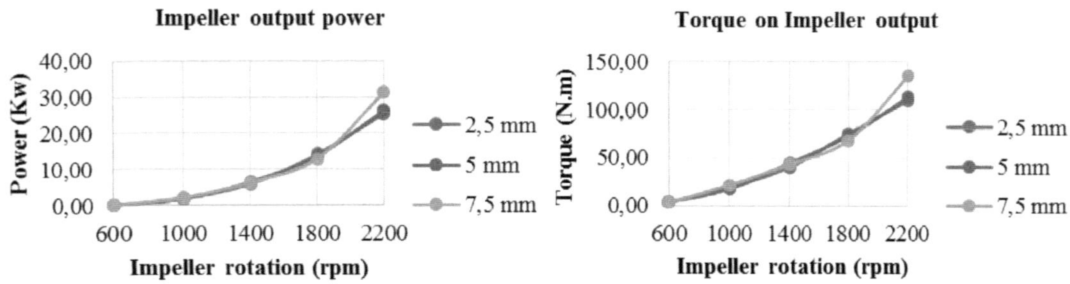

Figure 8. Power and torque output of torque converter

Correlation between torque and the power output on some impeller rotations on different impeller distances as shown in Figure 8 are tend to produce the same value on up to 1900 rpm impeller rotation. However, this rotation can be categorized as high for ship primer mover system. Hence, only the stress value in torque converter impeller that need to be considered.

6. Conclusion

The dimension of torque converter for fishing vessel in the area of South Sulawesi is 30 cm in diameter with the geometry area of impeller is 0.0176 m2. The length and width of impeller blades is 13.5 cm and 2.5 cm respectively with the number of blades is 16. The impeller distance has effect on the dynamic respond of torque converter including average pressure, shear stress, fluids speed, torque and power which can be transmitted on each distance/gap of impeller.

7. References

[1] Behrendt C and Rajewski P 2008 Analysis of propulsion system failures in the fishing fleet operating from Polish port Problemy Eksploatcji pp 147-152.

[2] Transportation Safety Board of Canada 2017 Catastropic failure of machinery Marine Investigarion Report M17A0039.

[3] Sutrisno R A and Triwilaswandio W P 2012 The production of traditional fishing vessel using laminated wooden deck and hull and using alluminium for frame and deck construction *J Teknik ITS* no. **1 1**pp G98-103 [in Bahasa].

[4] Prayitno M M E 2012 Optimalization of fishing vessel propulsion system using CVT gearbox *Kapal* **93** pp 116-122. [in Bahasa].

[5] Marathe BV, Lakshminarayana B and Maddock D G 1997 Experimental Investigation of Steady and Unsteady Flow Field Downstream of an Automotive Torque Converter Turbine and Inside the Stator: Part I—Flow at Exit of Turbine *Journal of turbomachinery***1193**pp624-633.

[6] Zarott GL and Nervegna N 1977 Hydraulic Torque Converters Design *Proc 6th Australa sian Hydraulics and Fluid Mechanics Conference* Adelaide (Australia).

[7] Asl H A, Azad N L and McPhee J 2012 Modeling torque converter characteristics in automatic drivelines: lock-up clutch and engine braking simulation*Proc ASME 2012 International Design Engineering Technical Conferences*.

[8] Guan P, Dong J C and Wu L G 2012 Secondary development of blades parametric design of hydraulic torque converter based on CATIA" *Proc Advanced Materials Research*.

IOP Conf. Series: Materials Science and Engineering **676** (2019) 012018 doi:10.1088/1757-899X/676/1/012018

Ultimate Strength Analysis of FPSO Hull Girder Under Longitudinal Bending

M Z M Alie[1]*, R Iriani[1], Juswan[1] and M I Ramadhan[2]

[1]Departement of Ocean Engineering/Ocean Structure Analysis Research Laboratory, Engineering Faculty, Hasanuddin University
[2]PT. PAL Indonesia, Surabaya, East Jawa, Indonesia

*E-mail: zubair.m@eng.unhas.ac.id

Abstract. Ship and offshore structures have become the key for the development of the offshore world. Production and processing equipment can be placed on a platform, ship and/or barge structure called FPSO (floating, production, storage, and offloading units). The FPSO's hull works like a beam and deforms in the vertical plane. In this case, weight and bouyancy are not normally distributed and wave loading. The most important aspect of FPSO structure is the materials and structural strength used in the design. The critical condition of FPSO is when the structure is under the operation so that the structure needs to be analysed for the design requirement. In the present study, the ultimate strength of FPSO hull girder under longitudinal bending moment is analysed using numerical approach. The nonlinear finite element analysis is adopted to handle the calculation. For the simple case, the one-frame space and the fully cross section of FPSO are modeled. The quadrilateral shell element of the nonlinear method is used for meshing. By performing the boundary condition with the Multi Point Constraint (MPC) is applied where the neutral axis located. The vertical bending moment is also placed at this point (MPC) and it is connected to the all nodes at both sides of the cross sections. The ultimate strength of FPSO is calculated only for intact in hogging and sagging conditions. The ultimate strength analysis is represented in terms of the moment-curvature relationship for hogging and sagging. The result obtained the nonlinear finite element analysis is compared with Smith's method including their collapse behavior. It is found that the comparison result of the ultimate strength is about 8.9% and 10.1% in hogging and sagging condition, respectively. It is shown that the nonlinear finite element analysis is in good aggrement with the analytical solution performed by Smith's method.

1. Introduction

The FPSO's system has become the principle method in offshore oil and gas production sites in the world. Offshore units in the form of ships having a variety of benefits when compared to other types of structures in terms of sufficient work areas, deck load, high storage capability, structural strengths, shorter waiting times, building costs, and suitability for conversion and reuse. However, similar to other floating platform types, the replacement volume below the water line is relatively large, and the response and structural failures under extreme environmental conditions are associated with waves, winds, and significant problems to consider in design and operation. These items must considered for the requirement of strucutral design.

The ultimate strength analysis of ship's hull has been done by many researchers. The asymmetrically damaged ship under sagging condition was investigated by Muis Alie [1]. A plate and/or stiffened plate element at the specified location so-called “critical element” reached its ultimate strength which represent that the hull girder strength also attained the ultimate strength. Tekgoz et al. [2] analyzed a container ship under asymmetrical bending taking the influence of structural damage and associated neutral axis translation and rotation of the residual load carrying capacity. The FE analysis was used and a formulation based on the Common Structural Rules (CSR). The ship was analyzed in intact and damaged condition. The assessment of the ultimate strength for Ro-Ro ship after damage was conducted by Muis Alie et al. [3]. The example of the calculation focused on the cross section. The side shell of the hull and bottom part were assumed to be damaged by simply removing those elements on that part. The result of analytical solution was compared for intact and damage under hogging and sagging conditions.

Kim and Paik [4] expanded a full automated method to optimize design for hull structural scantling of merchant cargo ships and the plate-shell were used for modelling. To minimize the structural weight and maximize structural safety, the technique for full optimization with multi-objectives was used based on the design constraint related to the ultimate limit states of the plate panels, support members and hull girders. The procedure of development was implemented to the hull structure of VLCC, the procedure's capacity is shown by this test for requirement of common structural rules. Gaspar et al. [5] evaluated the influence of the nonlinear vertical wave-induced bending moments on the ship hull girder reliability. A chemical tanker for which the nonlinearity of the vertical wave-induced bending moments was found to be significant was adopted as case study.

Muis Alie et al. [6] used Finite Element Method to analyze the ultimate strength of asymmetrically damaged ships. The collision damage was simply created and remove the elements on that part. The comparison between FE analysis and analytical solution was done including their collapse behavior of ship’s hull. Ultimate limit state-based ultimate longitudinal strength analysis was performed by Park et al. [7] to identify the operability of aged non-ice class ships in the Arctic Ocean considering aging. The hull girder ultimate strength was verified by Garbatov et. al. [8] based on the class society and the result obtained by experiment and dimensional theory.

An analytical method was proposed by Gao et. al. [9] for rapidly predicting response of FPSO side structures in case of being struck by a ship with rigid bulbous bow. The proposed was developed by combining several primary failure models of major double shell members, including the plate punching model, the plate perforating model, the plate denting model, the plate tearing model and the X-shaped structure crushing model. The residual strength analysis of ship with bottom damage was conducted by Muis Alie [10] taking the fully cross section into consideration and using the nonlinear finite element method.

In the present study, the ultimate strength analysis of FPSO hull girder is conducted using the nonlinear finite element method. The cross section and one-frame space of the FPSO are taken for the calculation, and those are modeled by 3D finite element method. The model is created by using shell element. Both two sides of the cross section are attached Multi Point Constraint (MPC) and at this point the moment is applied. The result obtained by nonlinear finite element analysis is compared to the analytical solution.

2. Method of Analysis

The shape configuration and properties of the FPSO’ cross section are summarized in Table 1, Table 2 and Figure 1, respectively, as follow:

Table 1. Sections properties of FPSO

Items	Unit (N,mm)
Breadth	49987,2
Depth	15849,6
Density	780000

Items	Unit (N,mm)
Yield Strength (*)	315
Elastic Modulus	210000
Poisson Ratio	0,3

Table 2. Stiffener dimensions of FPSO's cross section

Stiffener no.	Dimension	Type	Mark
1	125 x 75 x10	Flat Bar	*
2	250 x 90 + 12 x 16	Flat Bar	*
3	350 x 100 + 12 x 17	Flat Bar	*
4	300 x 90 + 11 x 16	Flat Bar	*
5	1524 x 304,8 + 15,875 x 25,4	Tee Bar	*
6	1270 x 203,2 + 14,288 x 25,4	Tee Bar	*
7	1524 x 304,8 + 15,875 x 25,4	Tee Bar	*

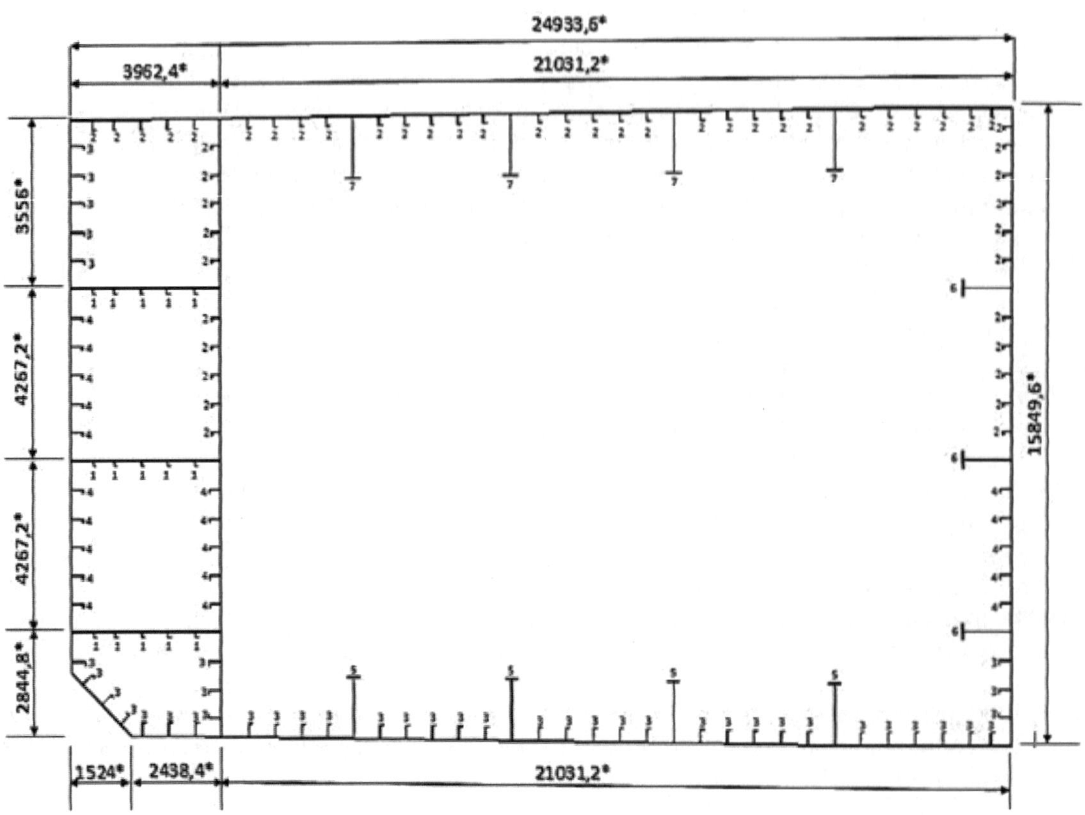

Figure 1. Cross section of FPSO

3. Finite Element Model

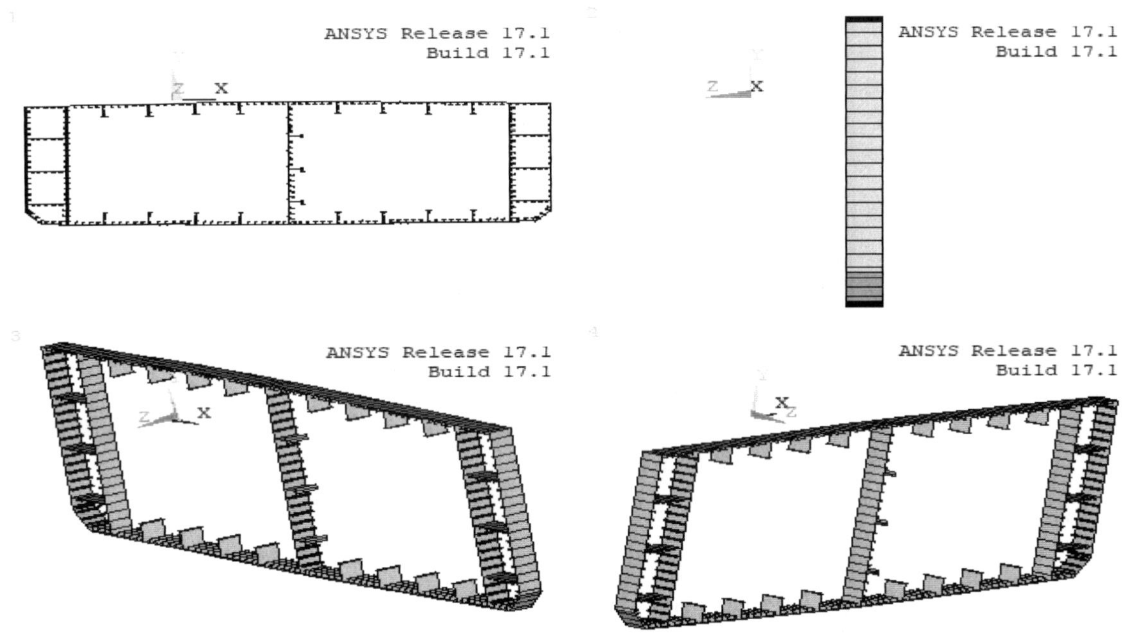

Figure 2. Finite element modeling

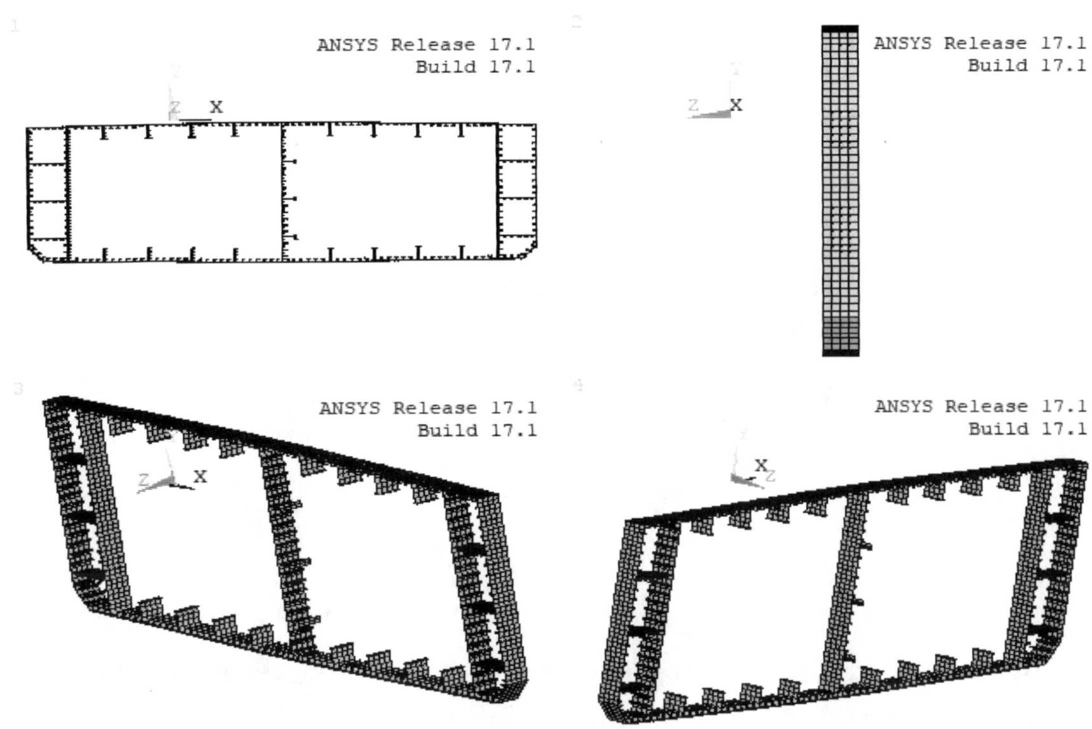

Figure 3. Meshing size

IOP Conf. Series: Materials Science and Engineering **676** (2019) 012018 doi:10.1088/1757-899X/676/1/012018

Figures 2 and 3 show the finite element modelling and meshing size together with their coordinate systems. In this model, the shell element type is used and generated to all area. The application of Multi Point Constraint (MPC) can be seen in Figure 4 as follow,

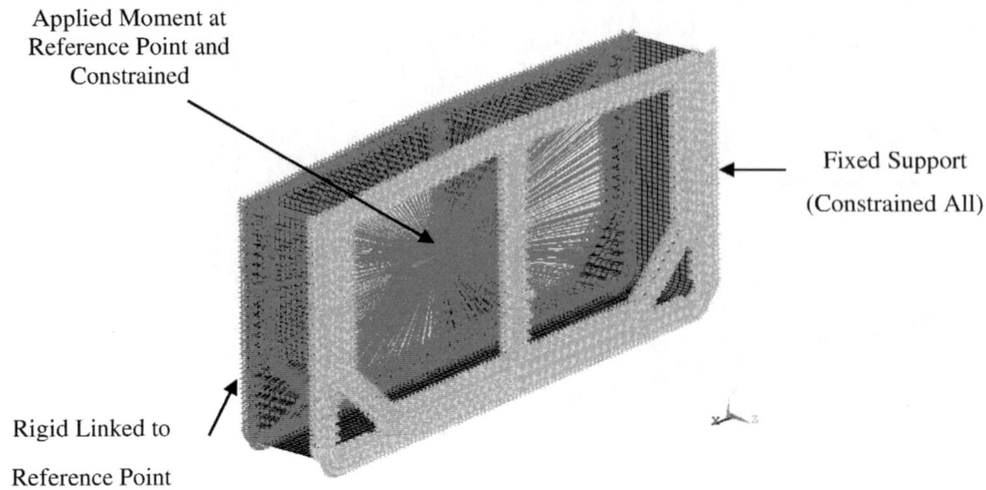

Figure 4. Boundary condition

4. Results and Discussions

Figures 5 and 6 show the stress distributions on the ultimate strength in hogging and sagging conditions, respectively. Tension occurs on deck part when the cross section is under hogging condition. While compression is placed on the bottom part when the cross section is under sagging. MX symbol indicates that value is maximum on that part.

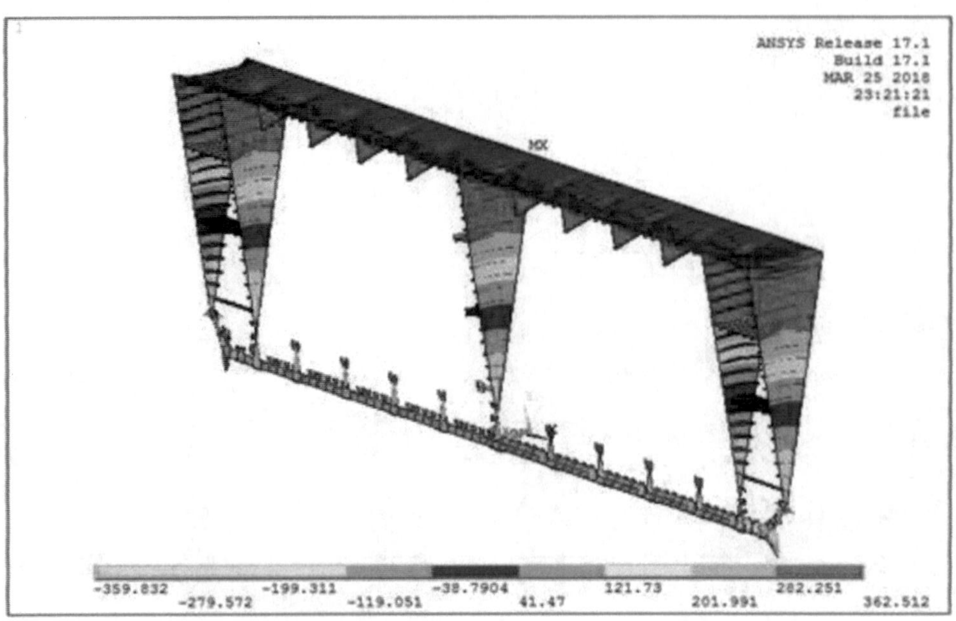

Figure 5. Ultimate strength in hogging condition

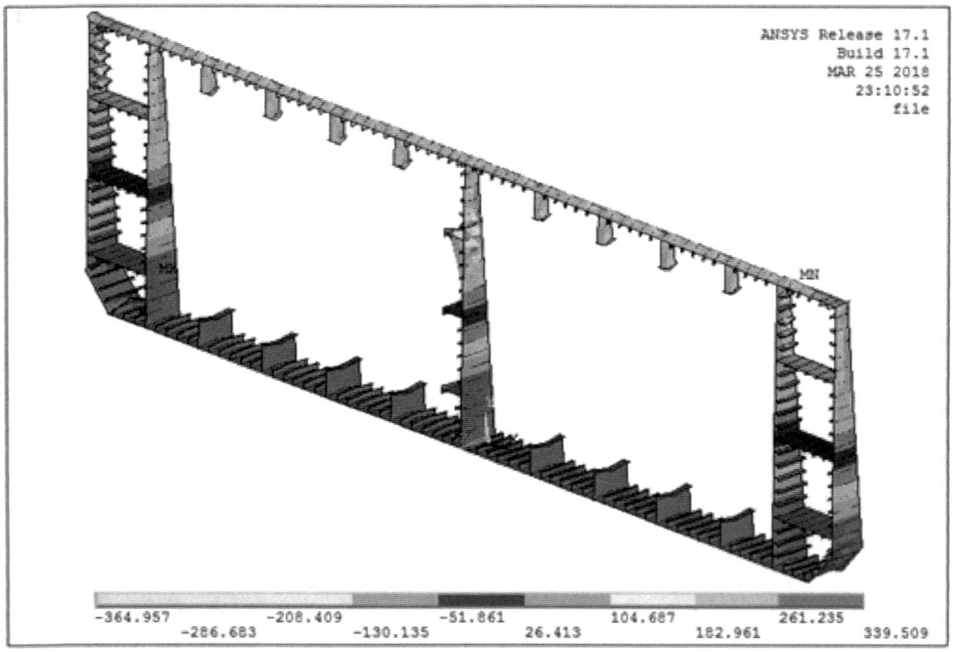

Figure 6. Ultimate strength in sagging condition

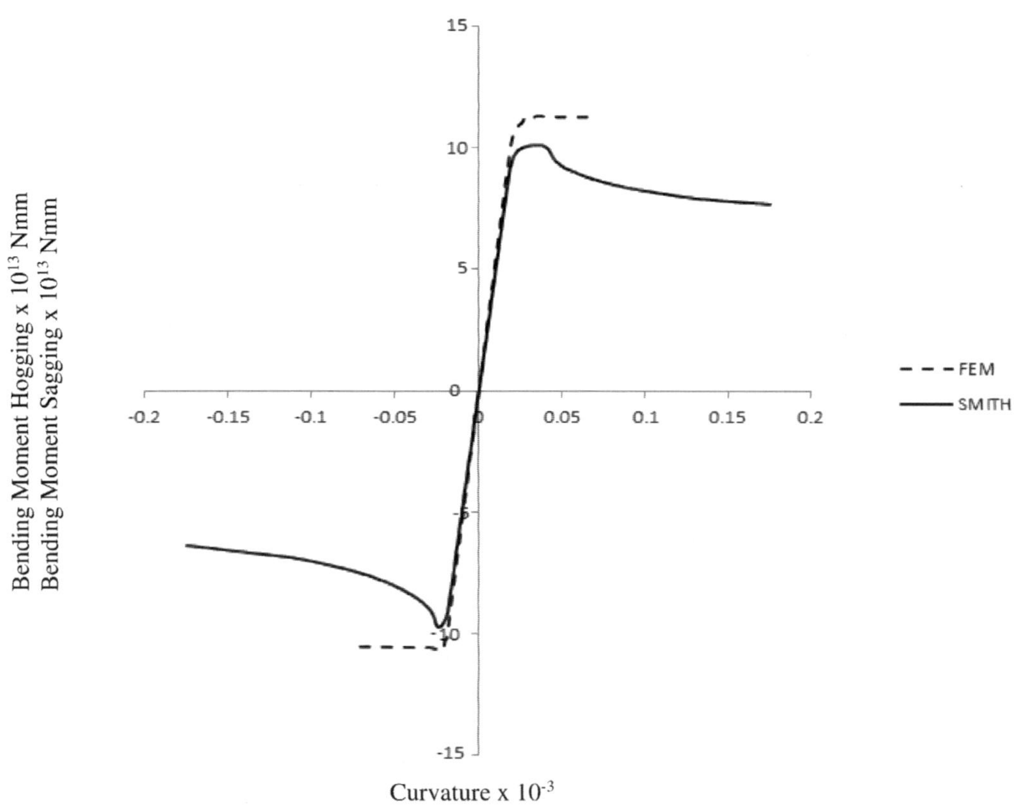

Figure 7. Comparison of the ultimate strength

Figure 7 shows the comparison of the ultimate strength obtained by nonlinear finite element analysis and Smith's method. The ultimate strength for the nonlinear finite element method is represented by the dot line and the solid line for Smith's method. It should be noted that the FE analysis considers many elements to construct the model. Therefore, the stress distribution spreads to all the elements. However, the bending stiffness both two methos are identic. The section modulus between deck and bottom is also different. This is marked by the distance from the neutral axis to deck and bottom. Therefore, generally, this behavior where the ultimate strength in hogging condition is always larger than sagging condition. It is also observed that the ultimate strength in hogging and sagging condition are different due to the redistribution of the stress concentration in the FE model.

5. Conclusion

The ultimate strength analysis of FPSO has been conducted using the nonlinear finite element method, the following conclusion is that the ultimate strength in terms of the moment-curvature relationship obtained by using nonlinear finite element analysis is in good aggrement with the analytical solution performed by Smith's method.

References

[1] Muis Alie, M.Z. (2018),' Simplified Approach on the Ultimate Hull Girder Strength of Asymmetrically Damaged Ships,' *International Journal of Offshore and Polar Engineering*, Vol. **28**(2), 200-205.

[2] Tekgoz, M., Garbatov, Y., Soares, C.G. (2018),' Strength Assessment of an Intact and Damaged Container Ship Subjected to Asymmetrical Bending Loadings,' *Marine Structures*, Vol. **58**, 172-198.

[3] Muis Alie, M.Z., Sitepu, G., Latumahina, S. I. (2017),'The Assessment of the Hull Girder Strength of a RO-RO Ship after Damages,' *Proceeding of the Twenty-seventh International Offshore and Polar Engineering Conference (ISOPE)*, Vol. **4**, 913-919.

[4] Kim, D.H., Paik, J.K. (201,' Ultimate Limit State-Based Multi-Objective Optimum Design Technology for Hull Structural Scantlings of Merchant Cargo Ships,' *Ocean Engineering*, Vol. **129**, 318-334.

[5] Gaspar, B., Teixeira, A. P., Soares, C.G. (2016),'Effect of the Nonlinear Vertical Wave-Induced Bending Moments on the Ship Hull Girder Reliability,' *Ocean Engineering*, Vol. **119**, 193-207.

[6] Muis Alie, M.Z., Sitepu, G., Juswan., Wahyuddin., Nugraha, A.M., Alamsyah. (2016),'Finite Element Analysis on the Hull Girder Ultimate Strength of Asymmetrically Damaged Ships,' *Proceeding of the 35th International Conference on Ocean, Offshore and Arctic Engineering (OMAE)*, Vol. **3**, V003T02A083.

[7] Park, D.K., Kim, D.K., Seo, J.K., Kim, B.J., Ha, Y.C., Paik, J.K. (2015),'Operability of Non-Ice Class Aged Ships in the Arctic Ocean-Part I: Ultimate Limit State Approach,' *Ocean Engineering*, Vol. **102**, 197-205.

[8] Garbatov, Y., Saad-Eideen, S., Soares, C.G. (2015). 'Hull Girder Ultimate Strength Assessment Based on Experimental results and the Dimensional Theory,' *Engineering Structures*, Vol. **100**, 742-750.

[9] Gao, Z., Hu, Z., Wang, G., Jiang, Z. (2014) 'An Analytical Method of Predicting the Response of FPSO Side Structures to Head-on Collision,' *Ocean Engineering*, Vol. **87**, 121-135.

[10] Muis Alie, M.Z. (2014),' Finite Element Analysis on the Ship Hull Girder under Longitudinal Bending with Bottom Damage,' *Proceeding of the Third International Conference on Ship and Offshore Technology (ICSOT)*, 1-4.

Investigation on Reflected Wave by Different Geometrical Ramp Shape of Overtopping Break Water for Energy Conversions Using Experimental and Simulation

M F Roslan[1], M F Ahmad[1]* and M A Musa[1]

[1]Programme of Maritime Technology, School of Ocean Engineering, Universiti Malaysia Terengganu, Malaysia 21030

*E-mail: fadhli@umt.edu.my

Abstract. Innovation trends on the development of breakwater have prolong its functionality from coastal protection to harness wave energy. One of the latest structural being study is the overtopping breakwater for the energy conversion which integrate breakwater with wave energy converter. An impact of linear ramps on reflection of the structural been studied in detail in some previous researches and create opportunity in exploring vast scope of study including the ramp shape. Therefore, intention of this study is to experimentally investigate the effectiveness of various geometrical ramp shapes of the structural toward the reflected wave. In addition, simulation was conducted and compare with the experimental result to ensure the agreement of the outcomes. This study used different wave parameters and effect of relative depth (H_{m0}/h) to acquire data on the behaviour of the reflected wave that describe the effectiveness of the structure. The findings of this study in the experimental work indicates that the cubic ramp shape precedes the linear ramp with 21.5% more efficient under certain wave conditions. Besides, this study additionally reveals similarities in both methods and conclude that ramp shape has significant influence on the reflection of the wave from the structure and the highest percentage of discrepancy for the CFD is 22.3%. However, the contribution of the different ramp shape is absent at the maximum depth but significantly affect as well as the depth decrease.

1. Introduction

Coastal area and the ecosystem offer many services to human society; residential, food supplies, recreational enchantment and industrial activities [1]. Rapid development of the coastal area transforms the landscape and exposed the vicinity toward pollution, erosion and habitat loss. Consequently, most of the world's shoreline are retreating in a landwards direction cause by erosion process [2] and the ecosystem experienced some of the most rapid environmental change but astonishingly the demographic tendencies increasing swiftly [3]. Hence, preservation and retaining the shoreline from nature characteristic of wave and current is essential.

Countermeasures that may be applied to the coastal area are beach nourishment, groins, sand bypass, detach breakwater and rubble mound breakwater [4]. For quite century, rubble mound breakwater is the most typically used to defend harbours and shore line [5]. It principally dissipated the incoming wave energy on the slope and porous structure, reflected partly to the ocean and partly transmitted into the harbour area because of penetration and overtopping [6]. Recently, the innovation of rubble mound breakwater have integrate the breakwater with wave energy convertor (WECs) recognized as overtopping breakwater for energy conversion (OBREC) [7]. The hybrid system harvest the wave energy using concrete ramp to increase overtopping phenomenon and store within the reservoir as potential energy which then turns the low head hydraulic turbines to generate electric [8].

The performance of OBREC rely on the overtopping activity that strongly related with the reflected wave; high overtopping creates low reflection due to more dissipation [9]. By changing wave climate and intense sediment scour, reflected wave result in structure destabilization, erosion at nearby beach and unstable ocean state at the entrance of harbor [10-12] Since OBREC used the concept of overtopping to capture wave energy, enhancement on the performance of OBREC are terribly useful. One approach to improvise OBREC is by optimize its geometrical shape or configuration [13,14]. Additionally, overtopping significantly relate with wave run up over the ramp surface; roughness, slope angle, length, crest freeboard permeability and shapes [15]; roughness, slope angle, length, crest freeboard permeability and shapes [16]. Different ramp shapes accumulate different energy for wave run up and have an impact on the reflection of the incoming wave [15].

Hence, this study aims to analyze the reflected wave for various geometrical ramp shapes of OBREC that quantified the amount of energy and determined the efficiency of the structure[17,18]. Reflection coefficient can be obtained using variety of methods that have been proposed by researchers and one of them is using the resolution technique [19,20].To attained the aims, this study performed the experimental work and used the CFD approach by referring previous researchers [21, 22].

2. Method and Materials

The design of the OBREC being improves through the geometrical ramp shape. The ramp designs are based on the previous study which are linear, concave, convex and this study introduce new idea to have cubic ramp that are expected to have better performance.

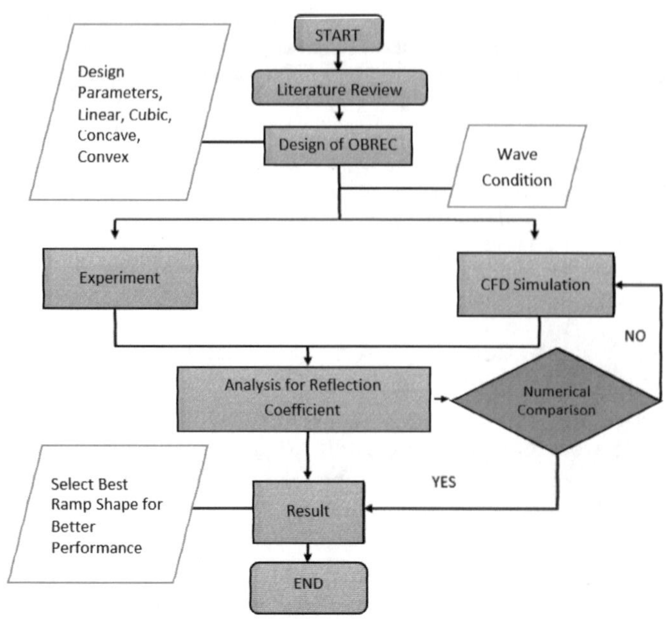

Figure 1. Flow chart of the study

2.1. Experimental setup

Experimental procedure has been carried out in Port and Harbour 3D wave basin at National Hydraulic Research Institute Malaysia, (NAHRIM). The basin is approximately 30m x 30m x 1.5m and is geared up with multi element wave maker with 30 paddles of the flat type to generate regular wave. and passive wave absorber to avoid reflection of wave. Couple of probes have been deployed in order to record wave elevation and used to calculate the reflection coefficient. Different type of wave parameters and water depth used to study reflection behaviour creates by OBREC model.

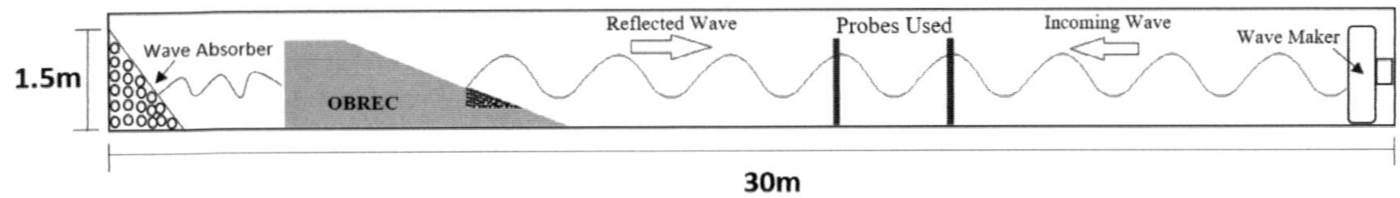

Figure 2. Illustration for experimental setup

Table 1. Wave parameter and depth

Type of Wave	Wave Period Tp (s)	Significant Wave Height Hs (m)	Depth(D) Min-Max (m)
R1	1.43	0.0527	0.35-0.45
R2	1.91	0.083	
R3	2.04	0.102	
R4	2.22	0.1173	

The OBREC model and ramp shapes are designed using AutoCAD software before being construct with the aid from CNC machine. Scale 1:15 (Froude scaling) was used based on previous model developed by Vicinanza [22]. The main structural of the experimental model is estimated to be 4m (wide) x 3.4m (length) x 0.7m (height).

(a) (b)

Figure 3. Design of OBREC and ramp shape using autoCAD

The ramp shapes are designed using polynomial equation with 1.8 m height of crest free board and Slope angle used is 19° for better performing of WECs [23].

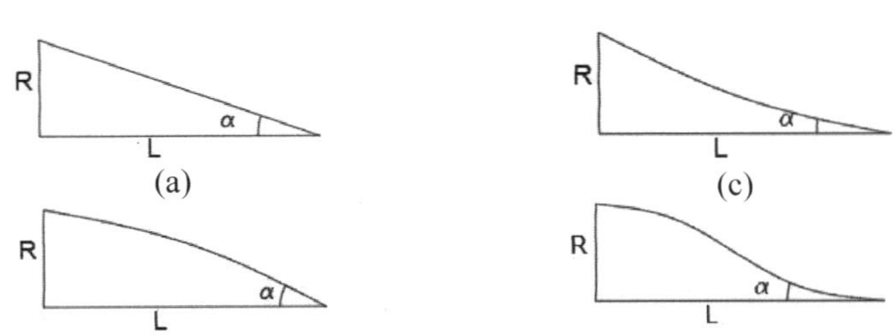

Figure 4. Ramp shape configuration: a) Linear b) Convex c) Concave d) Cubic

Numerous studies have attempted to separate components of incoming and reflected wave which is beneficial in determine the performance of many coastal structures [19]. This study used method proposed by Goda and Suzuki [20] where two history probes have been used to be function as the wave gauge by installing at least one λ from toe of the breakwater [24]. Fast Fourier Transform (FFT) being used to analyzed and estimate the amplitudes and by smoothing the estimated periodograms, wave spectra are presented to obtain the energies of resolve incident and reflected wave E_1 and E_2.

$$K_R = \sqrt{E_1/E_2} \tag{1}$$

2.2. Simulation setup

Computational fluid dynamic (CFD) of FLOW 3D provide advantages in time and cost consuming. In this study, FLOW 3D version 10.1 is used to simulate each of the conditions that run for the experiment. The result of both experiment and simulation then compare to validate the availability of the software for future work.

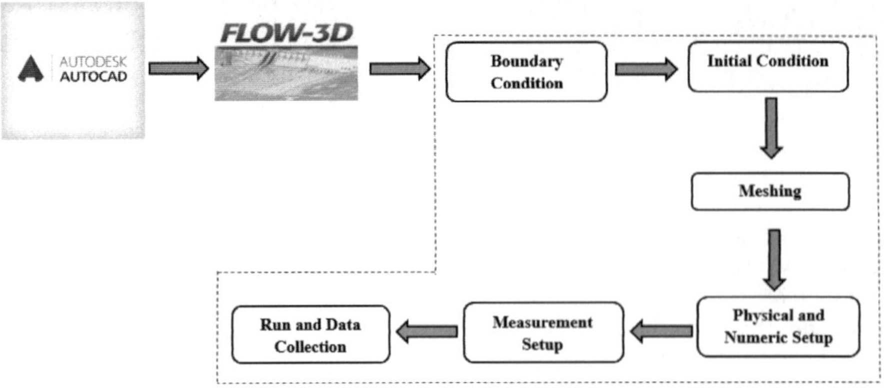

Figure 5. Flow diagram for simulation setup

Flow-3D has been widely tested and it is based on the RANS which designed for turbulence simulation and Volume of Fluid (VOF) for free surface computation methods. General equation of Nevier-Stoke and Continuity used in the simulation to discretize and solve each computational cell.

$$\frac{V_f}{\rho}\frac{\partial \rho}{\partial t}+\frac{1}{\rho}\nabla\cdot\left(\rho\vec{u}A_f\right)=-\frac{\partial V_f}{\partial t} \tag{2}$$

$$\frac{\partial \vec{u}}{\partial t}+\frac{1}{V_f}\left(\vec{u}Af\cdot\nabla\vec{u}\right)=-\frac{1}{\rho}\left[\nabla\rho+\nabla\cdot\left(\tau A_f\right)\right]+\vec{G} \tag{3}$$

$$\frac{\partial F}{\partial t}+\frac{1}{V_f}\nabla\cdot\left(F\vec{u}A_f\right)=-\frac{F}{V_f}\frac{\partial V_f}{\partial t} \tag{4}$$

In the simulation the basin used to be the major component in the simulation with two subcomponents, breakwater and the ramp. The boundary for the basin was once setup (4m width x 30m size x 1m height) without wave generator. It was setup to create actual condition where at origin point of y-min direction, wave boundary was declared to have linear wave and at the end of y-max direction outflow boundary was used. At z-max, the boundary has been integrated with pressure boundary to determine the time history of hydrodynamics parameters, Z-min is assigned as wall and the rest of the boundary (x-min – x-max) were setup as symmetry.

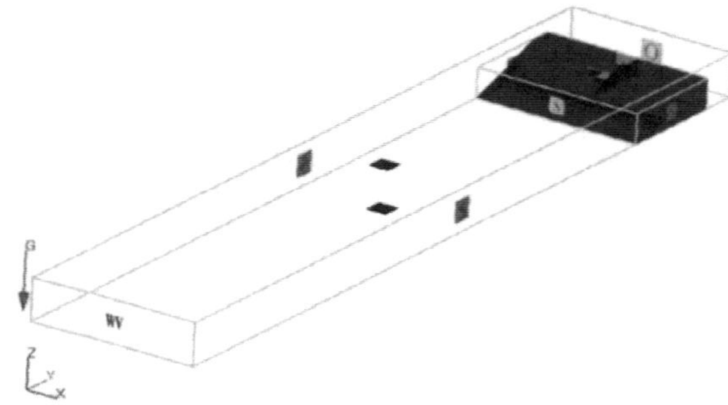

Figure 6. Boundary condition setup

Table 2. Boundary setup details

Boundary	Type of boundary condition
X_{min}	Symmetry
X_{max}	Symmetry
Y_{min}	wave
Y_{max}	Outflow
Z_{min}	Wall
Z_{max}	Specified pressure

Further, meshing block used to separate between the region where the fluid is flowing and location of reflected wave takes place using confirming mesh that fit to the geometry. To obtain best result for the simulation and minimize computational burden, ratio between fine and course mesh used is 1:2 as recommended by Flow 3D inventor [21].

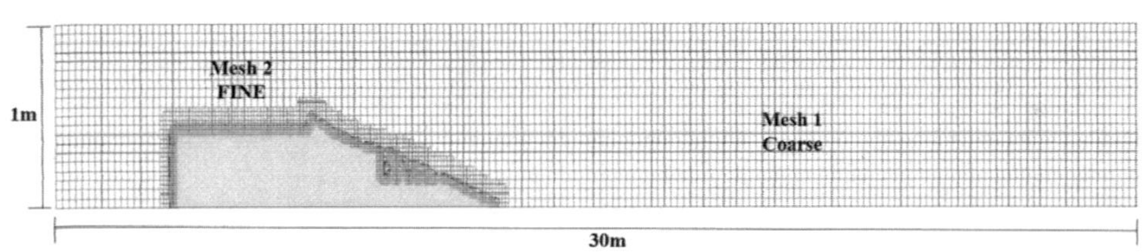

Figure 7. Mesh block setup

Fluid region has been set up to define fluid initial condition of the overall numerical set-up, the water level in front of the structure has been fixed at 0.4 m,0.35m and 0.45m. A RNG turbulence closure have been chosen. A time step of 0.01s has been employed and finish time has been made up to be 60s.

3. Result and Discussion

The discussion of the result begins with the comparison of experimental results for different ramp shape to the different wave parameter tested. Previous studies reveal that OBREC is able to reduce up to 22% reflection against traditional breakwater by using a linear ramp shape [22]. The most striking result to emerge from this study is that cubic ramp shape shows better performance in each test compared to linear, concave and convex. As Figure 7 shows, there is a significant difference between cubic ramp shape and linear ramp shape on the wave parameter R1 where the cubic precedes the linear ramp shape by 21.5%, Concave also precedes linear ramp by 10% better but linear lead convex shape by 13%. This result is due to the ability of each ramp shape to trap wave energy via overtopping. According to Eurotop, the concave shape has higher capabilities for overtopping compare to linear and convex, though this study has further enhanced OBREC's ability by introducing a new cubic ramp shape to increase overtopping and reduce reflection on coastal defense [13].

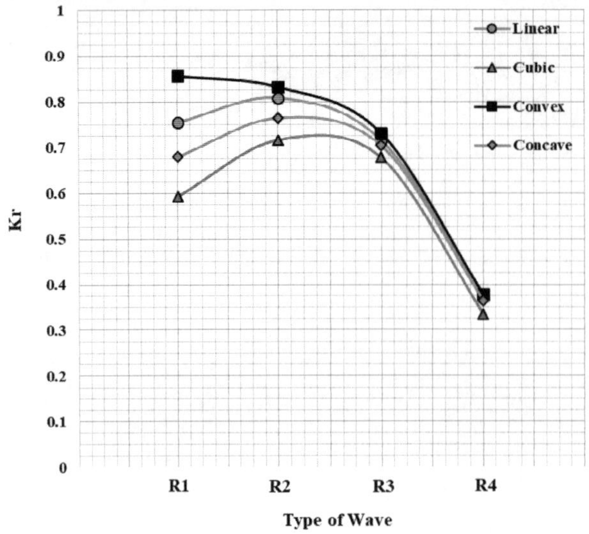

Figure 8. Experimental result for Kr against different wave

The result of comparison between experimental work and simulation presented in Figure 8 is interesting because both methods shows same trend for all tests.

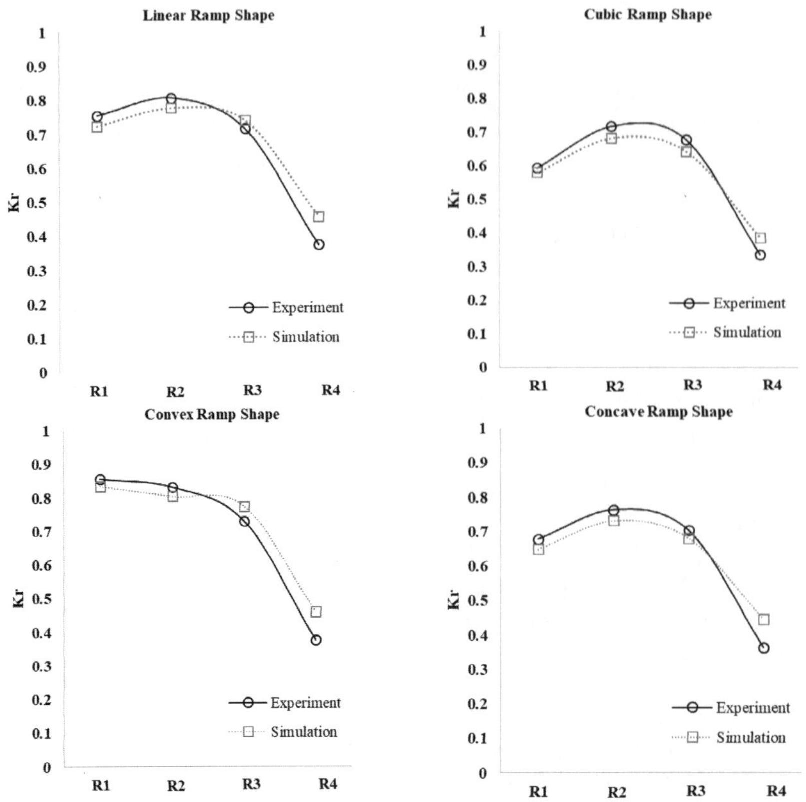

Figure 9. Comparison of experimental and simulation

Details for the results of the experimental work and simulation including the discrepancy percentage presented in Table 3. Obviously, the difference is not significant for R1, R2 and R3 wave parameter which is below 10% but the discrepancy percentage for R4 is 22%. This may be due to the splashing factor on the OBREC ramp or interpolation data error within the mashing block causing a certain volume of water miscalculated in the accumulation of energy wave run up and affect the reflection coefficient [25].

Table 3. Details on the result and discrepancy percentage of all tests

Type of Wave	Linear Ramp Shape		Discrepancy (%)
	Experiment	Simulation	
R1	0.7550	0.7236	4.1588
R2	0.8083	0.7778	3.7705
R3	0.7159	0.7403	3.4045
R4	0.3759	0.4588	22.0464

Type of Wave	Cubic Ramp Shape		Discrepancy (%)
	Experiment	Simulation	
R1	0.5921	0.5796	2.1129
R2	0.7161	0.6818	4.7870
R3	0.6772	0.6420	5.2060
R4	0.3349	0.3845	14.826

Type of Wave	Convex Ramp Shape		Discrepancy (%)
	Experiment	Simulation	
R1	0.8564	0.8345	2.5642
R2	0.8325	0.8048	3.3289
R3	0.7311	0.7754	6.0502
R4	0.3779	0.4621	22.2929

Type of Wave	Concave Ramp Shape		Discrepancy (%)
	Experiment	Simulation	
R1	0.6788	0.6484	4.4782
R2	0.7646	0.7327	4.1680
R3	0.7039	0.6809	3.2604
R4	0.3647	0.4450	22.0230

The correlation between reflected wave and relative depth (H_{m0}/h) presented in Figure 9. For cubic and concave ramp shapes, the reflection coefficient decrease with the depth, however linear and convex ramp shapes show differ trends. Among the plausible explanations for these findings is that when the water level decreases, the wave momentum flux decreases by decreasing its weight of water in the swash zone. Cubic and concave ramp shape indicates its ability when the curve angle affects greatly to reduce the reflection; Reflection coefficient tends to increase with the slope [8]. This study also indicates that at maximum depth, the difference for the reflection coefficients of each different shape is not significant because the area for wave run up has decreased. The ramp shape do not plays it roles to reduce the reflection at this situation although overtopping is expected to be higher; increase of the submerge ramp length causes an increase of reflection [8].

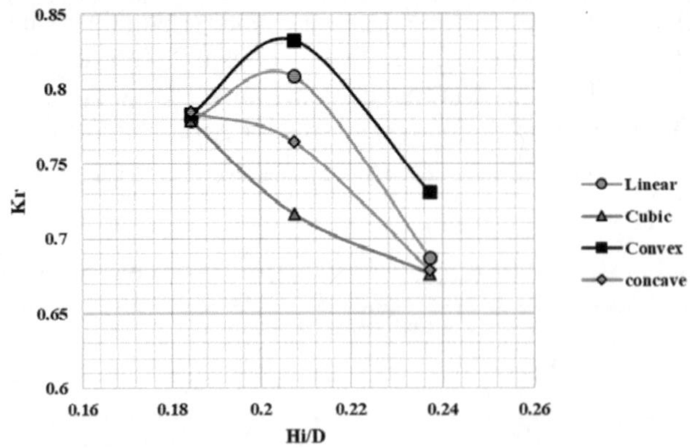

Figure 10. Reflection wave vs relative depth

4. Conclusions

In the present paper, different geometrical ramp shapes of overtopping breakwater for energy conversion being tested under circumstances of wave parameter and different depth using experimental work and CFD approach. Flow3D were properly carried out and compare with experimental result to ensure the agreement of the result and confirm the availability of the CFD. It can be conclude that geometrical ramp shape significantly influence the reflection coefficient where in this study reveal that cubic ramp shape has better performance compare to linear, concave and convex. Capability of the CFD in stimulating the experiment and study the behavior of the reflected wave are compatible and the ramp shape do not influence in reducing the reflected wave at maximum depth as the immerse ramp shape reduce the wave run up.

References

[1] C. Brown, E. Corcoran, P. Herkenrath, and J. Thonell, "Marine and coastal ecosystems and human well-being: a synthesis report based on the findings of the Millennium Ecosystem Assessment," *J. Bertrand Russell Arch.*, p. 80, 2006.

[2] O. H. Pilkey, W. J. Neal, and D. M. Bush, "Coastal Zones and Estuaries – Coastal Erosion," *Encyclopedia of Life Support Systems (EOLSS)*.

[3] T. Agardy *et al.*, "Coastal Systems," in *Ecosystems and Human Well-being: Current Status and Trends*, 2005, pp. 513–550.

[4] Paul D. Komar, *Beach Processes and Sedimentation*, vol. 79, no. 19. 1998.

[5] T. L. Andersen, "Hydraulic Response of Rubble Mound Breakwaters Scale Effects - Berm Breakwaters," 2006.

[6] H. Gravesen, "Hydraulic Engineering Rubble Mound Breakwaters," no. September 2008, 2014.

[7] D. Vicinanza, D. Stagonas, G. Müller, J. H. Nørgaard, and T. L. Andersen, "Innovative Breakwaters Design for Wave Energy Conversion," *Coast. Eng. Proc.*, vol. 1, no. 33, p. 1, 2012.

[8] C. Iuppa, P. Contestabile, L. Cavallaro, E. Foti, and D. Vicinanza, "Hydraulic performance of an innovative breakwater for overtopping wave energy conversion," *Sustain.*, vol. 8, no. 12, pp. 1–20, 2016.

[9] B. Zanuttigh, L. Margheritini, L. Gambles, and L. Martinelli, "Analysis of Wave Reflection from Wave Energy Converters Installed as Breakwaters in Harbour," *8th Eur. Wave Tidal Energy Conf. Uppsala, Sweden.*, 2009.

[10] R. M. Abernethy, N. W. H. Allsop, N. Poisson, and M. W. McBride, "Effects of Breakwater Reflections on Local Sea Steepness and Navigation," *Coast. Struct. 2003*, vol. 40733, no. July, pp. 777–788, 2004.

[11] A. S. Koraim, "Hydrodynamic Efficiency of Suspended Horizontal Rows of Half Pipes Used as A New Type Breakwater," *Ocean Eng.*, vol. 64, pp. 1–22, 2013.

[12] B. Zanuttigh and J. W. van der Meer, "Wave Reflection From Coastal Structures," pp. 1–12, 1992.
[13] J. Goggins and W. Finnegan, "Shape optimisation of floating wave energy converters for a specified wave energy spectrum," *Renew. Energy*, vol. 71, pp. 208–220, 2014.
[14] L. Margheritini, "R&D Towards Commercialization of Sea Wave Slot Cone Generator (SSG) Overtopping Wave Energy Converter," no. 24, p. 219, 2009.
[15] T. Pullen, N. W. H. Allsop, T. Bruce, A. Kortenhaus, H. Schüttrumpf, and J. W. van der Meer, "Wave Overtopping of Sea Defences and Related Structures: Assessment Manual," 2007.
[16] Pedersen.J, "Wave Forces and Overtopping on Crown Walls of Rubble Mound Breakwaters," no. August, p. (Series Paper; No. 12), 1996.
[17] W. P. Huang, C. R. Chou, and J. Z. Yim, "On reflection and diffraction due to a detached breakwater," *Ocean Eng.*, vol. 32, no. 14–15, pp. 1762–1779, 2005.
[18] S. K. Wang, T. W. Hsu, W. K. Weng, and S. H. Ou, "A three-point method for estimating wave reflection of obliquely incident waves over a sloping bottom," *Coast. Eng.*, vol. 55, no. 2, pp. 125–138, 2008.
[19] R. V Varghese, K. G. Shirlal, B. Hari, and P. S. Mohanan, "Review of Developments in Estimation of Wave Reflection from Coastal Structures," pp. 30–32, 2016.
[20] T. S. Yoshimi Goda, "Estimation of incident and reflected waves in regular wave experiments," *Ocean Eng.*, vol. 22, no. 1, pp. 77–86, 1995.
[21] A. Y. Maliki, M. A. Musa, M. F. Ahmad, I. Zamri, and Y. Omar, "Comparison Of Numerical And Experimental Results For Overtopping Discharge Of The Obrec Wave Energy Converter," vol. 12, no. 5, pp. 1337–1353, 2017.
[22] D. Vicinanza, P. Contestabile, J. Quvang Harck Nørgaard, and T. Lykke Andersen, "Innovative rubble mound breakwaters for overtopping wave energy conversion," *Coast. Eng.*, vol. 88, pp. 154–170, 2014.
[23] D. Vicinanza, L. Margheritini, J. P. Kofoed, and M. Buccino, "The SSG wave energy converter: Performance, status and recent developments," *Energies*, vol. 5, no. 2, pp. 193–226, 2012.
[24] G. Klopman and J. W. van der Meer, "Random Wave Measurements in Front of Reflective Structures," *Journal of Waterway, Port, Coastal, and Ocean Engineering*, vol. 125, no. 1. pp. 39–45, 1999.
[25] M. Barkhudarov, "Lagrangian VOF Advection method for FLOW-3D," *Flow Sci. Inc.*, vol. FSI-03-TN6, no. July, pp. 1–11, 2004.

Performance Comparison Between Ellipse and Circular Intake Shapes of Propeller Flow Cooling System (PFCS) with Numerical and Experimental Methods

F Mahmuddin[1]*, S Ramadhan[1], A Fitriadhy[2] and S Klara[1]

[1]Marine Engineering Department, Engineering Faculty, Hasanuddin University, Indonesia
[2] Programme of Maritime Technology, School of Ocean Engineering, Universiti Malaysia Terengganu, Malaysia

*Email: f.mahmuddin@gmail.com

Abstract. This research carried out an analysis on the use of high pressure flow behind propeller to cool the ship main engine known as Propeller Flow Cooling System (PFCS). The system is commonly found on relatively small fishing boats at South Sulawesi area of Indonesia. Using the system, the special pump for engine cooling purpose is not needed. The present research conducted an investigation on PFCS intake shape that would be optimal for delivering water to main engine which was arisen from the propeller rotation and to find the difference of water capacity produced from different shapes of intake. The implemented research methods were numerical and experimental methods that analysed the shape of two intake shapes namely, circular and elliptical shapes. In numerical computation, ANSYS CFX software was adopted. The result from numerical computations was an initial prediction to the tendency of the results. In order to verify the results, an experiment in a tank facility known as Circulating Water Channel (CWC) tank was conducted. In the experiment, measurements were performed at 2 (two) different locations which were at the center of propeller and at 0.7R of propeller blade. The results of the research found that the performance of the elliptical intake shape is more optimal compared to the circular one.

1. Introduction

Ship operation needs a large amount of energy which proportional to costs increase and pollution production. Therefore, energy efficiency in ship operation has been a great concern over the last decade [1]. Due to its importance, there are numerous research which have been performed related to ship energy efficiency such as the ones performed by Coraddu et. al [2], Armstrong et. al [3], Trodden et. al [4], Baldi et. al [5], and Mahmuddin [6].

Amongst equipment which consume energy aboard a ship, pumping system consumes approximately almost half of the electrical energy produced on a ship [7]. Thus, increasing the efficiency of pumping system will result in a significant increase of energy efficiency performance of a ship. One of the main pumps used in ship operation is the engine cooling pump. The pump is used to deliver water from open sea to the main engine as well as to discharge them after circulating inside the engine. In order to obtain an efficient engine cooling pumping system, Mrakovcic et. al [8] and Theotokatos et. al [9] performed modelling and simulation of the engine-cooling system. Su et. al [10] also conducted an investigation on a closed pump system for saving energy of marine power system.

2 (two) main categories of the engine cooling system are sea water system (open system) and central cooling system (using a combination of seawater and freshwater). The number of pumps used in an engine-cooling system depends on the type of the engine-cooling system chosen. However, central cooling-system is usually more complex so it needs more number of pumps compared to the sea water system. Another difference of those systems is that the central cooling system is usually installed in large size ships while small ships or boats usually use sea water or open system.

Especially in South Sulawesi area of Indonesia, the number of fishermen who use small ships or boats for fishing is quite common. These fishing boats usually adopt an open cooling system where water is flowed directly to cool the engine and then discharged outboard. However, in order to save space inside the ship, a special pump for engine cooling purpose is not installed inside the boat. Instead, they utilize the flow in front of the propeller to cool the engine. The system takes advantage of the high pressure flow behind propeller which is directed to circulate inside the engine. After circulating inside the main engine, the water is then discharged outboard. The system will be referred to Propeller Flow Cooling System (PFCS) in the present study. Using this system, not only initial and operational costs can be reduced but also some spaces inside the ship can be saved to store other essential equipment and loads.

Example of a fishing boat which implements the PFCS is shown in Fig. 1.

Figure 1. A fishing boat that implements Propeller Flow Cooling System (PFCS)

The performance of the (PFCS) highly depends on the propeller flow pressure. Therefore, it is important to determine and choose the optimal dimension and shape of the system components. As shown in Fig. 1, the main components of the PFCS are intake, distribution pipe, and outtake. The present study will evaluate and compare the performance between circular and elliptical intake shape. The shape of the intake is very significant in determining the flow capacity of the PFCS. In analysis and investigation, numerical methods using ANSYS CFX Software and direct measurement by experiment were performed.

2. Propeller Flow Water Cooling System

The main principle of the Propeller Flow Cooling System (PFCS) is using the pressurized flow in front of propeller to be directed and circulated in the main engine. In order to use the flow, an intake is placed in front of propeller. When the propeller rotates, some amount of water will get into the intake. Due to continuous propeller rotation, water will also continuously stream inside the distribution pipe which is directed to the main engine. The water will then circulate inside and cool the main engine. After the circulation, the hot water is then discharged outboard. The main components and concept of PFCS are illustrated in Fig. 2.

As shown in Fig. 2, the main components of the PFCS are intake, distribution pipe (hot water and cold water pipes), and outtake. In its application, several variations of PFCS can be found to be used by fishermen on their boats. The variations are mainly on the position and shape of the intake. For the intake shape, the most common adopted shapes are circular and elliptical shapes. Thus, performing an investigation on the performance comparison of these intake shapes in order to determine the best intake shape is necessary.

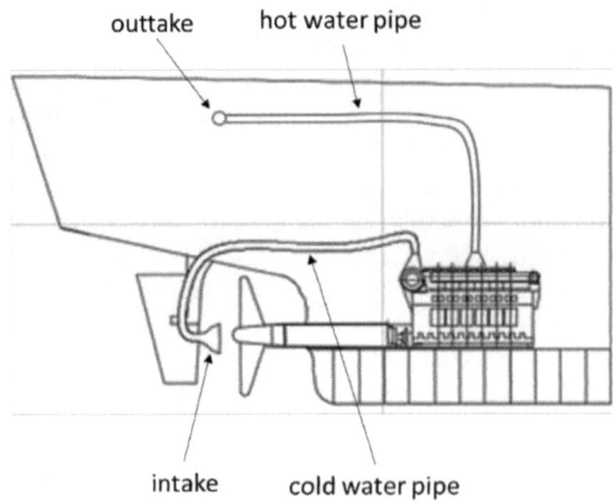

Figure 2. The main components and concept of PFCS

3. Research Methodology

For the analysis purpose in the present study, numerical analysis using software and an experiment in a water tank were conducted. The employed software is Ansys CFX software while the tank experiment was conducted in a tank facility known as Circulating Water Channel (CWC). The tank is located at Marine Structure System Laboratory, Marine Engineering Department, Engineering Faculty, Hasanuddin University, Indonesia.

The concept of the CWC is shown in Fig. 3.

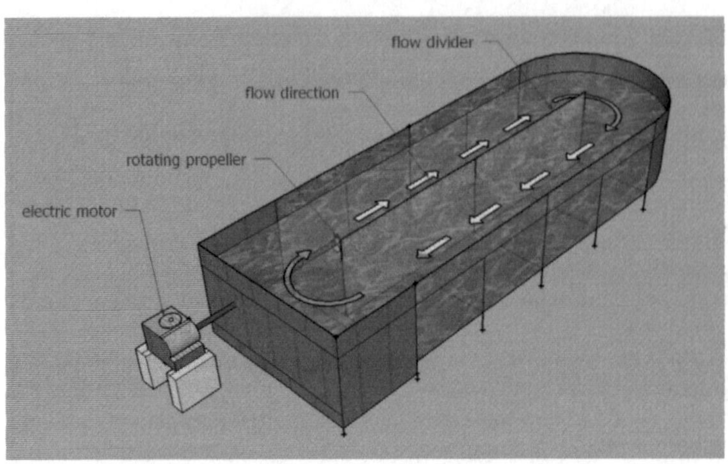

Figure 3. The sketch of CWC

As shown in Fig. 3, CWC is a tank where the water inside the tank can circulate horizontally which enables a convenient data collection for a certain period. The water circulation is made possible due to a rotating propeller connected to an electric motor. The motor is placed outside the tank as shown in Fig. 3. The speed of the flow is adjusted by regulating the speed of the motor (motor RPM) using an Automatic Voltage Regulator (AVR).

The main parameters of the CWC are shown in Table 1.

Table 1. CWC dimension

Parameter	Value
Length (L)	9,0 m
Breadth (B)	2,4 m
Height (H)	1,2 m
Water weight (M)	21 ton

In the present study, the investigated intake was placed in front of the rotating propeller of the CWC to simulate the inlet pipe position of PFCS on a fishing boat described in the previous sections. There are 2 (two) shapes which were investigated which were circular and elliptical shapes. Moreover, there were 2 (two) different vertical positions where measurements were performed which were at the center of propeller (Case 1) and at 0.7 x Radius (0.7R) of propeller blade (Case 2).

More detail about intake shapes and vertical measurement positions are described below.

3.1. *Shape description*

The sketch and dimension of each shape are shown in Fig. 4 and Table 2.

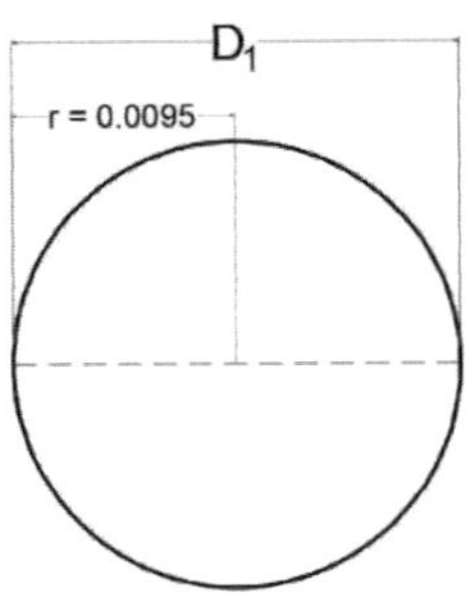

(a) circle

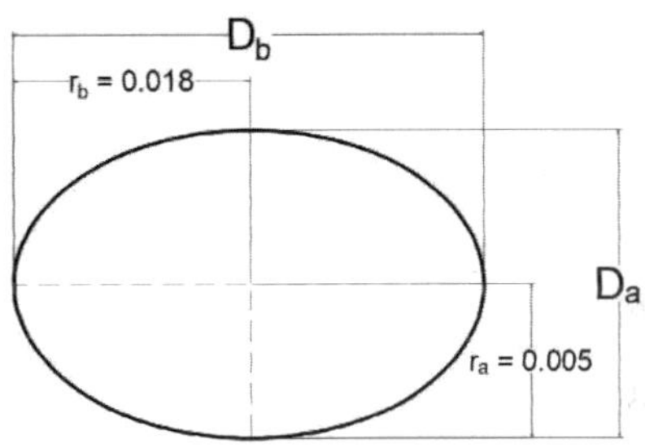

(b) ellipse

Figure 4. Shape dimension

Table 2. Dimension of shapes

Parameter	Intake Shape	
	Circle	Ellipse
Diameter	$D_1 = 0.019$ m	$D_a = 0.01$ m, $D_b = 0.036$ m
Surface area	0.00028 m^2	
Horizontal distance to propeller	0.1 m	

3.2. *Vertical position*

In order to verify the tendency, beside at the center of propeller, the data collection using these shapes also was conducted for the case when the intake was placed at location of 0.7R of the propeller vertically. The sketch of the location where the intake was placed is shown in Fig. 5.

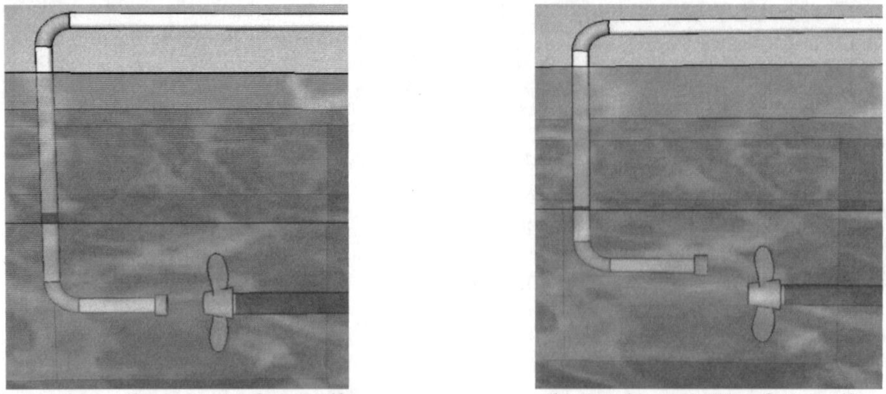

(a) Intake at the center of propeller (Case 1) (b) Intake at 0.7R of propeller (Case 2)

Figure 5. Vertical position of the intake

As shown in Fig. 5, there are 2 (two) cases of vertical intake placement. Case 1 is when the intake was placed exactly at the center of the propeller while Case 2 is when the intake was placed at 0.7R of propeller. The horizontal distance of intake to propeller in each case was same which equals to 0.1 m as described in Table 2.

4. Results Analysis and Discussion

In order to compare the performance, in the present study, a numerical simulation and an experiment were performed.

4.1. Ansys CFX simulation

Before conducting the experiment, a numerical analysis using ANSYS CFX software was performed. The main aim of the software analysis is to compare the flow velocity between circular and elliptical shapes. It become an initial prediction of the analysis results. However, only the case when the inlet pipe was placed at the center of the propeller (Case 1) which was simulated because the input velocity flow was just assumed for the software simulation. Thus, the second case will obtain the same result tendency.

In this analysis, the fluid source was considered to have a box shape. The domain of the intake and flow source were then meshed using unstructured meshes. The shape and their meshes are shown in Fig. 6.

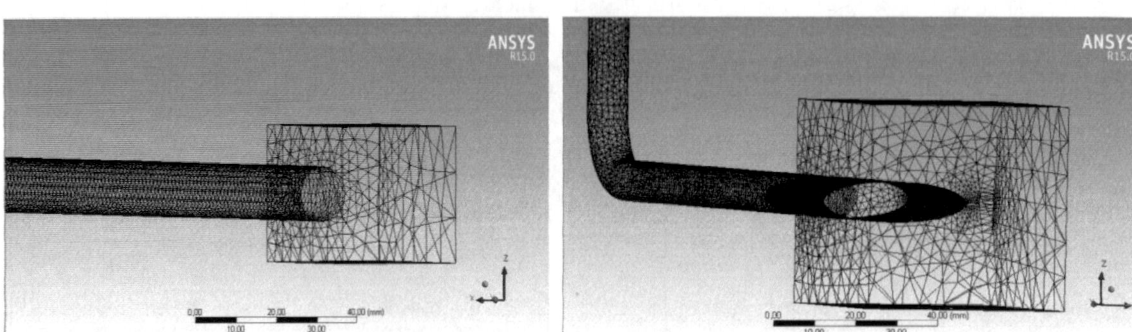

(a) Circular shape (b) Elliptical shape

Figure 6. Domain meshing in ANSYS CFX

IOP Conf. Series: Materials Science and Engineering **676** (2019) 012020 doi:10.1088/1757-899X/676/1/012020

After the meshing was converged, the computation was run. The computation results are shown in Fig. 7.

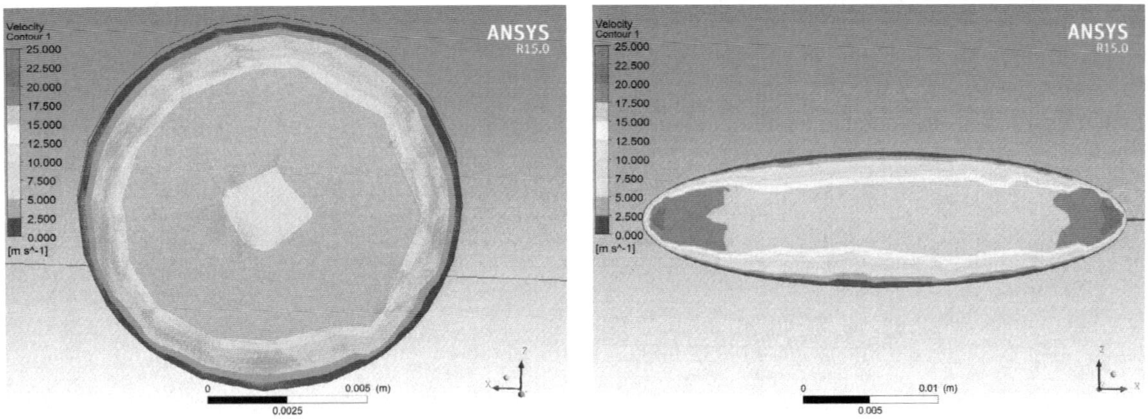

(a) (b)

Figure 7. Velocity contour of circle (a) and ellipse (b) intake

The computation results shown in Fig. 7 clearly shows that the flow velocity for the case of elliptical shape is higher than the one in circular shape. The main difference of velocity is mainly around the corner area where the flow velocity is higher on that area. Around the corner area of elliptical shape, the flow velocity could reach 23 m/s while on circular shape only reached 18 m/s. The obtained results are due to flow separation on the corner area which makes the flow could pass through with higher velocity.

4.2. Experimental results

Based on the experimental setup described in the previous section, several direct measurements were conducted. The main measured parameters in the experiment were motor shaft revolution (motor *RPM*) and water capacity (*Q*). The first data collection was conducted for the case when the intake was placed in front of the center of propeller as shown in. Fig. 5(a). The experimental results for this case are shown in Fig. 8.

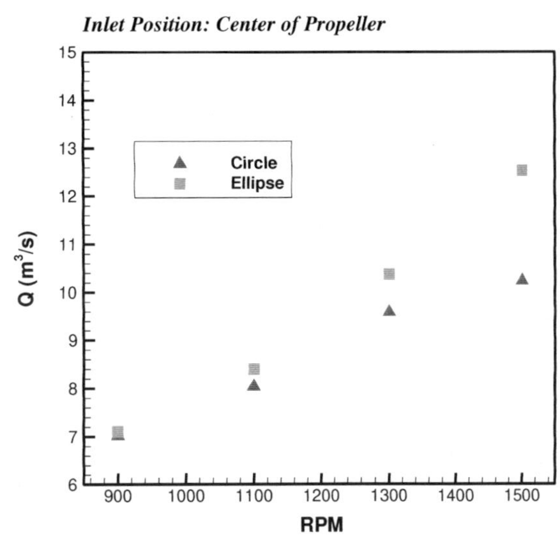

Figure 8. Flow capacity of circle and ellipse shape at the center of propeller position

In Fig. 8, the horizontal axis is motor revolution (motor RPM) while the vertical axis is the water flow capacity (Q) in m^3/s. The figure compares the flow capacity over a certain range of rpm for circular and elliptical shapes. The measured rpm is 900, 1100, 1300, and 1500. It can be seen from the figure that flow capacity is almost same for rpm equals to 900. However, the flow capacity for elliptical shape case is higher as the rpm become higher.

Moreover, to verify the tendency at a different location, the measurement was also performed at the location of 0.7R of propeller blade as illustrated in Fig. 5(b). The measurement results are shown in Fig. 9.

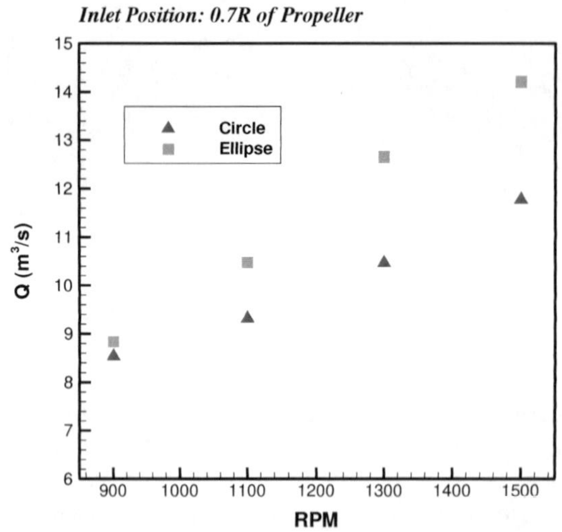

Figure 9. Flow capacity of circular and elliptical intake shapes at 0.7R of propeller blade position

From Fig. 9, the same tendency can be observed where elliptical intake shape has a better performance. The maximum flow velocity in circular shape is 11.8 m^3/s while in elliptical is 14.2 m^3/s. Moreover, it can also be seen that the flow capacity difference between circular and elliptical shapes are larger than the one in the previous case. As a result, a higher maximum flow capacity can be observed for this case.

5. Conclusion

A numerical analysis and an experiment have been conducted to investigate the performance of 2 (two) intake shapes which are circular and elliptical shapes. The main measured parameters in the experiment were motor rotation and intake flow capacity. Measurements at two different positions of the intake shapes were also conducted which were when the intake at the center of propeller and when the intake is at 0.7R of propeller blade position. It was found that elliptical shape has a higher flow rate which can be attributed to the flow separation around the corner area of the elliptical shape. More over, it was also found the measurement at 0.7R of propeller blade position has higher flow capacity compared to when the intake was placed at the center of the propeller.

Acknowledgement

The authors would like to express gratitude to JICA C-BEST project of Engineering Faculty, Hasanuddin University for funding this research. The authors also express gratitude to Professor Jun Ando from Kyushu University for the fruitful discussion and supervision of the project.

References

[1] E. K. Boulougouris, A. D. Papanikolaou, and A. Pavlou, "Energy efficiency parametric design tool in the framework of holistic ship design optimization," *Proc. Inst. Mech. Eng. Part M J. Eng. Marit. Environ.*, vol. 225, no. 3, pp. 242–260, 2011.

[2] A. Coraddu, M. Figari, and S. Savio, "Numerical investigation on ship energy efficiency by Monte Carlo simulation," *Proc. Inst. Mech. Eng. Part M J. Eng. Marit. Environ.*, vol. 228, no. 3, pp. 220–234, 2014.

[3] V. N. Armstrong and C. Banks, "Integrated approach to vessel energy efficiency," *Ocean Eng.*, vol. 110, pp. 39–48, 2015.

[4] D. G. Trodden, A. J. Murphy, K. Pazouki, and J. Sargeant, "Fuel usage data analysis for efficient shipping operations," *Ocean Eng.*, vol. 110, pp. 75–84, 2015.

[5] F. Baldi, H. Johnson, C. Gabrielii, and K. Andersson, "Energy and exergy analysis of ship energy systems - The case study of a chemical tanker," *Int. J. Thermodyn.*, vol. 18, no. 2, pp. 82–93, 2015.

[6] Mahmuddin, F., "Analysis of Wind Energy Potential with a Mobile Floating Structure around Sulawesi and Maluku Islands of Indonesia", Proceeding of the 34th International Conference on Ocean, Offshore, and Arctic Engineering (OMAE), St. John's, Canada, Vol. 9, 2015.

[7] G. Kocak and Y. Durmusoglu, "Energy efficiency analysis of a ship's central cooling system using variable speed pump," *J. Mar. Eng. Technol.*, vol. 17, no. 1, pp. 43–51, 2018.

[8] V. Medica, "Numeri ~ no modeliranje hladilnega sistema na motorju z notranjim zgorevanjem," vol. 50, pp. 104–114, 2004.

[9] G. Theotokatos, K. Sfakianakis, and D. Vassalos, "Investigation of ship cooling system operation for improving energy efficiency," *J. Mar. Sci. Technol.*, vol. 22, no. 1, pp. 38–50, 2017.

[10] C. L. Su, C. H. Liao, T. C. Chou, M. H. Chou, and J. M. Guerrero, "Variable flow controls of closed system pumps for energy savings in maritime power systems," *IEEE Ind. Appl. Soc. 52nd Annu. Meet. IAS 2016*, pp. 1–8, 2016.

IOP Conf. Series: Materials Science and Engineering **676** (2019) 012021 doi:10.1088/1757-899X/676/1/012021

Design of Software for Calculation and Analysis Engine Propeller Matching

Habibi[1]*

[1]Diploma Program of Maritime Academy, Hang Tuah University, Surabaya

*E-mail: habibi@hangtuah.ac.id

Abstract. The criteria of a good ship are not only could floating in the sea, but it should have a very complex construction development planning. Longitudinal and transversal strengths, stability, flood, bulkhead, ship motion etc. were interrelated criteria that should be calculated for safety and comfort of the ship, passenger and cargo passenger. Matching point is an operating point of main engine same with the load character of propeller. In another word, the operating point of main engine rotation, where absorbed power by propeller is equal by produced power by main engine, and delivery equal speed with planned service speed. One step in ship design evaluating is EPM ("Engine Propeller Matching"). The lack of available software to analyze relationship between characteristics of propeller with main engine and hull, causing the calculation of EPM (Engine Propeller Matching) takes more time. Java is a very popular application, especially for enterprise application. So, that Java applications can be used to build a programming language of Engine Propeller Matching.

1. Introduction

An efficient ship using main engine to generate power that is able to deliver a speed service as planned. Adji [1] define match point is rotation main engine operating point where absorbed power by propeller is equal with produced power by main engine and generate an equal speed with planned speed service.

Information technology development resulted in computer capability increasing, that allows ship designing process can be done in a relatively short period. This was supported by software development to running a complete numerical analysis such as C++, Java, Vb.Net, Delphi, Fortran, Matlab and others. Steps of design related to one another makes it possible to create a program that have integrated for design so that the evaluation can performed simultaneously.

Java is a programming language for all requirements, concurrent, class-based, object-oriented and designed independently. One of method in java language programming executes a block of statements repeatedly until the specified condition returns false. In general, statements are executed sequentially: The first statement in a function is executed, followed by the second, and so on.

Basicly, this study focused on how the combine method *for-loop* in java to find a *match point engine* and *propeller matching* [2].

1.1. Engine propeller matching

In general, ships that move in fluid with a certain speed, will make a resistance force opposite to the direction of ship movement. The amount of drag force that occurs should be able to be overcomed by the thrust force of the ship generated from propeller. The power delivered to propeller from the Shaft

Power, while the Shaft Power itself is sourced from the Brake Power (P) which is the output power of the ship's motor [3].

1.2. Characteristics of ships hull

One of the most influential stages in Engine Propeller Matching analysis process is the modeling stage of the ship's design. This is because *Ship Body Characteristics* have a direct effect on the characteristics of the propeller. This ship's resistance is the inhibitory force of the fluid medium through which the vessel operates at a certain speed. The total inhibitory force is the sum of all components drag force working on the ship.

1.3. General characteristics of propellers in the open water curve

Forces and moments produced by propellers can be described in the most basic form that presented in a series of non-dimensional characteristics. This characteristic is used to describe in general the performance of a propeller [4]. These characteristics are:

Thrust Coefficient:

$$KT = \frac{T}{\rho n^2 D^4} \tag{1}$$

Torque Advance Coefficient:

$$K_Q = \frac{Q}{\rho n^2 D^5} \tag{2}$$

Advance Coefficient:

$$J = \frac{Va}{nD} \tag{3}$$

Cavitation number:

$$\sigma = \frac{P_0 - e}{\frac{1}{2}\rho V^2} \tag{4}$$

where:

D	=	Diameter of propeller
Va	=	Advanced velocity of fluid flow
n	=	Propeller rotary speed
P	=	Fluid density
$P_0 - e$	=	Static pressure fluid around propeller

1.4. Hull and propeller interaction

Hull and Propeller Interaction is approach to obtain the performance characteristics of propeller for condition behind the ship. The method is by calculation as following [5]:

$$Tship = \frac{\alpha V_{A^2}}{(1-t)(1-w)^2} \tag{5}$$

$$Tship = Tpropeller$$

$$Tpropeller = KT \rho n^2 D^4 \tag{6}$$

$$KT = \frac{\alpha V_{A^2}}{\rho n^2 D^4 \left(1-t\right)\left(1-w\right)^2} \tag{7}$$

where:

α = Constant
t = Thrust deduction factor

1.5. Java programming desktop

The Java programming language was originally created by James Gosling in 1995 as part of the Sun Micro system Java Platform. Java syntax is much derived from C and C ++ but is simpler and has access to a more limited OS. This is because Java have been intended as a programming language that is simple enough to learn and easy to read. The Java application is written as a *.java extension* that is compiled into file *.class*. This *.class* file is a byte code that works in all Java Virtual Machines, regardless of its OS or processor architecture. Java is a language intended for all needs, concurrent, class based, object oriented and designed to be independent of the environment in which the application is run.

2. Design of System

System was designed to obtain a match point between the hull, propeller and main engine. It is the point where power absorbed by propeller is equal to power produced by main engine and produces the ship speed approaching speed planned. The system of application can be seen in Figure 1. The data that have been collected will be processed, as for the stages in the data analysis include: Presentation of Ship Data, Steps of workmanship, Open water diagram readings, Calculation of total ship resistance, Calculation of power main engine, Propeller characteristics calculation, Engine rotation calculation, Load propeller calculation, Determination engine propeller matching and Engine graph propeller matching [6].

3. Methods

This study begins by entering the type of propeller into the database which will be used in this software. Database storage using Xamp™ and PHPmyadmin™. Coding process method *for-loop* java using NetBeans™. Result of running process can be seen in the Figure 1. *For-loop* method process contained in the *data input form*. The data will be *looping*are propeller-type, propeller diameter, and speed service. Figure 2 shows *data input form*.

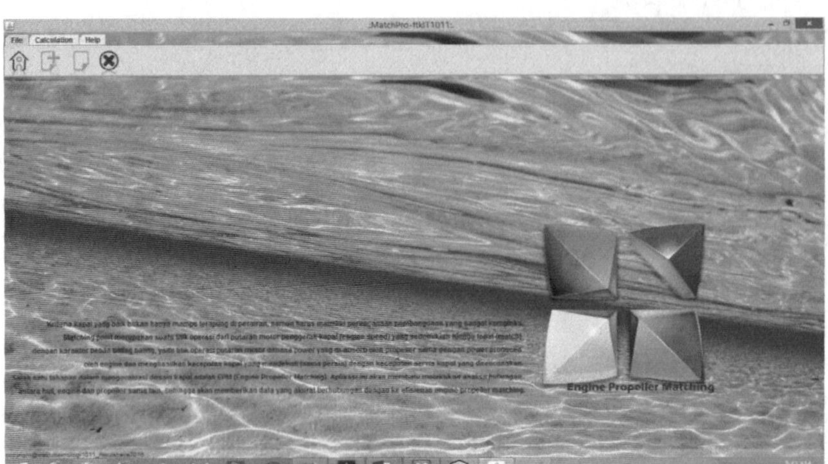

Figure 1. Main form

4. Testing Application/Software

4.1. Data collection

Testing application/software engine propeller matching begins with input data dimensions of the ship and the selection of methods of calculation of ship resistance to be used. The steps can be seen in Figure 2.

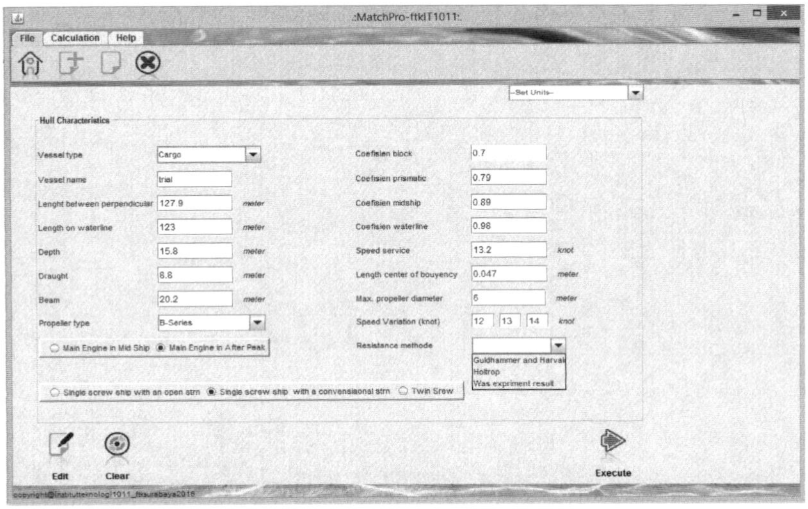

Figure 2. Form dimension of the ship and the selection of ship counting method of ship resistance

The process of calculating the total resistance of the ship by the system is pressing execute button and will end with the main machine input form as shown in Figure 3.

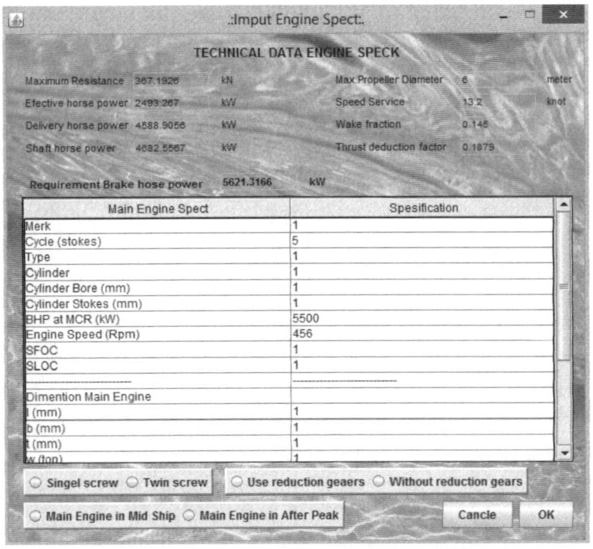

Figure 3. Machine input form

On the input form, machine will calculate the characteristics of the main engine to be used during the sailing vessel. The input form of this machine in addition to the characteristics of the engine that was also provided tables for input data gear reduction.

IOP Conf. Series: Materials Science and Engineering **676** (2019) 012021 doi:10.1088/1757-899X/676/1/012021

4.2. Output testing results applications software

Output of application/software testing results include:

The calculation of total ship resistance barriers. The total vessel barrier will be displayed by the application in the form of a table based on the variation of the speed covered in the input form. The results of the total ship resistance calculation can be seen in Figure 4.

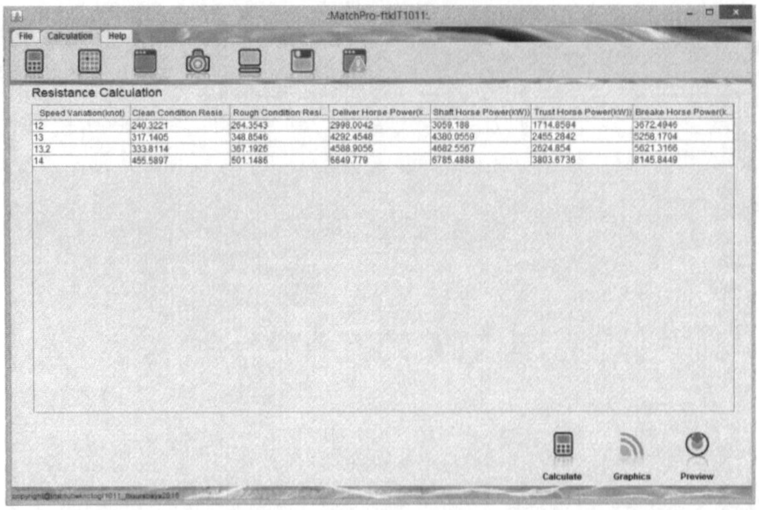

Speed Variation(knot)	Clean Condition Resis...	Rough Condition Resi...	Deliver Horse Power(k...	Shaft Horse Power(kW))	Trust Horse Power(kW))	Breake Horse Power(k...
12	240.3221	264.3543	2998.0042	3059.188	1714.8584	3672.4946
13	317.1405	348.8546	4292.4548	4380.0559	2455.2842	5258.1704
13.2	333.8114	367.1926	4588.9056	4682.5567	2624.854	5621.3166
14	455.5897	501.1486	6649.779	6785.4888	3803.6736	8145.8449

Figure 4. Form calculation of total ship resistance

Calculation of propeller characteristics. In the form of propeller characteristics will show the type of propeller that can be installed on the ship. Database application provides 8 (eight) types of propeller type B-series that can be selected that is B335, B350, B365, B380, B440, B455, B470 and B485. The type of propeller to be displayed by the application is a non-cavitation propeller type, have a maximum diameter and has a power that is smaller than the main engine power at the time of operation at the speed service.

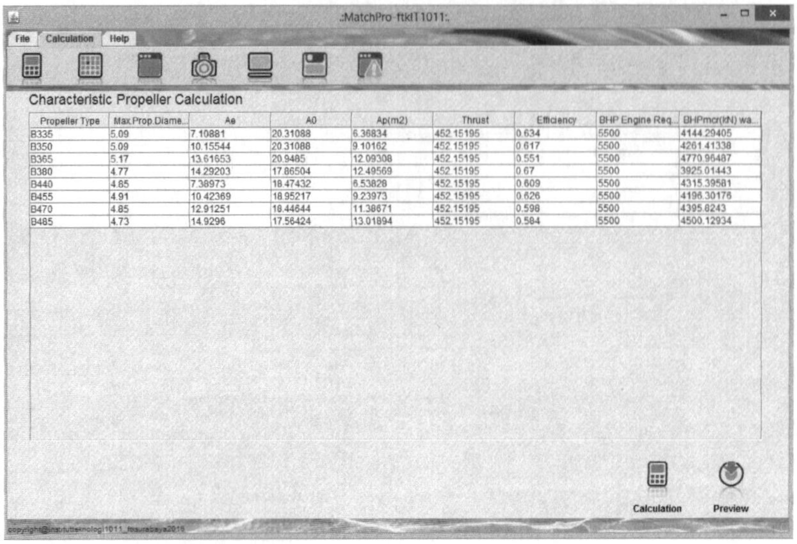

Propeller Type	Max.Prop.Diame...	Ae	A0	Ap(m2)	Thrust	Efficiency	BHP Engine Req...	BHPmcr(kN) wa...
B335	5.09	7.10881	20.31088	6.36834	452.15195	0.634	5500	4144.29405
B350	5.09	10.15544	20.31088	9.10162	452.15195	0.617	5500	4261.41338
B365	5.17	13.61653	20.9485	12.09308	452.15195	0.551	5500	4770.96487
B380	4.77	14.29203	17.86504	12.49569	452.15195	0.67	5500	3925.01443
B440	4.85	7.38973	18.47432	6.53828	452.15195	0.609	5500	4315.39581
B455	4.91	10.42369	18.95217	9.23973	452.15195	0.626	5500	4196.30176
B470	4.85	12.91251	18.44644	11.38671	452.15195	0.598	5500	4395.8243
B485	4.73	14.9296	17.56424	13.01894	452.15195	0.584	5500	4500.12934

Figure 5. Form characteristic propeller

Engine and propeller matching calculation form displays a match at the operating point where operation of engine power rotation was absorbed by propeller same with power produced by the main engine. The result of calculation is seen in Figure 6.

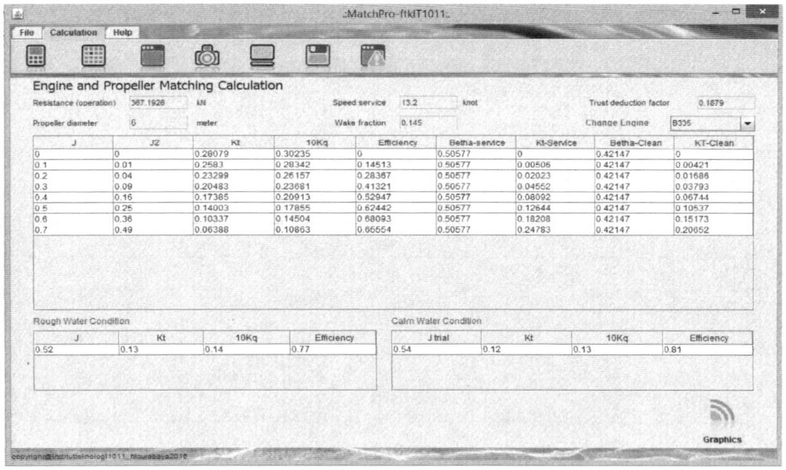

Figure 6. Engine and propeller matching.

Form of the Engine and Propeller matching, there are graphics buttons displaying KT-KQ-J chart which will automatically determine the point J (advanced coefficient) which is the intersection between KT propeller and KT ship. The view of graphical can be seen in Figure 7

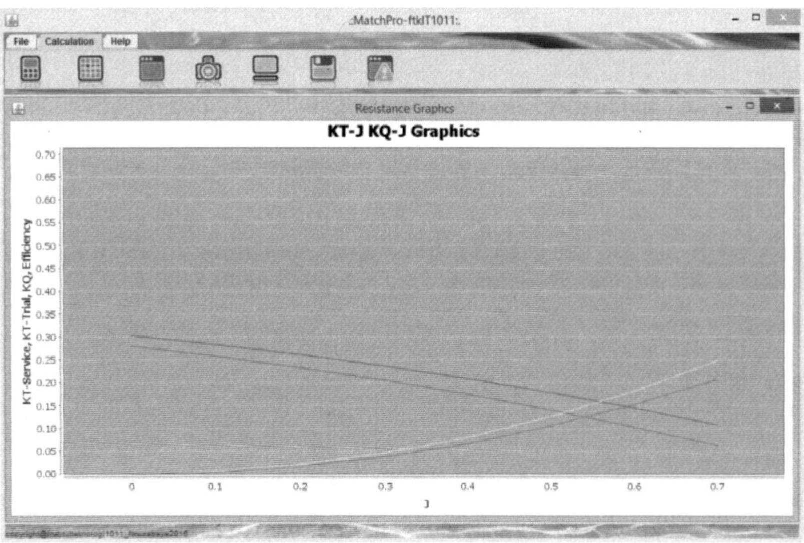

Figure 7. KT-KQ-J Graph form

5. Comparative Result

To obtain the validation of the calculation results, a comparison of the calculations results is made. The results of the calculation from application compared with the results of the calculation of ship have been operating. The data can be seen in the Table 1.

IOP Conf. Series: Materials Science and Engineering **676** (2019) 012021 doi:10.1088/1757-899X/676/1/012021

Table 1. Ship data

Ship Name: KM. Sungai 75	
Ship type	Car ferries
Resistance Method	experiment result from towing tank
Propeller type	B-Series
Lwl	20.453 m
Lbp	18.95 m
H	2.7 m
T	1.9 m
B	8 m
Cb	0.6
Cm	0.929
Cw	0.9
V	11.22 Knot
Lcb	-0.965
D	1 m

5.1. EHP calculation

Result of comparison of EHP calculation can be seen in Table 2 as follows:

Table 2. EHP Calculation

Speed (Knot)	Experiment	Software	Difference	Percentage
10.13	124.14152	131.3765	7.23	0.07
11.22	247.1784	232.8634	14.32	0.14
11.63	302.69488	320.3255	17.63	0.18
12	365.48036	386.6642	21.18	0.21

From the calculation result in Table 2 the percentage difference in vessel service speed is 0.14%. This is due to differences in formulas used by the shipyard and application. Application gives a value of 1 + 20% Resistance on sea margin, while testing shipyard adds 15% EHP to get BHP on service condition.

5.2. Propeller calculation

The result of comparison of propeller calculation can be seen in Table 3 as follows:

Table 3. Propeller Calculation

Calculation	Prop.Rps	Ae/A0	Eff. (%)	Trust (kN)	Cavitation
Ship yard	10.6900	0.7000	0.5311	21.8390	No
Software	10.6909	0.6999	0.4600	28.3733	No
Difference	0.00070	0.0000	0.0711	6.53426	

From the result of calculation of propeller in Table 3, the percentage of difference that is higher is in the thrust generated by propeller that is reaches 0.06%, while the rotation and propeller efficiency is close to 0%. Based on the calculation Propeller type B-470 can be used because of both calculations of shipyard and application does not occur cavitation.

6. Conclusion

Conclusion that can be taken from this research is as follows:

- Application/software has been able to produce match point between propeller and main engine mounted on ship.
- Using this application in the analysis of engine and propeller matching can reduce errors in manual readings.
- This application can accelerate the calculation and analysis of engine and propeller matching in which 8 (eight) types of B-Series propellers can be calculated/running for ± 60 seconds.

Acknowledgements

This Research was supported by Research And Community Service Institutions Hang Tuah University (LPPM). We thank our colleagues from Mr. I Made Aryana who provided insight and expertise that greatly assisted the research.

References

[1] Adji, S.W., 2005. Engine Propeller Matching. ITS., Surabaya.
[2] Bima.2003. Java Foundation. Ganesa. Bandung.
[3] H, Otto, Kristensen. 2012. Prediction of Resistance and Propulsion Power Of Ships. DTU.
[4] Harvald, Sv. Aa. 1992. *Tahanan dan Propulsi Kapal Edisi Terjemahan*. University Press. Airlangga . Surabaya.
[5] Carlton, JS.2007. Marine propellers and propulsion. Butterworth-Heinemann.
[6] Habibi. 2016. *Pembuatan Aplikasi Engine Propeller Matching*, Thesis, ITS, Surabaya.

IOP Conf. Series: Materials Science and Engineering **676** (2019) 012022 doi:10.1088/1757-899X/676/1/012022

The Effect of Hanging Sheet Pile Breakwater's Draft Relative on Wave Transmission Coefficient (K_t) in Irregular Wave

C Paotonan[1]*, H Umar[1], A Y Baeda[1], T Rachman[1] and W Hasan[1]

[1]Ocean Engineering Department, Faculty of Engineering, Universitas Hasanuddin, Indonesia

*Email: paotonan_ch@yahoo.com,

Abstract. Breakwater is a structure that builds parallel to the coast, and located within some distance from the coastline. The breakwater that is currently and widely used is Rubble Mound Breakwater; a large dimension coastal structure with also come in high price. Due to this matter, a Hanging Sheet Pile type breakwater can be one of good solution. This research was conducted at Ocean Technology Laboratory, Faculty of Engineering, Universitas Hasanuddin, in a flume thank with 1:25 model scale, and equipped with wave generator and probes as measurement apparatus. The purpose of this study is to identify the effect of wave parameters (H_i/gT^2) and structure parameters (draft relative, s/d) on fluctuated wave transmission coefficients produced. The waves are in the form of irregular waves that generated using JONSWAP spectrum. There were 5 variations in wave height (H) and wave period (T), while variations of structure in the form of draft (s) are in 3 variations. The results of this study indicate that the value of K_t will decrease when H_i/gT^2 value increase; so that the greater the value of s/d, the smaller value of K_t will be.

1. Introduction

Indonesia as an archipelago located between two continents and two oceans, definitely have several problems due to its very long coastline. For that fact, beach protection against wave energy is a must. Beach protection system often used in Indonesia is breakwaters. The most important sea wave parameters commonly used in ocean wave analysis are wave height and wave period. Based on the wave height, the probability distribution of ocean waves can be determined; while with the wave period, the shape of the ocean wave spectrum can be determined.

Wave transmission (K_t) is the ratio between the passings through sheet piles breakwater wave parameters to the incoming wave parameters. The wave height and period imposed on the model are used as variables that determine the magnitude of the wave transmission coefficient [1].

That parameter is very important in the planning of sheet pile breakwater. Therefore, it is necessary to conduct studies related to the effects of waves through the sheet pile structure before being applied to the actual conditions as breakwaters on the coast and harbor.

The subject of the study is to find out the influence of wave parameters such as wave height (H) and wave period (T) on sheet pile in a certain draft; by determining the effect of H_i/gT^2, as wave steepness, on the magnitude of the wave transmission coefficient (K_t) and the effect of draft relative (s) to the value of the transmission coefficient (K_t) which occurs on the hanging sheet pile.

IOP Conf. Series: Materials Science and Engineering **676** (2019) 012022 doi:10.1088/1757-899X/676/1/012022

2. Literature Review

Waves that travel from the deep to shallow water will experience a change in shape due to changes of sea depth. The more towards the beach, the wave crest will be sharper and the valley will be flatter. This phenomenon causes waves to break. Waves can be divided into several types. Among them are wind waves generated by the blowing of the wind at sea level, tidal waves that generated by the attraction of celestial objects, especially the sun and moon to the earth, and tsunami waves that occur due to volcanic eruptions or earthquakes in the sea [2].

2.1. Wave spectrum

The nature of ocean waves is random, both magnitude and direction, so that the nature of random wave energy is difficult to measure. Random waves are a combination of several sinusoidal waves with varying length and periods. The intensity of random wave components is generally expressed in the form of an amplitude density spectrum, wave energy density or commonly abbreviated as a wave energy spectrum.

In wave energy spectrum analysis, it is necessary to record wave data for 15-20 minutes. The principle of wave spectrum analysis is to describe an irregular wave into an arrangement of regular waves of various frequencies and wave heights [3].

The mostly single parameter spectrum that most commonly used is the Pierson-Moskowitz [4] which is based on significant wave height or wind speed. In addition there are several commonly used dual parameter spectrums.

2.2. Wave transmission

Wave transmission is the residual wave energy that occur after passing through the wave restraint structure [5]. The parameters are expressed as a comparison between the transmitted wave height (H_t) and the incoming wave height (Hi) or the root of the transmission wave energy (E_t) with the incoming wave energy (E_i).

$$H_t = \frac{H_{max} + H_{min}}{2} \quad (1)$$

$$K_t = \frac{H_t}{H_i} = \left(\frac{E_t}{E_i}\right)^{1/2} \quad (2)$$

The transmission coefficient is the ratio between the transmission wave height and the incoming wave height. From the energy function, the transmission energy is the root function of the comparison spectral wave energy transmission with the spectral wave energy coming. The smaller the wave reduction produced the greater the transmission coefficient [6].

2.3. Sheet Pile

Sheet pile is a relatively thin vertical wall that serves to hold the soil and to hold water into the excavation hole. Sheet pile is currently widely used in jobs such as retaining walls, excavation of excavation cliffs, buildings in ports, and dam construction.

Sheet pile is not suitable for holding very high soil because it will require a large area of large pile material. In addition sheet pile is also not suitable for use in soils that contain lots of rock, because it makes it difficult to erect. Sheet pile types can be distinguished according to the material used. The sheet pile material is various for example wood, reinforced concrete and steel.

3. Results and Discussions

After all wave spectrum variations and model scenarios are tested on wave flume, the results are then analysed in order to get an idea of what and how the effectiveness and response of various wave spectrums due to the reduction of transmission and wave reflection. For this research, JONSWAP wave spectrum will be use due to its commonly use in Asia, mainly in Indonesian waters and in some other places which are very similar surroundings [7].

1. Effect of Wave Steepness (H_i/gT^2) on the Transmission Coefficient (K_t) for Draft Variation (s), with the provision of a gap width of 0.005m. To present the magnitude of the draft (s) relationship with the transmission coefficient value (K_t), the H_i/gT^2dimensionless parameter is used which represents the wave characteristics. If the H_i/gT^2 value is the X-axis and the transmission coefficient (K_t) is the Y-axis variable, then the graph will be obtained as follows:

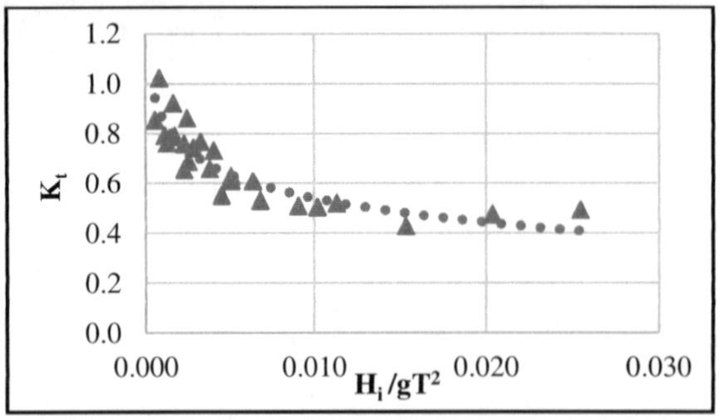

Figure 1. The relationship of H_i/gT^2 with K_t for draft 0,50m

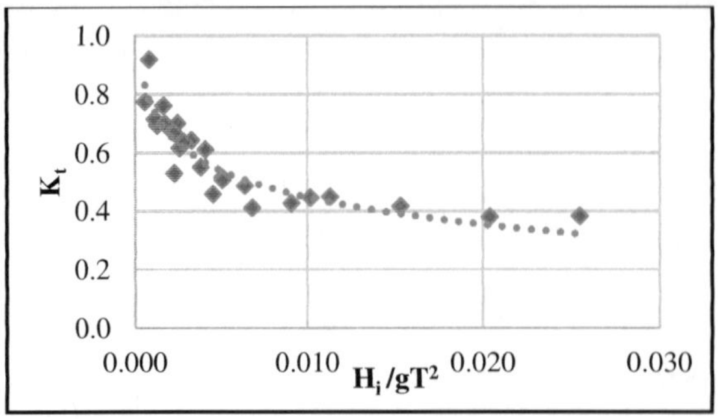

Figure 2. The relationship of H_i/gT^2 with K_t for draft 0,75

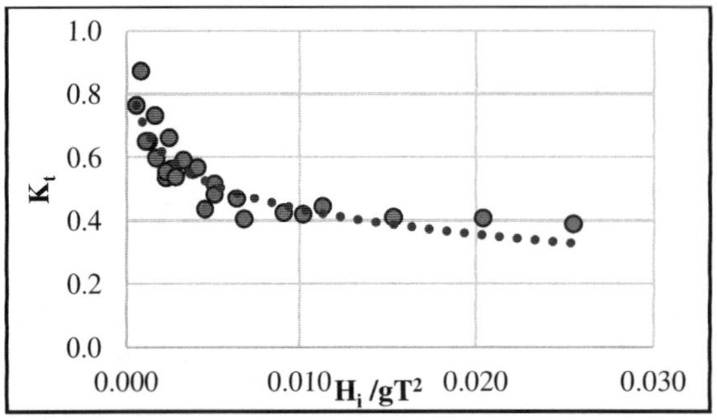

Figure 3. The relationship of H_i/gT^2 with K_t for draft 1,00

Figures 1 to 3 shows that an increasing wave steepness, (H_i/gT^2), will cause a decreasing value of the transmission wave energy coefficient, (KE_t). This indicates that the steeper the waves come in front of the sheet pile breakwaters model, will cause a shrinking ratio of wave height behind and the front of the model. This phenomenon occurs because a wave with a large steepness occurs significant turbulence in front of the model and causing a decrease in wave height passing the model relative to the incident wave height. But note that the distribution of data for all is relatively large. For all the values of s/d are shown in Figures 5 to 7, for the H_i/gT^2 less than 0.01, the decreasing of KE_t due to the increasing of H_i/gT^2, is significant. Meanwhile on the value of H_i/gT^2 greater than 0.01, the increasing of H_i/gT^2 is relatively have a small effect on the changing of KE_t.

Based on the graphs on the pictures above, it can be concluded that with the increase of H_i/gT^2, the K_t will decreases; this will consistently occur in each test with different variations of draft (s). For draft 0.50, wave steepness (H_i/gT^2) values ranging from 0.000 – 0.0250 with a range values of K_t between 0.8 – 0.40. For draft 0.75, wave steepness (H_i/gT^2) values ranging from 0.000 – 0.0250 with range values of Kt is between 0.84 - 0.31 and for 1.00, wave steepness (H_i/gT^2) values ranging from 0.000 - 0.0250 with range values of K_t between 0.79 - 0.31. This means that the steeper the incoming wave, the better the sheet pile will be in reducing the wave; or the smaller the wave transmission value (K_t) will be. The influence of wave steepness (H_i/gT^2) on large K_t in the range $H_i/gT^2 < 0.005$, can be seen from the significant change in K_t. Whereas for the H_i/gT^2 range > 0.005, the effect of H_i/gT^2 is relatively insignificant so that the K_t value tends to shrink gradually followed by the enlargement value of the H_i/gT^2.

2. The effect of s/d on the transmission coefficient (K_t). To presents the relationship of draft (s) and the relative depth (d) then the variable is based on s/d to K_t, then the graph are as follows:

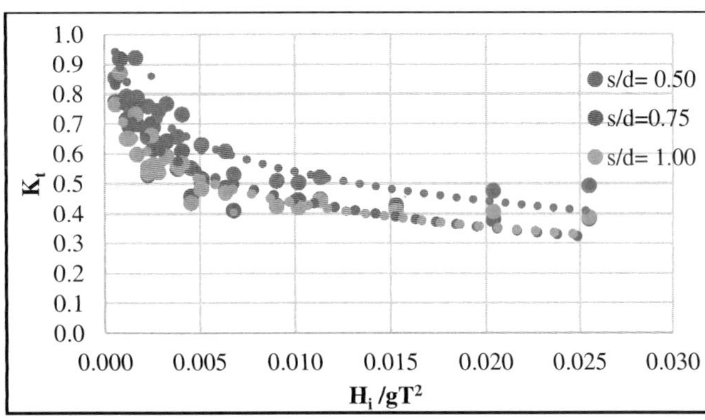

Figure 4. The relationship of H_i/gT^2 with K_t on each draft variations

Based on the graphs on previous pictures, it can be concluded that the relationship pictured in draft (s) and relative depth (d) as dimensionless numbers (s) to the value of K_t. Where the greater the value of s/d, the more declined the trend of K_t. This can be seen in the value up to 1.00 giving the smallest trend of K_t compared to the value up to 0.50 and up to 0.75. The decrease in the value of up to K_t tends to be drastic in the range $H_i/gT^2 < 0.005$, then in the range of $0.005 < H_i/gT^2 < 0.025$, the decreases tends to be gradual.

But from these experiments, there are some things that make the value of s/d against K_t, not always have a constant value, while H_i/gT^2 increases. This can be seen from the trend lines that coincides one another.

1. For 0.75 and 1.0 that coincide, the trend in which this event can be caused by several things, including: The model approach was not perfect so that the results obtained did not match expectations.

IOP Conf. Series: Materials Science and Engineering **676** (2019) 012022 doi:10.1088/1757-899X/676/1/012022

2. The wave damping behind the model is imperfect, so that the transmitted wave can be reflected back towards the model. This is what causes wave resonance events between the incoming wave and the transmitted wave.

For the range $0.020 < H_i/gT^2 < 0.025$ where the frequency of waves are large due to the small wave periods and higher wave heights, causes the generated waves to be more steep. So that in this range the incoming waves will be more dissipated, both in partially and whole, due to the instability of the water particle movement during wave propagation. This can be seen from the tendency of the K_t line to move almost in straight line, since most of the incoming wave energy has been dissipated.

4. Conclusions

1. The greater the wave steepness (H_i/gT^2), the smaller the transmission coefficient (K_t) will be, where the effect of wave steepness (H_i/gT^2) on K_t are significant on $H_i/gT^2 < 0.005$.
2. The transmission coefficient (K_t) will decrease by the increasing value of s/d equal to 1.00. This means that the greater the draft relative (s), the better structure functioning in transmitting waves.

References

[1] Murali, K., & Mani, J. (1997). *Performance Of Cage Floating Breakwater*. ASCE.
[2] Triatmodjo, B. 1999. *Coastal Engineering*. Beta Offset: Yogyakarta.
[3] Yuwono, N. 1982. *Coastal Engineering*. Teknik Sipil UGM: Yogyakarta.
[4] Chakrabarti, 1987, *Hydrodinamics of Offshore Structure*, Computational Mechanics, Cornell University.
[5] USACE, 2002, *Coastal Engineering Manual,* Washington, D.C.
[6] Bleck, M., & Oumeraci, H. (2001). Wave Dampening And Spectral Evolution At Artificial Reefs. 2, pp. 1062-1071. Proceedings of The International Symposium on Ocean Wave Measurement and Analysis (WAVES 2001), ASCE.
[7] Muis Alie, M., 2016. The Effect of Symmetrical and Asymmetrical Configuration Shapes on Buckling and Fatigue Strength Analysis of Fixed Offshore Platforms. s.l., International Journal of Technology (IJTECH).

Study on the Characteristics of Maneuvering Ferry Vessel as Effect of the Sea Waves

M Hasbullah[1]*, D Paroka[1], Rosmani[1] and Hanisa[1]

[1]Departemen of Naval Architecture, Faculty of Engineering, Hasanuddin University

E-mail: *mansyur_hasbullah@yahoo.com

Abstract. Wave is one of the natural phenomena that gives effect to the cruise ship. The energy produced by the wave is very large that makes the ship turn directions. Therefore the turn has an error coordinates of the starting point of a ship. Deviation of the coordinates has made a further distance and increase fuel consumption, so the operational costs being higher. The research would predict the drift angle, the ship speed and rudder angle to be used after the ship got the influence of wave by taking into account the motion of surge, sway, yaw. Simulations are performed by using Delphi software, with variations of the wind height of 0.5 m, 1 m, 1.5 m, and 2 m, the wave flatness of 0.01, the speed of the wind of 0 and 10 m/s and incoming wave direction 0° - 180° on the side of the ship. It is obtained that the biggest of drift angle happened on 50° wave direction. The biggest of rudder angle happened on 110° wave direction while that fastest ship speed happened on 160° wave direction.

1. Introduction

Ship maneuvering characteristics are influenced by several external factors which include sea state. This needs to be understood in view of the limited ability of the vessel to face the weather and sea waters as well as the movement of the ship in the water which also requires sufficient space. One of the external factors that influence the ship maneuvering is the sea wave. Meanwhile, influencing internal factors (ship) is the weight of the vessel, the dimensions of the vessel, hull, engine (propeller), and rudder system (steering system). They are instrumental in steering the ship when doing different movements (maneuver) in water [1].

Moreover, the influence of the waves against the yaw motion stability is very important to be analyzed, especially while the ship to be operating in wave conditions at far greater length than the length of the ship and the ship's position is on the slopes of the wave. This condition is known by the phenomenon of surf – riding [2, 3]. In this condition, the yaw motion of the vessel can be unstable despite steering control is given. The vessel will turn and move with the waves, the ship's position will ultimately sideways against the waves and eventually sink or reversed.

2. Theoretical Basis

2.1. Ships motion

On the principle of the motion, behavior of the ship is divided into six-degrees of freedom (six-degree of freedom), namely surge, sway, yaw, heave, roll, and pitch which are shown in Figure 1.

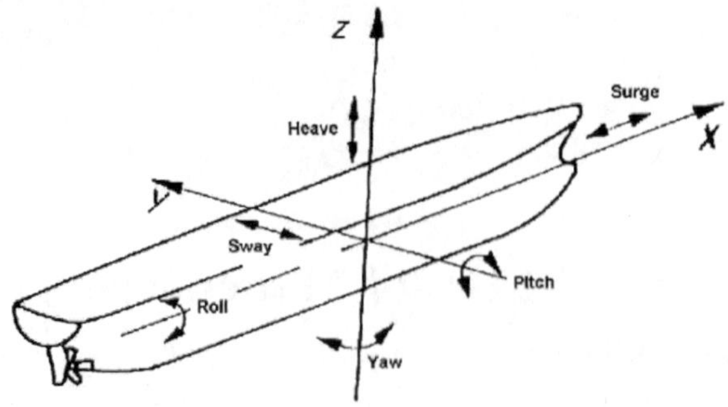

Figure 1. Six degrees of freedom of movement of ships

From Fig. 1, it can be understood as follows

1) *Surge* is one of the translational motion where the vessel experiences displacement on the *x*-axis (*x*-axis direction moving). The motion is not the case of mass change of the vessel, so there is no point bouyancy protection force.
2) *Sway* is one of the translational motion of the ship experienced displacement in the *y*-axis (move sideways), on the motion of the ship's center of gravity has not changed so there is no change of the mass of the ship and the bouyancy point is also does not move and no protection force.
3) *Yaw* is one of the rotational motion of a moving ship spinning on the *z*-axis so that if observed from the above, it can be seen that ship rotation moving.
4) *Heave* is the ship movement up and down vertically.
5) *Roll* is the rotational motion of the ship on the *x*-axis.
6) *Pitch* is the angular movement of the rotating fore and aft of the transverse axis of the ship, along the *y*-axis.

2.2. Steering ship

The rudder is a device to change direction ship by changing the direction of fluid flow that results in changes of ship direction. The steering wheel is placed at the threshold of the back hull (stern). The propeller is driven mechanically or hydraulically from the bridge by moving steering wheel. Because of the important role of steering in sailing, then SOLAS International Convention set rules in Chapter II-1 Construction regarding subdivision and stability, machinery and electrical installations [4, 5].

Rudder has the ability to maintain direction in accordance with the command. Since the requirements for passenger ships, steering wheel which is widely used is the type of Van Amorengen with specifications between -35° to 35° and rudder employment rate of 2.3° - 7° each second. The minimum requirement for an average rate of rudder is determined by the classification society. It is required that the rudder can be moved 35° from the port towards 35° to starboard in no more than 30 seconds.

Table 1. Standard maneuverability ship by IMO (Resolution MSC 137 (76) 2002)

Ability	Test	Criteria
Turning ability	Turning test with max. Rudder Angle (35 deg.)	Advance <4.5 L Tactical diameter <5.0 L
Initial turning ability	10^0 / Z-test10^0	Distance ship rudder run before execution <15 L2^{nd}

The maneuver used in the experiments at sea following a recommendation from the trial maneuvering code of ITTC (1975) and the IMO circular MSC 389 (1985). IMO also determine the appearance of some of the results on posters, bucklet and maneuvering bucklet in IMO Resolution A.601 (15) (1987) [6-8].

3. Methodology

3.1. Location and Research Time

This research was conducted in the laboratory of hydrodynamics, Naval Architect Department, Faculty of Engineering, University of Hasanuddin.

The vessel which was used as a sample of this research is Ro-Ro Ferry KMP SULTAN Murhum 300 GT. With the primary measure of data as shown in Table 2.

Table 2. Dimensions of the vessel

Main Dimensions			Form Coefficients			Propulsion System Parameters		
LOA	=	36.4 m	Cb	=	0.63	Hr	=	1.559 m
LWL	=	35.73 m	cm	=	0.986	lR	=	1.331 m
LBP	=	31.5 m	Cw	=	0.886	$N_{propeller}$	=	2
B	=	8.7 m	cp	=	0.804	n	=	8.578
H	=	2,65 m				xr	=	-15.75 m
T	=	1.65 m				D	=	1.1 m
V	=	10.5 Knot				N_{blade}	=	4

Source: PT IKI (Shipbuilding Industry Indonesia)

4. Results and Discussion

4.1. Coefficient of hydrodynamics

For calculating the coefficient of hydrodynamic hull according to Yoshimura and Ning Ma with a semi-empirical method, which considered three (3) movement of ships that are surge, sway, and yaw. To run the program, required input data are the main parameters of the vessel which is the object of research. The steps or display the data input on the Delphi program is as follows:

- Input ship's main dimensions

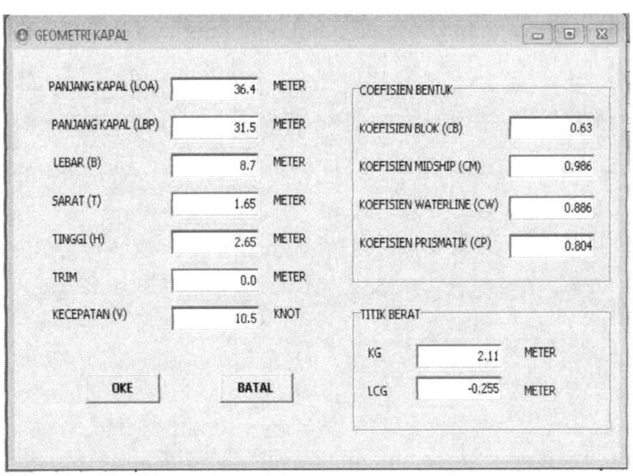

Figure 2. Display form of input data of the ship's main dimensions

- Once the input data is completed, then input the data section (section width, height and breadth section) as well as waveform data (height and flatness of the wave)

IOP Conf. Series: Materials Science and Engineering **676** (2019) 012023 doi:10.1088/1757-899X/676/1/012023

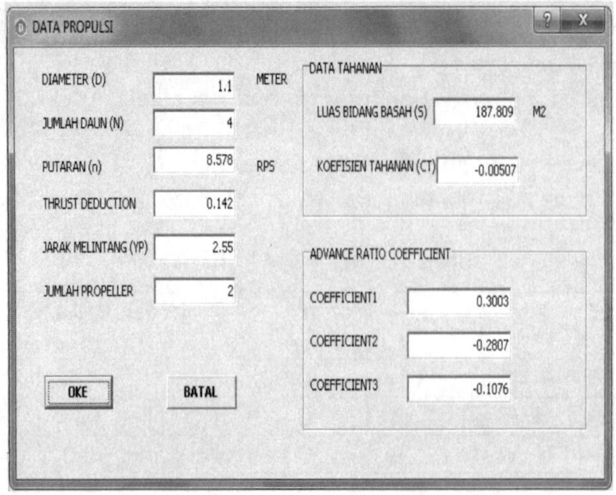

Figure 3. Display form of input data of section and wave

- The next step is propulsion data input

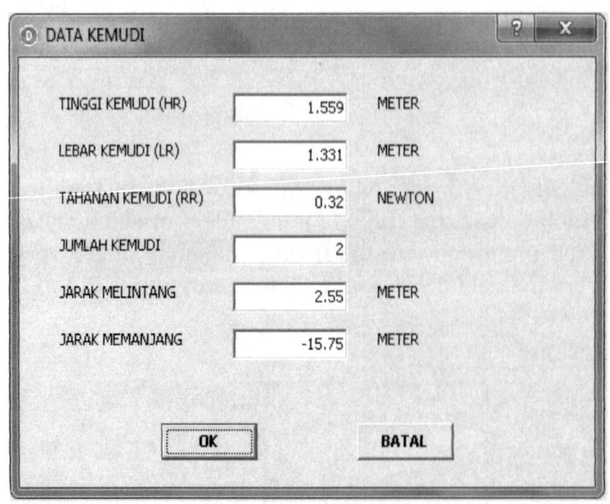

Figure 4. Display form of input data of propulsion and ship resistance

- Furthermore, the input data of the steering of the boat

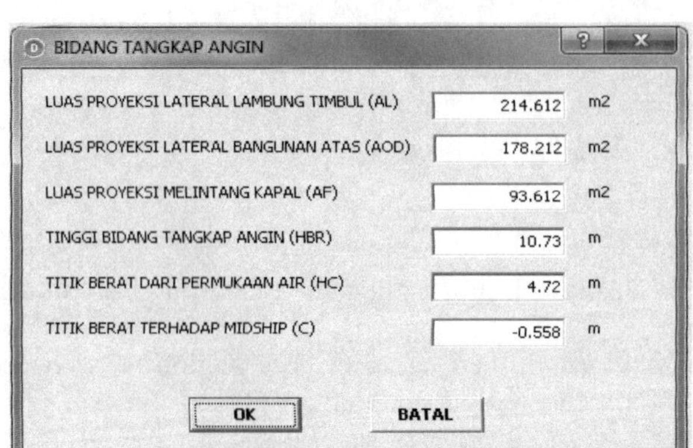

Figure 5. Display form of input data of steering of the boat

- Input field catch of the wind

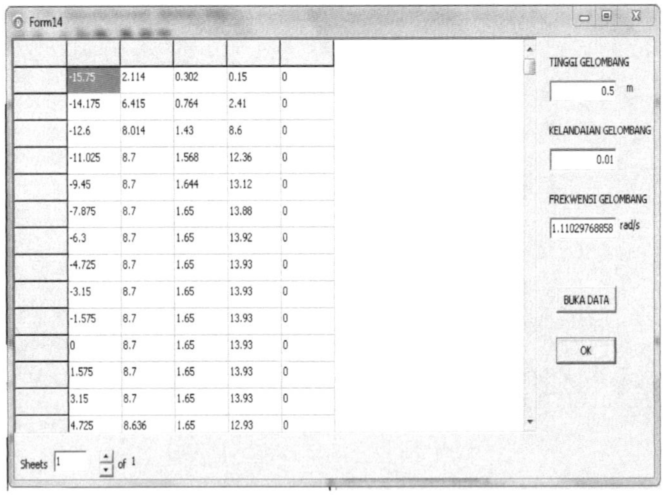

Figure 6. Display form of input data of capture wind field

- To get the predictions of the hydrodynamic coefficients, the used formulas are according to Yoshimura and Ning Ma which are conducted by a semiempirical method. At Delphi software, open window of hull hydrodynamic coefficients, press okay then the Fig. 7 should be filled.

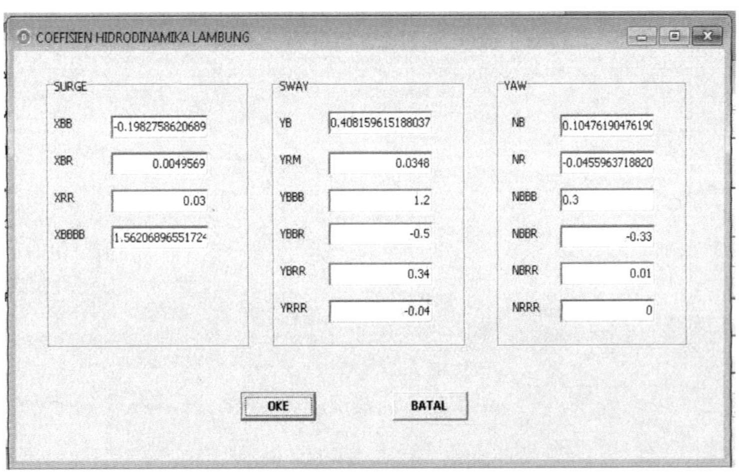

Figure 7. Display form of calculation results of hydrodynamic forces and moment of hull coefficients

- The result of the calculation of force and moment coefficients in the direction of the wind surge, sway, and yaw as a function of the direction the wind comes against the hull are given in the form of a curve as shown in Figure 8.

Figure 8 shows the curve of the relationship between the direction of the incoming wind with the wind force and moment coefficients. The coefficients in question are the coefficient of wind forces and moments on the motion surge (CAX), coefficient of wind forces and moments on the motion sway (GEO) and the coefficient of wind forces and moments on the motion of yaw (CAN). They were calculated using the equation of Fujiwara with a higher degree of accuracy than previous methods. Using physical components, the flow resistance longitudinal, cross flow resistance, lift, and induced obstacles.

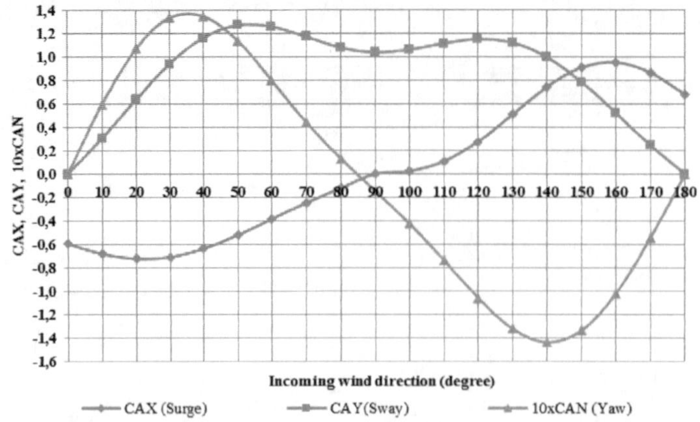

Figure 8. Relationship of incoming wind direction with wind force and moment coefficients on the motion of surge, sway, and yaw.

4.2. Wave effects against drift angle

Based on the purpose of this study which is to know the effect of the waves against the ship maneuvering characteristics and get the value of drift angle and rudder angle which affect the speed of the ship. To obtain these results, wind speed and wave height are varied and simulations using the Delphi program are carried out. Wave height (Hw) used were 0.5 m, 1 m, 1.5 m, and 2 m. The wind speeds used are 0 m/s and 10 m/s while the direction of the incoming wind together with the wave direction ranges from 0°-180° with 10° intervals.

The values of drift angles are obtained by simulation with the results shown in Figure 9.

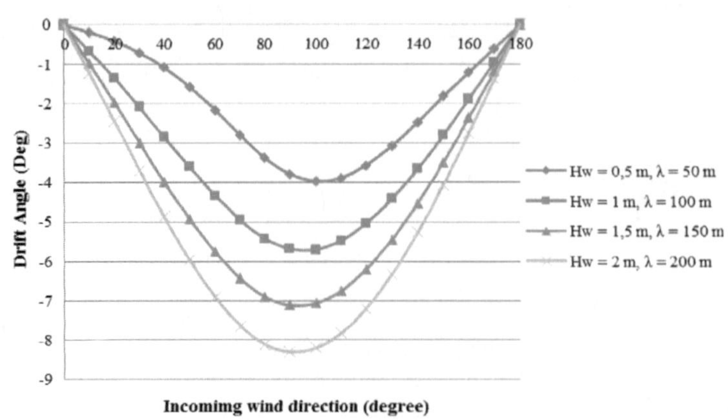

Figure 9. Drift angle at a various incident angle of wind for a wind speed of 0 m/s

Figure 9 compares the magnitude of the drift angle and incoming wind angle ranging from 0 - 180° for a wide variety of wave height and wind speed of 0 m/s. The blue curve illustrates the wave height (Hw) of 0.5 m and the flatness of 0.01 wave.

Drift angle changes because of the influence of the surge, sway, and yaw moment can be seen in Figs. 10-12.

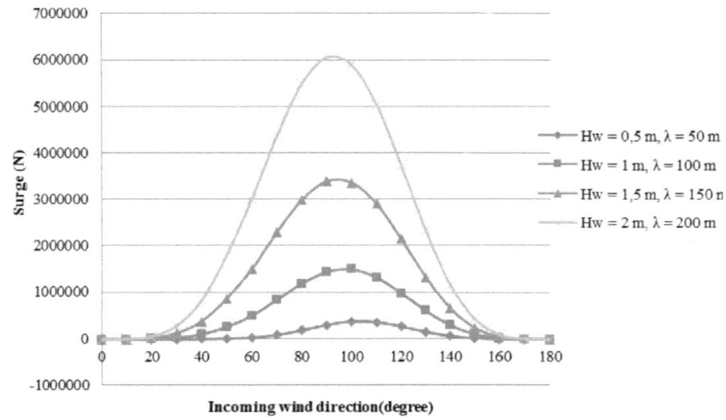

Figure 10. Surge force at various angles of wind for a wind speed of 0 m/s

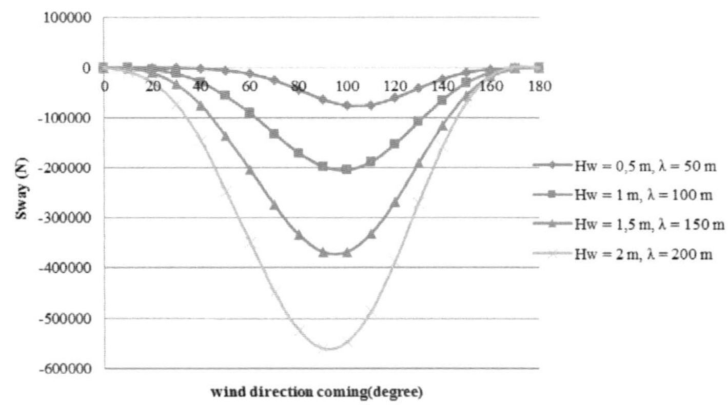

Figure 11. Sway force at various angles of wind for a wind speed of 0 m/s

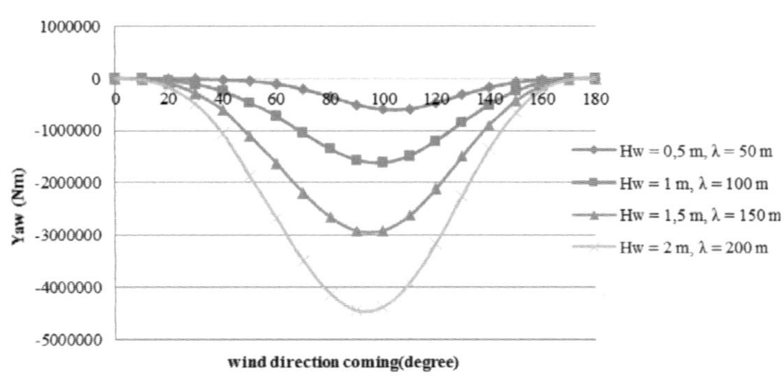

Figure 12. Yaw moment at various angles of wind for a wind speed of 0 m/s

Figures 10-12 indicates that the value of the surge, sway and yaw moments which have influence to change the drift angle on the ship. Sway force and yaw moments have an enormous influence on the change of angle of drift that consists of several elements of the hull, steering, wind, and waves. The greater the value of the sway force and yaw moment of the magnitude of drift angle deviation is also getting larger. This can be seen in a value of sway force and yaw moment when the wave heights are 0.5 m and 1 m primarily at an incidence wave angle of 100º.

IOP Conf. Series: Materials Science and Engineering **676** (2019) 012023 doi:10.1088/1757-899X/676/1/012023

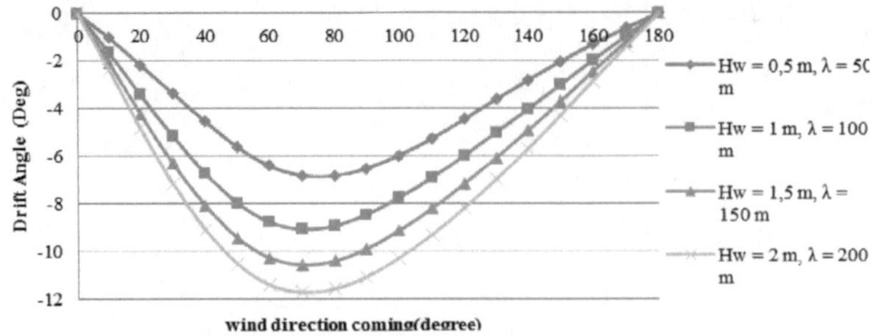

Figure 13. Drift angle at various incident wins angles for a wind speed of 10 m/s

Figure 13 shows that the effect of high waves against the ship drift angles is greater. The higher the waves, the vessel drift angle will be even greater. At the time of wave height (Hw) of 0.5 m and at an 80° angle of incidence, wind drift angle is -6,803°. For the wave height of 1 m, the drift angle when the angle of incidence of wind is 70°, is -9,059°. Drift angle changes because of the influence of the surge, sway, and yaw moment when the wind speed is 10 m/s, can be seen in Figs. 14 – 16.

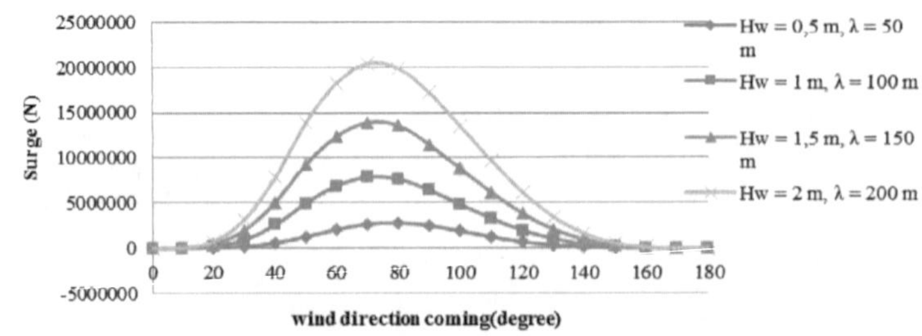

Figure 14. Surge force at various incident wind angles for a wind speed of 10 m/s

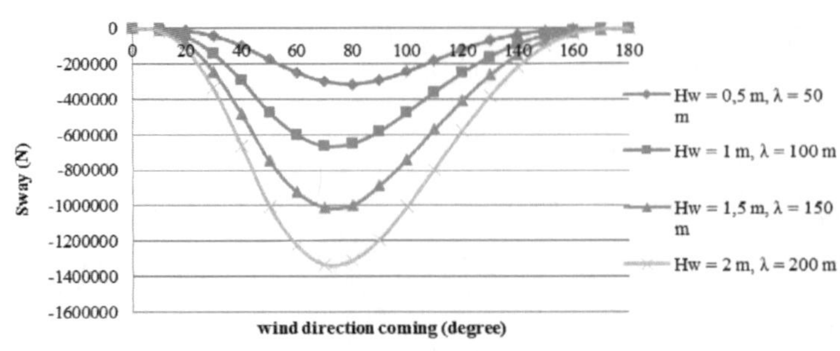

Figure 15. Sway force at various wind angles for a wind speed of 10 m/s

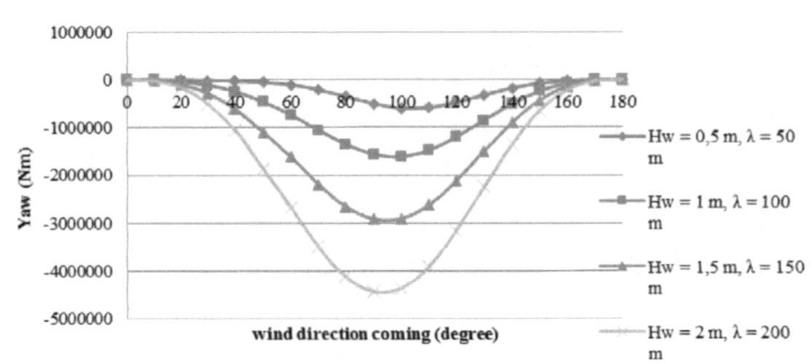

Figure 16. Yaw moment at various wind angles for the wind speed of 10 m/s

Figures 14-16 indicates that the value of the sway force and yaw moment of the direction of the wave direction, where the force of sway and yaw moment consists of elements (hull, propeller, rudder, wind, and waves). Sway force and yaw moments affect the drift angle is evident from the above picture is good value sway force and moments yaw, the greater the incidence angle of the wind against the ship, then the value of the sway force and moments yaw getting bigger until the angle of incidence of wind equal to 800 for the high waves of 0.5 m. In the incident angle is greater than 900 wind and wave height of 1 m, 1.5 m, and 2 m, the value sway force and yaw moment is getting smaller with increasing incidence angle of the wind against the ship. When the wind incidence angle equal to 180°, the value sway force and yaw moment is equal to 0. And the value of sway force and yaw moments occur in the largest wave direction 80° to wave height of 0.5 m. While the greatest value sway force and yaw moment to wave height of 1 m, 1.5 m, and 2 m occurs when the incident angle of 70° wind. So it can be concluded that the value of drift angle and value sway force and yaw moment proportional. The greater the value of the sway force and yaw moments then drift angle will also increase and vice versa the smaller the value sway force and yaw moments then drift angle will also be smaller.

4.3. Effect of waves against rudder angle

The amount of rudder angle obtained by simulation performed in the Delphi program. As a result of the simulation rudder angle when the wind speed of 0 m/s is shown in Figure 17.

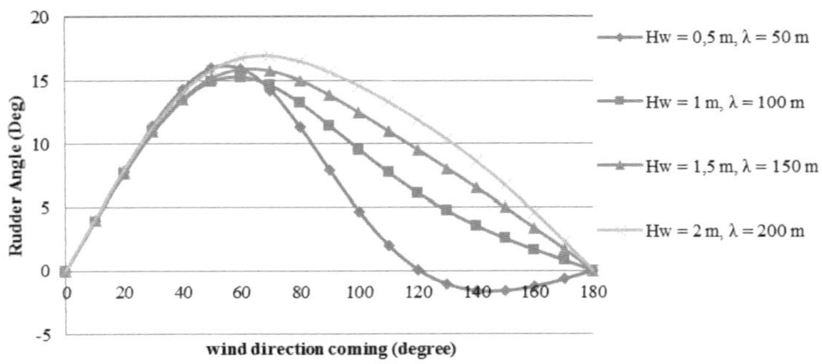

Figure 17. Rudder angle at different angles to the incoming wind speed of 0 m/s

To maintain the trajectory of the ship due to the drift angle that occurs due to the waves, the wheel should be tilted up to certain angle so that the movement of the ship remains on the specified track or path. Figure 17 shows the comparison of the amount of rudder angle and wind incidence angle ranging from 0 - 180° degrees for varying wave height of 0.5 m, 1 m, 1.5 m, and 2 m and at a constant wind speed which is equal to 0 m/s. The blue curve illustrates the wave height of 0.5 m, then red 1 m, green and purple 1.5 m and 2 m, respectively and the wave flatness of 0.01. From Fig. 17, it can be seen that the larger the incident angle of the wind, the greater the value of rudder angle ranging from 0-60°. For the incoming wind direction of 70-180°, rudder angle values tend to be smaller. For the wave height of 1 m, 1.5 m and 2 m, the direction of the rudder and the direction the incoming wind of 0-180° have an opposite direction to the incoming wind.

From Figure 18, it can be seen that the larger the incident angle of the wind, the greater the value of rudder angle ranging from 0-60° for a wave height of 0.5 m. For the wave direction of 70-180°, rudder angle values tend to be smaller. For the wave height of 1 m, 1.5 m, and 2 m, rudder angle direction coming from the direction of wind 0-80° has been larger. Meanwhile, when the direction of the wind coming from angle 90-180°, rudder angle direction gets smaller.

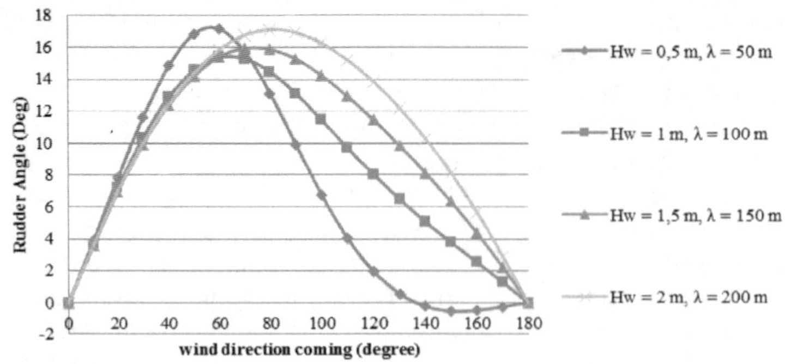

Figure 18. Rudder angle at various angles to the incoming wind speed of 10 m/s

4.4. Effect of waves against ships free

The speed and wave direction have a very significant influence on the operational speed of the ship.

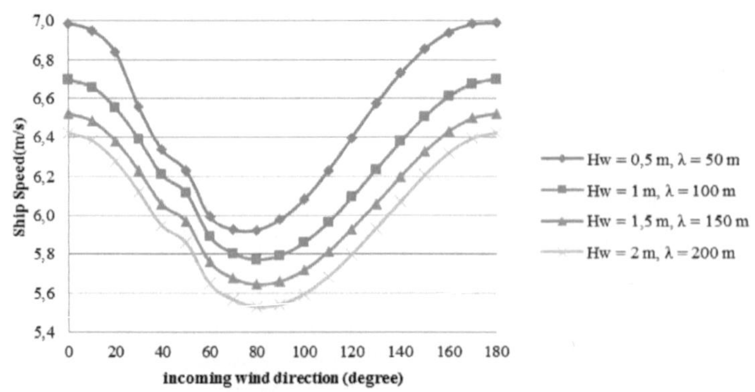

Figure 19. Vessel speed at various incoming wind angles for the wind speed of 0 m/s

Figure 19 illustrates that on the wind direction comes from angle 0 - 80°, the greater the incidence angle of the wind, the speed of the ship will be smaller. For the incoming wind direction greater than 80°, the speed of the ship was also greater with increasing incidence angle of the wind. When the wind speed ratio and the speed of the ship are equal to 0, the ship's speed increases from the with 5.4 m/s. When there is a change in the drift angle, the speed of the vessel decreased where the gradient of the decline getting smaller. The speed reduction occurs because when the ship began to turn, drift angle and vessel speeds increase, so the velocity of the ship in the direction of sway also increases.

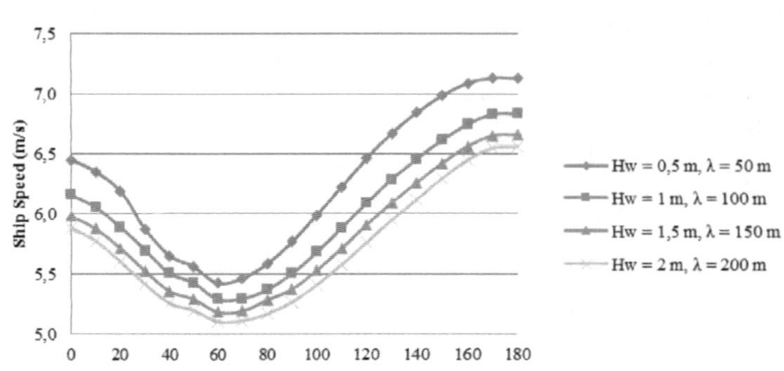

Figure 20. Vessel speed at various incoming wind angles for the wind speed of 10 m/s

IOP Conf. Series: Materials Science and Engineering **676** (2019) 012023 doi:10.1088/1757-899X/676/1/012023

Figure 20 depicts the wind speed ratio and the speed of the vessel which result in the drift angle and yaw rate is never stable. Changes of drift angle and yaw speeds cause the vessel speed will always vary in each direction of the incoming wind direction towards the ship. When the wind direction is less than 60° or in other words the ship operates on the bow wave, ship resistance increases due to surges that can cause the ship speed is significantly reduced. Meanwhile, when the incoming wind direction is greater than 60°, the ship's speed greater with increasing incidence angle of the wind on the ship. Ship resistance reduces due to the diminishing waves so that the ship's speed increases with increasing angle of incidence of the wind.

Factors that affect the speed of the ship are surge and sway forces arising from the hull, propeller, rudder, wind, and waves. The value of the surge and sway forces is inversely proportional to the speed of the ship. In this case, if the value of surge and sway forces are large, the speed of the ship will be smaller, and if the value of surge and sway forces are small, the speed of the ship will be even greater. The values which have a great influence on the value of the surge force are the values of the forces caused by the hull, while the forces and moments due to the propeller, rudder, wind, and waves.

5. Conclusions

From analysis results, it can be concluded that:

- Wave height and length wave gives the influence to characteristics of maneuvering of ferry ship. Large length and height of wave result in the small value of surge. On the contrary, the motion of sway and yaw are bigger because of the larger wave length.
- The largest change of drift happened on the incoming wave direction of 50° and high of wave 2 m. Angle corner happened on the incoming direction of the wave of 110° and wave height of 2 m, where angle corner gyrates 22.382° or more.

References

[1] Bhattacharya, Rameswar. 1978. Dynamic of Marine Vehicles. Jhon Wiley & Sons. New York.

[2] Djatmiko, E. B., and Murdijanto. 1993. Seakeaping: Behavior of Floating Structures in Waves. Ocean Engineering Faculty, ITS. Seakeeping dan Hidromechanics Short Course, Surabaya. (in Bahasa)

[3] Loyd, A. R. J. 1989. Seakeaping Ship Behavior in Rough Water. Ellis Hordword Ltd. New York.

[4] Makmur, Irfan. 2011. Analysis of Heaving Motion of Ferry Ships at Sabalana Water Area, Selayar. Departemen of Naval Architect, Faculty of Engineering, Hasanuddin University. Makassar. (in Bahasa)

[5] Manik Parlindungan 2007. Analysis of Ship Motion and Seakeaping on the Regular Wave. Naval Architect Departemen. Diponegoro Univ. Semarang.

[6] Pratama, Laode Abdul Haslan. 2013. Numerical Simulation of Heaving dan Pitching of Ferry Ships by Strip teory. Departemen of Naval Architecture, Faculty of Engineering, Hasanuddin University, Makassar. (in Bahasa)

[7] Sitepu, Ganding, 2001. The motion of ship. Department of Naval Architecture, Faculty of Engineering, Hasanuddin University. Makassar.

[8] Sulhan, Harjuni. Analysis on motion Crossing Ferry Vessel 600 GT, Department of Naval Architecture, Faculty of Engineering, Hasanuddin University. Makassar.

Feasibility Study of Biogas from Banana Peel Waste and Livestock Manure Mixture as Renewable Energy Source

I R Rahim*[1], A T Lando[1], K Sari[1], E Asriyanti[1], M Ihsan[2]

[1]Department of Environmental Engineering, Engineering Faculty, Hasanuddin University, Makassar, Indonesia
[2]Department of Civil Engineering, Sekolah Tinggi Teknik Baramuli, Pinrang, Indonesia

*E-mail: irwanrr@yahoo.co.id

Abstract. Biogas energy can be obtained from waste of domestic, sewage, cattle, market, food industry etc. Banana peels contain chemical compounds such as cellulose, hemicellulose which contain acidic organic compounds that can be used as substrate for biogas production. Purposes of this study are to analyse the characteristics, the composition and the production cost of biogas made of mixtures of banana peel and livestock manure. The study used 6 biogas reactors where each reactor has 19 L capacity with ratio of 1:2.5 for solid and wastewater. The result shows that reactor E2, containing 7.1% of cow and chicken manure and 14.3% of banana peel, produced the highest rate of biogas (91.4%) when compared to other reactors. E4 reactor produced 209 ml of biogas as effective composition, consisting of 2.4% of cow manure, 7.15% of chicken manure, and 19.0% of banana peel. In terms of production cost, E2 reactors have the most effective composition, with the ability to produce 0,000174 m3 of methane by Rp. 5,747 per kg. Biogas made of organic material and livestock manure is relatively less expensive compared to 3 kg of LPG gas with a price of Rp. 18,000. E4 reactor biogas could produce 209 ml with a composition of 357 g of cow manure, 1,071 g of chicken manure, 2850 g of banana peel, 5.355 l of water for livestock manure and 5.355 l of water for banana peel.

1. Introduction

According to Indonesian's Ministry of Energy and Mineral Resources [1], fossil energy in the form of oil reserves in Indonesia is estimated to run out in the next 12 years. Data of Indonesia's oil reserves shows that oil stock currently remains 3.65 billion barrels, with a production level of 800 thousand barrels per day [2]. In order to reduce dependence on fossil fuel, The Government of Indonesia has issued national energy policy to developed alternative source of energy in (Indonesian Government regulation No. 5, 2006) [3]. Renewable and environmentally friendly energy is therefore becoming the main option to overcome crisis of fuel where one of the sources is biogas. Due to its combustibility, this type of energy has the potentials to replace gasoline, kerosene, and diesel oil. The good point is, biogas energy can be obtained from domestic waste, manure of livestock, as well as organic waste from fresh markets, food industries and other sources [4].

Several previous studies have reported experimental results from the mixtures of cattle manure and banana peels. Banana peels contain chemical compounds such as cellulose and hemicellulose. These compounds consist of acidic organic compounds that can be used as substrates for biogas production. Labiba et al reported that a mixture of cow dung with banana peel could produce biogas of 3.6 L [5]. Yusiati et al found additional cattle manure in banana peel waste influenced concentration of methane gas production [6]. Bahite et al reported that volume of biogas production was higher with the addition of livestock manure [7].

IOP Conf. Series: Materials Science and Engineering **676** (2019) 012024 doi:10.1088/1757-899X/676/1/012024

2. Equipment and Materials

2.1. Equipment

This study used four (4) 19L reactor and 300 ml drinking plastic bottle as shown in Figure 1 and Figure 2. Beside the equipment, present study used also aquarium hose, rubber tire, balloons, plastic jars, measuring cup, pH meter, digital thermometer, modelling clay, rubber band, half-meter nets, and compressor faucet to support main equipment.

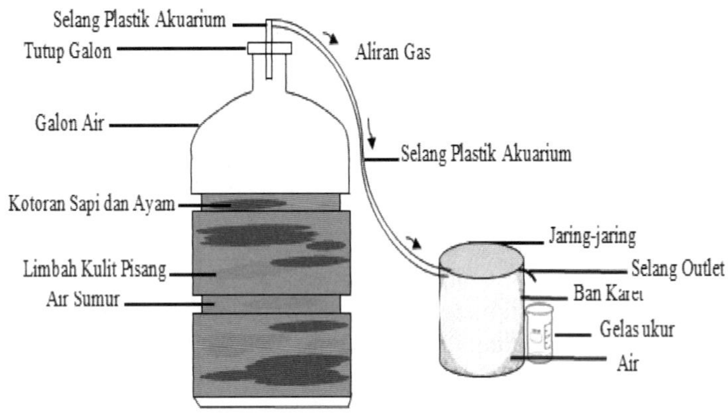

Figure 1. Sketch of 19 L - Biogas reactor

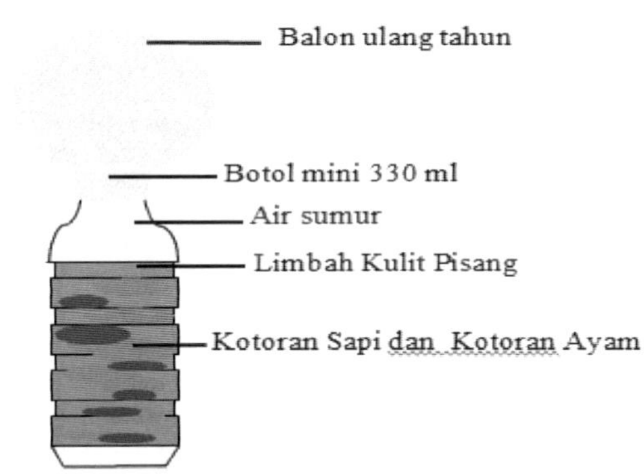

Figure 2. Sketch of 330 ml-plastic bottle as Biogas reactor Biogas

2.2. Material used

This research was conducted at Sanitation and Solid Waste Laboratory of the Faculty of Engineering, Hasanuddin University, Indonesia. Banana peel was collecting from banana sellers while livestock waste was taken from animal husbandry in Pangkep area. Composition of livestock manure and organic material to water was 1:2.5, which is also described in Table 1.

3. Laboratory Testing

3.1. Preparation, Mixing and Fermentation

After biogas reactor preparation completed, mixture material of livestock manure and banana peel is prepared based on designed composition by watered in separated containers and properly stirred as designed before mixed in larger media. Final mixture was placed into 330L-plastic bottles and 19L-reactors, while ensuring the reactors and bottle is properly confined and left to be fermented during observation. Values of pH, water content, solid total, temperature and COD of the waste mixture were recorded every day for 35 days.

Tabel 1. Material composition

Number of Sample	Type of Material				
	Cow manure (gr)	Chicken manure (gr)	Banana peel (gr)	Water for waste (L)	Water for banana peel (L)
E1	1608	536	2145	4.290	6.435
E2	1073	1073	2145	4.290	6.435
E3	1073	0	3216	4,.290	6.435
E4	357	1071	2850	5.355	5.355
E5	714	714	2850	5.355	5.355
E6	0	714	3564	5.355	5.355

3.2. Initial composition design

Several steps were conducted to find the effective composition of biogas mixture.

- Literature review on effective volume biogas produced from each material in previous studies. There are several sizes of reactors used in biogas production which are 2.8L, 4L, and 14L for P1, P2 and P3 respectively. Volume of biogas production from each study is divided by 75% of reactor volume, as shown in Table 2. According to Nurdimansyah [8], biogas processed in reactor works effectively when substrate level is 75% of reactor volume.
- Define the percentage of composition for biogas based on previous studies. The percentage was then calculated by the amount of composition and divided by total composition input and then multiplied by one hundred. Calculation result is presented in Table 3.

Table 2. Volume biogas from previous studies

No. research	Time (day)							
	7	10	14	20	21	28	30	35
	Volume of Biogas (ml)							
P1-1 (Cow manure)*	-	2	-	7	-	-	10	-
P1-2* (Horse manure)	-	3	-	4	-	-	43	-
P2-1 (A)**	3050	-	3425	-	3600	3750	-	3725
P2-2 (B)**	3450	-	3550	-	3775	3800	-	4100
P2-3(C)**	2900	-	3000	-	3025	3075	-	3275
P2-4 (D)**	3100	-	3475	-	3575	3750	-	3675
P2-5 (E)**	3450	-	3550	-	3775	3825	-	4075
P3-1 (A)***	2192	-	-	-	-	-	-	-
P3-2 (B)***	998	1689	3932	-	-	-	-	-

*Sources : *Yusiati et.al., [6] **Bahite et.al. [7],***Labiba, [5]*

Note :

*P1 = Effect of Addition of Banana peel Waste on Gas Production of Methane in Methanogenic Fermented Livestock Manure.

**P2 = Mixture of laying hens and banana peels.

***P3 = Banana Skin Mixture and Cow Manure Using Anaerobic Bioreactors.

Table 3. Mixture composition of previous studies

Material	Composition of previous studies (%)								
	P1-1*	P1-2*	P1-3*	P2-1**	P2-2**	P2-3**	P2-4**	P2-5**	P3-2***
Cow manure	11	9	8	0	0	0	0	0	14
Horse manure	11	9	8	0	0	0	0	0	0
Chicken manure	0	0		100	0	50	67	33	0
Banana Peel	0	3	7	0	100	50	33	67	14
Waste water	33	33	33	0	0	0	0	0	29
Banana peel mixed with water	44	44	44	0	0	0	0	0	43

*Sources : *Yusiati et.al., [6] **Bahite et.al. [7],***Labiba, [5]*

- In Figure 3 it can be seen that the result of P2-5 and P3-2 were the most effective of results on previous study. Based on these results, the present study is following their initial composition of the abovementioned study

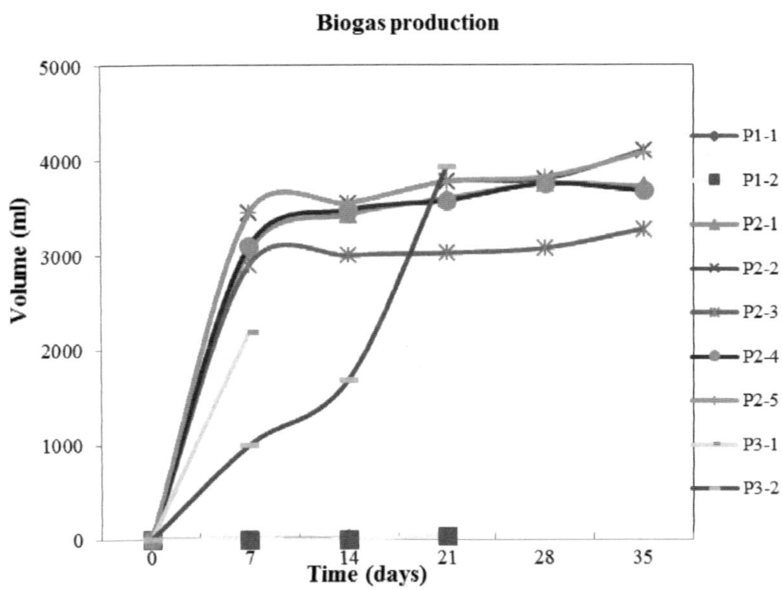

Figure 3. Biogas production of several studies

- The next step is to calculate the density of banana peel waste and cattle manure. As mention earlier, the most effective reactor volume was in P2-5 dan P3-2, therefore, by referring to the volume, this study applied 1:2.5 ratio of solid waste to water.

Table 4. Initial composition

Sample	Weight		%	Density (gr/cm^3)	Vol. (cm^3)
Cow manure	0	gr	0	1,07	0
Horse manure	0	gr	0	1,07	0
Chicken manure	714	gr	4,76	0,91	785
Banana Peel	3564	gr	23,8	0,48	7425
Sewage	5355	gr	35.7	1,00	5355

Sample	Weight		%	Density (gr/cm^3)	Vol. (cm^3)
M	5355	gr	35.7	1,00	5355
TOTAL	14988		100		18920
18,92	Litre				
Waste to sewage	1 :	7,5			14,3 lt
Banana peel to sewage	1 :	1,5			19,0
Solid waste to sewage	1 :	2,5			75%

- Final mixture composition is plotted in Table 5.

Table 5. Determination of initial composition

Material	Initial composition (%)					
	E1	E2	E3	E4	E5	E6
Cow manure	11	7	7	2	5	0
Chicken manure	4	7	0	7	5	5
Banana peel	14	14	21	19	19	24
Wastewater	29	29	29	36	36	36
Wastewater and banana peel	43	43	43	36	36	36

4. Result and Discussions

Data collecting was conducted by direct measurement comprising of volume of biogas, temperature of COD reactor, total solid (TS), water content, initial and final pH of mixture, as presented in Table 6.

Table 6. Parameter of research

Parameter	Time of sampling	Analysis method	Authors
pH	O	indicator of pH	-
Temperature	Everyday	Digital Thermometer	-
Content of Biogas	15th day	*GCMS* type *Gas Chromatography*	Lab. of Chem. Eng. UII
Volume of Biogas	Everyday	Archimedes law	-
Water content	Once in 7 days	Gravimetry	Abdullah, [9]
Total Solid	Once in 7 days	Gravimetry	Abdullah, [9]
COD	Once in 7 days	Closed reflux by titrimetry	SNI 6989.73:2009

4.1. Analysis of density and water content

Density measurement on investigated material as preliminary research, conducted in sanitation and waste laboratory in Engineering Faculty of Hasanuddin University, is presented in Table 7.

Table 7. Density of material

Materials	Average (gr/cm^3)
Banana Peel	0,5
Chicken manure	0,9
Cow manure	1,0

Density is a measurement of mass per unit volume. Accounting the special characteristics of material by density, the parameter can be used to determine substance type. The density is a very important property of material to determine the amount of composition to be incorporated into the reactor.

Table 8. Analysis of water content

No. Sample	Water content %	Average %
15 (K.P)	62	
15 (K.P)	61	61
15 (K.P)	61	
10 (K.S)	72	
10 (K.S)	71	72
10 (K.S)	72	
14 (K.A)	36	
14 (K.A)	35	35
14 (K.A)	34	

While water can increase the degradation of organic material, its excessive portions can inhibit the activity of methanogen bacteria, since addition of water will increase the concentration of oxygen which is toxic to anaerobic bacteria. Conversely, disproportionally low water content will prevent acetic acid accumulation [10].

4.2. Temperature

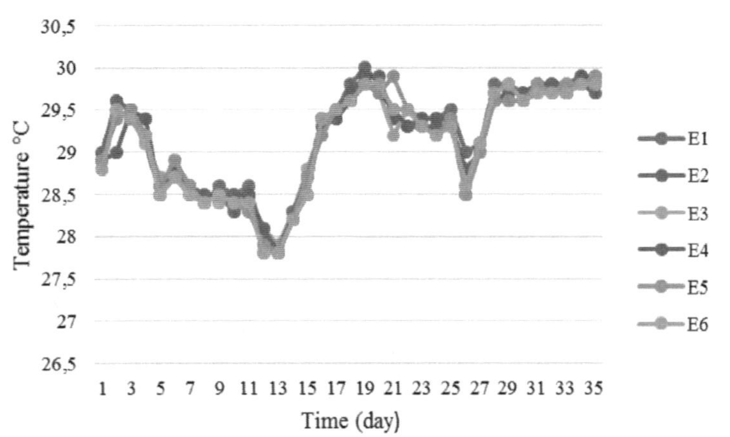

Figure 4. Temperature

Temperature in reactors is ranging between 27° to 30 °C. Recent study was conducted in mesophilic conditions where temperature is ranging between 25° to 30 °C, despite its fluctuation following changes in environmental condition due to adaptation of microorganism [11]. According to Figure 4, temperature inside reactor tends to rise as seen in the day 19th observation although for day 5th to day 15th the temperature was very low, which is probably influenced by rainy weather at the time of observation.

4.3. Value of pH

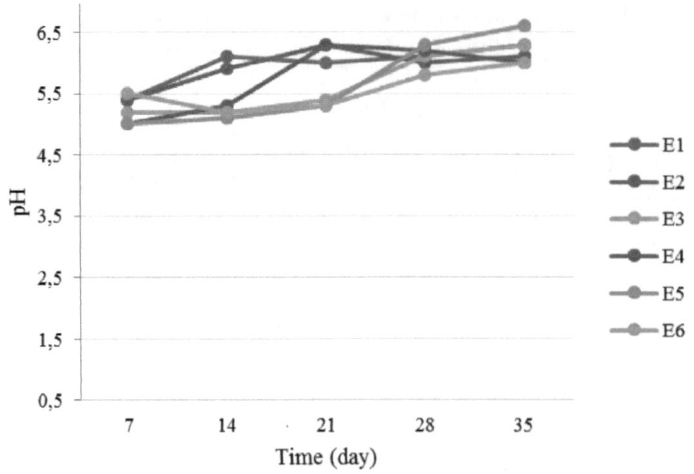

Figure 5. Measurement of pH

Results of pH analysis can be seen in Figure 5. At the beginning of fermentation process, pH was constantly decreasing due to the alteration of organic waste into organic acids by microorganisms. However, pH value increased again at the 5th week, ranging from 6 to 6.6. This is because organic acid products were used as a source of nutrition by microorganisms.

4.4. Water content

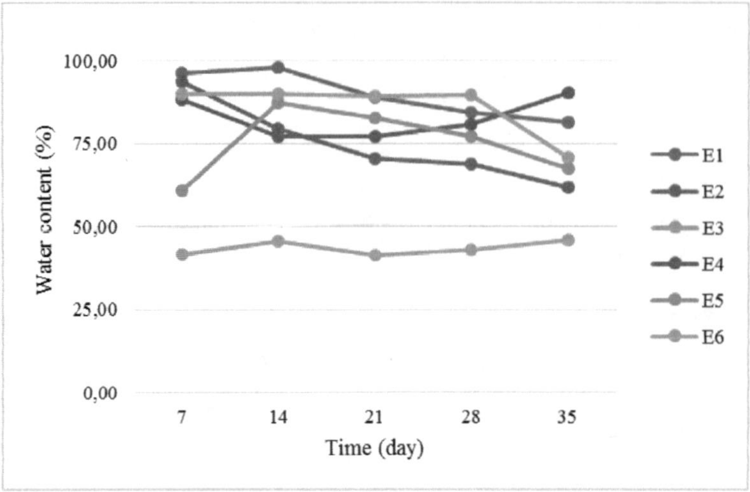

Figure 6. Determination of water content

As can be seen in Figure 6, water content of E3 reactor was decreasing by 42% in the 1st week. This is probably caused by the adaptation process of bacteria in E3 into new environment, causing slow degradation of organic materials. Water content was constantly decreasing to the range of 60% to 90% to day 5th. This shows that the bacteria contained in each reactor have undergone an adaptation process, leading faster degradation of organic materials [12].

4.5. Total Solid

Figure 7 shows differences of mean value of total solid reduction. While hydrolysis converted into water soluble compound, it was then used in acidogenesis reaction, causing the total dissolved solids to fell down. Data of E3 indicated a low decrease in TS because biomass contained not only lignin, but also other elements such as carbohydrates and cellulose [13].

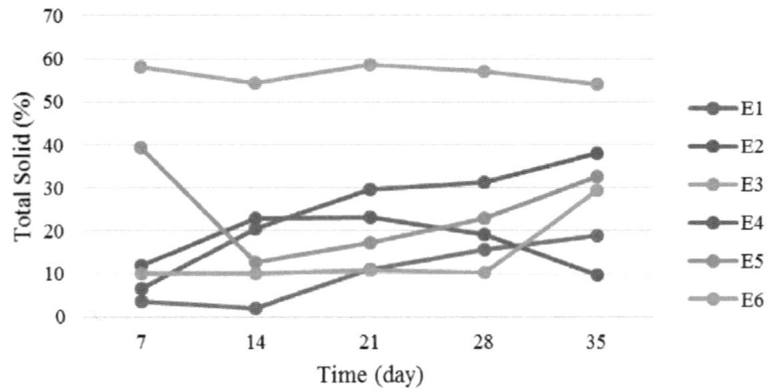

Figure 7. Total Solid measurements results

4.6. Value of COD

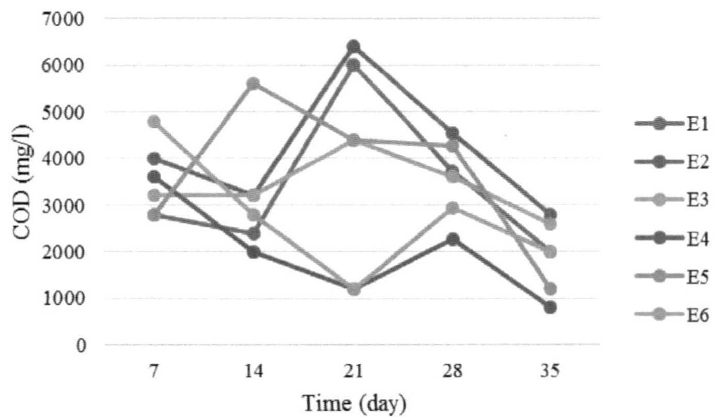

Figure 8. Measurement of COD

It can be seen in Figure 8 that COD was fluctuating, which is probably due to unstable COD formation process which influenced by several factors. When the value of COD decreases, there is a hydrolysis process and when the value of COD increases, the substrate decomposition occurs. The low efficiency of COD reduction may be due to high content of organic substances, which indicates dominant waste contains complex organic compounds with high levels of pollution. The second possibility was some chemical material was taken during collection of sewage and organic waste for solid sludge mixture which have not been decomposed and therefore affecting the value of COD concentration.

4.7. Volume of biogas

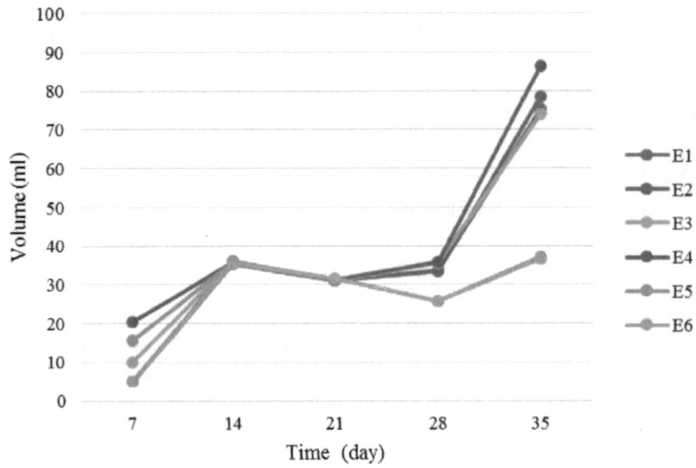

Figure 9. Measurement of biogas volume

Volume of biogas was measured every day for 35 days duration as depicted in Figure 9. Gas production was stored in balloons which were contained in bucket filled with water. The balloon was pushed against water in the bucket using nets then the water will come out from bucket as the measurement of the volume of gas production. Result show the highest volume and the lowest were 209.4 ml of E4 reactor and 134.5 ml of E6 reactor, respectively.

4.8. Biogas content

On the 15th day of observation, samples stored in gas-filled rubber tires were sent to Chemical Engineering Laboratory of the Islamic University of Indonesia, in Yogyakarta to be tested for biogas content using GCMS type gas chromatography. The results can be seen in Table 9.

Table 9. Biogas content

Samples	**Biogas content**	**Percent (%)**
E1	O_2	15,14
	$C_2H_6N_2O$	29,88
	C_4H_6O	29,45
	$C_{12}H_{10}N_4O_3$	25,54
E2	$C_2H_6N_2O$	91,73
	$C_{16}H_{22}O_2$	5,02
	$C_{12}H_{10}N_4O_3$	3,24
E3	$C_8H_8O_4$	2,63
	$C_2H_6N_2O$	90,84
	$C_{16}H_{22}O_2$	6,5
E4	$C_{16}H_{22}O_2$	49,85
	$C_2H_6N_2O$	50,15
E5	$C_{16}H_{22}O_2$	48,78
	$C_2H_6N_2O$	51,22
E6	$C_2H_6N_2O$	51,43
	$C_8H_8O_4$	48,57

Source: Result from laboratory in Chemical Engineering University of Islamic Indonesia, Yogyakarta

Furthermore, there were two components detected in the samples which were Oxygen (O_2) and Acetyl hydrazide (C2H6N2O). Oxygen (O_2), a chemical element, has no odour, colour, taste, and not flammable but it can help combustion by oxidizing. Acetyl hydrazide or acetyl hydrazine (C2H6N2O) is a compound of aromatic hydrazides or hydrazine. Hydrazine can be used as gas fuel due to its flammability, and it can be easily obtained from synthesis. This compound is a flammable, colorless liquid with a smell like ammonia. In addition, the hydrazine reaction produces a large amount of heat energy compared to the reaction of propane (LPG gas material in general) with oxygen because the hydrazine property is a good reductor agent.

4.9. Value of pH

Initial and final pH values were plotted in Table 10.

Table 10. pH

Sample	Initial pH	Final pH
E1	6	6,3
E2	6	6,1
E3	6	6,3
E4	6	6
E5	6	6,6
E6	6	6

Source: Measurement result in Environmental laboratory of Environmental engineering UNHAS

Reduction of initial pH value indicates the action of microorganisms due to process of acid formation or acidogenesis, while in the stage of methane formation (methanogenesis), bacteria which not produce methane and not sensitive to the changing of pH can work at a pH between 5 and 8.5. Based on this fact, the best composition was in E2 reactor, comprising 7% cow manure, 7% chicken manure and 14% banana peel.

4.10. Water content

Table 11. Condition of water content

Sample	Initial water content (%)	Final water content (%)
E1	96,3	81,2
E2	94,2	61,9
E3	46,5	45,8
E4	89,9	90,2
E5	65,0	67,4
E6	91,2	70,6

Source: Measurement result in Environmental laboratory of Environmental engineering UNHAS

Result analysis indicated water content was high at the initial stage but low at the final process as seen in Table 11. This is because at the initial stage it was still in the hydrolysis stage. Normal activity of methane microbes requires about 90% water. The high moisture content will facilitate the decomposition process, while the low water content results in the accumulation of acetic acid which inhibits the growth of methanogenic bacteria. Based on this explanation, composition of E2 reactor, 7% cow manure, 7% chicken manure, 14% banana skin is the closest composition.

4.11. Value of total solid

Decrease in evaporating solids occurs because of rapid formation of acid. A few days before the end of the fermentation process, condition would be stable and lead to increase of the rate of cumulative gas production. At the state of anaerobic fermentation, total solid values would decrease between 3.1-3.5% during biogas production process.

Table 12. Condition of total solid

Sample	Initial total solid (%)	Final total solid (%)
E1	3,68	18,8
E2	5,76	38,1
E3	53,53	54,19
E4	10,17	9,77
E5	34,97	32,62
E6	8,97	29,41

Source: Measurement result in Environmental laboratory of Environmental engineering UNHAS

Inconsistent decrease of total solid affects gas production. Result of total solid value ranging from 13% to 57% with fluctuation of percentage decrease as depicted in Table 12. Based on this explanation, the closest mixture composition was 7% cow manure, 7% chicken manure and 14% banana peel of E2 reactor.

4.12. Value of COD

Table 13. COD conditions

Sample	Initial COD (ppm)	Final COD (ppm)
E1	4400	2000
E2	5600	2800
E3	8000	2600
E4	4000	800
E5	3200	1200
E6	8000	2000

Based on observation results of COD value (Table 13), it appeared that the value rose in the beginning and diminished at the end. There was an indication of substrate breakdown by bacteria or other microorganisms during reducing of value. On the contrary, increasing of COD is due to presence of simple organic compounds which led to hydrolysis of organic polymers, but the compound was not further overhauled by bacteria into biogas. Based on this fact, the closest mixture composition was from E2 reactor with 7% of cow manure, 7% of chicken manure, and 14 % of banana peel.

4.13. Cost Analysis

The volume of produced methane (CH4) can be seen in Table 14.

Referring to Table 15, the usage of biogas as source of energy in the household is considerably more efficient compared to subsidized LPG or kerosene. It costs Rp. 5,747 to produce biogas while it needs Rp. 18,000 to have a 3 kg subsidized LPG. However, investment costs for installing biogas equipment can vary depending on the selected material and the volume of the reactor.

Table 14. CH4 content (% methane)

Sample	Organic material			Water		Measurement result		
	Cow manure (%)	Chicken manure (%)	Banana Peel (%)	Water mixed with livestock manure (L)	Sewage mixed with banana peel (L)	Volume (m^3)	CH_4 (% methane)	Volume of methane product (m^3)
E1	11	4	14	4	6	0,00018	29,5	0,000053
E2	7	7	14	4	6	0,00019	91,7	0,000174
E3	7	0	21	4	6	0,00018	90,8	0,000163
E4	2	7	19	5	5	0,00021	50,2	0,000105
E5	5	5	19	5	5	0,00015	51,2	0,000076
E6	0	5	24	5	5	0,00013	51,4	0,000067

Table 15. Consumption comparation between Biogas and LPG

Sample	Methane production (m3)	1 m^3 = 0,50 kg LPG	Cost of LPG 3 kg (Rp)	Cost of biogas (Rp)
E1	0,000053	0.50	18000	18868
E2	0,000174	0.50	18000	5747
E3	0,000163	0.50	18000	6135
E4	0,000105	0.50	18000	9524
E5	0,000076	0.50	18000	13158
E6	0,000067	0.50	18000	14925

5. Conclusion

Based on experimental results, current study can conclude some points as follows:

1) There are several characteristics which could be defined from biogas produced from a mixture of banana peel and livestock manure, such as, pH, water content, COD, and total solid (TS). Initial pH condition was 6, but it increased up to 6.1 after 35 days of fermentation in E2 reactor.
2) Water content of biogas at the beginning was 94.2% before decreasing to 61.9% after 35 days in E2 reactor. Total solid was 5.76% and 38.1% at the beginning and the end of observation, respectively. Initially, COD of mixture in E2 was 5600 ppm but reduced into 2800 ppm after 35 days. The highest biogas content was 91.4% (E2 reactor), composed of cow manure (7.1%), chicken manure (7.1%), and banana peel (14.3%).
3) The most effective composition for reforming biogas was mixture in E4 reactor. The mixture was composed of banana peel waste (19.0 %), cow manure (2.4%) and chicken manure (7.15%). This mixture could produce 209 ml of biogas. The lowest production was 135 ml which found in E6 reactor composed of 0% of cow manure, 4.76% of chicken manure, and 23.8% of banana peel waste.
4) Based on cost analysis, compared to other source of energy, mixture composition in E2 was more efficient since it costs Rp. 5,747 while LPG costs Rp. 18,000 for a 3 kg tube.
5) Compared to previous research, the highest volume of biogas produced in the P3-2 study was 3932 ml with a composition of 1000 grams of cow manure, 1000 gr of banana peels, 2000 ml of water for cow manure and 3000 ml of water for banana peels, while for research done to produce biogas volume of 209 ml in the E4 reactor with a composition of 357 grams of cow manure, 1071 grams of chicken manure, 2850 grams of banana peel waste, 5,355 L of water for livestock manure and 5,355 L of waste water for banana peel waste.

IOP Conf. Series: Materials Science and Engineering **676** (2019) 012024 doi:10.1088/1757-899X/676/1/012024

6. Recommendation

It is necessary to conduct further research on usage of waste material as biogas to find effective composition. It is better to check the C/N ratio of samples, since it can affect substances in samples. It is recommended to stir properly all material in reactor, because some raw materials are difficult dilute and lead to form clumping or deposition of organic matter and finally inhibit biogas formation.

References

[1] Directorate General of Oil And Gas, 2017. *Oil and Gas Statistic 2016*, Ministry of Energy and Mineral Resources, Jakarta.

[2] Setiawan rakhmat, 2017. *Cadangan energi indonesia menipis, saatnya melek energi terbarukan* (*Indonesian energy reserve running out, time to renewable energy literacy*) (https://www.kompasiana.com/cakmat/599aefc15af02c183e6ca1d2/cadangan-energi-indonesia-menipis-saatnya-melek-energi-terbarukan, accessed 4th February 2018.

[3] The Republic of Indonesia, 2006, *Indonesian Government Regulation No. 5, 2006 on National Energy Policy*, Jakarta.

[4] Azhari, F, 2015. Kualitas Biogas yang Dihasilkan dari Substrat Kotoran Sapi dan Penambahan Starter Buah-Buahan dengan Menggunakan Reaktor Kubah. (Biogas Quality Produced from Cow Substrates and Addition of Fruit Starters by Using a Dome Reactor) *Jurnal Wahana-Bio* Vol. 14 (2-2) pp. 69-77.

[5] Labiba, Q. and Wulandari, C., 2017. *Pembuatan Biogas dari Campuran Kulit Pisang dan Kotoran Sapi Menggunakan Bioreaktor Anaeobik.* (*Biogas production from banana peel and cow manure mixture using anaeobic bioreactor).* Institut Teknologi Sepuluh Nopember, Surabaya.

[6] Yusiati, L.M. et al, 2012. Pengaruh Penambahan Limbah Kulit Pisang (Musa sp) terhadap Produksi Gas Metan dalam Fermentasi Metanogenik Kotoran Ternak. (The Effect of Addition of Banana Skin Waste (Musa sp) on Methane Gas Production in Methanogenic Fermented Livestock Manure) *Buletin Peternakan* Vol. 36(2): 87-94.

[7] Bahite, R. et al, 2014. *Pengaruh Campuran Kotoran Ternak Ayam Petelur dan Kulit Pisang Kepok Terhadap Volume Biogas*. (*The influence of mixture of chicken manure and kepok banana peel on biogas volume*) Universitas Negeri Gorontalo, Gorontalo.

[8] Nurdimansyah, et al, 2015. Analisis Pengaruh Level Substrat Pada Digester Anaerob Skala Laboratorium terhadap Produksi Metana. (Analysis of the Effect of Substrate Level on the Anaerobic Digester on the Laboratory Scale of Methane Production) *E-Proceeding of Engineering* : Vol.2, No.2 p. 3260.

[9] Abdullah, N.O., 2017. *Peningkatan Produksi Biogas Sampah Pasar dengan Penambahan M-A6 dan Pengadukan Menggunakan Reaktor Anaerobik. (Increasing Biogas Production of Market Waste by Adding M-A6 and Stirring Using Anaerobic Reactors).* Institut Teknologi Sepuluh November. Surabaya.

[10] Lestari, D.I., 2016. *Efektivitas Rumput Laut Sargassum sp. sebagai Sumber Alternatif Penghasil Biogas. (Effectiveness of Sargassum sp. as an Alternative Source of Biogas Productio*n) Universitas Airlangga. Surabaya.

[11] Ramdiana, 2017. Pengaruh Variasi Komposisi pada Campuran Limbah Cair Aren dan Kotoran Sapi terhadap Produksi Biogas (Effect of Composition Variations on Palm Sugar Liquid Waste and Cow Manure on Biogas Production) *Eksergi*, Vol 14, No.2. ISSN : 1410-394X.

[12] Mariyani, S., 2016. Potensi Campuran Sampah Sayuran dan Kotoran Sapi sebagai Penghasil Biogas (Potential of Vegetable Waste and Cow Manure as Biogas Producer). Universitas Islam Negeri Maulana Malik Ibrahim, Malang.

[13] Saktiyudha, R.P., 2014. Produksi Biogas dari Pencernaan Anaerob Serasah dan Eceng Gondok (Eichornia Crassipes) dengan Sumber Inokulum Kotoran Sapi dan Kotoran Ayam. (Biogas Production from Litter and Hyacinth Digestive Anaerobic Digestion (Eichornia Crassipes) with Inoculum Source of Cow and Chicken Manure) *Bioteknologi* vol. 11 (2): pp. 23-27 EISSN: 2301-8658, DOI: 10.13057/biotek/c110201.

IOP Conf. Series: Materials Science and Engineering **676** (2019) 012025 doi:10.1088/1757-899X/676/1/012025

An Evaluation of Carrying Capacity of Jack-in Piles with Base Enlargement in Soft Clay

G Aponno[1]* and M Sholeh[1]

[1]Politeknik Negeri Malang, Jl. Soekarno-Hatta POBox 09 Malang, Jawa Timur 65141, Indonesia

E-mail: *gaponno@gmail.com

Abstract. The enhancement of the carrying capacity of the single pile can be achieved by improving the roughness of the pile shaft and/or enlarging the tip of the pile. This paper presents the results of a laboratory experiment of a modeled steel pipe pile with and without the enlargement of the pile base. The pile base was made separately from the pile shaft and was inserted using jacking pressures into an artificial and homogenous soft clay soils. The soil media used is classified as high plasticity clayey-silt is compacted in the concrete container with 600 mm in diameter and 1000 mm in depth. This study shows that the technique of pile tip enlargement works properly, and the carrying capacity of the pile is higher than a pile without base enlargement by increasing of up to ± 270%.

1. Introduction

The foundation of a structure is defined as an integrated part of the structure in direct contact with the ground and which transmits the load of the structure into the ground. There are two types of foundation namely, shallow foundation and deep foundation, where the shallow foundation is used when the soil near the ground surface has an adequate bearing capacity, and the working load is not too heavy. In some cases, shallow foundations are inadequate to support the structural loads, so deep foundations or pile foundations are required. Piles are columnar elements in a foundation which have the function of transferring load from the superstructure through weakly compressible strata or through water, onto stiffer or more compact and less compressible soils or onto a granular rock. The mechanism of bearing capacity of the pile foundation consisting of, the friction of the surrounding soils along the pile and the bearing capacity under the tip of the pile. To increase the carrying capacity of the single pile, various efforts have been done including increase the carrying capacity of the tip and or around the surface of the pile skin, by means of modifying the physical shape of the pile, for example, increase the pile roughness and or enlarge the dimensions of the pile tip cross-section.

There are several types of pile foundation with base enlargement, two of them are Franki Piles and Mac-Arthur Piles. Was developed in 1909 by Belgian Engineer Edgard Frankignoul, the Franki piling system (also called pressure-injected footing) is a method used to drive expanded base cast-in-situ concrete piles (Figure 1), where the installation method of this piling system is explained clearly in ref. [1]. This method can be applied to different site conditions and is still widely used due to its high tensile load capacity, and relatively low noise and ground vibration levels.

Franki piles can be used as high-capacity deep foundation elements without the necessity of excavation or dewatering. They are useful in conditions where a sufficient bearing soil can only be reached deeper in the ground, and are best suited to the granular soil where the rising bearing is primarily achieved from the densification of the soil around the base. They are not recommended for use in cohesive soils where compaction of the base is not possible. The Franki piling system is the quietest of the driven cast-in-place systems, and so is used in conditions where high noise levels could cause environmental problems.

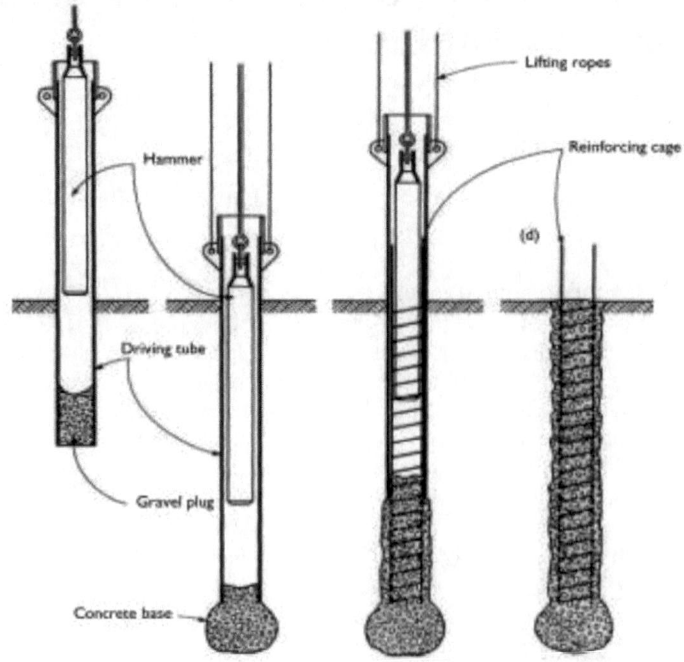

Figure 1. Stages in installing an open-ended Franki pile

The next type of pile with the enlargement on the base was developed by the MacArthur Concrete Pile Corp. (1950). A temporary steel tube is driven into the ground to the desired depth, then fill the tubes with concrete one-third the length of the tube, followed by raising and lowering the tube to produce bulb at the pile base. Once the bulb is formed, then drawn the tube gradually while filling the hole with concrete until completely fill the hole (Figure 2).

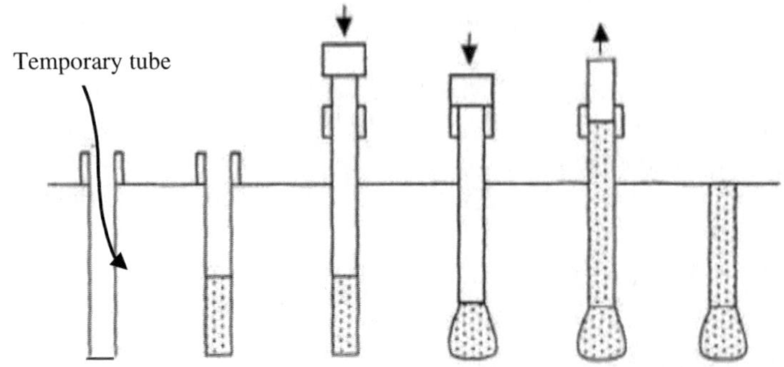

Figure 2. Stages in installing a Mac-Arthur pile

Both methods described above have a limitation if applied on a layer of soft clay because the bulb at the base of the pile cannot be formed perfectly. Therefore, to overcome this drawback, researchers from University of Christian Petra Surabaya has proposed a new method namely Preloaded Bulb Cast in-situ [2]. Although their research was successful to increase the carrying capacity of the pile tip, the bulb produced by this method is still less than the maximum. This paper describes a new method proposed by the authors which deal with the techniques to create a pile bulb in soft clay layer and will explain more detail in the next sections.

2. Model Description

In general, this research was a laboratory experiment consist of a model of steel tube in a scale of 1:10, that pushed into soft clays soil that compacted on the concrete container with a diameter of 600 mm and a height of 1000 mm.

2.1. Characteristics of soil media

A laboratory testing program was designed to determine the geotechnical and physical properties of the artificial clay soil as shown in Table 1. All of the tests are carried out based on Indonesian Standards SNI [3]. The grain-size distribution curves of the studied soils indicated that it is composed of predominantly clay and silt-sized particles. The Atterberg limit test revealed that examined soil would be classified as high plasticity clayey-silt.

Table 1. Properties of soft clay soil

Parameters	Values
The depth of sampling (m)	1.00
Colour	dark grey
Specific gravity	2.55
Natural water content (%)	42,23
Bulk density (kN/m^3)	1.56
Percent smaller than 75μm	85.12
Liquid limit (%)	68.00
Plastic limit (%)	42,66
Plasticity Index (%)	25,34
Unified soil classification	CH
Compaction (Proctor) test	
$\gamma_{d(max.}$(kN/m^3)	11.50
OMC (%)	34
Internal friction angle (ϕ),	0,0
degree	6,0
Cohesion (c), (kN/m^2)	

2.2. Pile model design

In this experiment, the author introduces two type of model piles, namely standard pile (pile without base enlargement) and pile equipped with a mechanical system to create the bulb at the pile base. To fit the scale 1:10 ratio conversion, therefore the dimension of model piles is designed with its external diameter by 40 mm, with a wall thickness of 2.0 mm, and a total pile length of 600 mm. Furthermore, in order to form the enlargement of the pile base, then a detachable steel shoe is designed at the end of the pile. For a standard pile, a steel shoe diameter was designed as large as the diameter of the pile. Since for the pile with base enlargement, the diameter of the steel shoe is slightly larger than the diameter of the pile, in this case, is determined by 60 mm (Figure 3). The main function of this pile shoe as an artificial supporting layer (artificial bearing stratum) to ensure that the enlargement of the pile base can perfectly be formed, through concrete pressure transferred by the internal pressing rod (internal rod pusher).

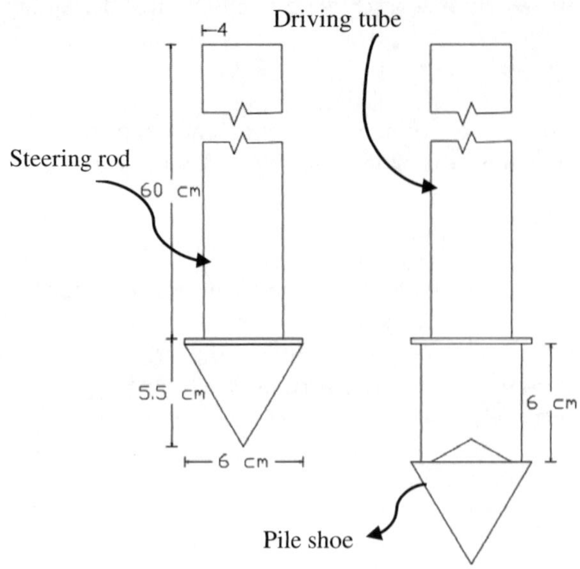

Figure 3. Components of detachable of pile shoe

2.3. Pile installation method

This type of pile may be driven (jack-in) through a soft clay soil. Furthermore, a steel tube having an external diameter of 40 mm and 2 mm in thickness, is driven into the ground. To facilitate driving of the pile, the steel tube is fitted with a detachable steel shoe with the diameter of 60 mm that completely closes at the bottom of the tube. When the tube has been driven to the required depth, a number of dry concrete is poured at its base, filling the bottom of the tube. Dry concrete is then pushed out of the tube together with a detachable steel shoe until it stops at a certain distance and forming a bulb which is desired.

Immediately after the bulb has been perfectly formed, then the tube is gradually withdrawn leaving the concrete below. Thus, reinforce by alternately pouring the fresh concrete and withdrawing the tube, the pile is constructed to its full length. The pile shoe will remain in place and hence a new one is needed for each pile installation. In case the pile is required to be reinforced, the reinforcement cage is lowered into the steel tube prior to the pouring of fresh concrete. Step by step in installing this piling system is depicted in Figure 4.

Furthermore, the pile is loaded statically using hydraulics jack and proving ring to measure the load, which gives a couple of load and the settlement data. To determine the ultimate load of the tested pile is done by analyzing the curve shape of the relationship load-net settlement, which is described in [4] as follows: net settlement is equal of total settlement minus the elastic settlement. In the special case when the relationship between the load and displacement curve does not show a clear ultimate capacity of the pile, then in practice the ultimate pile capacity can be defined as the load which causes the pile settlement of 10% times the diameter of the pile [5].

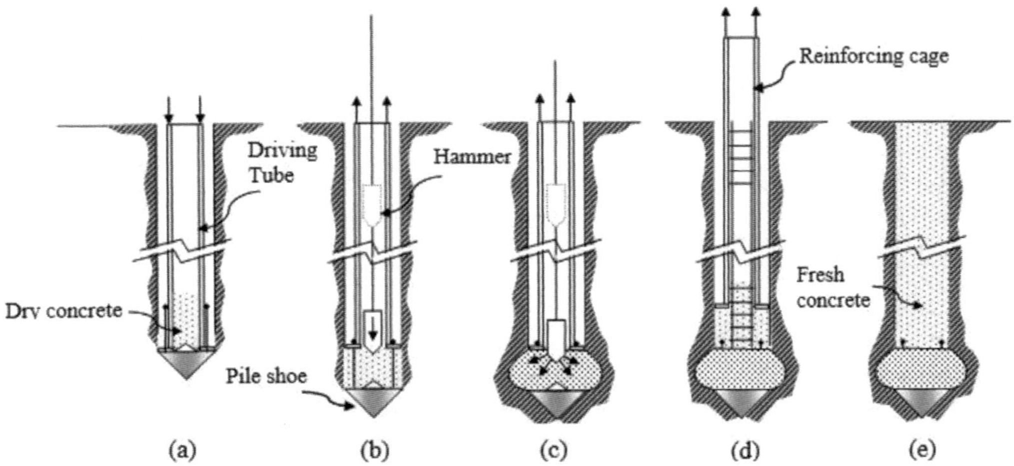

a. Push the tube down to its design depth.
b. Holding the top part of the tube and followed by pushing the dry concrete inside the tube consequently, the pile shoe will move down and terminate to its desire distance.
c. Create the bulb (base enlargement) at the pile base by continually pushing the dry concrete, add some dry concrete if needed.
d. Once the bulb was perfectly formed, then start to release the temporary tube from the pile shoe. Insert the reinforce cage according to the design requirements.
e. Lift up the tube gently, while pouring fresh concrete and compacted by using tremi pipe.

Figure 4. Stages in installing an artificial hard stratum

3. Result and discussion

Figure 5 below shows the relationship between the load and the total displacement curve of a standard pile (curve 1), while the curve 2 and 3 are the two series of testing of the pile with the enlargement at the pile base. In the curve 1 the peak value of the ultimate bearing capacity of the pile is not realized, therefore the ultimate bearing capacity is determined based on the settlement of the pile by 10% times the diameter of the pile, that is equal to 0.75 kN.

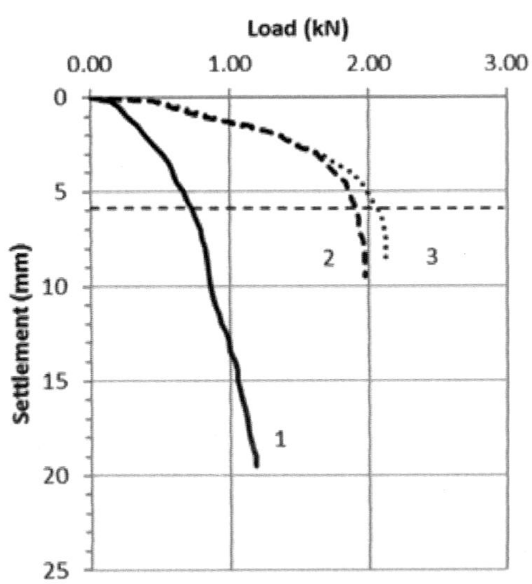

Figure 5. Load settlement curve

In the same way, the piles 2 and 3 successively obtained the ultimate load capacity of 1.95 kN and 2.05 kN, it means that carrying capacity has increased by 270%. In addition, from the visual observation of the shape of the enlargement of the pile base which is produced by using a system of pile shoe separate from the body of the pile, give a very effective result. Additional research still needs to be done especially to improve the method to separate the pile shoe from the temporary steel pipe after the formation of the enlargement of the pile base, to be more practical and effective.

4. Conclusion

The paper deal with the improving method to create a bulb or pile base enlargement in soft to very soft clay deposit. The result indicated that the proposed method works properly, where the ultimate pile capacity is increase up to 270%. It is noticed we cannot get all information with small-scale models, therefore in compliment to model test, it is necessary to continue to carry out tests on large-scale models and real structures.

Acknowledgement

This research was supported by The State Polytechnic of Malang through the swadana research program DIPA 2015.

References

[1] M. Tomlinson, M. & J. Woodward, Pile Design and Construction Practice, 5th Ed. Taylor & Francis. (2008).
[2] E. Abadi & A. Budiman, Penggunaan Preloaded Bulb untuk Meningkatkan Daya Dukung Ujung Tiang Cast in-situ, FTSP Universitas Kristen Petra Surabaya (Unpublished) (2007).
[3] G. Aponno, Petunjuk Praktikum Mekanika Tanah, Polinema Press. (2005).
[4] C. Liu, and J. Evett, Soils and Foundations, 7th ed., Prentice Hall, Upper Saddle River, NJ. (2008).
[5] B. Das, Principles of Foundation Engineering, 8th ed., New York: Cengage Learning. (2016).

Optimization of Coagulation-Flocculation Process for Tello River Water Treatment Using Poly Aluminum Chloride and Aluminum Sulfate

S E Widiyanti[1]*

[1]Chemical Engineering Department, State Polytechnic of Ujung Pandang, Makassar 90245, Indonesia

*E-mail: setyoernawidiyanti@poliupg.ac.id

Abstract. Tello River Water was utilized by PT. PLN SULSERABAR as Boiler feed water. Total Suspended Solid (TSS) in river water can be removed by the coagulation-flocculation process. Improved performance of coagulation process can be done by optimizing coagulant concentration and pH on coagulation. This research focused on optimization of pH and percent (%) removal of TSS using Poly Aluminum Chloride (PAC) and Aluminum Sulfate (Alum). The concentration of coagulant was 25% of the initial TSS of the sample for PAC and 35% of the initial TSS of the sample for Alum. The sample used in this research is Tello River Water which has initial TSS concentration of 195 mg/L with rapid mixing (350 rpm) for 1 minute, slow mixing (50 rpm) for 20 minutes and 30 minutes of sedimentation. The coagulation-flocculation process was carried out in the pH range of 4-10 and the TSS concentration was analyzed using the Gravimetric method. The optimum pH on coagulation and percent (%) TSS removal obtained from this study were 7, 99.23% for PAC and 6, 89.48% for Alum. These results indicated that PAC was the perfect coagulant for the removal of TSS because it was able to eliminate TSS nearly 100%.

1. Introduction

Tello River Water was utilized by PT. PLN SULSELRABAR as boiler feed water. Total Suspended Solids (TSS) from river become deposits inside the boiler. This deposit cause by decreased of heat transfer and reduced of the efficiency of the boiler [1]. TSS can be removed by a coagulation-flocculation process [2, 3]. Types of coagulants, coagulant concentrations, initial pH [4], speed of rapid mixing [5-7], speed of slow mixing [8] settling time [9] and additional aid coagulants [10] are several factors that influence the efficiency of the coagulation-flocculation process. Coagulant concentration and pH are the most important factors for efficiency of the coagulation-flocculation process [11]. The addition of coagulant is not always directly proportional to the percent of TSS removal. Particle restabilization can occur due to the addition of excess coagulants [12]. Stabilization of suspended particles and coagulant surface charge are affected by the pH of the process. A suitable pH can help neutralizing the negative charge on the particles and help floc formation [13].

Commonly used chemical coagulants such as calcium, aluminum, and iron. Based on the results of the experiment, Aluminum Sulfate and Poly Aluminum Chloride coagulants were selected coagulants [11]. Those coagulant with Aluminum basis, metal ion hydrolized to Aluminum hydroxide and Hydrogen ion. Hydrogen ion will be reacting within water alkalinity and it process, decrease of water pH can be shown from equation below:

For Alum [13]

$$Al_2(SO_4)_3.16H_2O \rightarrow 2Al^{3+} + 3SO_4^{2-} + 16H_2O \rightarrow 2Al(OH)_3 + 6H^+ + 3SO_4^{2-} + 10H_2 \quad (1)$$

For PAC [14]

$$Al_2(OH)_3Cl_3 \rightarrow Al_2(OH)_3^{3+} + 3Cl^- + 3H_2O \rightarrow 2Al(OH)_3 + 3H^+ + 3Cl^- \quad (2)$$

Based on the theory of coagulation-flocculation, the surface of precipitate from PAC with positive charges can neutralize the charges on the surfaces of primary particles suspended in the water [15]. After neutralization of the particle charge, microflocs is formed which has not been seen by eyes about 1 - 100 μm. Due to Brown's forces, microflocs collide and bind to each other to form larger aggregates (perikinetic flocculation). Slow mixing is done to give microflocs a chance to bind to each other so that a larger floc is formed until it is steady and settles [12].

Research on optimization of coagulant concentration and pH of coagulation-flocculation process has been carried out. Coagulant concentration, pH of process, and percent removal of TSS were 300 mg/L, 6, 97% for Poly Aluminum Chloride (PAC) and 500 mg/L, 6, 92% for Aluminum Sulfate (Alum) respectively [16]. Optimization of coagulant concentrations with Tello River Water samples was carried out, the result was 25% of initial TSS for PAC and 35% of initial TSS for Alum with percent removal of TSS were 89.6% and 86.67% [17]. Percent removal from the study has not been maximized so that in this study aims to improve the efficiency of the performance of the coagulation-flocculation process by optimizing the pH of the process using PAC and Alum coagulants through same operating conditions.

2. Methods

The water sample used in the research from Tello River Water Makassar on April 2018. Initial TSS concentrations and pH of Tello river water were 195 mg/L and 5. The conditions of the coagulation-flocculation process in various research can be seen in Table 1.

Table 1. Summary of coagulation-flocculation operating conditions from various research

Parameter	Range	Reference
Speed of rapid mixing (rpm)	80 - 350	[3, 17 – 21]
Duration of rapid mixing (min)	1 - 4	[3, 17 – 21]
Speed of low mixing (rpm)	30 - 50	[3, 17 – 21]
Duration of low mixing (min)	15 - 20	[3, 17 – 21]
Settling time (min)	15 - 45	[3, 17 – 21]

PAC with a concentration of 25% of the initial TSS was added to the sample. Based on Table 1 can be seen a summary of the conditions of the coagulation-flocculation process from various research. In this research used rapid mixing at 350 rpm for 1 minute, slow mixing at 50 rpm for 20 minutes, after 5 minutes from the beginning of the flocculation process measurements and adjustments pH were made (pH adjusted by adding NaOH or H_2SO_4) and sedimentation for 30 minutes. The supernatant was taken 2 cm below the upper limit of the sample solution, then TSS was analysed using the Gravimetric method. Variations in process pH 4-10 and coagulant Aluminum Sulfate were carried out with a concentration of 35% from the initial TSS.

3. Results and Discussion

Coagulant concentration and pH of process are the most important factors of the efficiency of the coagulation-flocculation process. When the coagulation-flocculation process works in optimal conditions, the use of coagulants can be minimized as needed. The required coagulant concentration increases in proportion to the amount of suspended particle concentration to complete destabilize through charge neutralizing and bridging. However, the addition of excess coagulant makes particle restabilization and the suspended particle concentration increases again. pH can affect the stabilization of suspended particles and coagulant surface charge. Suitable pH can help neutralize the negative charge on the particles and help the formation of the floc.

TSS removal resulting from the coagulation-flocculation process will not optimal when optimizing the coagulant concentration without observing to the pH process. Research on optimization of coagulant concentration without observing to pH gets the optimum results of PAC and Alum coagulant concentrations at 25% and 35% from initial TSS concentrations with TSS removal percent were 89.6% for PAC and 86.67% for Alum [17]. This study optimizes the coagulation-flocculation process by optimizing the pH of the process using coagulant concentrations obtained from Widiyanti research [17]. There was an increase in percent TSS removal resulting from the coagulation-flocculation process in this research. TSS removal percent using PAC and Alum coagulants increased from 89.6% to 99.23% for PAC and 86.67% to 89.48% for Alum. Figure 1 shown the relationship between pH and percent TSS removal using PAC and Alum coagulant results from this research.

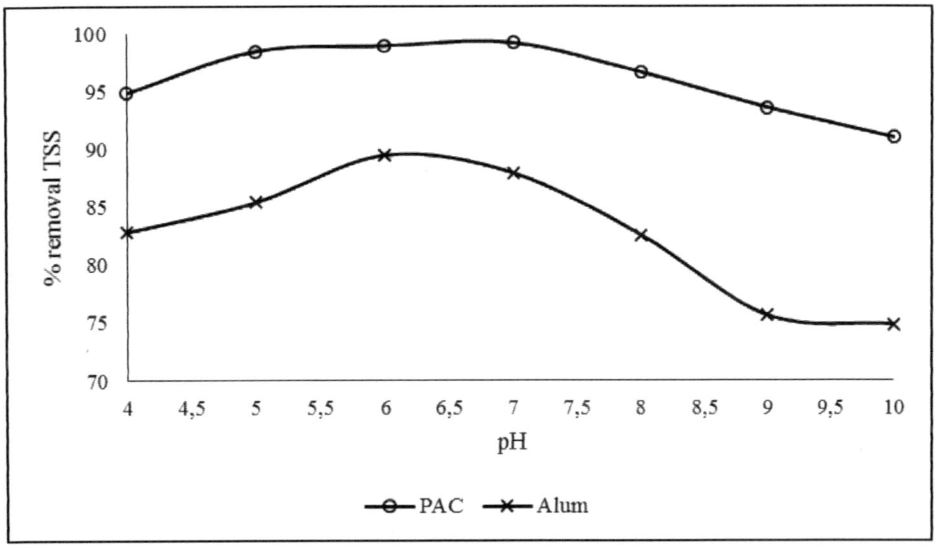

Figure 1. Chart of the relationship between pH and percent TSS removal using PAC and Alum coagulant

The highest TSS removal percentage was obtained at pH 7 for PAC and pH 6 for Alum based on Figure 1. Percent removal of TSS results from the coagulation-flocculation process in this research is greater using PAC than Alum. This is cause by the PAC coagulant is a coagulant resulting from polymerization of metal coagulants which is used to improve the performance of conventional coagulants such as Alum [22].

The conditions of the coagulation-flocculation process chosen are crucial in the success of removing TSS in water. Comparison of coagulation-flocculation process conditions between percent TSS removal during pH optimization using the PAC coagulant from this research and Nasir and Daud research [16] can be seen in Table 2 and Figure 2.

Figure 2 shown that the percent TSS removal results from this research are greater than the previous research which obtained the highest percent removal of 77%. This is because when determining the optimal pH of the process, the concentration of PAC coagulant used is not optimal. Since of coagulant concentration and pH of the process in optimal conditions, there was an increase in percent removal from 77% to 97% [16]. Due to the coagulant concentration and pH of the process was the most important factor in the success of the coagulation-flocculation process [11].

Table 2. Comparison of operating conditions of the coagulation-flocculation process of this research and Nasir and Daud research [16]

Parameter	Reference	
	This research	[16]
Initial of TSS sample (mg/L)	195	348
Speed of rapid mixing (rpm)	350	150
Duration of rapid mixing (min)	1	4
Speed of low mixing (rpm)	50	20
Duration of low mixing (min)	15	20
Settling time (min)	30	30
Concentration of coagulant (mg/L)	48,75	100
pH	4-10	2-12

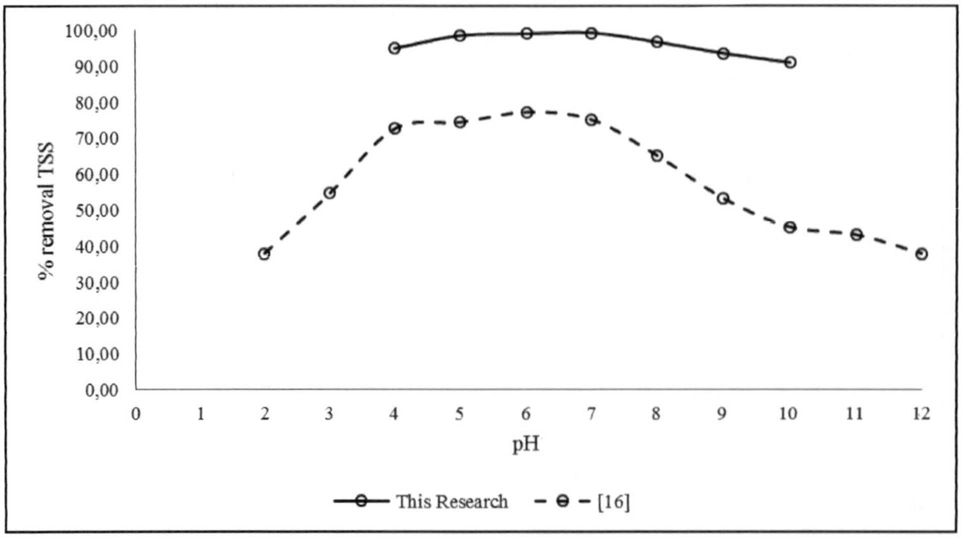

Figure 2. The relationship between pH and percent TSS removal results from the coagulation-flocculation process of this research and Nasir and Daud research [16] using PAC coagulant

Table 3. Comparison of operating conditions of the coagulation-flocculation process of this research, Nasir and Daud research [16] and Asadollahfardi *et al.* research [20]

Parameter	Reference		
	This research	[16]	[20]
Initial of TSS sample (mg/L)	195	348	287
Speed of rapid mixing (rpm)	350	150	100
Duration of rapid mixing (min)	1	4	3
Speed of low mixing (rpm)	50	20	30
Duration of low mixing (min)	15	20	15
Settling time (min)	30	30	30
Concentration of coagulant (mg/L)	68,25	100	1000
pH	4-10	2-12	2-10

Comparison of coagulation-flocculation process conditions and percent TSS removal during pH optimization using Alum coagulant from this research, Nasir and Daud research [16] and Asadollahfardi *et al.* research [20] can be seen in Table 3 and Figure 3.

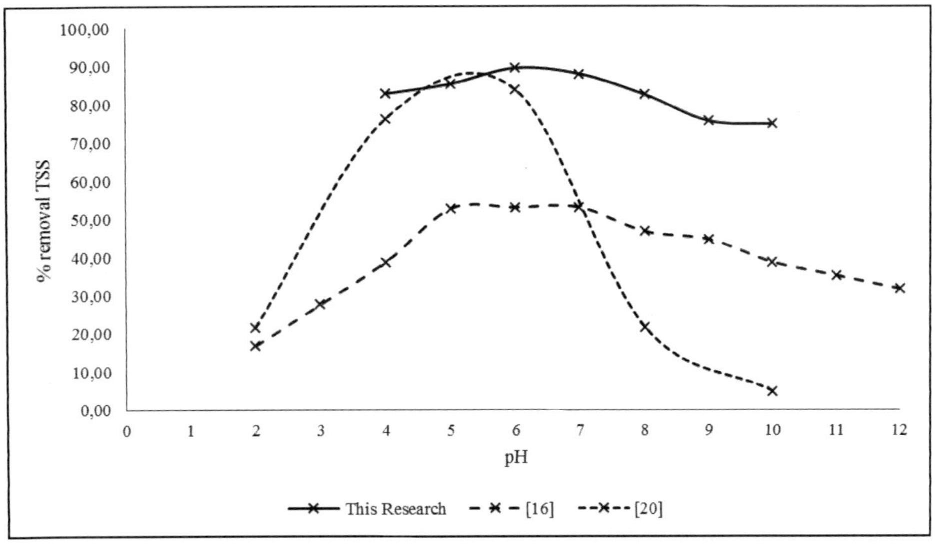

Figure 3. Chart of the relationship between pH and percent TSS removal results from the coagulation-flocculation process of this research and Nasir and Daud research [16] and Asadollahfardi et al. research [20] using Alum coagulant

Figure 3 shown that the pH optimum of the three research is 6. The percentage of TSS removal from this research is greater compared to the research of Nasir and Daud[16] and Asadollahfardi *et al.*[20]. This is caused when determining the optimal pH of the process, the concentration of Alum coagulant used is not optimal. At the time of coagulant concentration and pH of the process at optimal conditions, there was increasing removal percent from 53% to 92% [16] and 83.9% to 91.6% [20].

Based on the results of the coagulation-flocculation process from several research it can be seen as the most important factor to improve the performance of the coagulation-flocculation process are the optimal coagulant concentration and pH of coagulation. The highest percentage of coagulant PAC when compared with from Nasir and Daud research [16]. Whereas the percent uses the lowest Alum coagulant when compared with results from Nasir and Daud research [16] and Asadollahfardi *et al.* research [20], but the differences is not too large, which is around 3%. The advantage of optimizing the coagulation process of this research is that it can overcome the problem of TSS whereas water is always fluctuating depending on the season.

4. Conclusion

The optimum pH on coagulation and percent (%) TSS removal obtained from this study were 7, 99.23% for PAC and 6, 89.48% for Alum.

Acknowledgment

The author would like to thank the State Polytechnic Of Ujung Pandang for financially supporting this research

References

[1] Hassan F, Aftab T, Sfafique T and Mubasshir A 2010 *Asian Journal of Chemistry* **22** 5913-18.

[2] Hassan A, Ariffin M, Tan P L and Noor Z Z 2009 *Journal of chemical and natural Resources Engineering* **4** 43-53.

IOP Conf. Series: Materials Science and Engineering **676** (2019) 012026 doi:10.1088/1757-899X/676/1/012026

[3] Chuan C B, Yu G B, Hua X C, Ying F and Xin L 2010 *Chinese Sci Bull* **55** 1382-87.

[4] Aygun A and Yilmaz T 2010 *International Journal of Chemical and Environmental Engineering* **1** 97–101.

[5] Zhou Y, Liang Z and Wang Y 2008 *Desalination* **225** 301–11.

[6] Ramphal S and Sibiya S M 2014 *Procedia Engineering* **70** 1401–10.

[7] Ahmed S B, Ayoub G, Al-Hindi M and and Azizi F 2015 *Desalination and Water Treatment* **53** 3388–96.

[8] Ebeling J M and Ogden S R 2004 *North American Journal of Aquaculture* **66** 198–207.

[9] Rattanapan C, Sawain A, Suksaroj T and Suksaroj C 2011 *Desalination* **280** 370–7.

[10] Saraswathi R and Saseetharan M K 2012 *Water Science and Technology* **66** 1231–8.

[11] Kun Y 2012 *International Journal of Digital Content Technology and its Applications (JDCTA)* **6** 429-36.

[12] Choy S, Prasad K, Wu T and Ramanan R 2013 *International Journal of Environmental Science and Technology* **12** 367-90.

[13] Mbaeze M C, Agbazue V E and Orjioke N M 2017 *Modern Chemistry & Applications* **5** 1-14

[14] Gebbie P 2005 *68th Annual Water Industry Engineers and Operators Conference* Schweppes Centre – Bendigo 75-83.

[15] Wei N, Zhang Z, Liu D, Wu Y, Wang J and Wang Q 2015 *Chinese Journal of Chemical Engineering* **23** 1041-46.

[16] Nasir N and Daud Z 2014 *Journal of Mechanical Engineering and Sciences (JMES)* **7** 1189-95

[17] Widiyanti S E 2018 Konversi **7** 1-5.

[18] Ghafari S, Aziz H A, Isa M H and Zinatizadeh A A 2009 *Journal of Hazardous Materials* **163** 650–6.

[19] Zand A D and Hoveidi H 2015 *Journal of Applied Biotechnology Reports* **2** 287-92.

[20] Asadollahfardi G, Zangooei H, Motamedi V and Davoodi M 2018 *Advances in Environmental Research* **7** 1-11.

[21] Xie J, Wang D, Leeuwen J V, Zhao Y, Xing L and Chow C W K 2012 *Journal of Environmental Sciences* **24** 276–83.

[22] Sinha S, Yoon Y, Amy G and Yoon J 2004 *Chemosphere* **57** 1115–22.

IOP Conf. Series: Materials Science and Engineering **676** (2019) 012027 doi:10.1088/1757-899X/676/1/012027

Natural Frequencies of Laminated CFRP Square Plates with Slightly Different Material Properties

Y Narita[1] and N Innami[2]

[1]Hokkaido University, Sapporo, Japan
[2]Hokkaido Polytechnic College, Sapporo, Japan

ynarita@eng.hokudai.ac.jp,

Abstract. The influence of using different material properties is studied on the natural frequencies of laminated composite square plates composed of recently developed carbon fiber reinforced plastic (CFRP) materials. The plate is symmetrically laminated by thin layers with each layer reinforced by parallel straight carbon fibers. Using the Ritz method in conjunction with the thin plate theory and the lamination theory, the displacements are assumed as finite polynomial functions with the boundary index to accommodate any combination of classical boundary conditions. The material properties are expressed by a set of four elastic constants, and some typical constants are referred from the recent literature. In addition to these sets of constants, a new standard set of discretized constants is proposed to consider the underlying characteristics of the existing constants. In numerical experiment, the convergence study is carried out, and the lowest five natural frequencies are presented in frequency parameters for five sets of classical boundary conditions. Next, for better insight, a new definition of frequency parameters is introduced to promote more physically meaningful comparison, and the influence of using slightly different constants in frequency calculation is clarified for the purpose of deriving approximate frequency formula.

1. Introduction

Plate components are found practically in all fields of industry, and the literature related to free vibration of flat plates is voluminous. Monograph "Vibration of Plates" compiled by Leissa [1] in 1969 is still now quite often cited in many papers, and for *isotropic* rectangular plates he published a comprehensive set of natural frequencies for all possible combinations of classical boundary conditions [2]. Since in the 1970's, advanced fibrous composite materials have been developed and served as plate components in many industrial applications, such as airplane, automobile and marine structures. Generally, fiber reinforced plastics have microstructures made of reinforcing fiber and matrix material. There are various fibers, ranging from natural fibers to chemical fibers such as glass, boron and carbon fibers. When such constituents merge into one material, it shows anisotropic characteristics. Among various types of fibers, the use of carbon fibers in reinforced plastics (CFRP) is becoming more increasingly dominant in weight-sensitive structures. For the reason, some books [3-5] dealing with mechanics of the composites have been published to serve for analysis and design, and review papers have appeared, for example [6].

Starting the publication of journal papers in the 1970's, researchers such as Bert [7] published papers on vibration of *laminated composite* plates, and it has led to improvement of lamination theories to the first order and higher order plate theories [5, 8]. For rectangular plates with arbitrary edges, combination of classical boundary conditions became possible to be analyzed [9]. Application of the finite element method has been active in analyzing laminated plates [10, 11].

For defining elastic problem of laminated CFRP plates, four independent elastic constants are necessary, unlike isotropic plates with only two independent constants being needed. Previous literatures have used different values of the constants, because the improvement of fiber stiffness is in progress, and many past researchers have used elastic constants obtained from measurement tests supplied from different chemical companies. This fact makes direct comparison difficult among the published results, and was also obstacles preventing from compiling design date book.

Despite such practical needs, however, there is no literature to discuss the influence caused by using different elastic constants. The present study takes up this problem for studying the influence when slightly different constants are used in the vibration analysis of laminated CFRP plates. In doing so, numerical experiments are conducted to calculate natural frequencies of laminated CFRP square plates by using various sets of constants, and for effective insights, new standard constants are proposed to understand vibration behaviors in the comprehensive way.

2. Analytical Method Used in Numerical Experiment

Figure 1 shows a laminated rectangular plate in the coordinate system and in each layer the major and minor principal axes are denoted by the L and T axes. The dimension of the whole plate is given by $a \times b \times h$ (thickness). The plate considered is limited to symmetric laminate, and the total number of layers is defined as $2N$ (i.e., N layers in the upper(lower) half cross-section).

Free vibration of a macroscopic model for such thin symmetric plates is governed in the classical lamination (plate) theory by

$$D_{11}\frac{\partial^4 w}{\partial x^4}+2(D_{12}+2D_{66})\frac{\partial^4 w}{\partial x^2 \partial y^2}+D_{22}\frac{\partial^4 w}{\partial y^4}+4D_{16}\frac{\partial^4 w}{\partial x^3 \partial y}+4D_{26}\frac{\partial^4 w}{\partial x \partial y^3}+\rho h\frac{\partial^2 w}{\partial t^2}=0 \tag{1}$$

where ρ is a mean mass per unit volume of the plate. The D_{ij} (i, j=1, 2, 6) are the bending stiffness of the symmetric laminate defined by

$$D_{ij}=\frac{2}{3}\sum_{k=1}^{N}\bar{Q}_{ij}^{(k)}\left(z_k^3-z_{k-1}^3\right) \tag{2}$$

with z_k being a thickness coordinate measured from the mid-surface and $\bar{Q}_{ij}^{(k)}$ being elastic constants [3, 4] in the k-th layer, obtained from

$$Q_{11}=\frac{E_L}{1-\nu_{LT}}, Q_{12}=\frac{E_L \nu_{TL}}{1-\nu_{LT}}, Q_{22}=\frac{E_T}{1-\nu_{LT}\nu_{TL}}, Q_{16}=\frac{G_{LT}}{1-\nu_{LT}\nu_{TL}} \tag{3}$$

(superscript (k) is omitted) by considering an fiber orientation angle $\theta^{(k)}$ in the layer. The E_L and E_T are moduli of longitudinal elasticity in the L and T directions, respectively, G_{LT} is a shear modulus and ν_{LT} is a Poisson's ratio.

For the small amplitude (linear) free vibration, the deflection w of a thin plate may be written by

$$w(x,y,t)=W(x,y)\sin\omega t \tag{4}$$

where W is the amplitude and ω is a radian frequency. Natural frequency is normalized as a frequency parameter

$$\Omega=\omega a^2\left(\rho h/D_0\right)^{1/2} \text{ with } D_0 \text{: reference bending stiffness} \tag{5}$$

where ω is a radian frequency of the laminate plate.

Then, the maximum strain energy due to the bending is expressed by

$$U_{max}=\frac{1}{2}\iint_A \{\kappa\}^T \begin{bmatrix} D_{11} & D_{12} & D_{16} \\ D_{12} & D_{22} & D_{26} \\ D_{16} & D_{26} & D_{66} \end{bmatrix} \{\kappa\}\, dA \tag{6}$$

and $\{\kappa\}$ is a curvature vector

$$\{\kappa\}=\left\{-\frac{\partial^2 W}{\partial x^2} \quad -\frac{\partial^2 W}{\partial y^2} \quad -2\frac{\partial^2 W}{\partial x \partial y}\right\}^T \tag{7}$$

The maximum kinetic energy is given by

$$T_{max}=\frac{1}{2}\rho h\omega^2\iint_A W^2 dA \tag{8}$$

In the Ritz method the amplitude is assumed in the form

$$W(x,y)=\sum_{m=0}^{M-1}\sum_{n=0}^{N-1}A_{mn}X_m(x)Y_n(y) \tag{9}$$

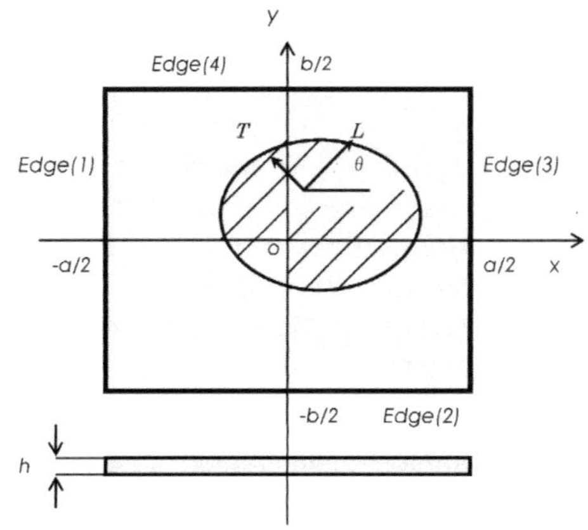

Figure 1. Laminated rectangular plate

Where A_{mn} are unknown coefficients, and $X_m(\xi)$ and $Y_n(\eta)$ are the functions modified so that any kinematical boundary conditions are satisfied at the edges with applying "boundary indices" [9] on each of four edges.

After substituting Eq. (9) into the sum of energies (6) and (8), the stationary value is obtained by Fig. 1 Laminated rectangular plate

$$\frac{\partial}{\partial A_{\overline{m}\overline{n}}}(T_{max}-U_{max})=0 \quad (\overline{m}=0,1,..,M-1;\overline{n}=0,1,..,N-1) \tag{10}$$

The minimizing process gives a set of linear simultaneous equations in terms of the coefficients A_{mn}, and the eigenvalues Ω may be extracted by using existing computer subroutines. This analytical procedure is a standard routine of the Ritz method but the special form of polynomials can satisfy any kinematical boundary conditions [9].

3. Numerical Examples and Accuracy of Solution

3.1 Numerical examples

Symmetrically laminated square plates (a/b=1) are considered as numerical example. The number of layers is eight and a lamination sequence is denoted by $[\theta_1/\theta_2/\theta_3/\theta_4]$s, where θ_1 is a fiber orientation angle of the outer most layer measured from x axis (as shown in Fig.1) and θ_4 is the angle of inner most layer. In the examples, the lamination sequence is limited to two typical types:

Cross-ply plate [0°/90°/0°/90°]s and Angle-ply [30°/-30°/30°/-30°]s (s: symmetric lamination) (hereafter, "°" is omitted). The boundary conditions are denoted by four capital letter, such as CSFF labelling counter-clockwise starting from Edge (1) in Fig.1. Here, "C" denotes a clamped edge (i.e., deflection and rotation are both rigidly constrained), "S" does a simply supported edge (deflection is constrained but bending moment is zero) and "F" does a free edge (bending moment and shear force are both zero). By using this notation, natural frequencies for SSSS (totally simply supported plate) are presented in both Table 3 and 4, and those for FFFF (totally free plate), CFFF (cantilever plate), CSFF (combination of C,S,F) and CCCC (totally clamped plate) are presented in Tables 5, 6, 7 and 8, respectively.

3.2 Elastic constants of different CFRP materials

For a fix geometry (a,b,h) and boundary condition (C,S,F) of plate, it is easily understandable that frequency parameters are determined only by the values of elastic constants, and it is the main topic of this work to study the influence from the constants. As shown in Eq.(3), there are four independent elastic constants in Q_{ij}: E_L and E_T being modulus of longitudinal elasticity (Young's modulus) in the L (fiber) and T directions, respectively, G_{LT} being a shear modulus and a Poisson's ratio ν_{LT}, and there is a relation of $E_L/\nu_{LT}=E_T/\nu_{TL}$ among the constants.

The survey by the authors showed that a little different value of elastic constants has been employed previously, as listed in Table 1. It is observed that the difference is mainly caused by values of E_L but not by E_T and G_{LT}, because Young modulus of the fibers is a key factor in controlling the stiffness in the longitudinal direction within lamina, while the matrix material is almost the same, mostly epoxy material. In the table, Mat.1 indicates values used by Stanford and Jutte [12] in NASA and shows the lowest anisotropy rate E_L/E_T=11.6 among the listed sets of constants. Mat.2 and 3 are medium values listed in [3] and [13], respectively. Mat.4 indicates those used by Panesar and Weaver [14], a group in Bristle University, and it shows the highest anisotropy E_L/E_T=18.7. Finally, Mat.5 is introduced to represent the mean values of recently used CFRP materials and used as the standard reference for comparison. This set of discretized values is fictious and not measured in experiment, but it will be shown that it is useful as the representative constants for CFRP. When one measures deviation of listed materials from Mat.5, the differences of the constants differ by -15% (Mat.1)≦0% (Mat.5)≦ 15%(Mat.4) for E_L, -12%(Mat.3)≦0%(Mat.5)≦11%(Mat.1) for E_T, -10%(Mat.1)≦0%(Mat.5)≦ 42%(Mat.2) for G_{LT}, and -17%(Mat.1)≦0%(Mat.5)≦7%(Mat.3) for ν_{LT}.

Table 1. Elastic constants of carbon fiber reinforced plastic (CFRP) materials.

	E_L [Gpa]	E_T [Gpa]	(E_L/E_T)	G_{LT} [Gpa]	ν_{LT}
Mat. 1 [12]	128	11	(11. 6)	4. 5	0. 25
Mat. 2 [3]	138	8. 96	(15. 4)	7. 1	0. 30
Mat. 3 [13]	139	8. 76	(15. 9)	4. 57	0. 32
Mat. 4 [14]	168. 980	9. 050	(18. 7)	5. 00	0. 288
Mat. 5	150. 0	10. 0	(15. 0)	5. 00	0. 30

Table 2. Convergence of the solution for cross-ply and angle-ply square plates (Mat.5).

Number of term					
m×n	Ω_1	Ω_2	Ω_3	Ω_4	Ω_5
Cross-ply plate [(0/90)2]s, CSFF					
6x6	12.55	40.36	73.06	90.49	119.9
8x8	12.55	40.36	73.06	90.48	118.7
10x10	12.55	40.36	73.06	90.48	118.7
Cross-ply plate [(0/90)2]s, SSSS					
6x6	42.55	102.1	134.5	170.2	217.2
8x8	42.55	102.1	134.5	170.2	213.9
10x10	42.55	102.1	134.5	170.2	213.8
Exact sol.	42.55	102.1	134.5	170.2	213.8
(m,n)	(1,1)	(1,2)	(2,1)	(2,2)	(1,3)
Angle-ply plate [(30/−30)2]s, CCCC					
6x6	88.50	143.7	209.1	225.8	273.9
8x8	88.50	143.7	209.1	225.6	273.8
10x10	88.50	143.7	209.1	225.6	273.8

The following comparison is made by a frequency parameter $\Omega = \omega a^2 \left(\rho h / D_0\right)^{1/2}$ defined in Eq. (5), but two types of reference bending rigidity are used as

$$D_0 = E_T h^3 / 12\left(1-\nu_{LT}\nu_{TL}\right) \text{ with } E_T, \nu_{LT} \text{ for each material in Table 1} \quad (11)$$

$$D_0^* = E_T^* h^3 / 12\left(1-\nu_{LT}^*\nu_{TL}^*\right) \text{ with } E_T^*=10 \text{ GPa}, \nu_{LT}^*=0.3 \text{ (note } \nu_{TL}^* = \left(E_T^* / E_L^*\right)\nu_{LT}^* \text{)} \quad (12)$$

of Mat.5, for global comparison among different materials listed in the table.

3.3 Convergence and comparison of the solution

Before the natural frequencies are compared to discuss the discrepancy stemming from different CFRP materials, validity of the solution should be established. Table 2 presents frequency parameters of laminated square plates with Mat.5 (i.e., $D_0 = D_0^*$ in Es.(11)(12)). Number of terms M, N in Eq.(9) is increased from six to ten for cross-ply (CSFF, SSSS) and angle-ply (CCCC) plates. They converge well within four significant figures.

For a special case of cross-ply (i.e., $D_{16} = D_{26} = 0$) simply supported square plate ($a=b$, SSSS), the exact solution can be written in the form

$$\Omega = \pi^2 \left[\left(\frac{D_{11}}{D_0}\right) m^4 + 2\left(\frac{D_{12}+2D_{66}}{D_0}\right) m^2 n^2 + \left(\frac{D_{22}}{D_0}\right) n^4 \right]^{1/2} \quad (13)$$

with m and n being half wave numbers in x and y direction, respectively. The results from this formula exactly agree with the present solution within four significant figures in Table 2. All the frequency parameters listed hereafter will be obtained by using 12×12 solution.

4. Results and Discussions

Table 3 presents lowest five frequency parameters Ω, defined in Eq.(11) by using E_T of each material, for simply supported (SSSS) square plate. The results are given for Mat.1～Mat.5, and the differences (%) in frequency measured from Mat.5 are also written in the table. The frequency parameter Ω is non-dimensional and it excludes influence of plate geometry (a, b, h) and mass density ρ. So for the fixed geometry and density, values of the frequency parameters depend only ω itself and E_T and ν_{LT}.

From observation in the table, no significant differences (%) exist among the frequency orders (i.e., from the first to fifth frequency) and between the lamination schemes of cross-ply and angle-ply. The increasing (or decreasing) relation of the frequencies is unclear with respect to the constant E_L, which should be the most influential factor on the frequency values. Later in Fig.2, the degree of influence by E_L will be quantitatively focused. Therefore, use of Ω* defined in Eq. (12) is considered next to make effective comparison and to provide with clear physical interpretation. Table 4 is in the same format as Table 3 for the same square plate, except that a new reference stiffness $D_0^*=E_T^* h^3/12(1-\nu_{LT}^*\nu_{TL}^*)$ (E_T^*, ν_{LT}^* for Mat.5) is used for all the materials (i.e, even for Mat.1,2,3 and 4). Therefore, for Mat.5, the frequency parameters are identical as those in Table.3.

The idea here is that by using the identical elastic constants E_T^* and ν_{LT}^* in the frequency parameter, the differences are caused only by ω in the new frequency parameters. Actually, it is observed in Tables 4-8 that the increasing order of E_L (Mat.1<Mat.2≒Mat.3<Mat.4) among the materials basically reflects the order in the difference, except for FFFF plate, and the differences only for Mat.4 are positive Ω* (Mat.5)< Ω* (Mat.4) due to E_L (Mat.5)< E_L (Mat.4).

The frequency parameters listed in Tables 5 (FFFF), 6 (CFFF), 7 (CSFF) and 8 (CCCC) are based on the same idea, and Ω* uses the identical values of E_T^*=10GPa and ν_{LT}^*=0.3. In Table 5 (FFFF), it is seen that the differences for Mat.2 take alternatively positive and negative values, but those of Mat 3 show only all negative differences. This unpredicted behavior may be caused by the fact that in the FFFF case, there are three rigid body motions (elastic frequencies become zero) and the vibrations take complicated mode shapes to satisfy the self-equilibrium. Such strange behavior is found when the plate edges involve a number of free edges, such as FFFF, CFFF and CSFF, and it decreases as the number of free edges (F) diminished.

For Mat.1 with E_L being the smallest in Table 1, all the differences are negative in Tables 4-8 for ten cases (cross-ply and angle-ply plates with 5 different combinations of boundary conditions), and

IOP Conf. Series: Materials Science and Engineering **676** (2019) 012027 doi:10.1088/1757-899X/676/1/012027

the average differences of the lowest five Ω for the ten cases stay within the range of -6.8~-4.6%. Similarly, for Mat.5 with E_L being the largest in the table, all the differences are positive for the ten cases.

Table 3. Frequency parameters Ω (Eq.(11)) of symmetric 8-layer square plates (SSSS).

	Ω_1	Ω_2	Ω_3	Ω_4	Ω_5
Cross-ply [(0/90)$_2$]s					
Mat. 1	37.92	91.53	119.2	151.7	191.8
(%)	-10.9	-10.3	-11.4	-10.9	-10.3
Mat. 2	44.31	105.3	137.8	177.2	218.5
(%)	4.1	3.2	2.5	4.1	2.2
Mat. 3	43.67	104.6	138.2	174.7	219.1
(%)	2.6	2.5	2.7	2.6	2.5
Mat. 4	46.76	111.9	149.0	187.0	234.9
(%)	9.9	9.7	10.8	9.9	9.9
Mat. 5	42.55	102.1	134.5	170.2	213.8
Angle-ply [(30/-30)$_2$]s					
Mat. 1	46.13	91.04	127.7	156.8	185.4
(%)	-11.0	-10.0	-11.4	-9.2	-10.9
Mat. 2	52.71	103.7	147.0	178.4	212.2
(%)	1.8	2.5	2.0	3.3	2.0
Mat. 3	53.16	103.60	148.1	176.6	213.5
(%)	2.6	2.4	2.7	2.3	2.6
Mat. 4	57.28	110.7	159.9	187.6	229.9
(%)	10.6	9.5	10.9	8.6	10.5
Mat. 5	51.80	101.1	144.1	172.7	208.1

Table 4. Frequency parameters Ω* (Eq.(12)) of symmetric 8-layer square plates (SSSS).

	Ω_1^*	Ω_2^*	Ω_3^*	Ω_4^*	Ω_5^*
Cross-ply [(0/90)$_2$]s					
Mat. 1	39.76	95.97	125.0	159.0	201.1
(%)	-6.6	-6.0	-7.1	-6.6	-5.9
Mat. 2	41.94	99.70	130.4	167.7	206.8
(%)	-1.5	-2.3	-3.0	-1.5	-3.3
Mat. 3	40.89	97.96	129.4	163.5	205.1
(%)	-3.9	-4.0	-3.8	-3.9	-4.0
Mat. 4	44.45	106.4	141.7	177.8	223.3
(%)	4.5	4.2	5.3	4.5	4.4
Mat. 5	42.55	102.1	134.5	170.2	213.8
Angle-ply [(30/-30)$_2$]s					
Mat. 1	48.37	95.45	133.8	164.4	194.4
(%)	-6.6	-5.6	-7.1	-4.8	-6.6
Mat. 2	49.89	98.11	139.2	168.8	200.8
(%)	-3.7	-3.0	-3.5	-2.3	-3.5
Mat. 3	49.77	96.96	138.6	165.4	200.0
(%)	-3.9	-4.1	-3.8	-4.3	-3.9
Mat. 4	54.45	105.2	152.0	178.3	218.5
(%)	5.1	4.1	5.5	3.2	5.0
Mat. 5	51.80	101.1	144.1	172.7	208.1

Table 5. Frequency parameters Ω (Eq.(12)) of symmetric 8-layer square plates (FFFF).

	Ω_1^*	Ω_2^*	Ω_3^*	Ω_4^*	Ω_5^*
Cross-ply $[(0/90)_2]s$					
Mat. 1	15. 76	48. 72	58. 25	67. 64	74. 65
(%)	-5. 2	-6. 0	-5. 7	-7. 3	-6. 9
Mat. 2	19. 66	49. 62	63. 55	69. 93	80. 21
(%)	18. 3	-4. 2	2. 9	-4. 1	0. 0
Mat. 3	15. 89	49. 71	59. 23	70. 17	77. 06
(%)	-4. 4	-4. 1	-4. 2	-3. 8	-3. 9
Mat. 4	16. 64	54. 28	63. 89	77. 12	84. 04
(%)	0. 1	4. 80	3. 4	5. 7	4. 8
Mat. 5	16. 62	51. 82	61. 79	72. 93	80. 18
Angle-ply $[(30/-30)_2]s$					
Mat. 1	23. 67	33. 64	60. 88	70. 96	73. 45
(%)	-1. 4	-6. 4	-6. 9	-2. 7	-5. 9
Mat. 2	25. 90	35. 14	64. 11	74. 19	77. 31
(%)	7. 9	-2. 3	-1. 9	1. 8	-1. 0
Mat. 3	22. 78	34. 52	62. 83	69. 40	74. 89
(%)	-5. 1	-4. 0	-3. 9	-4. 8	-4. 1
Mat. 4	23. 64	37. 79	68. 58	73. 23	81. 65
(%)	-1. 5	5. 1	4. 9	0. 5	4. 6
Mat. 5	24. 00	35. 95	65. 37	72. 90	78. 09

Table 6. Frequency parameters Ω* (Eq.(12)) of symmetric 8-layer square plates (CFFF).

	Ω_1^*	Ω_2^*	Ω_3^*	Ω_4^*	Ω_5^*
Cross-ply $[(0/90)_2]s$					
Mat. 1	10. 63	14. 34	53. 23	66. 61	71. 40
(%)	-7. 3	-6. 4	-6. 0	-7. 3	-7. 0
Mat. 2	10. 99	16. 22	55. 72	68. 86	75. 94
(%)	-4. 1	5. 9	-1. 6	-4. 1	-1. 1
Mat. 3	11. 02	14. 70	54. 29	69. 10	73. 80
(%)	-3. 8	-4. 0	-4. 1	-3. 8	-3. 9
Mat. 4	12. 12	15. 83	59. 03	75. 95	80. 66
(%)	5. 8	3. 4	4. 3	5. 8	5. 1
Mat. 5	11. 46	15. 31	56. 60	71. 82	76. 77
Angle-ply $[(30/-30)_2]s$					
Mat. 1	9. 212	22. 31	46. 28	60. 04	78. 55
(%)	-7. 3	-6. 7	-5. 5	-6. 9	-6. 2
Mat. 2	9. 740	23. 25	48. 83	62. 89	82. 95
(%)	-1. 9	-2. 7	-0. 4	-2. 4	-1. 0
Mat. 3	9. 553	22. 96	46. 94	61. 96	80. 35
(%)	-3. 8	-4. 0	-4. 2	-3. 9	-4. 1
Mat. 4	10. 46	25. 16	50. 92	67. 71	87. 45
(%)	5. 3	5. 3	3. 9	5. 0	4. 4
Mat. 5	9. 933	23. 90	49. 00	64. 46	83. 76

Table 7. Frequency parameters Ω (Eq.(12)) of symmetric 8-layer square plates (CSFF).

	Ω_1^*	Ω_2^*	Ω_3^*	Ω_4^*	Ω_5^*
Cross-ply $[(0/90)_2]s$					
Mat. 1	11.68	37.95	67.81	84.51	111.6
(%)	-6.9	-6.0	-7.2	-6.6	-6.0
Mat. 2	12.54	39.93	70.64	91.10	114.7
(%)	-0.1	-1.1	-3.3	0.7	-3.4
Mat. 3	12.06	38.72	70.28	86.91	113.9
(%)	-3.9	-4.1	-3.8	-4.0	-4.1
Mat. 4	13.16	42.08	77.12	94.42	124.1
(%)	4.9	4.3	5.6	4.4	4.6
Mat. 5	12.55	40.36	73.06	90.48	118.7
Angle-ply $[(30/-30)_2]s$					
Mat. 1	14.42	36.15	66.51	80.68	110.1
(%)	-6.7	-5.5	-6.6	-4.6	-6.6
Mat. 2	15.03	37.82	70.11	83.64	114.7
(%)	-2.8	-1.2	-1.6	-1.1	-2.7
Mat. 3	14.85	36.67	68.40	80.90	113.3
(%)	-3.9	-4.2	-4.0	-4.3	-3.9
Mat. 4	16.27	39.79	74.56	87.13	123.9
(%)	5.2	4.0	4.7	3.0	5.1
Mat. 5	15.46	38.27	71.23	84.56	117.9

Table 8. Frequency parameters Ω* (Eq.(12)) of symmetric 8-layer square plates (CCCC)

	Ω_1^*	Ω_2^*	Ω_3^*	Ω_4^*	Ω_5^*
Cross-ply $[(0/90)_2]s$					
Mat. 1	85.47	154.8	196.1	240.3	277.2
(%)	-6.7	-6.2	-7.1	-6.7	-6.0
Mat. 2	88.65	159.7	203.7	250.9	284.5
(%)	-3.3	-3.2	-3.5	-2.6	-3.6
Mat. 3	88.08	158.4	203.0	247.5	283.0
(%)	-3.9	-4.0	-3.8	-3.9	-4.0
Mat. 4	96.31	172.6	222.6	270.0	308.4
(%)	5.1	4.6	5.5	4.9	4.6
Mat. 5	91.64	165.0	211.1	257.5	294.9
Angle-ply $[(30/-30)_2]s$					
Mat. 1	82.66	135.8	194.2	214.9	255.8
(%)	-6.6	-5.5	-7.1	-4.7	-6.6
Mat. 2	85.84	140.2	202.4	221.4	265.2
(%)	-3.0	-2.4	-3.2	-1.9	-3.2
Mat. 3	85.04	137.7	201.1	215.9	263.1
(%)	-3.9	-4.1	-3.8	-4.3	-3.9
Mat. 4	92.85	149.2	220.4	232.4	287.3
(%)	4.9	3.8	5.4	3.0	4.9
Mat. 5	88.50	143.7	209.1	225.6	273.8

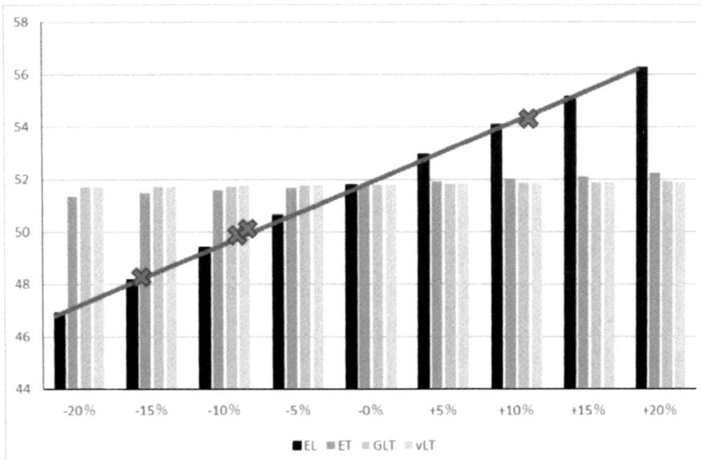

Figure 2. Variations of frequency parameter Ω_1 with change (-20%～+20%) of the four elastic constants (E_L, E_T, G_{LT}, ν_{LT}) for square plate (SSSS) with $[(30/-30)_2]$s lamination.

From the results in Tables 4-8, it is obvious that Young's modulus E_L in the fiber direction is the most decisive controlling factor among the four constants to determine the frequency parameters. Therefore, numerical experiment is done in Fig.2 to calculate Ω_1 for the angle-ply square plate (SSSS) [30/-30/30/-30]s by changing the four elastic constants with the range of 80 %～120 % from the reference value of Mat.5 (i.e., E_L=150GPa, E_T=10GPa, G_{LT}=5GPa, ν_{LT}=0.3). In the figure, the values with change of E_L are denoted by black solid columns and those with other three constants E_T, G_{LT}, ν_{LT} are by lighter grey columns. As clearly seen, the frequency values increase significantly in linear fashion with change of E_L, while those values for change of the other three constants stay almost unchanged. The cross marks (×) in the figure are the frequency values Ω_1 plotted from Table 4 (Ω_1 for Mat.1-4). Thus, the straight line can be used to approximate the frequency values for laminated plates with different CFRP materials under specified lamination condition.

5. Conclusions

It has been demonstrated that, for symmetrically laminated CFRP plates having slightly different elastic constants, there is a underlying relation among values of frequency parameters when the parameters are properly defined. The Ritz method was used as an analytical tool in numerical experiments for solving free vibration of the plates. In numerical results, examples are given for cross-ply and angle-ply square plates. The frequency parameters for four different sets of materials are calculated and compared to those of new hypothetical material with the averaged and discretized constants. It turned out that, despite the difference of CFRP materials, the frequency parameters show unified behaviors with the change of Young's modulus in the fiber direction. It was suggested that it is feasible to derive approximate formulas to simultaneously predict the frequency parameters for laminated plates composed of different CFRP materials.

References

[1]. Leissa, A.W. 1993 *Vibration of Plates*, Acoustical Society of America; previously, 1969 NASA SP-160, U.S. Government Printing Office, Washington D.C.
[2]. Leissa, A.W. 1973 The free vibration of rectangular plates, J. Sound Vib., vol.31, pp.257-293.
[3]. Vinson, J.R., Sierakowski R.L. 1986 *The Behavior of Structures Composed of Composite Materials.* Martinus Nijhoff Publishers, Dardrecht, The Netherlands.
[4]. Jones, R.M. 1999 *Mechanics of Composite Materials*, 2nd ed. Taylor & Francis.
[5]. Reddy J.N. 1997 *Mechanics of Laminated Composite Plates: Theory and Analysis*, CRC Press, Boca Raton, FL.
[6]. Sharma A.K., Mittal N.D. 2010 Review on stress and vibration analysis of composite plates, J.

of Appl. Sci., vol.10(23), pp.3156-3166.

[7]. Bert C.W., Mayberry B.L. 1969 Free vibration of unsymmetrically laminated anisotropic plate with clamped edges, J. Compos. Mater. Vol.3 pp.282-293.

[8]. Reddy, J.N. 1984 A simple higher-order theory for laminated composite plates, Trans. ASME J. Appl. Mech., vol.51, pp.745-752.

[9]. Narita, Y. 2000 Combinations for the free-vibration behaviors of anisotropic rectangular plates under general edge conditions, Trans. ASME J. Appl. Mech., vol.67, pp.568-573.

[10] Narita, Y. 2006 Maximum frequency design of laminated plates with mixed boundary conditions, Int. J. Solids Struct., vol.43, pp.4342-4356.

[11] Pandit, M.K., Haldar, S., Mukhopadhyay, M. 2007 Free vibration analysis of laminated composite rectangular plate using finite element method, J. of Reinforced Plastic Compo., vol.26, p.69.

[12] Stanford B.K., Jutte C.V. 2017 Comparison of curvilinear stiffeners and tow steered composites for aeroelastic tailoring of aircraft wings, Comput Struct, vol.183, pp.48-60.

[13] Ogasawara T., Ishikawa T. 2010 Proposal of a convenient compressive test method for carbon fiber reinforced plastics composites (in Japanese), J. Japan Soc. Comp. Mat., vol.36, pp.33-40.

[14] Panesar A.S., Weaver P.M. 2012 Optimisation of blended bistable laminates for a morphing flap, Compos Struct, vol.94, pp.3092-3105.

IOP Conf. Series: Materials Science and Engineering **676** (2019) 012028 doi:10.1088/1757-899X/676/1/012028

The Effectiveness of PAC and Chitosan Usage in Jeneberang River Raw Water Treatment

B Bakri*[1], M Selintung[2], R N Hamdani[2], M Ihsan[3], Y Arai[4]

[1]Civil Engineering Department, Hasanuddin University, Makassar, Indonesia
[2]Environmental Engineering Department, Hasanuddin University, Makassar, Indonesia
[3]Civil Engineering Department, Sekolah Tinggi Teknik Baramuli, Pinrang, Indonesia
[4]Department of Civil and Environmental, Tokyo Metropolitan University, Tokyo, Japan

*E-mail: bambangbakri@gmail.com

Abstract. Raw water treatment commonly uses effective coagulants with the purpose to reduce the concentration of particles causing turbidity. One of the most popular coagulants is poly aluminum chloride (PAC) with some advantages. It has low corrosiveness. Formed floccules are relatively easier to detach and pH result is in moderate level. Another alternative chemical substance as a coagulant is Chitosan, a high degradability, non-toxic substance, with high market availability. Chitosan usage produces small amount of sludge and no cationic residue. This research intended to analyze the comparison of effectiveness of PAC and chitosan as coagulants for raw water treatment in Jeneberang River. Samples were taken in two locations; at the long storage (first location) and at the ground sill (second location) in Jeneberang River. Samples were analysed at Water Quality Laboratory of Environmental Engineering Department, Engineering Faculty, Hasanuddin University. For first location, PAC has 94.57% effectiveness to reduce turbidity when compared to as for chitosan which gained 69.67% effectiveness in reducing turbidity. Therefore, the results show that PAC is more effective and efficient as a coagulant for raw water treatment in Jeneberang River when compared to Chitosan.

1. Introduction

In the processing raw water into clean water, several methods are carried out namely physical, chemical and biological processing. In physical processing, the methods include filtration and sedimentation. Biological processing is usually conducted to kill pathogenic microorganisms by introducing disinfectant. In chemical processing, the processing is carried out by adding a chemical compound called coagulant and flocculent to function as a water purifier [1-3]. In general, the most commonly used coagulant is aluminum sulphate or colloquially named as alum. However, Poly Aluminum Chloride (PAC) has been found to perform better than aluminum sulphate. In addition to PAC, an alternative material that can be used is chitosan. Chitosan is a chitin deacetylation product, which is a long chain polymer from glucosamine (2-amino-2-deoxiglucose). The molecular weight of chitosan depends on the degradation that occurs in the process of its production. Chitosan has free amino groups as polycationic, chelating and forming dispersions in acetic acid solutions [4]. This study aims to compare the effectiveness of raw water treatment using two alternative materials, which are Poly Aluminum Chloride (PAC) and Chitosan. Samples of water to be used are sourced from the Jeneberang River as one of the sources of raw water in Makassar City's clean water production. The objectives of this research are to analyze the effectiveness of the results of processing using Poly Aluminum Chloride (PAC) and chitosan as coagulants for raw water treatment in Jeneberang River and to compare the effectiveness and efficiency of both materials.

2. Literature Review

If the solution is added with a solution with different charge in colloid system, the colloid system will pull the other to form a double layer. The first layer is a solid layer where the charge of the colloidal particle attracts ions with the opposite charge from the dispersing medium. The second layer is a diffusion layer where the charge from the dispersing medium diffuses into the colloidal particles. This multiple layer model is called the Strern double layer. The existence of this layer causes overall solution to be neutral. If the colloidal particles are neutral, there will be clotting and precipitation due to the influence of gravity. This process of clotting and precipitation is called coagulation [5]. PAC is a special aluminum chloride base salt designed to provide coagulation and flocculation that is stronger and better than ordinary aluminum and iron salts. Chitosan ($C_6H_{11}NO_4$) is a polymer with the chemical name 2-amino-2-deoxy-D-glucose, containing free amino groups in its carbon chain and positively charged. This free amine group has many functions for chitosan. Chitosan in an amorphous solid is one of the few natural polymers in the form of cationic polyelectrolytes in organic acid solutions [6].

3. Research Method

The type of research that will be carried out is an experimental development research conducted to prove the hypothesis in order to compile a generally accepted conclusion. This research was conducted in October 2017 at the Water Quality Laboratory of the Faculty of Engineering, Hasanuddin University which lasted in 3 days. Material collection was carried out in the Jeneberang River, namely Long Storage and Groundsill. The locations were chosen because it is a source of raw water for water treatment plant in its vicinity. Equipment in this research includes a set of jar test, turbidimeter, and pH meter. Raw water from the Jeneberang River was then mixed with 1% coagulant solution, and 10% PAC solution.

Testing the effectiveness of the coagulant solution is carried out in three stages. The first stage is the dissolution of the coagulant in the sample water, stirred in 150 rpm speed for 1 minute. The second stage is 50 rpm slow stirring to form floccules for 10 minutes. The third stage is the sedimentation process for 10 minutes. Determination of the dose of coagulant that will be used is determined based on the jar-test test where a dose of 10 ppm of coagulant solution can reduce turbidity in the sample water. To investigate the effectiveness of the two types of coagulants, the dosage was varied to 10 ppm, 20 ppm, and 30 ppm.

The turbidity measurement tool used is Lutron TU 2016 Turbidity Meter which has been calibrated to a standard of turbidity values in accordance with the standard of 0 NTU. Turbidity measurements used homogenised 10 mL of sample water (raw water and jar test water) which was put into the cleaned cuvette and then put into a turbidimeter. The results will be displayed after 10 seconds on the turbidimeter. The turbidity of water in jar-test was then measured for every 2 minutes to see a reduction in turbidity for a period of 10 minutes.

Electrodes were installed at pH-meter and dipped in a pH 4 buffer and then into a pH buffer 9 then the electrodes were transferred and dipped at pH 7. After the pH-meter was calibrated, the electrode was dipped in the sample solution of mixed raw water and jar test water and the pH indicated on the pH-meter device was recorded.

The effectiveness of the coagulant in reducing the turbidity value is expressed in percentage (%) and calculated using the following equation

$$ef\,(\%) = \left(\frac{initial\ value - final\ value}{initial\ value} \right) \times 100$$

where,

ef (%)	= effectiveness in percentages
Initial value	= turbidity value before processing
Final value	= turbidity value after processing

4. Results and Discussion

4.1. Relationship of sedimentation time and turbidity values

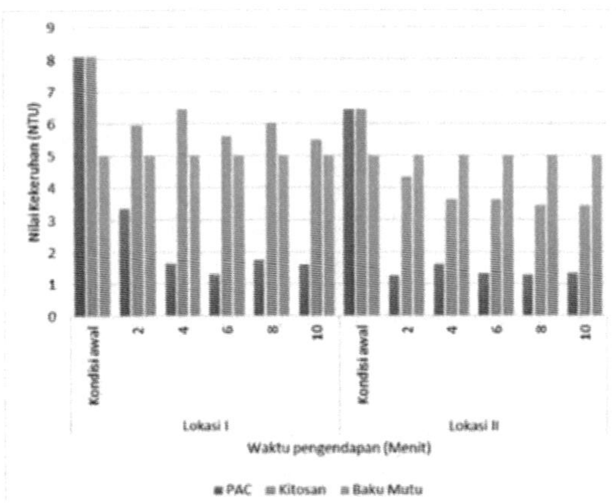

Figure 1. Results of 10 ppm measurements

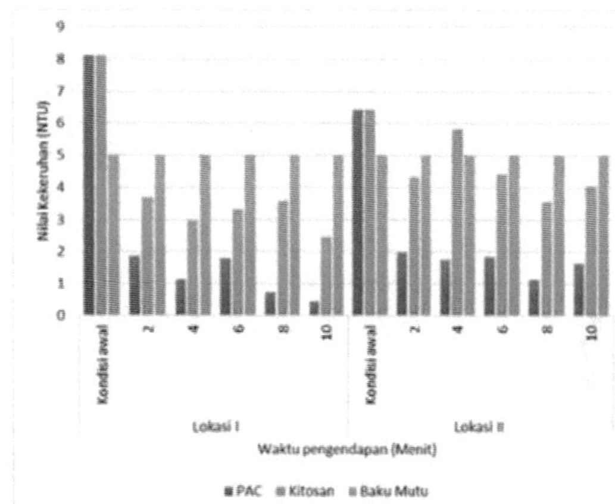

Figure 2. Results of 20 ppm measurements

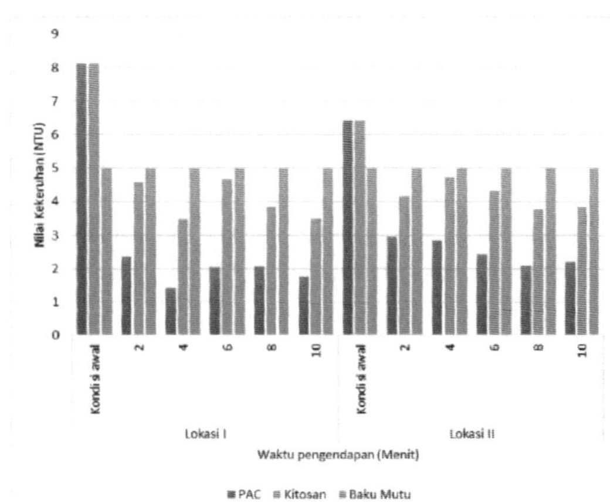

Figure 3. Results of 30 ppm measurements

Based on the graph above it can be seen that PAC coagulant is more effective in reducing turbidity values, and optimal precipitation occurs at a dose of 20 ppm. The highest effectiveness for location I is 94.575% using PAC coagulant as much as 20 ppm while the effectiveness for location II is 79.004% using PAC coagulant at a dose of 10 ppm.

4.2. Effectiveness of coagulants in reducing turbidity

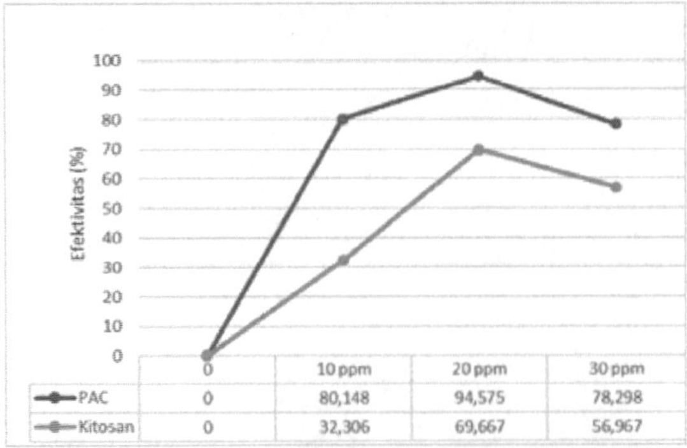

Figure 4. Effectiveness of coagulants in location I

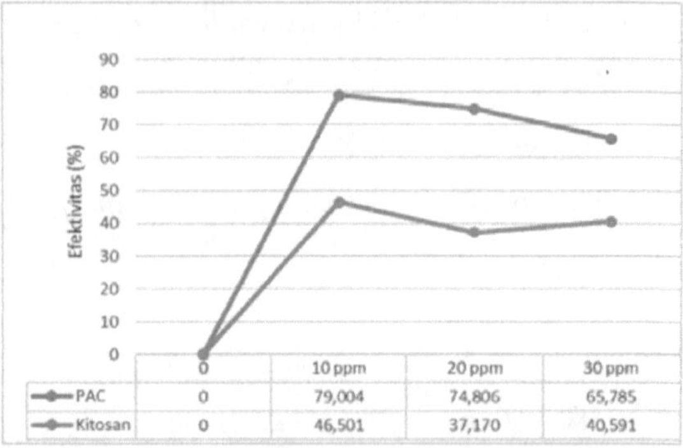

Figure 5. Effectiveness of coagulants in location II

4.3. Results of pH measurements

The pH value showed that the coagulation results with the addition of PAC coagulant were at the optimum concentration for 10 ppm, which are 6.8 at location I and 6.9 at location II. Whereas with the addition of chitosan, the pH value drops to 6.6 and has met the requirements of drinking water according to drinking water quality standards. The pH value decreases because both coagulants have acidic properties. In the sample using PAC, the pH will get more acid as the coagulant dose increases. This is because PAC contains free hydrogen ions produced from hydrolysis reactions. In general, the more coagulants used, the lower the pH will be higher [7].

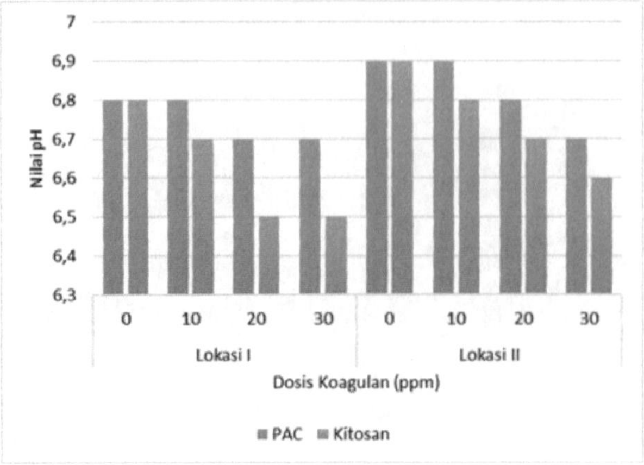

Figure 6. Results of pH measurements

4.4. Relationship of pH and the effectiveness of coagulants

In the graph below it can be seen that the highest effectiveness of PAC coagulant for location I uses a dose of 20 ppm with a pH value of 6.7. As for location II, a 10 ppm PAC dose provides the highest effectiveness with a pH value of 6.9. Based on this finding, it is known that PAC coagulants can reduce turbidity optimally at pH 6.7 - 6.9.

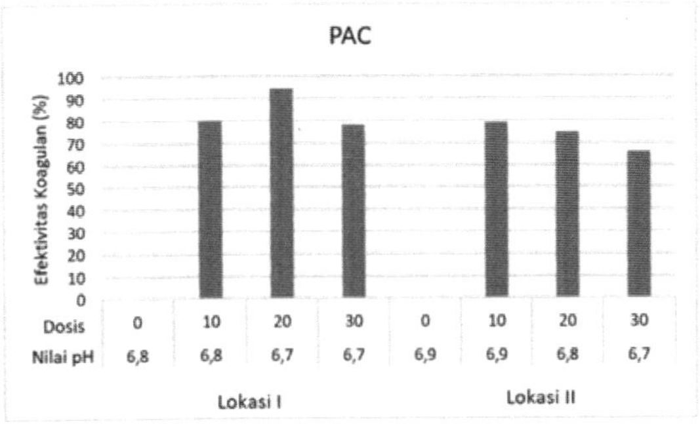

Figure 7. pH of PAC coagulant

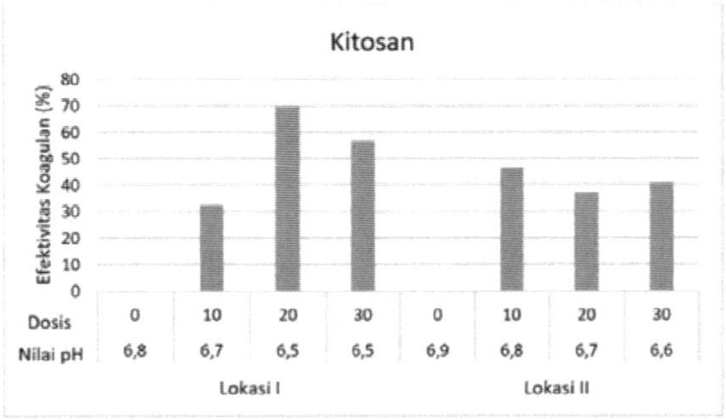

Figure 8. pH of Chitosan coagulant

The use of chitosan showed the highest effectiveness at a dose of 20 ppm at location I with a pH of 6.5 while at location II, the effectiveness of the highest coagulant occurred at a dose of 10 ppm with a pH value of 6.8. Based on this, it can be seen that chitosan coagulant can reduce turbidity optimally at pH 6.5 - 6.8. The initial pH value can affect the solubility of a coagulant. PAC has a large solubility in the pH range 5-9. While chitosan coagulant dissolves in acidic solutions with a pH range of around 4-6. The easier a coagulant to dissolve, the formation of aquometallic ions will be easier which eventually accelerate neutralized colloidal particles to form floccules. A higher pH gives smaller the solubility of the coagulant causing aquometallic ions to be increasingly difficult to form, which ultimately reduces the number of colloidal particles that can be neutralized to form floccules [8].

4.5. Production cost analysis

The calculation of production costs for both coagulants uses raw water discharge according to processing capacity in the treatment plant is as follows;

Table 1. Analysis of production costs

Coagulant	Usage (mg/L)	Hourly demand (kg/h)	Daily demand (kg/day)	Monthly demand (kg/mo.)	Price per kg	Production cost per month
PAC	20	3.6	86.4	2592	5,000	Rp 12,960,000
Chitosan	20	3.6	86.4	2592	54,000	Rp 1,399,680,000

A supporting factor causing PAC coagulant to more superior than chitosan coagulant is that PAC coagulant is in powder form which dissolved easily with water while chitosan can only dissolve with acidic solvents making its usage to be less efficient.

5. Conclusion

Based on the results of testing and data processing, it can be concluded that the PAC coagulant can reduce the turbidity of the Jeneberang River water with the highest effectiveness of 94.57%, at Location I with a dose of 20 ppm. For Location II the highest effectiveness was 79.01% at a dose of 10 ppm. Chitosan coagulant can reduce the turbidity of Jeneberang River water at Location I with the highest effectiveness of 69.67% at a dose of 20 ppm. For Location II the highest effectiveness is 46.5% with a dose of 10 ppm. The use of PAC coagulant is more effective than chitosan coagulant in reducing turbidity of Jeneberang River raw water in accordance with quality standards. In economic terms, the use of PAC coagulants is more efficient than chitosan coagulants.

IOP Conf. Series: Materials Science and Engineering **676** (2019) 012028 doi:10.1088/1757-899X/676/1/012028

References

[1] Ravina L 1993 *Coagulation and Flocculation*, (Virginia: Zeta-Meter, Inc).

[2] Gozan, Misri 2006 *Water treatment for plant utilities,* (Depok: Department of Chemical Engineering, University of Indonesia).

[3] Noviani H. 2012 *Analysis of the use of poly aluminum chloride coagulant (PAC) and Chitosan in the water purification process in PDAM Tirta Pakuan Bogor*. (Bogor: Pakuan University)

[4] Knorr D 1983 Dye binding properties of Chitin and Chitosan. *J. Food Sci*. 48. P: 36-41.

[5] Pararaja 2008 *Review: Process of Coagulation & Flocculation in a Water Treatment Plant*. Skima Madiun.

[6] Hirano S 1986 *Chitin and Chitosan,* In *Encyclopedia of Industrial Chemistry*, (Weinheim: New York).

[7] Yuliastri I R 2010 *Use of Moringa Oleifera seed as coagulant and flocculent in improving the quality of wastewater and groundwater,* (Jakarta: Syarif Hidayatullah State Islamic University).

[8] Karamah EF 2014 *Treatment of coagulation in process membrane water treatment: effect of slow stirring time aluminum sulfate coagulant against membrane performance*. (Depok: University of Indonesia).

IOP Conf. Series: Materials Science and Engineering **676** (2019) 012029 doi:10.1088/1757-899X/676/1/012029

Tidal Flood in Pekalongan: Utilizing and Operating Open Resources for Modelling

M B Pratama[1]*

[1]Faculty of Civil & Environmental Engineering, Institut Teknologi Bandung, Bandung, Indonesia

*E-mail: munawirbintang@gmail.com

Abstract. The open resource materials are used in this research to construct a numerical model and study the tidal flood in Pekalongan City, including the impact of land subsidence. The area of interest is in 6 villages in the north coast of Pekalongan namely, Jeruksari, Pabean, Bandengan, Kandang Panjang, Panjang Baru, and Panjang Wetan. Delft3D, an open ware, is used for numerical modelling. While the input data are obtained from GEBCO for bathymetry, SRTM for topography, and TPXO for tidal constituent. The resulting model shows a decent agreement with field tidal elevation data from Indonesian Navy Tide Tables and GLOSS. After applying the annual land subsidence, compare to flood in May 2018, it is found that the inundation height increases for around 11, 36, and 132 cm in 2019, 2025, and 2050. In 2050, an extreme inundation over 2.5 m is found. Similar with Jakarta flood, to protect the land area, coastal dykes may be proposed. However, for long-term resilience, it is recommended to emphasize efforts in decreasing the land subsidence rate. This study indicates the threat of tidal flood in Pekalongan, thus further studies followed with field survey and more constituents in the modelling are strongly recommended.

1. Introduction

Land subsidence has been an important subject of studies in the past decades. Researcher across the globe has studied the subsidence rate [1], causes [2], impacts [3], and how to adapt to the threat especially in a coastal-front area at various countries [4]. In Indonesia, one of the sexiest topics is the capital city, Jakarta. Northern Jakarta has been regularly flooded in a certain time due to the tidal propagation intrusion and high precipitation. The land subsidence worsens the condition more significant than the sea level rise since it is declining the land elevation constantly so then the land is lower than the mean sea level [4].

However, Jakarta is not the only place suffered. Until the present time, some regions at the Northern Coast of Central Java are also threatened by the tidal flood, Pekalongan City for an instance [5]. National and local news reported that the city was severely flooded in the second half of May 2018 [6-9]. Tracking the history, the flood has been happening regularly years before. Meanwhile, Pekalongan can be described as one of the representatives of Indonesia batik culture [10]. Thus, to preserve human, culture, and assets in Pekalongan City, a study to understand and forecast the future scenarios of Pekalongan is needed.

This study aims to construct a reliable hydrodynamic model to simulate the tidal flood in Pekalongan City and its surrounding area. The highlight of this study is also the utilization of various kind of open data which freely accessible on the internet. Not to mention, the tool for numerical

IOP Conf. Series: Materials Science and Engineering **676** (2019) 012029 doi:10.1088/1757-899X/676/1/012029

modelling itself is also an open ware. In the end, this study will present the inundation forecast at 2019, 2025 and 2050.

2. Methodology

Study location is in the northern part of Java Island as shown in Figure 1, there are global domain, local domain and main area. Global domain covers West Java Province and Central Java Province. Local domain covers Pemalang Regency, Pekalongan Regency, Pekalongan City and Batang Regency, in Central Java Province. The main area is in the northern region of Pekalongan City. The illustration of study locations is presented in Figure 1.

The numerical modelling takes the main role in the study which performed to construct a reliable simulation of tidal propagation at Java Sea and its intrusion to the urban residential area in the northern coast of Pekalongan City. In this study, the only constituent included is the tidal, however, in the future, the involvement of wind and other forcing is promising. The modelling time frame is specified at the third to fourth week in May which is known as the worst period of the tidal flood in Pekalongan in 2018.

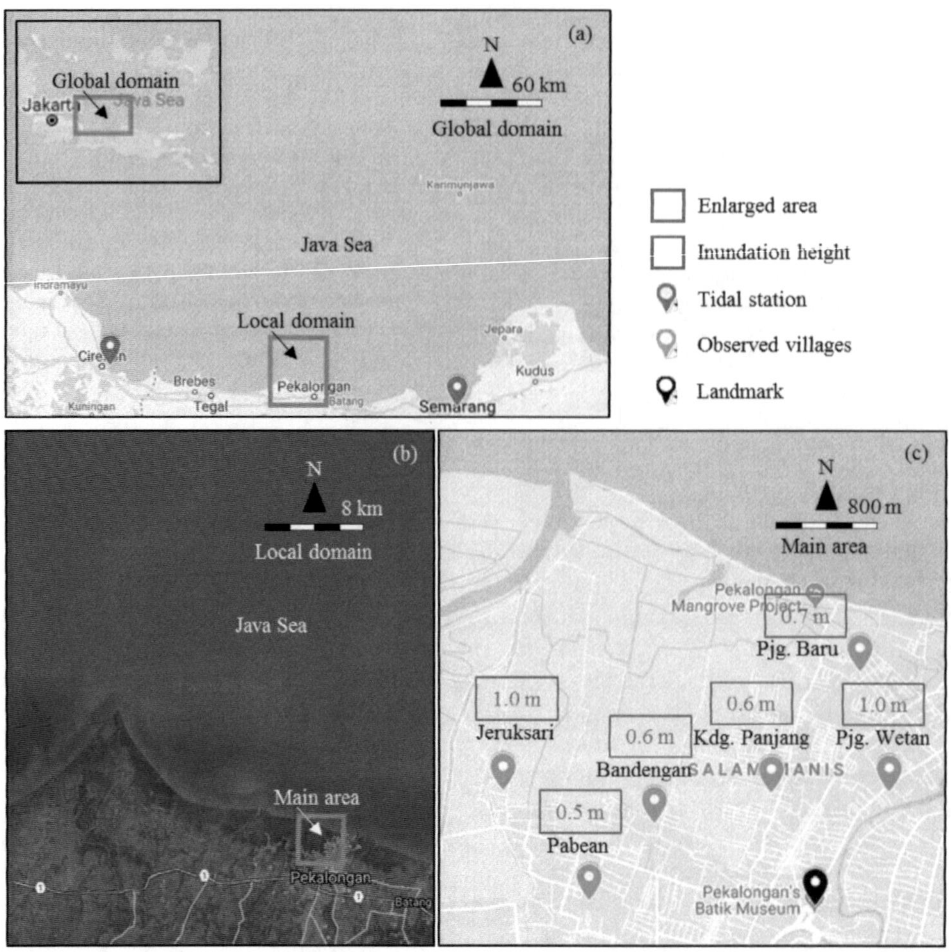

Figure 1. Study locations and the three categories of area of interest, the (a) global domain, (b) local domain, and (c) main area

The study framework is given in Figure 2. The numerical model is set up few times before result in a forecast model. The steps are rough, initial, benchmark, and forecast model. The rough model is a model defined by default value in the modelling tool and default input from open data. The initial model is a rough model validated with tidal elevation record. The benchmark model is a model which reliable to simulate the tidal propagation to land area, resulting in a correct value of inundation level. Finally, the forecast model is the future inundation model in Pekalongan City, set in 2019, 2025, and 2050.

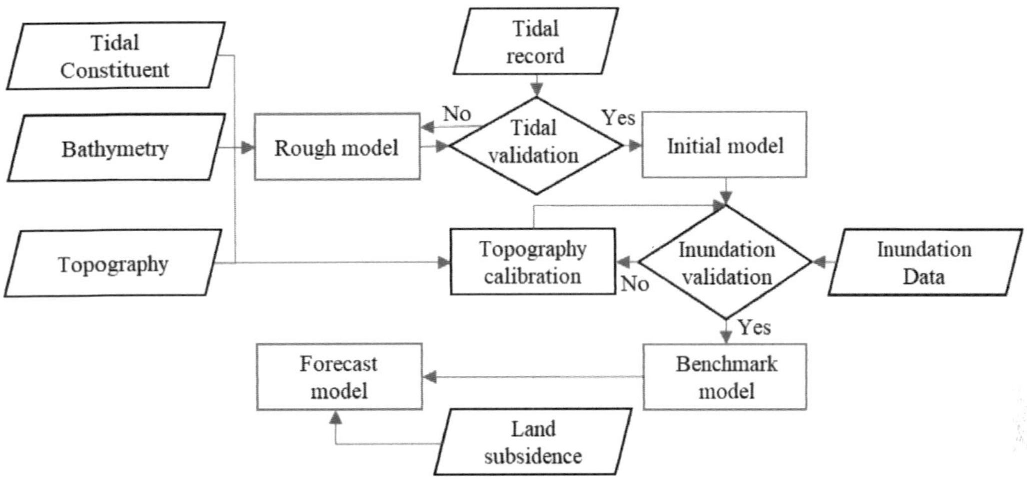

Figure 2. Framework of numerical modelling.

2.1. Data compilation

As mentioned, open data and open ware combination is the highlight of this study. The numerical model is constructed using this widely available open resources. As in present, a large number of researcher have also applied these resources in their studies.

As shown in Figure 2, the required input to assemble the model are bathymetry, topography, and tidal boundary. Bathymetry data are essential for hydrodynamic modelling in the coastal environment since the water depth is one of the main variables in the shallow water equation. The bathymetry data used in this study is obtained from the GEBCO, which operates under the joint auspices of the International Hydrographic Organization (IHO) and the Intergovernmental Oceanographic Commission (IOC) [11]. While for land elevation, topography data is taken from the Shuttle Radar Topography Mission (SRTM) provided by the National Geospatial-Intelligence Agency (NGA) and National Aeronautics and Space Administration (NASA) [12]. Data resolutions for each data are per 30 and 1 arc second for the bathymetry and topography respectively. Tidal boundary is generated using TPXO 7.2, a tool for generating tidal constituent [13]. The program is developed by the Oregon State University in the United States.

Further for model validation, tidal elevation and inundation height record are gathered. There are two sources of tidal elevation data, they are the Indonesia Navy Tide Tables 2018 and GLOSS database. Tidal data at Semarang and Cirebon Station are obtained from the forecasted tide tables [14] which have been widely used in Indonesia. However, to present a recorded tidal data including the occurring abnormal tidal behaviour, GLOSS database is used which provides a dataset of tidal observation in Indonesia including in Semarang Station [15]. The comparison of tidal elevation at Cirebon and Semarang Station on May 15th – 30th, 2018 are shown in Figure 3 while the station locations are denoted in Figure 1.

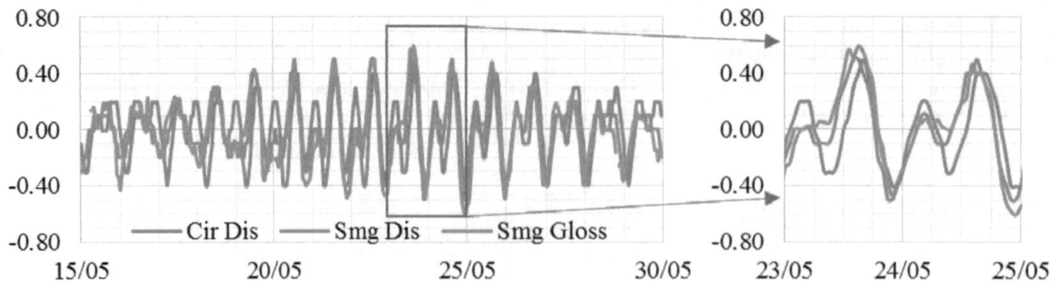

Figure 3. Comparison between the three obtained tidal data, Cirebon Tide Tables (blue line), Semarang Tide Tables (orange line), and Semarang Tidal Record (grey line).

For inundation data of tidal flood, since there is not any document yet which provide solid data of inundation height and location, national and local newspaper are chosen as the sources. The data is obtained at the 6 villages around North Pekalongan District, in Pekalongan City, the inundation level is varied from 0.5 – 1 m on May 22nd and 23rd 2018 [6-9]. The location and inundation level are illustrated in Figure 1.

3. Numerical Model

Numerical model in coastal environment has been widely used by researcher around the world. The range of its utilization is about ocean renewable energy [16], coastal process [17], contaminant transport [18], and even for port planning [19] or other coastal infrastructure as well. For this study, the numerical modelling is proposed to show the impact of tidal propagation and intrusion in the north coast of Pekalongan City.

Some of well-known modeling tools at coastal environment are the MIKE [16], Delft3D [20], Telemac [20], SMS [18], and etc. Particularly in this study, Delft3D is chosen since it provides the open ware version for research activity and has a user-friendly interface.

3.1. Model description

Delft3D is one of the leading tools in hydrodynamic modelling for coastal, river, and estuarine areas. The tool is capable to simulate phenomena of flows, sediment transport, waves, water quality, and etc. Delft3D-FLOW which used in this study is a module in the Delft3D package. It uses and solves the Navier Stokes equations for an incompressible fluid, within the shallow water and the Boussinesq assumptions [21].

The modelling is set as a 2-dimensional model using flood scheme to able the tidal propagate into land area. The model is designed into two stages, the global and local domain. The global domain covers a large water area at northern part of West and Central Java. The local domain covers Pekalongan City and 3 other regencies. The global and local domain are presented in Figure 4. Each of Figures 4a and 4b correspond to Figures 1a and 1b.

The mesh of numerical model, both in the global and local model is design as a uniform rectangular grid from the boundary to the interest area. The mesh size of global domain is 1 x 1 km while the local domain is 5 times finer [20], which designed to be smaller to provide a more accurate model result. Overall, the model only applies the offline nesting, leaving the online nesting out. The transition of global and local domain is driven by the nesting 1 and 2 features in Delft3D [21]. Further, to assure period for model to be stable, 2 – 3 days spin up time is applied to both domains.

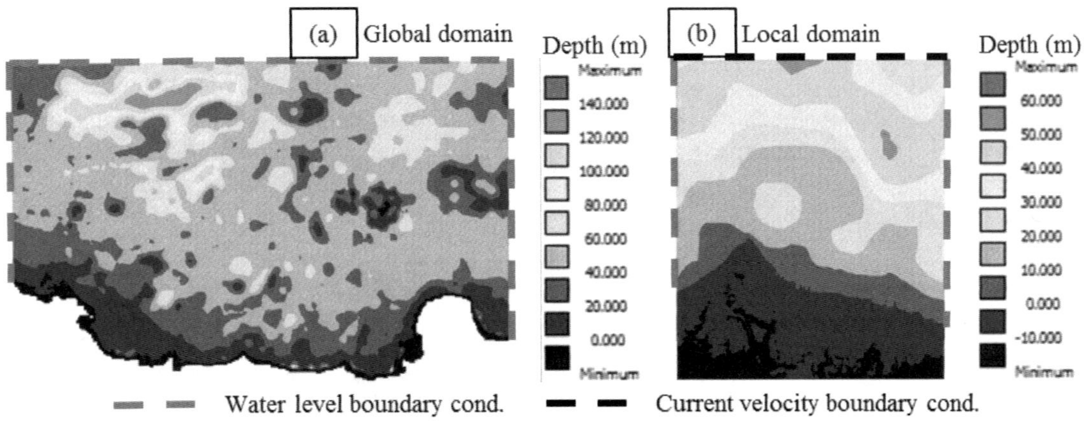

Figure 4. Overview of (a) The global and (b) local domain, including bathymetry, land elevation, and boundary condition setting

The seabed and land elevations are obtained from the GEBCO and SRTM combination which also have been used in Takagi et. al. (2016) for a similar study in Jakarta, the capital city of Indonesia [3]. The resulted bathymetry and land elevation are shown in Figure 4. The tidal constituents at boundary conditions of global domain are generated using TPXO 7.2 [13] while the boundaries of local domain are from global domain output translated using nesting features. The northern boundary is set in current velocity (black line in Figure 4b) and western and eastern boundaries (red lines in Figures 4a and 4b) are in water level. This arrangement is adapted from A. A. Rahman et al. (2017) to result in a more stable model [20].

To result in a reliable rough and initial model (stated in Figure 2), the tidal constituent and topography data are required to be calibrated, respectively. Particularly for the topography calibration, an effort to correct the data is needed since the land elevation is the main variables to determine the inundation height. In addition, the modelling which including land area, the manning value used is 0.02 and 0.04 for water and land respectively [22]. The summary of the 4 models set up is presented in Table 1.

Table 1. Set up of the rough, initial, benchmark, and forecast model.

Parameters	Rough	Initial	Benchmark	Forecast
Domain	Global	Local	Local	Local
Grid size	331 x 216	159 x 188	159 x 188	159 x 188
Grid resolution	1 x 1 km^2	0.2 x 0.2 km^2	0.2 x 0.2 km^2	0.2 x 0.2 km^2
Time step	4 minutes	1 minute	1 minute	1 minute
Time frame	May 9th – 30th, 2018	May 11th – 30th, 2018	May 11th – 30th, 2018	Nov. 12th – 30th, 2019 Nov. 20th – Dec. 8th, 2025 Oct. 28th – Nov 17th, 2050
Roughness (Manning's n value)	Uniform 0.02	Uniform 0.02	Water 0.02 Land 0.04	Water 0.02 Land 0.04

3.2. Scenarios

There are 4 modelling steps as presented in Figure 2. In the forecast model, land subsidence parameter is applied. The used annual subsidence rate is shown in Figure 5 which is adapted from Chaussard et. al. [2], varying spatially along Pekalongan City and Pekalongan Regency [2]. The minimum and maximum values are approximately 0 and 8 cm/year at the highland area and Bandengan Village respectively. Models with different land elevation affected by the land subsidence are run in 2019, 2025, and 2050. The result is inundation level at the 6 investigated villages and the spatial distribution.

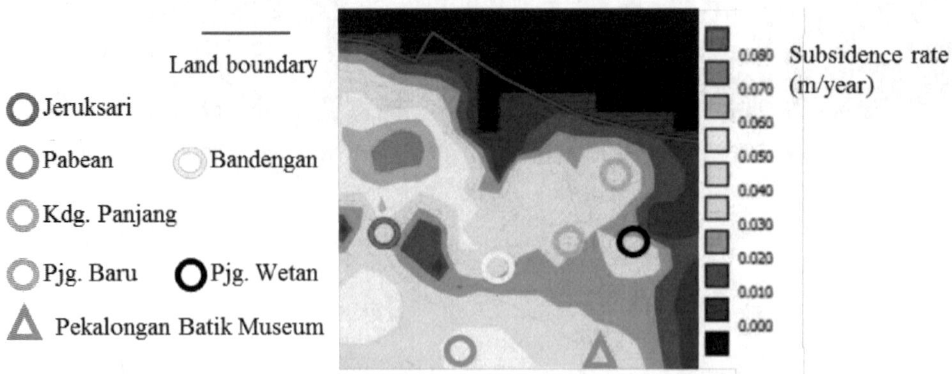

Figure 5. Annual land subsidence rate, adapted from Chaussard, et al. [2]

The forecast model is set in the highest tidal elevation to represent the most extreme inundation occurrence. To know the time frame of this high tidal elevation at those years, ErgTide, a software to predict tidal elevation is used. Ergtide also has been used in few studies [23-24]. ErgTide builds the tidal constituents from a known tidal dataset (at least for 14 days) using least square analysis and converts it again into elevation data at any time. As shown in Figure 6d, the predicted elevation by ErgTide is showing a decent agreement to both model and field data. After generated for full year in 2019, 2025, and 2050, the obtained maximum tidal elevations are 46 cm (November 22nd 2019), 48 cm (November 30th 2025), and 40.5 cm (November 9th 2050). Considering the spin-up time and to show a cycle of flood-ebb tide, the time frame for forecast models are determined as shown in Table 1.

4. Result and Discussion

4.1. Tidal elevation validation

The resulting model data are compared with the field data at the corresponding locations. The errors of model and field data are calculated using mean absolute error formula as defined in Eq. 1. There are two sites, Cirebon and Semarang. In Semarang, there are the Indonesian Navy (Dishidros) and the Gloss data. The comparison is given in Figures 6a – 6c for Cirebon Dishidros, Semarang Dishidros, and Semarang Gloss respectively. The obtained value are 6.4, 4.1, and 5.8% respectively. Thus, the simulation is concluded to be reliable to simulate the tidal propagation decently.

$$Error = \frac{|Model\ data - Field\ data|}{Tidal\ range} \quad (1)$$

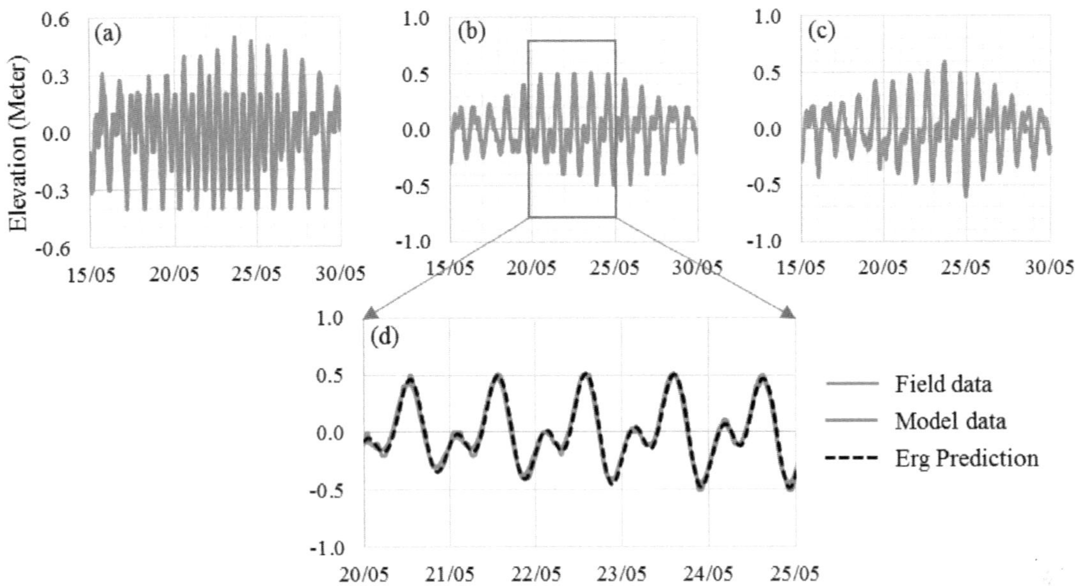

Figure 6. Tidal elevation validation result for (a) Cirebon Tide Tables, (b) Semarang Tide Tables, and (c) Semarang Tidal Record also (d) the predicted tidal from Ergtide.

4.2. Topography calibration

Previously, inundation depths at 6 villages (see Figure 1) have been introduced as summarized in Table 2. With the validated tidal elevation model (in the initial model), the resulting inundation depth is found to be 0 meter at all districts (as given in Table 2). It indicates that the land elevation is overestimated which is reasonable since SRTM data accuracy is around 10 meters [25] and taken in 2,000 [12]. To correct the data, elevation of 5 rail stations [26] in the global domain are taken as a benchmark. Further, the land elevation data in a region is reduced with the difference between the SRTM data with the rail stations and inundation level data in the corresponding region. Model equipped with land elevation corrections which resulting in a closer inundation value as shown in Table 2, is taken as a benchmark model.

Table 2. Inundation height comparison for field data vs initial and benchmark model.

No	Villages	Longitude (E)	Latitude (S)	Dates (May 2018)	Inundation (Meter)		
					Field Data	Initial Model	Benchmark Model
1	Jeruksari	109.65426	6.86999	22nd	1.0 [6]	0	1.02
2	Pabean	109.66034	6.88022	22nd	0.5 [8]	0	0.52
3	Bandengan	109.66449	6.87221	23rd	0.6 [6]	0	0.62
4	Kandang Panjang	109.67237	6.87015	22nd	0.6 [6]	0	0.67
5	Panjang Baru	109.67846	6.86198	22nd	0.7 [9]	0	0.72
6	Panjang Wetan	109.68042	6.87012	22nd	1.0 [7]	0	0.97

4.3. Future inundation and Solutions

As the forecast models have run in the mentioned time frame in Table 1, the resulting inundation height is given in Figure 7, which taken at the highest tidal condition. Clearly seen that in 2019 the inundation values around 1 – 1.5 meter then in 2025 and 2050 the range grows to 2 and 3 meters respectively.

IOP Conf. Series: Materials Science and Engineering **676** (2019) 012029 doi:10.1088/1757-899X/676/1/012029

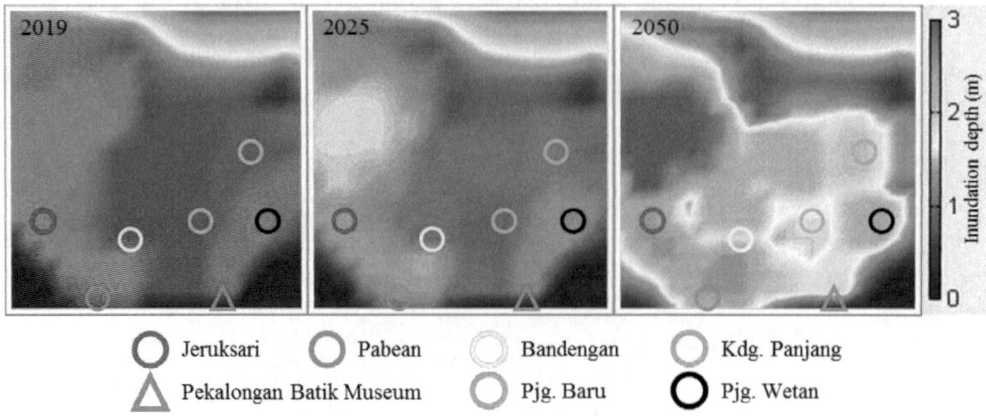

Figure 7. Forecasted spatial inundation at 2019, 2025, and 2050, including the effect of land subsidence

To summarize the specific inundation heights at 6 observation point (marked as coloured dots), Table 3 is presented. Compare to inundation in 2018 (given in Table 2), 2019's model shows an increment around 8 – 14 cm. While in 2025, the increment rises to around 31 – 41 cm. It is also shown that if the existing rate of land subsidence is continuing to 2050, inundation in an extreme level is inevitable, over 2.5 meters at worst.

Table 3. Forecasted inundation height in 2019, 2025, and 2050

No	Villages	Inundation height (meter)		
		2019	2025	2050
1	Jeruksari	1.14	1.40	2.51
2	Pabean	0.64	0.91	2.08
3	Bandengan	0.73	0.93	1.82
4	Kandang Panjang	0.76	0.92	1.59
5	Panjang Baru	0.83	1.06	2.04
6	Panjang Wetan	1.08	1.31	2.28

As observed in Figure 6, the maximum tidal elevation at Pekalongan Coast is around 45 cm or 60 cm (in the abnormal condition). Considering the existing inundation given in Table 2, it is certain that some area is already below the mean sea level (MSL). For instance, Jeruksari and Panjang Wetan are approximated to be 55 cm below the MSL since the 45 cm tide elevation resulting 100 cm inundation.

Exactly similar occurrence takes place in few cities around the world, namely Jakarta, the Capital City of Indonesia where 2 meters coastal dyke is built to protect the land area from tidal inundation. This coastal dyke can also be applied in Pekalongan. However, it is only for short term since the coastal dyke will subside along the land in the meantime caused by land subsidence. To fight tidal inundation, the core problem is the excessive groundwater extraction. So, it is strongly recommended to manage the groundwater consumption in Pekalongan coastal area and another coastal-front city in Indonesia.

5. Conclusion

Numerical hydrodynamic modelling using open ware Delft-3D is performed in this research. The input data are based on open data which are the GEBCO for bathymetry, SRTM for land elevation, and

TPXO 7.2 for tidal constituent. Open access tidal record data of GLOSS is also used to validate the constructed model. In addition, the Indonesian Navy's tidal elevation data is used to ascertain the validity. The utilization of GEBCO and TPXO 7.2 is decent enough to result in a good tidal propagation model, as validated with the tidal prediction and record data. While SRTM data requires to be corrected to be used for tidal inundation model. Some landmarks elevation can be used to formulate the correction. With open data utilization, model accuracy may be an issue in this research. However, the result of this research can be taken as a basic evidence and in the future, it is crucial to conduct field survey (such as topography and tidal elevation) and compose an advance model to present the behaviour of flood in Pekalongan better. The involvement of other constituents like wave and river discharge is promising.

Model with corrected land elevation model shows that some area in Pekalongan is already below the mean sea level, most probably as an impact of land subsidence. Thus, after the land subsidence rate is applied in the model, it is shown that after yearlong, the tidal flood inundation height is increasing in the range of 8 – 14 cm. The further projection shows that in 2025, inundation height in Jeruksari closing down to 1.5 meters while in 2050, it may reach 2.51 meters. At this level, the flood may severely damage urban living in Pekalongan. As happened in Jakarta, the coastal dyke guarding coastal-front area is one of the proposed solutions. However, it is only for short term since the land subsidence will lower the dyke elevation. It is highly recommended to control the groundwater consumption to fight the land subsidence and the threat of tidal flood as well.

References

[1] Abidin H, Andreas H, Gumilar I, Sidiq T P and Fukuda Y 2012 Land subsidence in coastal city of Semarang (Indonesia): Characteristics, impacts and causes *Geomatics, Natural Hazards and Risk* 1–15.

[2] Chaussard E, Amelung F, Abidin H and Hong S H 2013 Sinking cities in Indonesia: ALOS PALSAR detects rapid subsidence due to groundwater and gas extraction *Rem. Sensing of Env.* **128** 150–61.

[3] Takagi H, Esteban M, Mikami T and Fujii D 2016 Projection of coastal floods in 2050 Jakarta, *Urban Climate* **17** 135-45.

[4] Takagi H, Fujii D, Esteban M and Yi X 2017 Effectiveness and limitation of coastal dykes in Jakarta: The need for prioritizing actions against land subsidence *Sustainability* **9** 1-15.

[5] Sarah D and Soebowo E 2018 Land subsidence threats and its management in the North Coast of Java *IOP Conf. Ser.: Earth Environ. Sci.* 118.

[6] Bernardi Robby, 22nd May 2018, High waves, thousand houses in Pekalongan are flooded, Detik, news.detik.com/jawatengah/4033546/gelombang-tinggi-ribuan-rumah-di-pekalongan-terendam-rob [In Bahasa].

[7] Ridlo Muhamad, 26th May 2018, Caused by flood, hundred convicts are moved to Nusakambangan, Liputan 6, liputan6.com/regional/read/3539076/gara-gara-banjir-ratusan-napi-lapas-pekalongan-dipindah-ke-nusakambangan [In Bahasa].

[8] Radar Pekalongan, 23rd May 2018, Flood inundation is getting higher, Radar Pekalongan, radarpekalongan.co.id/31620/genangan-rob-makin-tinggi/ [In Bahasa].

[9] Isnawati, 23rd May 2018, Flood, local people sahur in the evacuation site, Suara Merdeka, suaramerdeka.com/news/baca/88722/rob-warga-sahur-di-pengungsian [In Bahasa].

[10] Rukayah R S, Wibowo A A, Wahyuningrum S H 2015 Public participation in branding road corridor as shopping window or batik industry at Pekalongan *Procedia - Social and Behavioral Sciences* **168** 76-86.

[11] Weatherall P, et al. 2015 A new digital bathymetric model of the world's ocean *Earth and Space Science* **2(8)** 331-45.

[12] Farr T G, et al. 2007 The shuttle radar topography mission *Review of Geophysics* **45(2)** 1-33.

[13] Egbert, Gary D and Svetlana Y E 2002 Efficient inverse modeling of barotropic ocean tides *Journal of Atmospheric and Oceanic Technology* **19(2)** 183-204.

IOP Conf. Series: Materials Science and Engineering **676** (2019) 012029 doi:10.1088/1757-899X/676/1/012029

[14] Dinas Hidro-Oseanografi TNI-AL 2018 Tide Tables of Indonesia (Indonesia: Dishidros) p 245, 252.
[15] Bradshaw E, Rickards L and Aarup T 2015 Sea level data archaeology and the global sea level observing system (GLOSS) *GeoResJ* **6** 9-16.
[16] Ajiwibowo H, Lodiwa K S, Pratama M B and Wurjanto A 2017 Field measurement and numerical modeling of tidal current in Larantuka Strait for renewable energy utilization *International Journal of Geomate* **13** 124–31.
[17] Hu K, Chen Q, Wang H, Hartig E K and Orton P M 2018 Numerical modeling of salt marsh morphological change induced by Hurricane Sandy *Coastal Engineering* **132** 63-81.
[18] Ajiwibowo H 2018 Numerical modeling for the selection of effluent outlet location *International Journal of Geomate* **14(45)** 192-201.
[19] Kurniawan A, Mujahidin B and Idris K 2018 Analyzing Harbor Location and Orientation Based on Waves Condition: A Case Study at West of Borneo, *6th Reg. Conf. on Natural Disaster.*
[20] Rahman A and Venugopal V 2017 Parametric analysis of three dimensional flow models applied to tidal energy sites in Scotland, *Est., Coastal, and Shelf Sci.* **189** 17-32.
[21] Deltares 2014 Delft3D-Flow, Simulation of multi-dimensional hydrodynamic flows and transport phenomena, including sediments (Netherlands: Deltares) p 186, 679.
[22] Brunner G W 2016 HEC-RAS River Analysis System, 2D Modeling User's Manual Version 5.0 (USA: US Army Corps of Engineers) p 3-26.
[23] Wisha U J, Tanto T A, Pranowo W S and Husrin S 2018 Current movement in Benoa Bay water, Bali, Indonesia: Pattern of tidal current changes simulated for the condition before, during, and after reclamation *Reg. Stud. in Marine Sci.* **18** 177-87.
[24] Wisha U J, Dhiauddin R and Kusumah G 2017 Remote estimation of total suspended solid transport affected by tidal bore "Bono" of Kampar Big River Estuary using Landsar 8 Oli Imaginery *Marine Res. in Indonesia* **42** 37-45.
[25] Mcleod E, Poulter B, Hinkel J, Reyes E and Salm R 2010 Sea-level rise impact models and environmental conservation: A review of models and their applications, *Ocean & Coastal Man.* **53** 507-17.
[26] Direktorat Jenderal Perkeretaapian 2015 The railway's information book year 2014 (Indonesia: Kementerian Perhubungan Republik Indonesia) p 115. [In Bahasa].

IOP Conf. Series: Materials Science and Engineering **676** (2019) 012030 doi:10.1088/1757-899X/676/1/012030

Flexural Strength Test For Concrete Beam With Mild Steel and Reinforcing Iron

M Z Umar*[1] and A Faslih[1]

[1]Architecture Study Program, Vocational Education Program, Haluoleo University, Kendari, Indonesia

*Email: muzakum.uho@gmail.com

Abstract. Mild steel in buildings tends to only be used as a roof frame. The phenomenon in the field that lightweight steel roof truss can be combined with mortar mixture. In general this union is found on the walls of the construction of the red stone horses. This phenomenon is captured and applied dynamically, creatively, and innovatively with new idioms. Mild steel can be used as a beam reinforcement. This research is important in order to find alternative reinforcement of beams other than iron and strong beam reinforcement. This study aims to determine the ratio of flexural strength between concrete from mild steel reinforcement and concrete from steel reinforcement. This study uses a concrete flexural strength test and simple test beam method. The test beam is loaded and centered directly. Press testing machine is used as a flexural test device. The machine is equipped with two support beams and one load beam, load construction, and support blocks. Six beams are used as specimens. Test objects are made of different reinforcement. Concrete beams are made from a cross section of 600 mm x 150 mm x 150 mm, used reinforcing iron diameter of 6 mm, and amounting to three (first test object). Concrete beams are made from a cross section of 600 mm x 150 mm x 150 mm, mild steel is used with the profile shape C73.35.10 mm, and there are three pieces (second test object). This study concluded that the test beam was made from a cross section of 600 mm x 150 mm x 150 mm. and mild steel (profile C73.35.10 mm) has twice the maximum flexural strength compared to 600 mm x 150 mm x 150 mm cross section beam and reinforcing iron (6 mm diameter).

1. Introduction

At present, the world of housing construction is developed rapidly through innovation. Innovation is expected to create efficient construction products. At present, housing construction innovation in Indonesia is considered to be immovable [1]. Reinforced concrete structures are planned based on weak reinforced structures (bending collapse). A collapse in the structure occurs due to bending failure. High compressive strength possessed by concrete and high tensile strength is owned by reinforcing steel material [2]. Beams are used to resist bending force [3]. Flexural strength is the tensile strength of indirect concrete. Flexural strength is caused by concrete moments. Maximum flexural strength occurs in fibers under concrete beams. Flexural collapse in reinforced concrete structures occurs in tensile collapse, draw collapse, and compressive collapse [4]. Beams are used to resist bending force. The bending force is caused by the load acting on the floor. The loads are distributed to the supporting columns. The flexural strength of the beam is obtained from beam bending testing. The beam is given two centralized loads with ½ P at the point 1/3 of the span of the pedestal (figure 1) [3]. Roofing material from wood in housing can be used as an alternative, such as mild steel. Mild steel in buildings tends to only be used as a roof frame. The phenomenon in the field that lightweight steel roof truss material can be combined with mortar mixture on the walls of the construction of horses made of red stone. Thus this phenomenon is captured and then applied dynamically, creatively, and innovatively with a new idiom as a beam reinforcement. This research is important in order to find alternative reinforcement of beams other than iron and strong beam reinforcement. This study aims to determine the ratio of flexural strength between concrete from mild steel reinforcement and concrete from steel reinforcement.

2. Literature Review

Concrete materials are reinforced with mild steel. Mild steel is the main material when bending occurs. The condition of the reinforcing steel is closely related to the magnitude of the flexural strength [4]. The crack condition is caused by three voltages, as follows: The first condition is called flexural cracks; The second condition is called flexural-shear cracks; and the third condition is called the diagonal tension cracks [5]. Flexural capacity and high ductility are owned in square columns. Square columns are created with circle circles. Flexural crack patterns occur in all column crack patterns [6]. Thus, bending patterns occur in all columns.

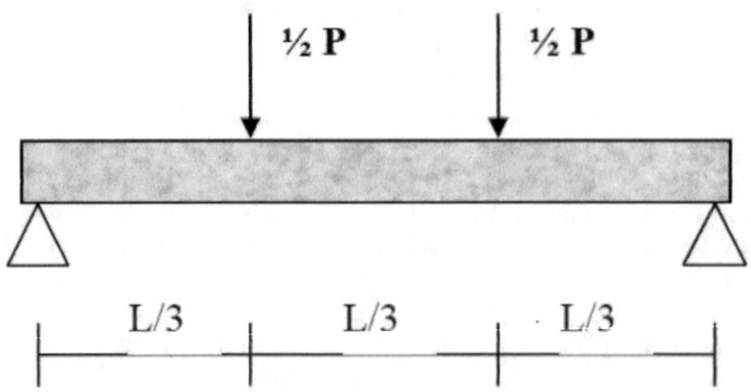

Figure 1. Testing of bending 2 centralized loads [2]

At present, the C channel profile is used not only roof beam *(gording)*, but is used as a home structural element such as columns [7]. C channel profiles are used as columns. The C channel profile is cast with fill concrete. This is proven to prevent local buckling. Profile stability is added with mild steel and cast concrete is added in the profile [8, 9]. The C profile filled with concrete can prevent shear failure and crack patterns [10]. C channel profile beam filled with lightweight concrete can increase flexural strength compared to C channel profile beam without lightweight concrete fillers [11]. C channel profiles have stability weaknesses. Therefore, the profile application needs to be given a stiffener and installed in double. Thus, the double C channel filled with cast concrete can prevent local buckling, increase load capacity, increase strength in the column, increase profile stability, and increase flexural strength [12].

3. Research Method

Test method of concrete flexural strength is used simple test beam. The test beam is loaded and centered directly. This method is intended as a reference for carrying out bending tests in the laboratory. This research was conducted on Saturday, August 19, 2017 at the Material and Construction Testing Laboratory, Department of Civil Engineering, Faculty of Engineering, Halu Oleo University. Test equipment is used, as follows: press testing machine (the machine must be equipped with two support beams and one load beam), load block construction, support blocks, and measuring instruments such as rulers, shovel runners, feelers, scales, cameras, stationery, and test results forms are used as flexural testing equipment. Test objects amounted to 6 pieces and made of different reinforcement. Three beams are made of lightweight steel reinforcement material. Three other blocks are made of reinforcing iron material. The test material must meet the provisions of SNI 03-2493-1991 concerning the Making and Maintenance of Concrete Test Objects in the laboratory. Length 600 mm, width 150 mm, and 150 mm are used in the size of the test beams. All surface areas are required to be flat, free from defects, holes, and curves. Side fields must be perpendicular to the upper plane and lower plane. Flexural strength is calculated by the equation, as follows:

$$flt = \frac{3PL}{2bd^2} \quad (1)$$

Information:

f_{lt}	=	Flexural strength (Mpa)
P	=	Maximum load resulting in test beam collapse (N)
L	=	The span length between the two support beams (mm)
b	=	The average beam width in the collapsed section (mm)
d	=	The average beam height in the collapsed section (mm)

4. Result and Discussion

Test method of concrete flexural strength is used simple test beam. The test beam is loaded and centered directly. Test object is 28 days old. The load when the first crack column when receiving the load (Pmax) is obtained by testing the direct flexural strength. Load data obtained maximum compressive strength and obtained by the compressive force of loading. Six specimens were used in each reinforcement variation. The flexural strength test results are presented in tables (1) and tables (2), as follows:

Table 1. Condition of test objects

Code	Date	Long (mm)	Cross section		Weight (kg)
			Wide (mm)	High (mm)	
KBT1	07/30/2017	600	150,00	150,00	32,70
KBT2	07/30/2017	600	150,00	150,00	32,50
KBT3	07/30/2017	600	150,00	150,00	31,00
KBR1	07/30/2017	600	150,00	150,00	31,75
KBR2	07/30/2017	600	150,00	150,00	33,00
KBR3	07/30/2017	600	150,00	150,00	32,60

Table 2. The results of the average flexural strength of the reinforcing material of beams from reinforcing iron and reinforced beam material of mild steel.

Beam Test Specimen						
Code	b.j. (kg/m^3)	Date	Age (day)	Max. Strength (kN)	Flexural Strenght (Mpa)	Average Flexural Strenght (Mpa)
KBT1	2,422	08/19/2017	28	17,50	3,889	
KBT2	2,407	08/19/2017	28	20,00	4,444	3,889
KBT3	2,296	08/19/2017	28	15,00	3,333	
KBR1	2,352	08/19/2017	28	30,00	6,667	
KBR2	2,444	08/19/2017	28	25,00	5,556	6,296
KBR3	2,415	08/19/2017	28	30,00	6,667	

Table 1 above explains that the condition of the test object before being tested is flexible. Table 2 above explained that the results of the maximum flexural strength test of the beam made from a cross section of 600 mm x 150 mm x 150 mm and steel bars with a diameter of 6 mm obtained a mean of 3.889 Mpa. The beam test results are made from a cross section of 600 mm x 150 mm x 150 mm and a mild steel reinforcement with a profile form C73.35.10 mm obtained a mean of 6.296 Mpa.

5. Conclusion

Based on the results of the test that the test beam is made of cross section 600 mm x 150 mm x 150 mm and mild steel in the form of profiles C73.35.10 mm has two times the maximum flexural strength compared to the test beam of the cross section 600 mm x 150 mm x 150 mm, and iron reinforcement with a diameter of 6 mm. This research can be continued to test concrete columns of mild steel with a profile form C73.35.10 mm to the maximum flexural strength.

References

[1] Buku Manual National Steel Truss Innovation Contest "Ritter" 1 2016 In National Steel Truss Innovation Contest Civil of Engineering Faculty Tadulako University.
[2] Widyawati R 2009 *J Rekayasa* **13** 285.
[3] Helmi M 2009 *J. Rekayasa* **13** 55-56.
[4] Cahyono T A 2015 Thesis (Undergraduate) Department of Civil Faculty of engineering Jember University p 10-11.
[5] Dewi I C, Dewi S M and Soehardjono A 2012 *J. Rekayasa Sipil* **6** 13-14.
[6] Busthamy I 2014 Thesis (Undergraduate) Civil Engineering Studies Program Department of Civil Engeneering Faculty of Engeneering Jember University p 9.
[7] Lisantono A, Santoso B H and Sugianto R 2013 In Konferensi Nasional Teknik Sipil 7 (KoNTekS 7) Sebelas Maret University Surakarta October 24-26 p S-171.
[8] Jiwandono D P P 2010 Thesis (Undergraduate) Civil Engineering Studies Program Faculty of Engineering Atmajaya University Yogyakarta Chapter **II.**
[9] Laksono D B 2009 Thesis (Undergraduate) Civil Engineering Studies Program Faculty of Engineering Atmajaya University Yogyakarta p xvi.
[10] Wigroho H Y 2008 *J. Teknik Sipil* **8** 265.
[11] Lisantono A and Sari M R 2009 In Konferensi Nasional Teknik Sipil 3 (KoNTekS 3, Jakarta, 67 Mei) p S-99.
[12] Lisantono A and Jiwandono P P 2011 In Prosiding Peningkatan Peran Riset dan Pendidikan Teknik Sipil dalam Penyelenggaraan Infrastruktur yang Efektif, Efisien, dan Berkelanjutan Sumatera Utara University Medan October 14 **2** S-271.

IOP Conf. Series: Materials Science and Engineering **676** (2019) 012031 doi:10.1088/1757-899X/676/1/012031

Subsurface Investigation of Freshwater-Seawater Interface on Gowa-Takalar Coastal Aquifer, Indonesia

M Ramli[1]*, Purwanto[1] and A V Anas[1]

[1]Mining Engineering Department, Engineering Faculty, Universitas Hasanuddin

*Email: ramli@unhas.ac.id

Abstract. Gowa-Takalar Coastal Aquifer is an unconfined aquifer located in the southward of Makassar City. The Gowa-Takalar area has a heavy groundwater exploitation particularly on dry season. In this area, the groundwater is utilized to support water need for agriculture activities, marine tourism facilities, and new settlement development. The highly economic activities in this area, it causes an intensive groundwater exploitation. Since this area is located around coastal area, this area has highly susceptibility to suffer from seawater intrusion. In order to maintain the continuity of groundwater supply, a special study is needed on hydrodynamics of fresh-sea water interface as the basis for the formulation of a sustainable management model. This study as an early step of groundwater management research emphasizing on basic of aquifer and groundwater data. The research method consisting of reconnaissance study for geological & hydrogeological mapping, groundwater chemistry laboratory analysis, and geo-electrical investigation. Geological condition of this area is composed of alluvial and coastal deposits, volcanic deposits, and marine sedimentary rocks interbedded with volcanics. These materials form three hydrogeological characteristics: extensive and moderately productive aquifer, locally productive aquifers, and poorly productive aquifers of local importance. The chemical composition of groundwater from dug wells show the chloride concentration vary from 3.55 ppm to 81.65 ppm, with chloride-bicarbonate ratio is 0.008 – 0.035 indicating aquifer without sea water invasion. However, based on vertical electrical sounding, the resistivity layers vary from 0.02 to 62.50 ohm-m indicating the present of seawater intrusion into coastal aquifer.

1. Introduction

Groundwater plays an important role for the environment and the economy. In many regions, groundwater is exploited in a big volume because ground water has several advantages compared to surface water. However, issues of groundwater quantity and quality have received far less attention than the surface water, particularly in some developing countries. Moreover, due to its existence in subsurface, the issues of groundwater are often not a problem arises until it comes to public attention such as quality and quantity degradation, land subsidence, etc. In the groundwater development area, a heavy groundwater exploitation will result in an uncontrolled decreasing of water table or groundwater piezometric level covering a very large area. This groundwater level change will be followed by number of environmental impacts, which are classified into three main categories; seawater intrusion, land subsidence, and the spread of other contaminants [1]. The seawater intrusion is one of the most common problems in almost all coastal aquifers around the world. It takes place when saline water displaces or mixes with freshwater in aquifers. This phenomenon can be attributed by a variety of conditions; hydrogeological conditions, hydraulic gradients, tidal and estuarine activity, sea level, infiltration, and groundwater withdrawal.

Coastal area of Gowa-Takalar, which is a buffer zone for Makassar City, has a heavy groundwater exploitation particularly on dry season. In this area, the groundwater is utilized to support water needs for agriculture activities, marine tourism facilities, and new settlement development. The uncontrolled exploitation leads the coastal aquifer to be a highly susceptibility of seawater intrusion. In order to avoid the seawater intrusion problem, it is required assessment activities. The most common methods for assessing seawater intrusion through an aquifer in coastal belts is a periodic analysis of groundwater chemistry [2]. The groundwater chemical composition generally exhibits high concentrations not only in total dissolved solids (TDS) but also in major cations and anions as well accumulation of selective trace elements [3]. Revelle recommended to calculate chloride-bicarbonate ratio (Cl^-/ (CO_3^{2-} + HCO_3^-) to identify seawater intrusion [4].

The extent of saline water intrusion is influenced by nature of geological settings, hydraulic gradient, rate of groundwater withdrawal and its recharge. In studying the thickness and geometry of depositional systems, a common procedure is to make use of information from geological research, drilling, and exploitation boreholes. However, these methods are expensive and time consuming. In contrast, geophysical measurements can provide a less expensive way to improve the understand of subsurface geological condition. For this reason, in many cases, geophysical prospecting techniques can provide complementary data that enable geological correlation, even in sectors where there are no data from boreholes. There are a variety of geophysical techniques for which it would be useful to know the subsurface resistivity in order to estimate the saline water intrusion. The large differences between the resistivity of saltwater saturated zones and the freshwater saturated zones have been used by number of investigations for determination of saltwater intrusion in many coastal areas [5].

Since the study of seawater intrusion problem in the Gowa-Takalar Coastal Aquifer has not been carried out intensively, this study has been done to provide a basic data of groundwater as interface of fresh-seawater. Therefore, the objective of this study is to provide information of general status of groundwater quality and salinization on unconfined aquifer. To achieve the above objective, the following tasks are to be carried out: (1) mapping of geological and hydrogeological condition; (2) analysis of groundwater hydro-chemical; (3) interpretation of resistivity to estimate position of freshwater-seawater interface.

2. Research Methodology

Reconnaissance of surface geological and hydrogeological mapping was done to have a complete understanding of geomorphology, litology/stratigraphy, and structural geology of the research area. This reconnaisance mapping was based on regional geological map of Ujung Pandang, Benteng, and Sinjai sheets, South Sulawesi published by the Geological Research and Development Center Bandung [6]. Moreover, the hydrogeological mapping using the Hydrogeological Map of Ujung Pandang, Benteng, and Sinjai Quadrangles, South Sulawesi published by Center for Environmental Geology Bandung [7] was a basic information. In the hydrogeological mapping, it was measured the depth of water table on dug wells and was took groundwater samples. The mapping was carried in August 2018 in which dry season and was collected samples of 25 dug wells. The groundwater samples were analyzed of chemical composition for chloride, carbonate, and bicarbonate element. Result of the chemical analysis is used to predict mixing seawater and freshwater by calculating chloride-bicarbonate ratio.

Overview the groundwater salinity of rock layer beneath surface was investigated by using geoelectrical prospecting. The geoelectrical surface applied Vertical Electrical Sounding (VES) by employing Schlumberger electrode array. The Schlumberger electrode configures current electrode spacing varying between 1 to 400 m. The VES was conducted using the resistivity meter for 10 measurement points. The VES field data was processed to determine layer parameters through 1-D inversion technique of IPI2Win software [8]. This software capable solving resistivity electrical prospecting ID forward and inverse problems for a variety of commonly used arrays for the cross sections with resistivity contrasts within the range of 0.0001 to 10000. The forward problem is solved using the linear filtering developed at the Near-surface Electrical Prospecting Lab. Geophysical Department, Moscow State University. The inverse problem is solved using a variant of the Newton algorithm of the least number of layers of regularized fitting minimizing algorithm using Tikhonov's approach to solve incorrect problems. The inverse problem is solved separately for each sounding curve. By using the IPI2win software, it will be estimated number of layers, true resistivity, thickness, depth, and altitude of layer contacts.

3. Geological and Hydrogeological Characteristics

The Gowa-Takalar coastal aquifer stretches from Pappa River in south to Jeneberang River in north with elevation vary from 0 – 25 m above mean sea level. The rainfall analysis was based on rainfall occurrence within the latest 5 years. The monthly rainfall that occurs varies with the lowest value of 0.70 mm in September and the highest value in January is 772.40 mm. Geological condition of the Gowa-Takalar coastal area was described on the published Geological Map of Ujung Pandang, Benteng and Sinjai Quadrangles [6]. The research consists of three rock formations: Alluvial and Coastal Deposits (Qac), Baturape-Cindako Volcanics (Tpbv), and Camba Formation (Tmc).

Alluvial and Coastal Deposits (Qac) is composed of gravel, sand, clay, mud, and coral limestone. Those materials were deposited in rivers, swamps, coast and delta environments. Around the Jeneberang River, the alluvial deposits consist mainly of detritus derived from volcanic materials of Lompobattang Mountain.

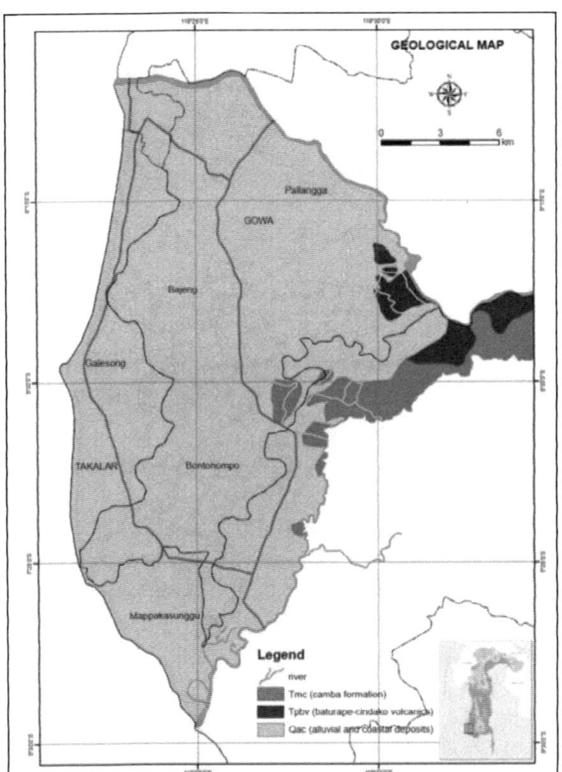

Figure 1. Geological map research area (modified from Ref. [6])

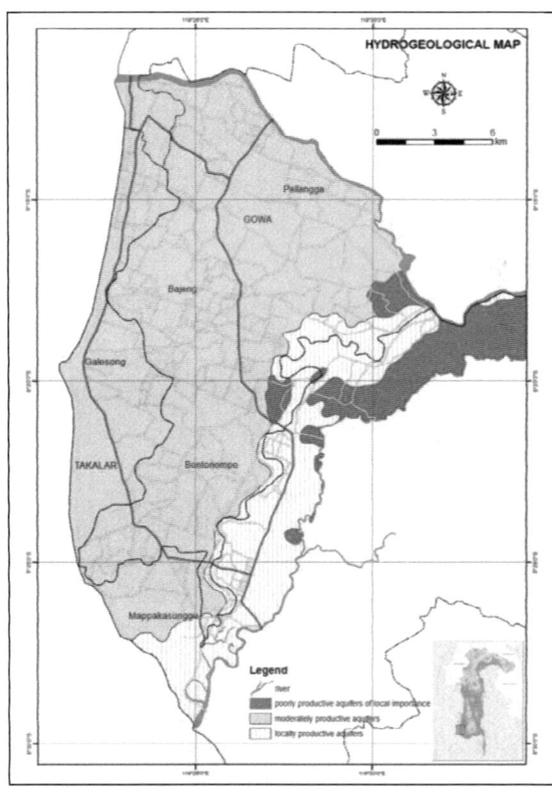

Figure 2. Hydrogeological map of research area (modified from Ref. [7])

Baturape-Cindako Volcanics (Tpbv) consist of lava and breccia, with the intercalation of some tuff and conglomerate. The lava composition is basaltic, mostly porphyritic, with big phenocrysts of pyroxene up to 1 cm, and partly aphanitic, with greenish dark gray to black in color. The lava commonly displays columnar and sheeted jointing. The breccia consists mostly of coarse components with size of 15 - 60 cm, mainly basaltic with small amounts of andesite mineral, cemented with coarse tuff to lapilli. Diorite intrusive consisting of stock and dyke at vicinity of Baturape-Cindako is thought to be the remnant of eruptive center (Tpbc). The surrounding rocks are highly altered, amygdaloidal with secondary minerals of zeolite and calcite. Ore mineral of galena in Baturape is possibly related to the dioritic intrusive. The area around Baturape and Cindako is dominated with lava (Tpbl).

Camba Formation (Tmc) is generally marine sedimentary rocks interbedded with volcanics. The formation consists of tuffaceous sandstone interbedded with tuff, sandstone, and claystone; with intercalation of marl, limestone, volcanic conglomerate and breccia, and coal. The rocks has varies in color from white, brown, red, light to dark grey. The materials are mostly well consolidated. Tuffs are fine to lapilli in size. Red clayey tuff contains abundant biotite. Conglomerate and breccia composed of andesite and basalt pebbles of 2 – 30 cm in size. Sandy limestone contains fragment of coral and mollusca. Dark gray claystone and marl contain small forams.

The hydrogeological conditions of Gowa-Takalar coastal area and its surroundings have been published by the Bandung Center for Environmental Geology on a Hydrogeological Map of Indonesia, Ujungpandang, Benteng, and Sinjai Quadrangles [7]. Hydrogeological conditions based on rocks properties are categorized into 3 categories, namely;

1. Extensive and moderately productive aquifer. The aquifer has low to moderate transmissivity. Water table on this aquifer lies at depth less than 5 meter below land surface, with wells yield is less than1.0 liter/second.
2. Locally productive aquifers. The material is mostly incoherent aquifers of low thickness and transmissivity. The positon of water table is 1 – 5 meter below land surface, and wells yield is less than 1.0 liter/second.
3. Poorly productive aquifers of local importance. Generally, aquifer has very low transmissivity, local, limited shallow groundwater resources. The aquifer can be obtained in weathered zones of solid rocks or in the valleys.

4. Freshwater and Seawater Interface

4.1. Chloride-bicarbonate ratio

The chemical composition of ground water is related to the soluble products of rock weathering and decomposition and changes with respect to time and space. Geochemical study can provide a complete knowledge of water resources of a hydrogeological system. Therefore, several hydro-chemical approaches have been published regarding different uses. To investigate seawater enroachment into coastal aquifer, Revelle, 1941 recommended to calculate chloride-bicarbonate ratio (Cl^-/ (CO_3^{2-} + HCO_3^-). Chloride is the dominant ion in seawater and it is only available in small quantities in groundwater while bicarbonate which is available in large quantities in groundwater occurs only in very small quantities in seawater [4]. An appropriate assessment of salt water intrusion aspect requires establishing the levels of Cl^-, CO_3^{2-} & HCO_3^- quantitatively [9].

On hydrogeological mapping, Groundwater samples were collected from 25 dug wells in August 2018 (dry season). The well's locations are shown on Figure 3. The sampling objectives to provide information groundwater quality laterally within the aquifer, so that the samples were taken from water table. The samples were sent to Laboratory of Water Quality, Faculty of Marine Science and Fisheries, Universitas Hasanuddin for chemical composition analysis. The analysis was carried for major chemical constituents; cations (Na^+, K^+, Mg^{2+}, and Ca^{2+}), and anions (Cl^-, HCO_3^-, SO_4^2, and CO_3^2). The sum of the cations and anions when expressed in milliequivalents per liter will be equal and may expected as total dissolved solid. The ionic balance was determined for the results of the major cations and anions calculated by Balance = ((Cations – Anions)/(Cations + Anions))*100. Where Cations is the total cation milliequivalents per liter. Anions is the total anions milliequivalents per liter.

Since this paper describes fresh-sea water interface, result of chemical quantity analysis of groundwater samples focuses on chloride (Cl^-), bicarbonate (HCO_3^-), and carbonate (CO_3^{2-}). Table 1 shows the concentration values for carbonate (CO_3^{2-}), bicarbonate (HCO_3^-), and chloride (Cl^-) and calculated of chloride-bicarbonate ratio. The results of the analysis show that concentration of chloride ion vary from 3.55 - 81.5 ppm, carbonate ion vary 588.0 - 2256.0 ppm, and bicarbonate ion vary from 434.0 - 2839.6 ppm. The calculated chloride-bicarbonate ratio varies from 0.001 to 0.045. All the samples showed the ratio less than 0.5 indicating that there is no salt water intrusion in the ground water of the study area.

Table 1. Chemical analysis results of collected groundwater samples.

No	SAMPLE CODE	CHEMICAL CONC. (ppm)			MOLARITY (m.eq/ltr)			CHLORIDE-BICARBONATE RATIO
		HCO_3^-	CO_3^{2-}	Cl^-	HCO_3^-	CO_3^{2-}	Cl^-	
1	OW-01	2021.2	1956.0	53.25	0.033	0.033	0.0015	0.023
2	OW-02	2070.8	2004.0	56.80	0.034	0.033	0.0016	0.024
3	OW-03	1190.4	1152.0	24.85	0.020	0.019	0.0007	0.018
4	OW-04	1959.2	1896.0	39.05	0.032	0.032	0.0011	0.017
5	OW-05	1438.4	1392.0	28.40	0.024	0.023	0.0008	0.017
6	OW-06	818.4	792.0	24.85	0.013	0.013	0.0007	0.026
7	OW-07	434.0	420.0	7.10	0.007	0.007	0.0002	0.014
8	OW-08	1500.4	1452.0	14.20	0.025	0.024	0.0004	0.008
9	OW-09	1562.4	1512.0	81.65	0.026	0.025	0.0023	0.045
10	OW-10	2331.2	2256.0	3.55	0.038	0.038	0.0001	0.001
11	OW-11	1971.6	1908.0	39.05	0.032	0.032	0.0011	0.017
12	OW-12	731.6	708.0	14.20	0.012	0.012	0.0004	0.017
13	OW-13	942.4	912.0	17.75	0.015	0.015	0.0005	0.016
14	OW-14	2170.0	2100.0	42.60	0.036	0.035	0.0012	0.017
15	OW-15	2157.6	2088.0	49.70	0.035	0.035	0.0014	0.020
16	OW-16	1748.4	1692.0	28.40	0.029	0.028	0.0008	0.014
17	OW-17	1426.0	1380.0	39.05	0.023	0.023	0.0011	0.024
18	OW-18	942.4	912.0	17.75	0.015	0.015	0.0005	0.016
19	OW-19	669.6	648.0	10.65	0.011	0.011	0.0003	0.014
20	OW-20	607.6	588.0	24.85	0.010	0.010	0.0007	0.035
21	OW-21	2839.6	2748.0	49.70	0.047	0.046	0.0014	0.015
22	OW-22	1103.6	1068.0	14.20	0.018	0.018	0.0004	0.011
23	OW-23	1860.0	1800.0	49.70	0.030	0.030	0.0014	0.023
24	OW-24	1450.8	1404.0	28.40	0.024	0.023	0.0008	0.017
25	OW-25	1624.4	1572.0	49.70	0.027	0.026	0.0014	0.027

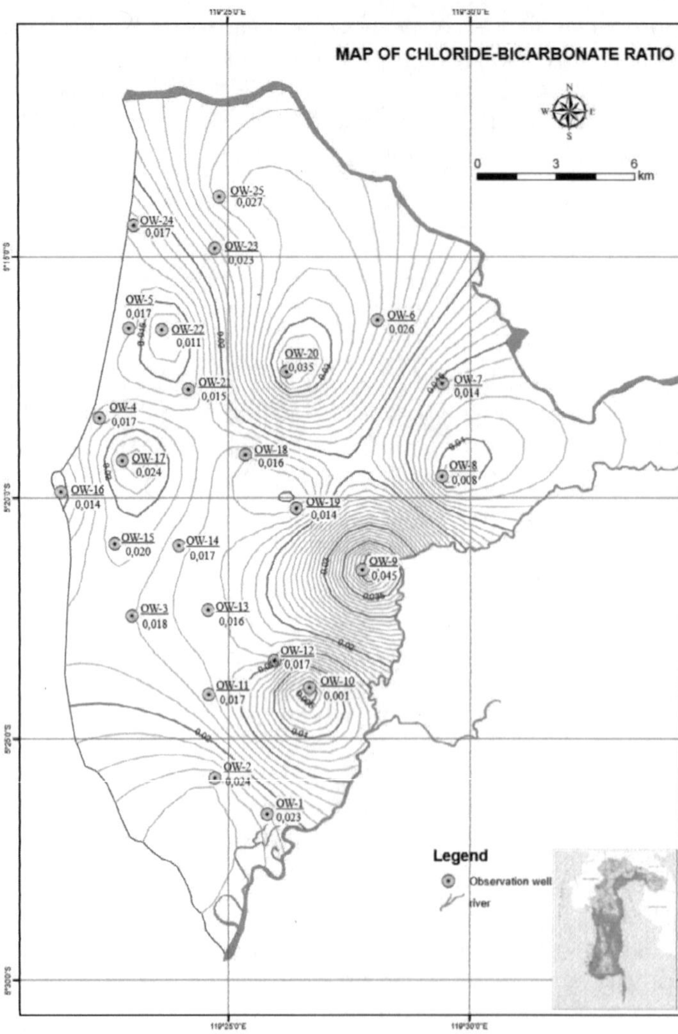

Figure 3. Map of chloride-bicarbonate ratio

4.2. Groundwater resistivity

The surface electrical resistivity method is a useful tool in determining seawater intrusion in coastal areas for its capability to discriminate the large resistivity contrast between the presence of seawater that strongly reduces the resistivity values and saturated freshwater layers [10]. The Vertical Electrical Sounding (VES) survey reported in this paper was carried out with the primary objective of obtaining information on the hydrogeological behavior of the aquifers, the subsurface lithology and groundwater salinity. Surface Schlumberger electrode array with electrode spacing varying between 1 to 400 m was applied for the VES survey to investigate subsurface condition for maximum depth of 150 – 200 m. Location and distribution of the VES stations are shown on Figure 4, for 10 VES using the resistivity meter.

The electrical resistivity of sediments depends on lithology, water content, clay content and salinity. Variations in rock type resistance values at a sounding point can be used to predict; (1) the position of the water bearing layer, (2) the thickness of the water bearing layer, (3) the quality of ground water, (4) the lateral spread of the water bearing layer, (5) the condition of the material below the surface. Some interpretation problems for the VES can occur when mapping subsurface salinity, since resistivity for salt water, saturated clay and sand overlaps. Both groundwater saline and saturated clay has low resistivity. It is, therefore, important to correlate the VES results with the lithological and hydrological information.

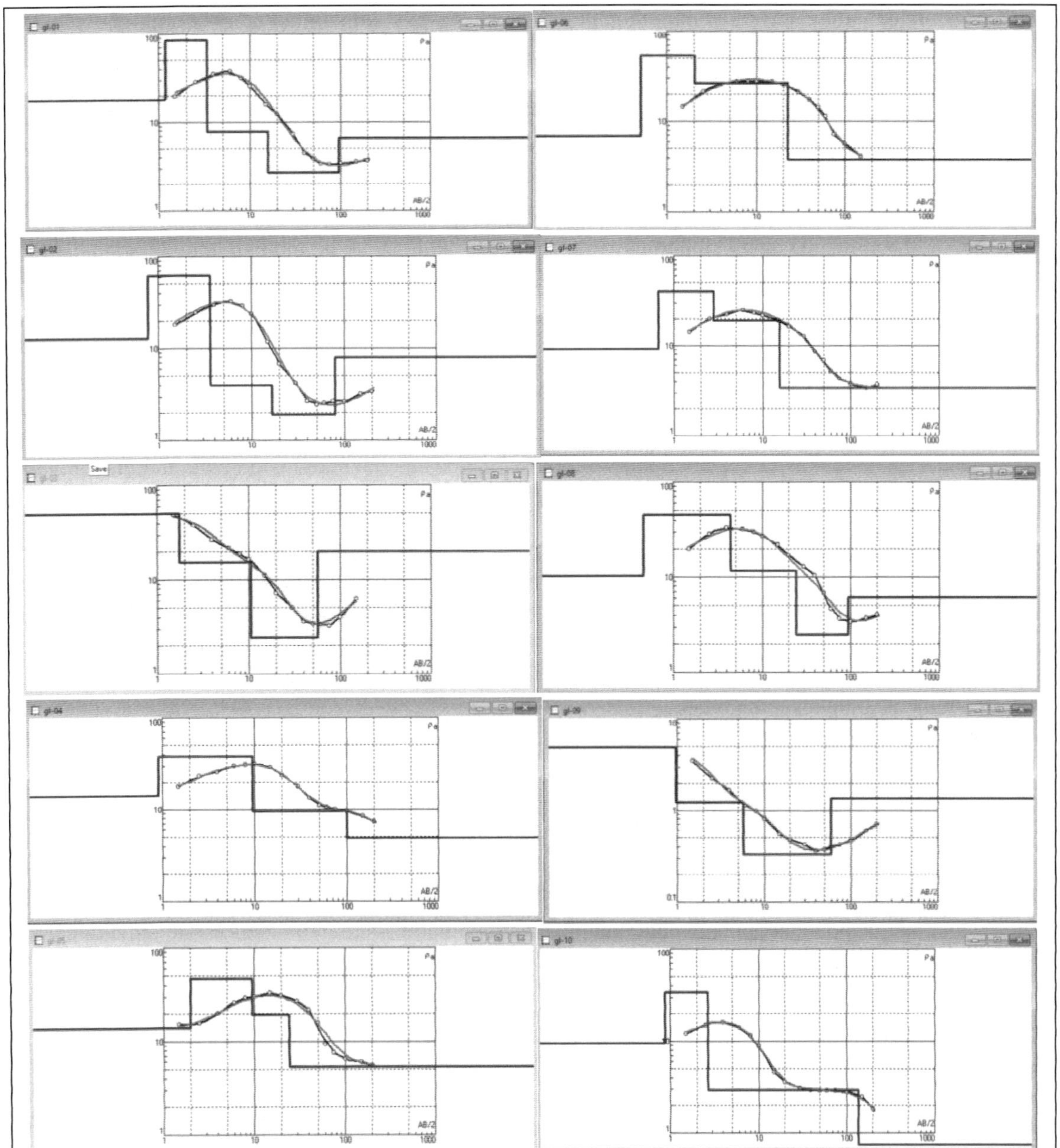

Figure 4. Matching field data curve with calculated IPI2Win curve.

Owing to the differing character of features in the apparent resistivity curves, the VES stations show different types of curves. The interpretation results reveal significant variations in subsurface resistivity of the coastal rock units in the study area. The types of VES curves were defined in terms of the number of geoelectrical layers and their resistivity relationship. If the lateral variations in resistivity in the shallow subsurface units as a result of heterogeneity of lithology and soil moisture content are ignored, most of the VES resistivity curves show two types of resistivity curve showed in

Figure 3. The first type with a trend of decreasing resistivity until about 5 ohm-m and then increasing in the last part of the cuver ($\rho1 > \rho2 < \rho3$) for VES-1, VES-2, VES-3, VES-8, and VES-9. The second type is decreasing curve to end of the curve (cuver ($\rho1 < \rho2 < \rho3$) for VES-4, VES-5, VES-6, VES-7, and VES-10). In general, the type of data obtained varies from 0.02 ohms-m to 62.50 ohms-m.

Tabel 2. Interpretation of resistivity layers into groundwater quality.

No	VES CODE	True Resistivity (ohm-m)	Layer Depth (m)	Groundwater Condition
1	VES_01	18.00	0,00 – 1,80	-
		5,85	1,80 – 20,71	Brackish water.
		1,76	20,71 – 106,90	Saline water
		7,14	>106,90	Brackish water
2	VES_02	12,90	0,00 – 0,80	-
		62,50	0,80 – 3,59	Fresh water
		3,97	3,59 – 16,53	Brackish water.
		1,89	16,53 – 79,04	Saline water
		7,90	>79,04	Brackish water
3	VES_03	50,20	0,00 – 1,74	-
		15,20	1,74 – 10,52	Fresh water
		2,44	10,52 – 56,50	Saline water
		19,90	>56,50	Fresh water
4	VES_04	14,30	0,00 – 0,90	-
		37,60	0,90 – 9,54	Fresh water
		9,56	9,54 – 100,30	Brackish water
		4,86	>100,30	Brackish water
5	VES_05	13,90	0,00 – 1,98	-
		47,30	1,98 – 9,51	Fresh water
		19,30	9,51 – 24,37	Fresh water
		5,37	>24,37	Brackish water
6	VES_06	6,91	0,00 – 0,53	-
		53,70	0,53 – 2,03	Fresh water
		26,50	2,03 – 22,31	Fresh water
		3,73	>22,31	Brackish water
7	VES_07	9,19	0,00 – 0,68	-
		40,40	0,68 – 2,79	Fresh water
		19,00	2,79 – 15,41	Fresh water
		3,43	>15,41	Brackish water
8	VES_08	10,40	0,00 – 0,48	-
		45,40	0,48 – 4,39	Fresh water
		11,60	4,39 – 24,37	Fresh water
		2,48	24,37 – 93,50	Saline water
		6,14	>93,50	Brackish water
9	VES_09	4,95	0,00 – 0,97	-
		1,22	0,97 – 5,74	Saline water
		0,33	5,74 – 59,18	Saline water
		1,36	>59,18	Saline water
10	VES_10	9,51	0,00 – 0,88	-
		34,10	0,88 – 2,65	Fresh water
		2,94	2,65 – 136,1	Saline water
		0,02	>136,10	Saline water

The results from the analysis of VES data is combined with the lithology data to interpretate the groundwater condition on the aquifer. The interpretation of the value of the prisoner into the condition of groundwater can be explained that:

a. Overburden/top soil layer: Layer with a resistivity value of 4.95 ohm-m - 50.20 ohm-m. This layer lies at a depth from 0.0 m to varies bottom layer at depth of 0.30 - 1.98 m from the top surface. This layer is the result of weathering of the rock beneath it which is clay-glazed material without groundwater.
b. The layer of fresh groundwater carrier with resistivity value varies from 11.60 ohm-m - 62.50 ohms-m. This layer composed of sandstones to sandstone-gravel layers. This layer lies at a very varied depth. In coastal areas found at a depth of 0.80 m - 20.71 m from the top surface, while in the areas far from the coast can be found deeper than 100 meters.
c. Brackish ground water is indicated with resistivity value of 3.0 ohm-m - 10 ohm-m. This layer presents at variable depths depending on the VES distance from coastline. Contact between brackish – saline water is shown in Figure 5 assumed as a fresh-sea water interface.
d. Saline groundwater carrier layer with a resistivity value of less than 3.00 ohm-m. This layer was found at the all VES measurement points except in VES-05, VES-06, and VES-07.

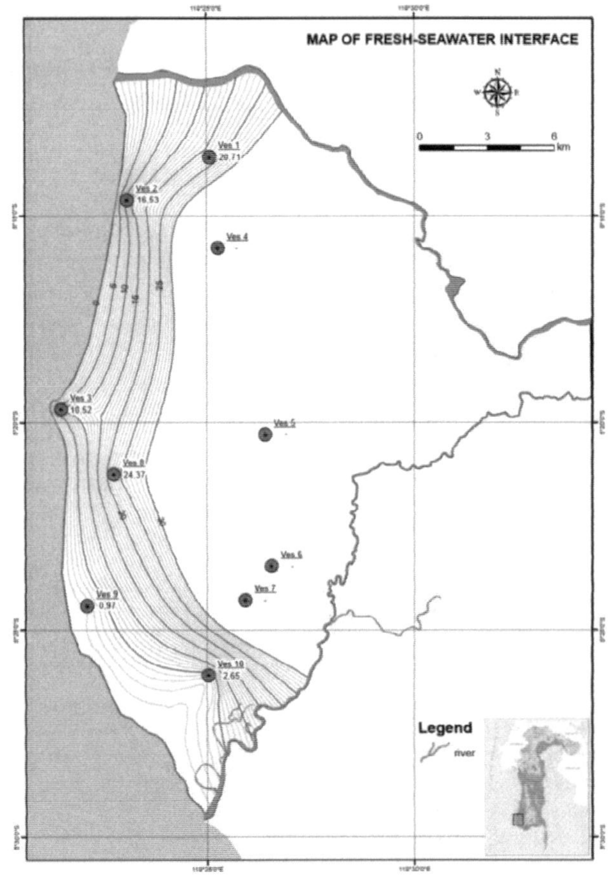

Figure 5. Interpreted fresh-sea water interface based on true resistivity of rock layers.

5. Conclusions

Geological condition of the Gowa-Takalar Coastal Aquifer, mainly, consists of Alluvial and Coastal Deposits (Qac) with material size of sand to boulder. Those materials form unconfined aquifer categorized as moderately productive aquifer. Water table on this aquifer lies at depth less than 5 meter below land surface leading to an easy condition to extract ground water form aquifer. The chemical analysis of ground water samples collected from water table of 25 dug wells shows the calculated chloride-bicarbonate ratio is far less than 0.50. The result indicates that here the shallow ground water aquifer has not contaminated yet with sea water intrusion. However, the VES results demonstrate that only three geoelectrical measurement location (VES-05, VES-06, and VES-07) have not indicated the existence of saline water. All VES has low resistivity value indicating the present of brackish water into aquifer.

References

[1] Foster S, Lawrence A, and Morris B 1998 *Groundwater in Urban Development* Assesing management needs and formulating policy strategies World Bank Technical Paper No. 390.
[2] Todd DK 1980 *Groundwater Hydrology* Second Edition John Wiley and Sons New York p. 535.
[3] Mondal NC, Singh VP, Singh S, and Singh VS 2011 Hydrochemical characteristic of coastal aquifer from Tuticorin, Tamil Nadu, India *Environ Monit Assess* 175: pp. 531–550.
[4] Ragunath HM 2003 *Groundwater* 2nd Edition New Age International Limited Publisher New Delhi.
[5] Rao VVSG, Rao GT, Surinaidu L, Rajesh R, and Mahesh J 2011 Geophysical and Geochemical Approach for Seawater Intrusion Assessment in the Godavari Delta Basin, A.P. *India Water Air Soil Pollution* Vol. 217 pp. 503–514.
[6] Sukamto R, and Supriatna S 1982 *Geological Map of Ujungpandang, Benteng, dan Sinjai Quadrangles, Sulawesi* Geological Research and Development Center, Bandung.
[7] Mudiana W, Mukna HS, and Soetrisno S 1984 *Hydrogeological Map of Indonesia*, Ujungpandang, Benteng, dan Sinjai Quadrangles, Sulawesi, Center for Environmental Geology Bandung.
[8] Moscow State University Geological Faculty Department of Geophysics 2001 WIN with IPI2Win Geoscan-M Ltd. Moscow.
[9] Sarada P, and Bhushanavathi P 2014 $Cl^-/(CO_3^{2-} + HCO_3^-)$ Ratio to Evaluate Salt Water Intrusion: A Case Study of Gnanapuram Area of Visakhapatnam, AP, India *International Journal of Science and Research* (IJSR) pp. 1311 – 1313.
[10] Mogren S 2015 Saltwater Intrusion in Jizan Coastal Zone, Southwest Saudi Arabia Inferred from Geoelectric Resistivity Survey *International Journal of Geosciences* Vol. 6 pp. 286-297.

IOP Conf. Series: Materials Science and Engineering **676** (2019) 012032 doi:10.1088/1757-899X/676/1/012032

Study on Sc-bearing Lateritic Ni deposits in Ultramafic Rock from Sulawesi: A New Paradigm in Indonesia Metal Mining Industry

A Maulana[1]*, Sufriadin[1], K Sanematsu[2], M Sakakibara[3]

[1]Geology Department, Engineering Faculty, Hasanuddin University, Indonesia 92171
[2]Japan Advanced Institute of Science and technology, Tsukuba, Japan
[3]Department of Geoscience, Ehime University, Japan

*Email: adi_maulana97@yahoo.com

Abstract. Scandium (Sc) and its compounds are used in many modern-technology applications such as in optical, electronic, aeronautical, and fuel-cell industries. However, Sc occurrence were only limited in a very specific area with limited deposit and little attention has been paid to the genesis of Sc-bearing deposits in the world. 6 (six) bedrock samples and 15 (fifteen) lateritic samples (saprolite to limonite) were collected and studied to investigate the distribution of Sc grades in lateritic Ni deposits and identify Sc-bearing minerals in laterites from ultramafic rocks in West Blok, East Block and Petea prospect in Soroako, South Sulawesi, Indonesia. The study suggests that scandium is enriched in limonite layer in weathering profile. In general, pyroxene bearing rich bedrock (harzburgite in composition) will contain higher Sc content in their weathering product. It is suggested that the Sc are associated with the occurrence of iron oxide from pyroxene. However, as the occurrence of Sc is very limited, the most promising result was shown by East Block profile in which the enrichment of Sc is also supported by enrichment of NiO content. Sc will be viable as by-product of nickel laterite.

1. Introduction

Scandium (Sc) is an important metal for various applications in globalization era and has been considered as one of the critical metals [1]. Scandium and its compounds have been found wider applications such as in optical, electronic, aeronautical, and fuel-cell industries [1-4]. The properties of Sc-strengthened alloys and Sc_2O_3-stabilised ZrO_2 materials are particularly promising. The Al–Sc alloys (0.35–0.4% Sc) have a number of superior properties including light weight, high strength, good thermal resistance and long durability [5]. The principal applications of the Al–Sc alloys were for sporting equipment and military demand [3]. For example, scandium alloys were applied in premium bicycle frames and in fighters and missiles. The application of Mg–Sc alloy to aircraft engines could save electricity and energy consumption [1, 6]. Scandia-stabilized zirconia has extremely high oxygen-ion conductivity for use as a high efficiency electrolyte in solid oxide fuel cells [1, 7].

New report suggests that Sc^{3+} enriched in mafic minerals such as pyroxene, amphibole and magnetite and poor in olivine [1]. These typical minerals are found in mafic to ultramafic rocks. Based on the most update investigation, scandium can be extracted from weathering crust of highly weathered nickel-contained ultramafic rock which is called as lateritic deposit. Scandium will be concentrated in limonite and saprolite layer, but the volume of enrichment is still unknown. Sulawesi Island is one of the islands where lateritic deposits from heavily weathered ultramafic are widely

IOP Conf. Series: Materials Science and Engineering **676** (2019) 012032 doi:10.1088/1757-899X/676/1/012032

distributed [8]. Currently, some mining companies extracted nickel ore from the lateritic deposit from ultramafic rocks, particularly from the olivine-rich parent rocks. However, the Sc occurrence and distribution in this deposit have not been studied in detail. The lateritic soil of the ultramafic rocks can be potential source of scandium. In this study, we investigate the distribution Sc grades of lateritic Ni deposits and identify Sc-bearing minerals (they may be amorphous materials) in laterites in ultramafic rocks from Indonesia, particularly in Sulawesi as one of the Sc-bearing deposits. As the price of Sc ore is significantly expensive along with demand, Sc will be a new target to intensify the mining exploration process

2. Regional Geology

Sulawesi consists of four major litho-tectonic units comprising the Eastern, Northern and Western Sulawesi Provinces and the Banggai-Sula Microcontinent (Figure. 1). Following is the regional geology of Sulawesi based on [8] and references there in. The Western Sulawesi Province is divided into three parts based on geomorphological and geological criteria, viz. Southwest (SW), Central West (CW), and Northwest (NW) Sulawesi. SW Sulawesi is different from the others being less tectonically deformed and less mountainous, and lacking CAK-type igneous rocks. As discussed by [8, 9] the Eastern Sulawesi Province consists of several quasi-concentric arcuate belts, which are composed of, from west to east: 1) sheared metamorphic rocks commonly referred to as the Central Sulawesi Metamorphic Belt; 2) a highly tectonized mélange of ophiolitic, metamorphic, and Mesozoic-Paleogene sedimentary rocks, and 3) predominantly ophiolite rocks. A fourth zone, consisting of imbricated Mesozoic and Paleogene rocks that fringes the southeast margin of the East Arm, belongs to the Banggai-Sula Province and marks the Neogene collision zone between the Banggai-Sula continental fragment and the ophiolite terrane. These four zones are unconformably overlain by syn-to post-orogenic sedimentary deposits ("Celebes Molasse").

The Northern Sulawesi Province occupies a large part of the North Arm. The E-W part of this arm is made up of Cenozoic arc volcanic and plutonic rocks and associated sedimentary rocks. The NE-trending part of the north arm is largely covered by Pliocene to recent volcanic rocks. The Banggai-Sula and Tukang Besi Continental Fragments are located in the eastern and southeastern parts of Sulawesi, respectively. The Bangai-Sula microcontinent is represented above sea level by a group of islands, including Peleng, Banggai, Taliabu and Mangole Islands whereas the Tukang Besi microcontinent comprises Buton, Muna and surrounding smaller islands. The mafic-ultramafic sequences formed ophiolite complex which are widely distributed in eastern Sulawesi Province (especially in Soroako area) (Figure 2) and small outliers in western Sulawesi Province (Bantimala and Barru areas). These complexes have been variously interpreted as members of ophiolites from different tectonic settings.

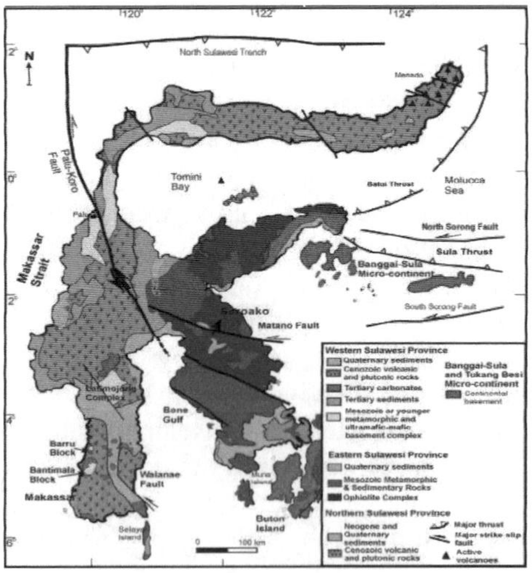

Figure 1. Regional geology of Sulawesi [8, 10]

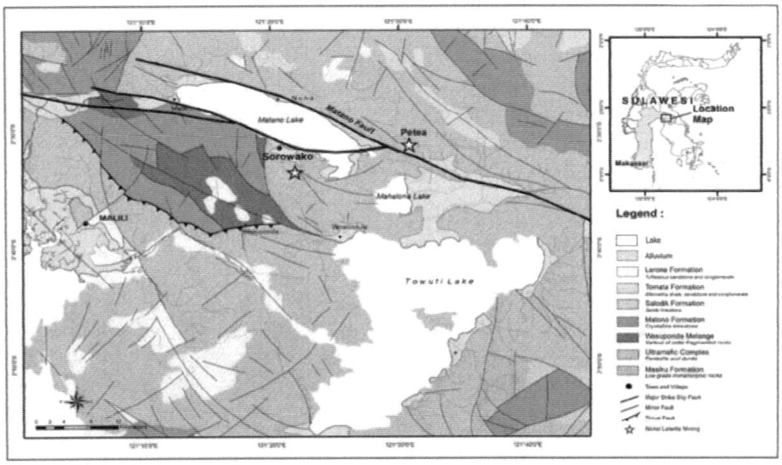

Figure 2. Local geology of Soroako (study area)

3. Methodology

The weathered ultramafic (lateritic) were analyzed for concentration of major elements using an X-ray fluorescence spectrometer (XRF) method in AIST Laboratory in Tsukuba and Department of Earth Resources Engineering, respectively. Rare earth element and trace element composition for both samples groups were determined by the ICP-MS method in Actlab, Canada. Mineral chemistry analysis will be conducted by electron microprobe analysis (EMPA) method using JEOL JXA-8530M in AIST laboratory in Tsukuba Japan as our international collaboration institution.

4. Result and Discussion

4.1. Sample description

We collected 6 bedrock samples and 15 lateritic samples (saprolite to limonite) from ultramafic rocks in West Blok, East Block and Petea prospect in Soroako. The bedrocks consist of peridotite or harzburgite (East Block), dunite (West Block) and Serpentinite (Petea). Weathering profile in Soroako showed a typical of lateritic profile consists of bedrock, saprolite, limonite and overburden as shown in Figure 3.

Figure 3. Weathering profile of lateritic deposit in Soroako area (upper). Limonite layer (red-brownish in color) in Soroako area (lower)

Harzburgites in the East Block (21-2 BR) are moderately to strongly serpentinized and consist of olivine (50–55%), relict orthopyroxene (20–25%), clinopyroxene (0–5%) and amphibole (5–10%) (Figure 4). Olivine varies in size, usually 0.4–1.2 mm, but sometimes up to 3 mm. Most olivines are altered, with serpentine veins replacing the olivine. Orthopyroxene occurs as subhedral to anhedral crystals 0.5–1.5 mm in size, sometimes up to 4 mm, and commonly replaced by amphibole, which is usually found in the contact between orthopyroxene and olivine, and is sometimes altered to

IOP Conf. Series: Materials Science and Engineering **676** (2019) 012032 doi:10.1088/1757-899X/676/1/012032

serpentine. Associated magnetite is presumed to be a by-product of serpentinization. A dunite (22-2 BR) is show low to moderate serpentinization in West Block (Figure 5). Olivine occurs as main minerals and made up 90–95% of the rock. Orthopyroxene show a distinct extinction and partly replaced by chlorite. Chromite abundance is 3–5%; it usually forms small grains 0.2–1 mm in size. Serpentinite (21-4 BR) shows a mesh texture and in places partially pseudomorphs olivine and orthopyroxene, along with spinel grains. The absence of evidence for clinopyroxene implies that it is a highly altered harzburgite. The serpentine is admixed with small amounts of chlorite. Individual serpentine pseudomorphs are generally fine-grained, with randomly oriented fibrous texture (Figure 6).

Figure 4. Outcrop of peridotite (harzburgite) bedrock in East Block (upper) Photomicrograph of peridotit showing olivine (OL), orthopyroxene (OPX), serpentine (SER) and opaque mineral (OM)

4.2. Geochemistry of Sc in weathering profile

Results of chemical analyses for peridotite are described in Table 1. The bedrock in Petea profile contains of 40.8 wt% SiO_2, 34.8 wt% of MgO, 8.7 wt% Fe_2O_3, 0.2 wt% Ni, 9.8 ppm Sc (Figure 7). In relation to this, the weathering profile show decreasing content of SiO_2 and MgO, and increasing trend of Fe_2O_3 and Sc. SiO_2/MgO ratio shows a slightly increasing trend from bedrock to soft saprolite but significantly enriched in yellow limonite layer.

Overall, West Block weathering profiles showing a somewhat similar pattern to Petea profile. SiO_2 and MgO are significantly decreased whereas Fe_2O_3, Sc and SiO_2/MgO ratio are enriched. This variation suggests the enrichment of Sc was concentrated in the limonite horizon.

Weathering profile of East Block also displays a similar trend in which SiO_2 and MgO are significantly depleted. Fe_2O_3 and Sc are enriched toward the upper part of the weathering profile whereas SiO_2/MgO ratio shows an extreme enrichment in yellow limonite, similar to Petea profile (Figure 8). These finding suggests that scandium is enriched in limonite layer in weathering profile from ultramafic rocks in Soroako.

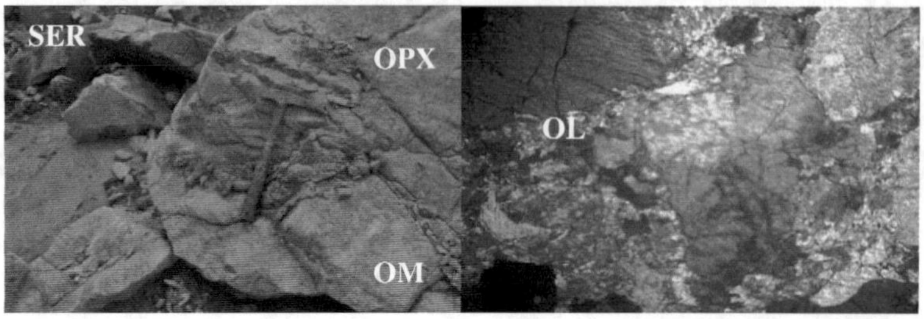

Figure 5. Heavily jointed dunite outcrop in West Block (upper) photomicrograph of dunite consists of olivine (OL) that has been serpentinized (SER), orthopyroxene (OPX) and opaque mineral (OM) (lower)

Figure 6. Outcrop of serpentinite in Petea (upper) and photomicrograph of serpentinite (SER) (lower)

It is also interesting to note that generally pyroxene bearing rich bedrock (harzburgite in composition) contain higher Sc content in their weathering product. The enrichment of Sc in limonitic layer is in a good agreement with the result of study in Eastern Australian [1]. The study suggests that scandium is enriched in limonite layer in weathering profile from ultramafic rocks in Soroako. In general, pyroxene bearing rich bedrock (harzburgite in composition) will contain higher Sc content in their weathering product. It is suggested that the Sc are associated with the occurrence of iron oxide from pyroxene. However, as the occurrence of Sc is very limited, the most promising result was shown by East Block profile in which the enrichment of Sc is also supported by enrichment of NiO content. Sc will be viable as by-product of nickel laterite.

Table 1. Geochemical composition of bedrock, saprolite and limonite

Sample name	Rocktype	Deposit	SiO2	Fe2O3 (T)	MnO	MgO	NiO	Co	Sc	SiO2/MgO
			(%)	(%)	(%)	(%)	(%)	(%)	ppm	%
140621-1RL	Red limonite	Petea (A)	3.16	50.09	0.06	0.7	0.83	0	82	4.514286
140621-1YL	Yellow limonite	Petea (A)	27.86	43.7	0.08	1.18	0.75	0	85	23.61017
140621-1SS	Soft saprolite	Petea (A)	36.32	26.55	0.98	17.3	1.55	0.1	32	2.096998
140621-1HS	Hard saprolite	Petea (A)	39.35	16.35	0.65	26.5	1.65	0	16	1.482668
140621-1BR	Bed rock (harzburqite)	Petea (A)	40.818	8.773	0.21	34.8	0.2		9.8	1.171954

Sample name	Rocktype	Deposit	SiO2	Fe2O3 (T)	MnO	MgO	NiO	Co	Sc	SiO2/MgO
			(%)	(%)	(%)	(%)	(%)	(%)	ppm	%
140621-2RL	Red limonite	Petea (B)	7.35	62.47	0.59	1.54	0.86	0.1	83	4.772727
140621-2YL	Yellow limonite	Petea (B)	20.74	45.29	0.73	10.6	1	0.1	58	1.965877
140621-2SS	Soft saprolite	Petea (B)	32.15	35.73	0.49	10.8	1.29	0	48	2.987918
140621-2BR	Bed rock (harzburqite)	Petea (B)	38.703	8.543	0.09	39.9	0.22		7.1	0.971168

Sample name	Rocktype	Deposit	SiO2	Fe2O3 (T)	MnO	MgO	NiO	Co	Sc	SiO2/MgO
			(%)	(%)	(%)	(%)	(%)	(%)	ppm	%
140621-3RL	Red limonite	East Block	12.82	47.41	0.96	4.97	1.77	0.1	72	2.579477
140621-3YL	Yellow limonite	East Block	39.35	45.67	0.65	0.83	1.03	0.1	40	47.40964
140621-3SS	Soft saprolite	East Block	34.46	18.63	0.25	27.6	3.6	0	20	1.249003
140621-3BR	Bed rock (peridotite)	East Block	39.21	9.054	0.1	40.6	0.22		8	0.966692
140621-4BR	Bed rock (peridotite)	East Block	42.999	10.047	0.11	44.1	0.24		9.3	0.975676

IOP Conf. Series: Materials Science and Engineering **676** (2019) 012032 doi:10.1088/1757-899X/676/1/012032

Sample name	Rocktype	Deposit	SiO2	Fe2O3 (T)	MnO	MgO	NiO	Co	Sc	SiO2/MgO
			(%)	(%)	(%)	(%)	(%)	(%)	ppm	%
140622-1RL	Red limonite	West Block	1.82	69.84	0.55	0.82	1.26	0.1	87	2.219512
140622-1YL	Yellow limonite	West Block	3.42	66.9	1.38	1.45	0.83	0.2	71	2.358621
140622-1SS	Soft saprolite	West Block	58.35	16.42	0.07	12.5	2.57	0	19	4.686747
140622-1BR	Bed rock (dunite)	West Block	42.439	9.746	0.1	44.8	0.25		9.4	0.948273

Sample name	Rocktype	Deposit	SiO2	Fe2O3 (T)	MnO	MgO	NiO	Co	Sc	SiO2/MgO
			(%)	(%)	(%)	(%)	(%)	(%)	ppm	%
140622-2RL	Red limonite	West Block	2.62	75.02	1.57	0.83	1.53	0	40	3.156627
140622-2SS	Soft saprolite	West Block	43.15	16.85	0.2	30.3	1.61	0	7	1.424092
140622-2BR	Bed rock (dunite)	West Block	42.386	9.735	0.1	44.9	0.23		8.4	0.944893

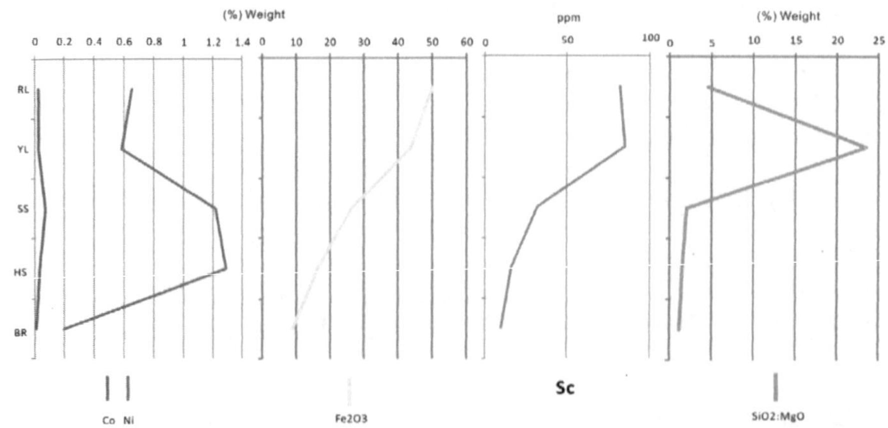

Figure 7. Diagram showing vertical distribution of some elements in Petea A

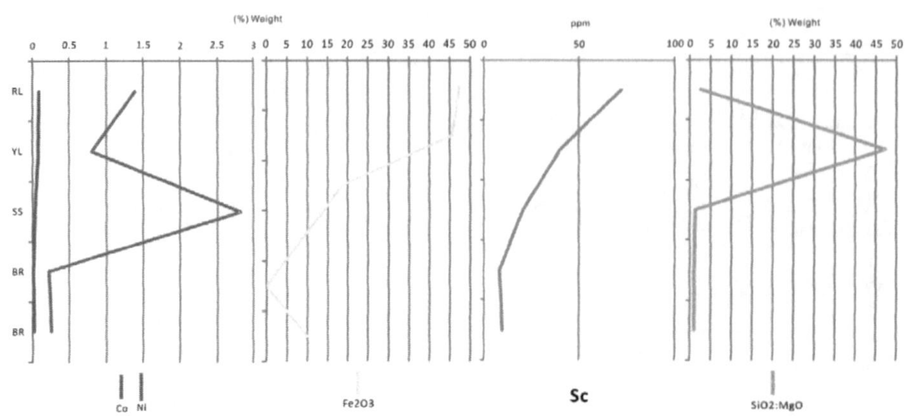

Figure 8. Diagram showing vertical distribution of some elements in East Block

IOP Conf. Series: Materials Science and Engineering **676** (2019) 012032 doi:10.1088/1757-899X/676/1/012032

5. Conclusions

- Scandium is enriched in limonite layer in weathering profile from ultramafic rocks in Soroako
- Pyroxene bearing rich bedrock (harzburgite in composition) will contain higher Sc content in their weathering product.
- Sc are associated with the occurrence of iron oxide from pyroxene and will be viable as by-product of nickel laterite

Acknowledgement

This research is financially supported by JICA C-BEST UNHAS research grant 2015-1016. We would like to express our gratitude to CoT member and staff for their generous support during the research. PT Vale Soroako, particularly Geology and Mining Department staff (Mr. Sahlan Suyuti and Mr. Jasman) is greatly acknowledged for permission and logistic support during our field work.

References

[1] Chasse, M., Griffin, W.L., O`Reilly, S.Y., Calas, G. Geochemical Perspective Letters, 105-114 (2017).
[2] G. Guo, Y. Chen, Y. Li, JOM **40** (7), 28–31 (1988).
[3] J.B. Hedrick. *Mineral Commodity Summaries 2008: Scandium*. US Geological Survey (2010).
[4] J.T.S. Irvine, T. Politova, N. Zakowsky, A. Kruth, S. Tao, R. Travis. *Proceedings of the NATO Advanced ResearchWorkshop on Fuel Cell Technologies*, State and Perspectives, Kyiv, Ukraine, 35–47 (2004).
[5] Z. Ahmad. J.of Material, **55**, 35–39 (2003).
[6] F.T. Ciacchi, S.P.S. Badwal, J. Drennan. J. Eur. Ceram. Soc. **7**, 185–195 (1991).
[7] V.A. Shalomeev, N.A. Lysenko, E.I. Tsivirko, V.V. Lukinov,V.V. Klochikhin. J. Metal Science and Heat Treatment, **50** (1-2), 34-37 (2008).
[8] A. Maulana. *M.Phil. Thesis*, Australian National University, Canberra, Australia (2013).
[9] W. Hamilton. *U.S. Geological Survey Professional Papers*, **1078**, 345 (1979).
[10] A. Kadarusman, S. Miyashita, S. Maruyama, C.D. Parkinson, A. Ishikawa. Tectonophysics, **392**, 55-83 (2004).

IOP Conf. Series: Materials Science and Engineering **676** (2019) 012033 doi:10.1088/1757-899X/676/1/012033

Hydrothermal Alteration Associated with Vein-Type Sulphide Mineralization at Lappadata Prospect, South Sulawesi, Indonesia: A Preliminary Study

I Nur[1]*, Sufriadin[1], A Ilyas[1] and U R Irfan[2]

[1]Mining Engineering Department, Hasanuddin University, South Sulawesi, Indonesia
[2]Geological Engineering Department, Hasanuddin University, South Sulawesi, Indonesia

*E-mail: irzal.nur@eng.unhas.ac.id

Abstract. In Lappadata area, Bone Regency, South Sulawesi Province, Indonesia, indication of vein-type sulphide mineralization is occurred. This paper describes a recent study of the prospect which focused on hydrothermal alteration associated with the mineralization. In this study, alteration and mineralization samples have been collected selectively and representatively from the field and then analyzed in laboratory using petrography, ore microscopy, and X-ray diffraction methods. The study resulted that host rocks of the mineralization are porphyritic basalt and limestone members of the Early Eocene to Late Oligocene Salo Kalupang Formation. Hydrothermal alteration mineral assemblages include quartz, chlorite, muscovite, sericite, calcite, actinolite, epidote, illite, kaolinite, and pyrite, which generally zoned from distal-propylitic mainly characterized by chlorite, calcite and epidote to proximal-argillic characterized by sericite and illite. The mineralization occurred in quartz vein and disseminated with identified hypogene ores of sphalerite, chalcopyrite, galena, and pyrite as well as supergene ores of covellite, chalcocite, and malachite. Study of the hydrothermal alteration stability temperatures suggested that the mineralization was formed in temperature range of about 200 to 300°C, from a near-neutral pH hydrothermal fluid. The whole characteristics of hydrothermal alteration, ore and gangue minerals, mineralization types, as well as range of formation temperature and composition of the responsible hydrothermal fluid indicated that the mineralization in Lappadata prospect is an epithermal type.

1. Introduction

In Lappadata area, Bone Regency, southern part of Sulawesi Island, Indonesia, indication of vein-type sulphide mineralization is occurred, where fine-grained sulphides observed in quartz veins hosted in volcanic and sedimentary rocks. The indication was previously reported and shown in the regional geologic map of the Pangkajene and western part of Watampone quadrangles, Sulawesi, which marked by Cu symbol in Salo Kalupang Formation [1] (Figure 1). Preliminary field observation resulted that around the outcrops of the mineralization, boulders of secondary manganiferous ironstones were found scattered in places. In the most southern part of Bone Regency, South Sulawesi, which famously known as the Biru area (e.g.,[2], [3,], [4]), located not far south of Lappadata area, the similar association of vein-type sulphide mineralization with secondary manganiferous ironstones (hematite and manganese) was also reported. Quartz-sulphide veins in the Biru area penetrate limestone in the same direction with hematite ore zone. The veins are discontinuous and steeply dipping with <5 m thickness, and composed of disseminations, small blebs and veinlets of argentiferous galena, pyrite, aresnopyrite, magnetite, and minor chalcopyrite and pyrrhotite. Galena also occurs as lenses and short veins. Quartz and carbonate are the dominant gangue minerals, and alteration in form of silicification only occurs around ore body [5-7]. Currently a domestic company, PT. Wijaya Eka Sakti, conducting an exploration program (including core drilling) to test the prospect.

This paper describes a preliminary study of hydrothermal alteration associated with the mineralization in Lappadata area, which includes alteration and ore mineral assemblages, alteration zones, and estimated formation temperature of the mineralization as well as its responsible hydrothermal fluid composition inferred from sensitive stability temperature of selective alteration minerals.

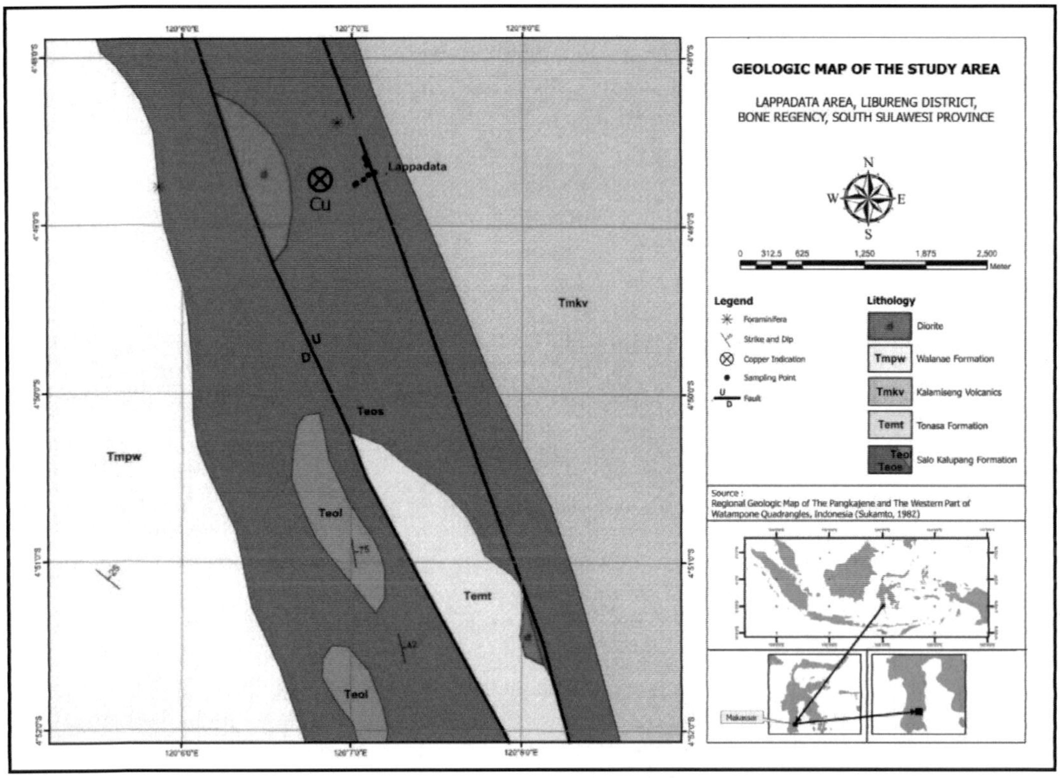

Figure 1. Geological map of Lappadata area and surroundings (cropped and modified from Ref. [1]). The sampling points near Lappadata village is the study area, which is situated around the indication of Cu-mineralization occurred in the Salu Kalupang Formation (Teos)

2. Geological Background

Regional geology of Lappadata area and surroundings is shown in Figure 1. The followings are description of each lithology units in the area, from the oldest to the youngest [1]:

a. Salu Kalupang Formation (Teos); arranged by sedimentary rocks interbedded with volcanics; composed of sandstone, shale, and claystone interbedded with volcanic conglomerate, breccia, tuff, with intercalations of lava, limestone and marl; the limestone and marl contain fossils of molluscs and foraminifera; calcite veins occurred in places; in the area south of Lappadata, members of the Salu Kalupang Formation (Teos) are bedded in nearly north-south strike which parallel to the bedding strike of its limestone member (Teol), the dips are 42° and 75° to the east, respectively (Figure 1); the foraminifera assemblages indicated age of Early Eocene to Late Oligocene; the thickness is not less than 4500 m.

IOP Conf. Series: Materials Science and Engineering **676** (2019) 012033 doi:10.1088/1757-899X/676/1/012033

b. Tonasa Formation (Temt); mostly composed by limestone which includes of massive coral-, crystalline-, and bioclastics- limestone, as well as calcarenite; well bedded and intercalated with marl; the bedded limestone and marl contain fossils of foraminifera, pelecypods and gastropods; the massive limestone is strongly jointed; fossills assemblages indicated age of Early Eocene to Middle Miocene; the thickness is not less than 3000 m; intruded by sill, dike, and stock of basalt, trachyte, and diorite; in the Biru area, the limestone member of this formation is the host rock of the quartz-sulphide veins described by Widi et al. [5] and Van Leeuwen and Pieters [6 ,7].
c. Kalamiseng Volcanics (Tmkv); composed by lava and breccia with intercalations of tuff, sandstone, claystone, and marl; mostly basaltic and partly andesitic; dark grey to black, generally altered, amygdaloid with secondary minerals of carbonate and silicate; parts of the lava showed pillow structure; intruded by stocks of basalt, andesite, and diorite; Early Miocene in age and not less than 4250 m in thickness.
d. Walanae Formation (Tmpw); consists of sandstone interbedded with siltstone, tuff, marl, claystone, conglomerate, and limestone; bedded in northwest-southeast strike and 25° dipping to northeast (Figure 1); fossils assemblage indicated age of Middle Miocene to Pliocene; the thickness is not less than 4500 m.
e. Diorite intrusive (d); intruded the Salo Kalupang and Tonasa Formation near Lappadata village (Figure 1); generally grey in fresh color, porphyritic with amphibole and biotite phenocrysts; based on K/Ar dating, the age is Late Miocene [1, 8].

3. Methods

This study generally consists of two stages, field works and laboratory works. The field works were conducted along Lappadata river, where fresh and altered rocks as wells as mineralization samples were collected selectively at each representative outcrops. During the field work, 13 samples have been collected from 8 outcrops. The samples points are shown in Figure 1.

The laboratory works performed include petrograhy, ore microscopy, and X-Ray diffraction (XRD) analysis. Sample preparations for microscopic analyses i.e., thin sections for petrography and polished sections for ore microscopy, as well as their observations using transmitted and reflected light polarizing microscope were performed in Optical Mineral Laboratory, Department of Geological Engineering, Hasanuddin University. For the XRD analysis, alteration and mineralization samples were pulverized to form powder and then scanned using an X-Ray diffractometer machine of Shimadzu Maxima_X XRD-7000. The diffraction patterns were recorded with scanning stage of 5-70° 2-theta. Results of the XRD scanning then subsequently analyzed using PDX-2 software from the Mineral Data Institute (MDI) combined with Impact Match! Version-3, to identify mineral species contained in the samples. The whole stages of the XRD works, from sample preparation, scanning, analysis, interpretation and identification of minerals were conducted in Laboratory of Mineral Exploration, Department of Mining Engineering, Hasanuddin University.

4. Host Rocks and Mineralization

The study resulted that host rocks of the vein-type and less disseminated mineralization in Lappadata area are porphyritic basalt and limestone (Figure 2). Field observation showed that porphyritic basalt exhibited grey to black in fresh colors and greenish in altered, fine hypocrystalline texture, and mostly composed by mafic minerals. Whereas the limestone was light grey in color, crystalline texture, and mainly composed by carbonate minerals. Based on the lithological characteristics and geographical distribution, both host rocks are the lava and limestone members of the Salu Kalupang Formation (Teos and Teol, respectively (Figure 1)). In addition, disseminated sulphides were also recognized occurred in the Diorite Intrusive (d). This unit intruded Salo Kalupang Formation along the north-northwest fault zone near Lappadata (Figure 1), thus it can be interpreted that the diorite intrusion is the hydrothermal source of the mineralizations.

Figure 2. (A) Outcrop of altered porphyritic basalt lava in Lappadata river, contains 0.1 to 0.2 cm quartz veinlets. (B) Outcrop of altered bedded limestone in Lappadata river. Argillic alteration strongly develops next to the limestone. (C) Mineralized quartz vein hosted in altered limestone, the thickness is about 2 to 3 cm. (D) Strongly altered and mineralized volcanic rocks that contains disseminated fine-grained sulphides and vuggy quartz. The fine-grained sulphides were also observed disseminated in the vuggy

During the field works, ore minerals identified in alteration and mineralization outcrops mostly include chalcopyrite and pyrite, as well as supergene mineral of malachite. Chalcopyrite and pyrite were identified both in altered volcanic rocks (porphyric basalt) including its vuggy quartz, as well as in altered limestone. In addition, malachite was also recognized in the field, stained particularly on surfaces of weathered limestone, indicating a supergene product. Outcrops of mineralization in the field are shown in Figures 2.C and 2.D.

Under the microscope, identified hypogene ore minerals include sphalerite, chalcopyrite, galena, and pyrite. Supergene mineral covellite was also recognized, mostly distributed at the fringe of chalcopyrite and some totally replaced chalcopyrite (Figure 3.A-D). In addition, chalcocite also identified through XRD analysis. Textural analysis resulted ore mineral paragenesis as follow: sphalerite and galena are the earliest minerals; followed by chalcopyrite and pyrite; and then covellite and chalcocite; and finally malachite that formed as supergene mineral stained on surfaces of weathered limestone host rock.

IOP Conf. Series: Materials Science and Engineering **676** (2019) 012033 doi:10.1088/1757-899X/676/1/012033

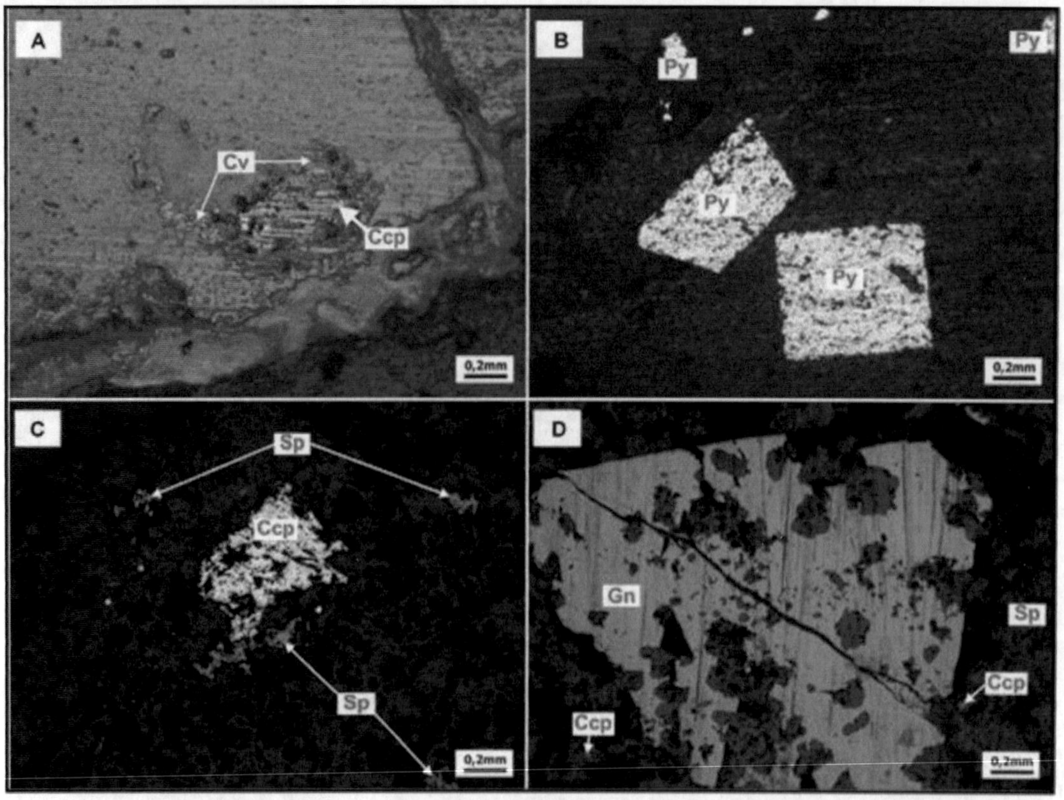

Figure 3. Photomicrographs of selected polished section samples from the study area, showing ore mineral assemblages and their textural relationships: (A) Chalcopyrite fringe replaced by covellite. (B) Euhedral cubic-prismatic crystals of pyrite. (C) Granular texture of anhedral chalcopyrite and sphalerite. (D) Intergrowth texture of sphalerite and galena; fine-grained chalcopyrite scattered on early sphalerite. Abbreviations: Ccp: chalcopyrite, Cv: covellite, Gn: galena, Py: pyrite, Sp: sphalerite

5. Hydrothermal Alteration

In the study area host rocks of the mineralization were intensively altered. The altered rock samples vary in colors such as grey, green, brownish yellow, orange and white. In weakly or selectively altered samples, the protolith still can be identified through field (macroscopic) observation, and under the microscope their primary minerals can be observed easily. Whereas in the strongly or pervasively altered samples, most of the primary minerals have changed or altered to secondary minerals, the composition of their primary minerals is minor, and some can not be identified anymore. Indentification of the pervasively altered samples in this study was conducted by XRD analysis.

Figures 4.A and 4.B showing two altered porphyritic basalt samples where their original colors (dark grey to black) have changed to greenish grey (Figure 4.A) and light green with brownish red stains on its surface (Figure 4.B). The green color on both samples indicated presence of chlorite as an alteration product of ferromagnesian minerals of the porphyritic basalts, mainly pyroxene. Petrographic study on the sample in Figure 4.A showed that the primary minerals still present, such as subhedral to euhedral plagioclase and pyroxene, but most of the pyroxene phenocrysts have altered to chlorite and epidote, and the groundmass altered to sericite (Figure 4.C). While sample in Figure 4.B under the microscope showed domination of chlorite and quartz, with relics of anhedral plagioclase phenocrysts; the pyroxene crystals were not observed since they have totally altered to chlorite (Figure 4.D).

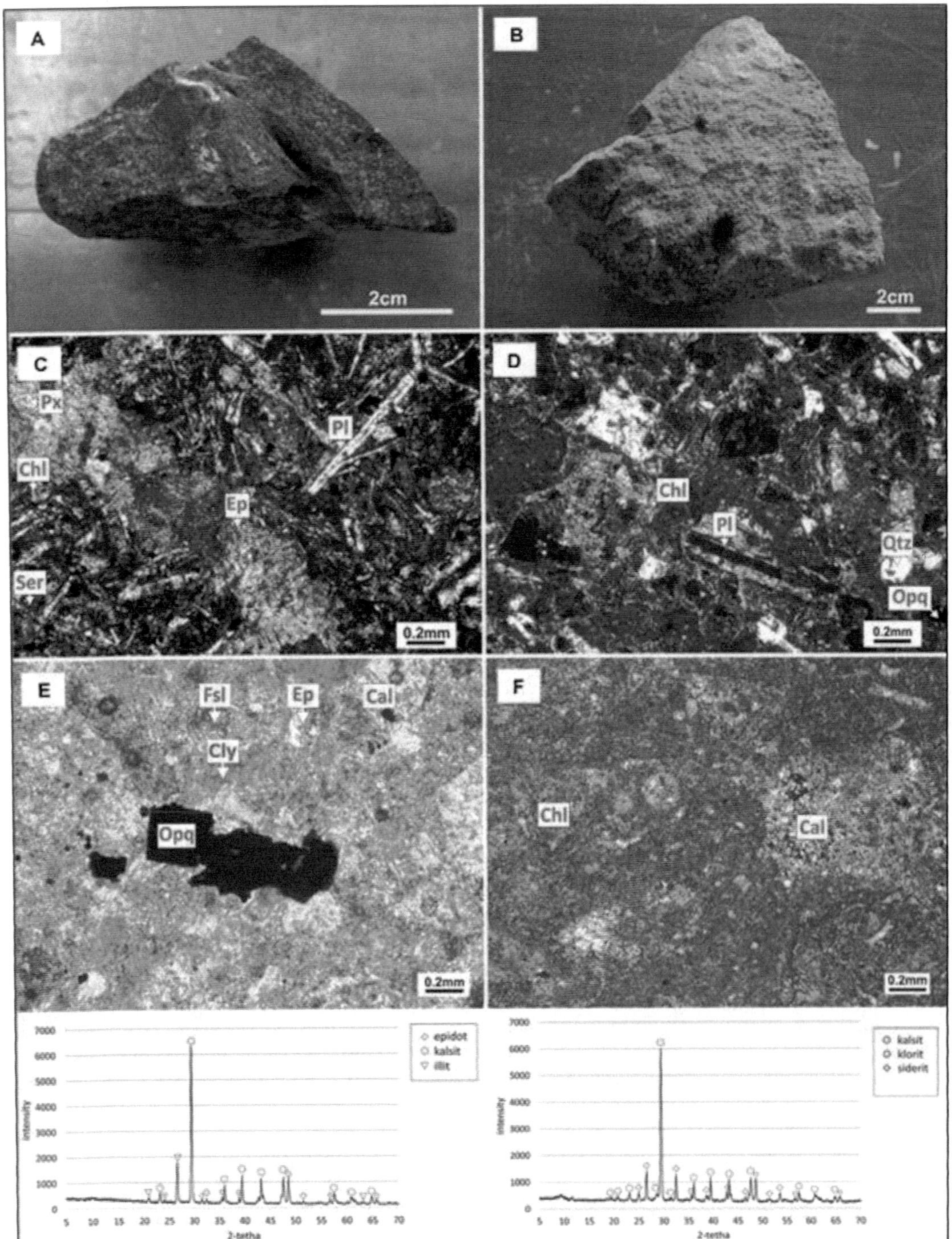

Figure 4. (A,B). Hand specimen samples of altered porphyritic basalt. (C,D). Photomicrographs of altered porphyritic basalt. (E,F) Photomicrographs and XRD diffractogram of altered limestone (sampel BN.IN.06.A and BN.IN.04.B, respectively). Abbreviations: Cal: calcite, Chl: chlorite, Cly: clay, Ep: epidote, Fsl: fossil, Opq: opaque, Pl: plagioclase, Px: pyroxene, Qtz: quartz, Ser: sericite

IOP Conf. Series: Materials Science and Engineering **676** (2019) 012033 doi:10.1088/1757-899X/676/1/012033

Limestone which hosts mineralized quartz veins are generally strongly altered as well. Minerals identified macroscopicly in the limestone include calcite, aragonite, and malachite. Petrography and XRD analysis indicated that mineral assemblages in the altered limestone samples include calcite, siderite, chlorite, epidote, and clay which identified as illite (Figures 4.E and 4.F).

In this study, 13 alteration samples have been analyzed from 8 locations. Alteration mineral assemblages identified from the whole samples by petrography and XRD analyses include: quartz, chlorite, muscovite, sericite, calcite, actinolite, epidote, illite, kaolinite, and pyrite. Using the hydrothermal alteration classification and terminology which generally applied in mineral exploration, e.g., [9], [10], [11], and by considering the distances of each sample locations to the mineralization zones, it was concluded that the hydrothermal alteration which relatively distal to the mineralization is propylitic type, which mainly characterized by the occurrences of chlorite, calcite, and epidote; whereas alteration proximal to mineralization is argillic type which indicated by dominant occurrences of sericite (muscovite) and illite. Mineral assemblages and alteration types as well as their relative distances to mineralization are described in Table 1. In the table, alteration mineral assemblages is arranged in order to their abundances, and the key minerals of each alteration type are marked by bolded font (Table 1).

Tabel 1. Mineral assemblages, types, and zones of hydrothermal alteration

Sample	Alteration minerals	Alteration types and zones
BN.IN.01	Quartz, **chlorite**, muscovite	Propylitic, distal
BN.IN.02	Sericite, quartz, **chlorite, calcite**, muscovite	Propylitic, distal
BN.IN.03	**Calcite**, quartz, **chlorite**, pyrite	Propylitic, distal
BN.IN.04.A1	**Sericite**, quartz, actinolite	Argillic (sericite-illite), proximal
BN.IN.04.A2	**Sericite**, quartz	Argillic (sericite-illite), proximal
BN.IN.04.B	Calcite, **chlorite**	Argillic (sericite-illite + chlorite), proximal
BN.IN.05	Chlorite, epidote, **sericite**	Argillic (sericite-illite), proximal
BN.IN.05.C	Calcite, chlorite, **illite**	Argillic (sericite-illite), proximal
BN.IN.06.A	Calcite, epidote	Argillic (sericite-illite), proximal
BN.IN.06.B1	Quartz, **illite**, kaolinite	Argillic (sericite-illite), proximal
BN.IN.07	Calcite, quartz, chlorite, **illite**	Argillic (sericite-illite), proximal
BN.IN.08.A	Quartz, **chlorite, calcite**	Propylitic, distal
BN.IN.08.B	Quartz, **chlorite, calcite, epidote**	Propylitic, distal

In epithermal environment, several epithermal minerals have a sensitive stability temperatures that can be used to estimate their formation temperature and interprete their responsible hydrothermal fluid composition (pH). In this study, chlorite, epidote, illite, sericite (fine-grained muscovite), and actinolite were selected to estimate and interprete the temperature and pH, since they are sensitive-temperature minerals. Quartz, calcite, and pyrite were not chosen since they formed in wide temperature ranges, from <100°C to >300°C [10, 11]. The presence of acid mineral kaolinite which not suitable to the other alteration minerals which are generally neutral to near-neutral pH, became a consideration why kaolinite was not used in the formation-temperature estimation in this study. The presence of kaolinite together with other neutral hydrothermal alteration minerals commonly formed by overprinting of acid and near-neutral fluids [12].

Studies on hydrothermal minerals stability in epithermal environment suggested that chlorite formed in temperatur range of about 200 to 300°C, epidote about 200 to 320°C, illite 200 to 300°C, sericite 200 to 250°C, and actinolite 280 to 300°C, all formed from near-neutral hydrothermal fluids [10, 11, 13, 14]. Thus, by combining formation-temperature ranges of the five minerals, it can be estimated that mineralization in Lappadata prospect was formed in a temperature range of about 200 to 320°C, from a near-neutral pH hydrothermal fluid.

6. Conclusions

Several conclusions defined from this study include, host rocks of the mineralization are porphyritic basalt and limestone members of the Early Eocene to Late Oligocene Salu Kalupang Formation. The mineralization also genetically related to Late Miocene diorite intrusion, which is inferred as the hydrothermal fluid source. The mineralization formed in quartz veins and disseminated, with hypogene ores of sphalerite, chalcopyrite, galena, and pyrite as well as supergene ores of covellite, chalcocite, and malachite. Hydrothermal alteration mineral assemblages include quartz, chlorite, muscovite, sericite, calcite, actinolite, epidote, illite, kaolinite, and pyrite, which are generally zoned from distal-propylitic mainly characterized by chlorite, calcite and epidote, to proximal-argillic characterized by sericite and illite. Based on study of stability temperature of the hydrothermal alteration minerals, it is concluded that the mineralization was formed in a temperature range of about 200 to 320°C, from a near-neutral pH hydrothermal fluid. The whole characteristics of hydrothermal alteration (mineral assemblage and its spatial distribution to the mineralization), ore mineral assemblage and *gangue*, mineralization types, as well as estimated formation temperature and pH of the responsible hydrothermal fluid, indicated that the mineralization in Lappadata area is epithermal type. The characteristics of hydrothermal alteration, mineralization types, and the formation environment can be used to develop the prospect to the next stage and more detailed exploration.

Acknowledgments

The authors wish to express a sincere gratitude to the Ministry of Research, Technology, and Higher Education, Republic of Indonesia, as well as Hasanuddin University Institution of Research and Community Service (LP2M UNHAS) for the financial support through “2018 Higher Education Excellent Basic Research Grant”.

References

[1] Sukamto, R., 1982. *“Geologic Map of the Pangkajene and Western Part of Watampone Quadrangles, Sulawesi”*, Geological Research and Development Center, Bandung, Indonesia.

[2] Van Leeuwen, T.M., 1981. *“The Geology of Southwest Sulawesi with Special Reference to the Biru Area”*. In Barber, A.J. and Wiryosujono, S. (Eds.), The geology and tectonics of eastern Indonesia, Geological Research and Development Centre, Bandung, Special Publication, 2, p.277-304.

[3] Elburg, M.A., Van Leeuwen, T.M., Foden, J. and Muhardjo, 2002. *“Origin of Geochemical Variability by Arc-Continent Collision in The Biru Area, Southern Sulawesi (Indonesia)”*, Journal of Petrology, 43, p.581-606.

[4] Jaya, A. and Nishikawa, O., 2011. *“Deformation Microstructures of Metamorphic Rocks in The Biru Area South Sulawesi, Indonesia”*, Proceedings of JCM Makassar 2011, The 36th HAGI and 40th IAGI Annual Convention and Exhibition, Makassar, 26-29 September 2011, p.10.

[5] Widi, B.N., Pardiarto, B. and Mulyana, 2007. *“Mineralization System of the Iron Ore Deposits in Bontocani District and Its Adjacent Bone Regency, South Sulawesi Province”*, Proceedings of HAGI-IAGI-IATMI Joint Convention Bali 2007, 13-16 November 2007, p.864-870.

[6] Van Leeuwen, T.M. and Pieters, P., 2011. *“Mineral Deposits of Sulawesi”*, Proceedings of the Sulawesi Mineral Resources MGEI-IAGI Seminar, Manado, North Sulawesi, Indonesia, 28-29 November 2011, p.78-79.

[7] Van Leeuwen, T.M. and Pieters, P., 2012. *“Mineral Deposits of Sulawesi”*, Geological Agency Publ., Ministry of Energy and Mineral Resources, Republic of Indonesia, Bandung, p.90-93.

[8] Sukamto, R. dan Supriatna, S., 1982. *“Geological Map of the Ujung Pandang, Benteng and Sinjai Quadrangles, Sulawesi”*, Geological Research and Development Center, Bandung, Indonesia.

[9] Thompson, A.J.B. and Thompson, J.F.H., 1996. *“Atlas of Alteration, A Field and Petrographic*

Guide to Hydrothermal Alteration Minerals", Geological Association of Canada, Mineral Deposits Div., p.118.

[10] Hedenquist, J.W., Izawa, E., Arribas, A. and White, N.C., 1996. *"Epithermal Gold Deposits: Styles, Characteristics, and Exploration"*, Resource Geology Special Publication, 1, Tokyo.

[11] Hedenquist, J.W., Arribas, R.A. and Gonzalez-Urien, E., 2000. *"Exploration for Epithermal Gold Deposits"*, Reviews in Economic Geology, 13, 245-277.

[12] White, N.C. and Hedenquist, J.W., 1995. *"Epithermal Gold Deposits: Styles, Characteristics and Exploration"*, Society Economic Geologist Newsletter 23, pp.1, and 9-13.

[13] Reyes, A.G., 1990. "Petrology of Philippine Geothermal Systems and the Application of Alteration Mineralogy to Their Assessment", Journal of Volcanology and Geothermal Research, 43, 279-309.

[14] Corbett, G.J. and Leach, T.M., 1998. *"Southwest Pacific Rim Gold-Copper Systems: Structure, Alteration, and Mineralization"*, Society of Economic Geologists Special Publication, 6, p.237.

IOP Conf. Series: Materials Science and Engineering **676** (2019) 012034 doi:10.1088/1757-899X/676/1/012034

Characteristic Sulfat Geochemistry of Barite (BaSO4) Marine Type on Marl Tonasa Formation Barru, South Sulawesi

A Tonggiroh[1]*, M Farida[1], H Sirajuddin[1] and D P Battu

Department of Geology, Engineering Faculty, Hasanuddin University, Makassar, Indonesia

*E-mail: adi_unhas@yahoo.com

Abstract. Barites can form in the marine environment as autogenic minerals, which are dominated by barium, carbonate and sulfate elements. Barite mineral distribution on Marl are occurred on two location which are Ralla and Barru Area (1) Ralla, irregular shape (2) Barru, barite concression, slightly rounded. The study aims to determine sulfat geochemistry as chemical bond and the influence of barium depositional environment. Applied Petrography, SEM and XRD Analysis on Marl, Andesite, and Barite mineral. Tabular rosset texture with minor Calcite are indicated that source of Sulfat material comes from Neritic environment (Barru) and penetration of hydrothermal solution.

1. Introduction

Barite is autogenic sediment mineral [1] formed by chemical mixing of unsaturated seawater (Ba^{2+}) dan (SO_4^{2-}) deposited or filled in cavity of hydrothermal rocks [2]. The chemical structure of barite is slightly simple and processing to obtain economically value through extracted of separation between sulfate and clay. Barite ($BaSO_4$) is mostly contain barium formed by various low temperature and pressure (25 °C, 1 atm) and very low solubility [3].

Chemical and physical properties of barite are common reason this minerals are used in various industries such as : (1) lubricating and solid material at drilling oil and gas industry to increase hydrostatic pressure (2) medical equipment industry; X-RAY filter in the field of radiology (3) mixture on plastic industry, rubber sludge, (4) automotive, clutch bearings, car sound absorbers, brake pads (5) electronic, radiation shield for television and computer (6) golf ball (7) paint industry(8) glass industry (9) mold coating, easily release of material from the mold.

Barium sulfat is formed from chemical mechanism of mixing interaction between dominated Ba solution and SO_4 as deposition material on shore and marine environment [3, 4].

Barite as marine deposition, characterized by occurrence on lithology with relative high calcium contain such as calcareous clay/marl, recognized on research area as nodul fragment concression on marl layer of Tonasa Formation (Figure 1).

Research was held by LBE (*Labo Based Education*) Program, Engineering Faculty of Hasanuddin University to analysis sulfat geochemistry that occurred as Barium element on Barite mineral that correlated with Sulphide Alteration Mineralization on Dacite igneous rock.

IOP Conf. Series: Materials Science and Engineering **676** (2019) 012034 doi:10.1088/1757-899X/676/1/012034

Figure 1. Barite on Marl Tonasa Formation

2. Regional Geology

Research area regional include on Tonasa Formation that formed by intercalation between Limestone and Marl abundant of fossil lies on Barru Area and Limestone contain with schis fragment and ultramfic occurred at Ralla Area (Figure 2). Tonasa Formation consist of marl with limestone as member, neritic environment depositional, Miosen age.

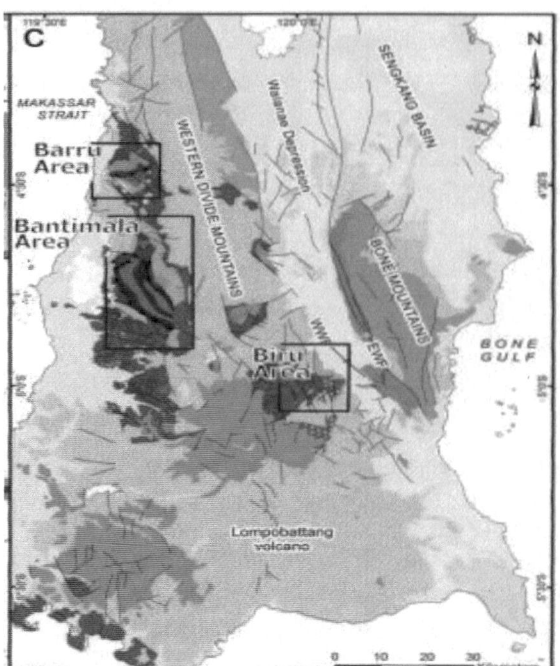

Figure 2. Regional Geology [5], research area Ralla and Barru

Abundantly of macro and micro foraminifera, shown if Tonasa Formation range Early Eosen (Ta.2) to Middle Miosen (Tf). Relative age of marl and limestone is Early Eosen to Late Eosen (P9-P16). Cycle of depositional environment between Inner Neritic-Middle Neritic-Outer Neritic occurred during sedimentation, medium water temperature (range 35°/oo), salinity on marine lagoon and carbonate platform [5].

3. Method

Barite mineral sample (ADB01; ADR02) convert to thin section with 0.8 – 1.2 cm thickness, SEM-EDX (Scanning Electron Microphobe-Emision Diffraction X-ray) Jeol-JSM 6610LV, Mineral Resources Laboratory Akita University (Figure 3).

Determined barium contain on Iodin solution 50 ml (0.5 N) and 1N hydrochlorid acid, followed by agglomeration and titration, conducted at Chemical Laboratory Engineering Faculty Unhas.

Igneous rock sample (ADB 03; ADB04), convert into thin section with thickness 1.0 cm polished till 0.03 and 0.05 mm thickness than observed under polarization microscope on Optical Mineral Laboratory of Technical Faculty of Hasanuddin University.

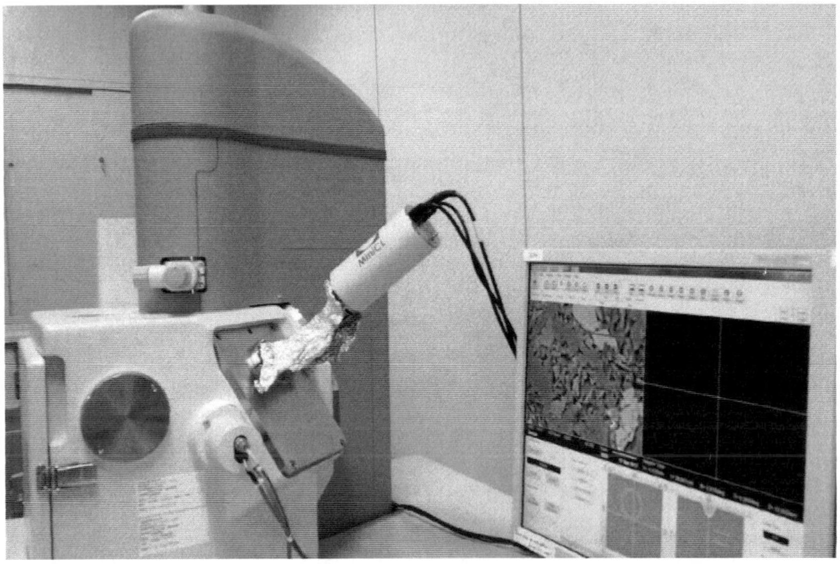

Figure 3. SEM EDX Jeol-JSM 6610LV

4. Field Situation

Research area is located on Barru and Ralla District Barru Regency South Sulawesi Province coordinate 119°37'00"–119°38'00"; 119°41'00"–119°42'00" East Longitude and 04°25'00"–04°26'00" South Latitude.

Field orientation shown if barite deposited has white clear brownish, rosset tabular structure, kohesif, slightly heavy and rounded with diameter < 6 cm, found as infiltration nodule on marl at Barru and slightly irregular shape at Ralla.

Intercalation between marl and limestone was shown changing on depositional environment that relative southwest – Northeast or from Barru to Ralla (Figure 4). Geological structure was developed as Normal Fault on marl with Middle Eocen – Post Middle Miocen.

Figure 4. Barite outcrop on Barru (a & c), Ralla barite (b & d)

5. Lithology

Field interpretation shown two outcrops were founded on research area; Andesite and Marl. Macroscopic description recognized blackish intercalation limestone on Marl, bioclastic texture and layering (Figure 5).

Figure 5. Outcrop (a) andesite (b) Prophylitic alteration

Marl megascopic description (AD01) grey, brownish grey, fine clastic, good sortation, fabric, silt to clay grain size and calcareous. Microscopic analysis shown; yellow brownish absorb color, brown interferen color, clastic with fossil (62%), micro foraminifera (Globigerina sp), calcite (20%), pore (3%), mud supported (15%), ***Grainstone***.

Limestone megascopic description (AD02) brown absorb, brown reddish interference, bioclatic texture with grain (35%), foraminifera fosil Actinocyclina radians (d'Archiac) and Dyscocyclina pratti (Michelin) and Alga (15%); Grain non skeletal litoclas (5%), sparite (10%), carbonate mikrit (35%), **Packstone** (Figure 6).

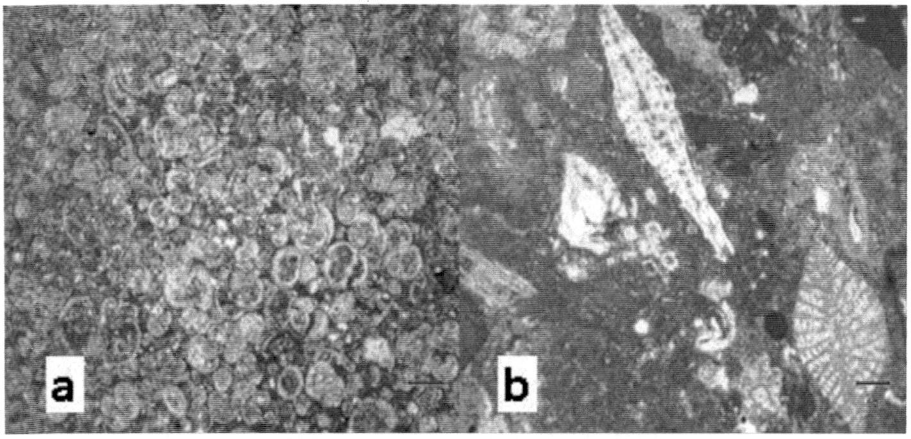

Figure 6. Thin section 50x, fragment : (a) grainstone (b) packstone

Andesite (AD03), grey, brownish grey, medium prophylitic alteration, patchy feldspar with pyrite lineation on oxide cracking (Figure 7).

Microscopic description, white greyih, brown, hypocrystalline, porfiritic, (Fenokris 40% ; Glass 60%), hyalopilitc, mikrolite feldspar subparalel on glass mineral. Fenokris Sanidine, plagioclas, feldspatoid, biotite and Hornblend (50 – 500 μm), euhedral - anhedral, **Trachyte Andesite**.

Field description dacite (AD04), grey to brown blackish, hipocrystaline, porfiritic, euhedral-subhedral, inequigranular, and masif. Composition mineral plagioclase, quartz, hornblende, and glass.

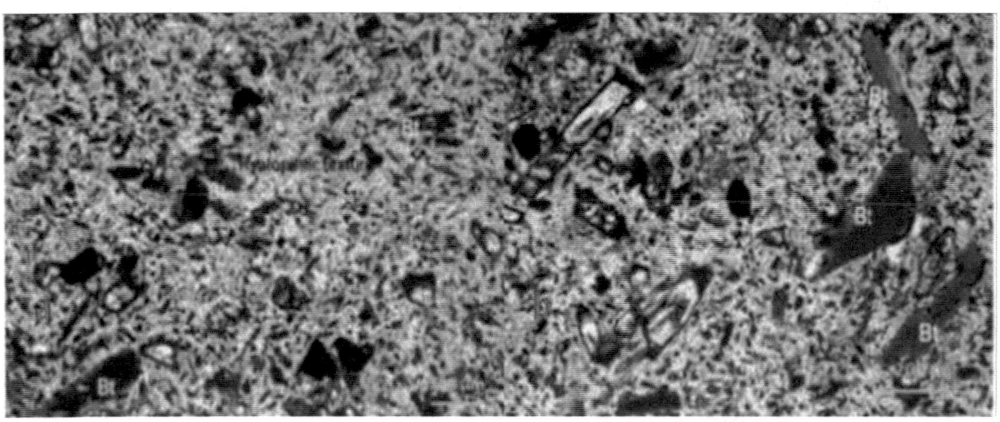

Figure 7. Thin Section 50x, (a,b) trachyte andesite

Mikroscopic, brown absorb color with grey blackish maximum interference, hipocrystalline, faneroporfitic, euhedral-subhedral, inequigranular, labradorit (35 - 40%) occasionally change to sericite, hornblende (15 - 20%) occasionally change to chlorite, quartz (10 – 15%), groundmass plagioclase crystalline (20 – 25%), groundmass as glass mineral (10 – 15%), indeks warna 20, **Dasit Porfiri** (Travis, 1955).

6. Barite Petrogrpahy

Petrography analysis for sample AD_05 shown if mostly barite mineral has transparent absorption, size (0,1 sampai 2,6 mm); calcite rhombohedral, size (0,1 sampai 0,875 mm), plaktonik fosil microcrystalline (0,125 sampai 2,75 mm). Orthorombic structure form pinacoid block, prismatic that contain with clay micrite and sronsium (Sr) minor (Figure 8).

Figure 8. Thin section 50x, mineral barit

Based on carbonate composition, microcrystalline and micrite, indicated if barite was deposited on lower part of Marl marine sedimentation; deep-sea marine with abundant of calcareous dan siliceous biological debris.

7. Crystal Chemistry

Barite has orthorhombic dipirimaidal crystal structure, one sulfur bond and two-oxygen bond as sulfat (SO_4^{-2}) tetrahedron [6]. Sulfur has chemical substitution affected by chemistry of depositional environment and hydrothermal solution. Number of sulfur and barium are linier as chemical reaction on environmental deposition (Table1).

Table 1. Percentage of Barium and sulfur

Sample ID	Ba (%)	S (%)	Si (%)
AD05	72.04	20.67	-
AD06	67.16	13.46	-
AD07	51.49	14.28	5.68
AD08	54.76	16.52	0.41
AD09	40.39	11.39	12.5
AD10	68.92	16.19	-

Minimum S (40,39; 11,39) and maximum (72,04; 20,67) are the hypothesis explain dominated Ba than S. Indicated if depositional environment of barite was dominantly effected by marine than shore sedimentation (Figure 9).

On marine environment, subtitution Ba^{2+} occurred as ion radius and anion of Sr^{2+} dan Ca^{2+} on particle phase with other element such as carbonate, organic, oval, ferromanganese hydroxide, mainly exist on marine and terrestrial environment that contain of silicate of detrital environment.

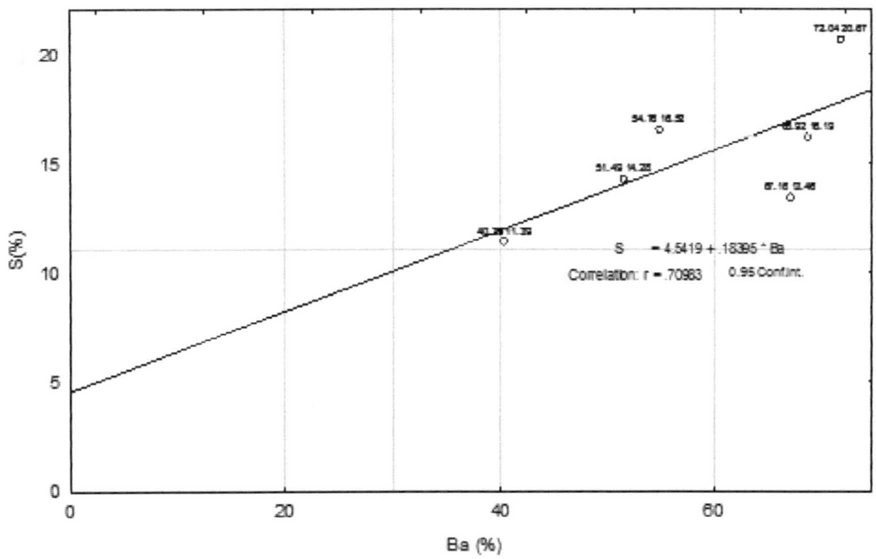

Figure 9. Correlated graphic element Ba and S

8. Discussion

According dominated of barium element (Ba), less number of sulfat and petrography analysis are shown the characterized of sedimentation process such as deposition, diagenetic and erosional to build texture and shape of lithology. During diagenetic process, mud material is fill into the cavity of barite, thus precipitation of marl formed the angular prismatic crystal as erosion effect. Occurrence of micrite and plactonic are shown condition wet paleoclimate where clay depositional exist on marine environment. Uneven distribution of silica on few samples and less number of sulfat are need further research as part of mixing system either hydrothermal or erosion of continent element.

References

[1] Boggs JR, 2009, Petrology of Sedimentary Rocks, Cambridge University Press, p.6.

[2] Ehya. F, Mazraei SM., 2017, Hydrothermal barite mineralization at Chenarvardeh deposit, Markazi Province, Iran: Evidences from REE geochemistry and fluid inclusions, Journal of African Earth Sciences, 134, p.299-307.

[3] Church TM, Wolgemuth K, 1972, Marine barite saturation, Earth and Planetary Science Letters,V.15, Issue 1, p.35-44.

[4] Griffith, E.M., Paytan, A., 2012. Barite in the ocean-occurrence, geochemistry and palaeoceanographic applications. Sedimentology 59, 1817-1835.

[5] Jaya, A, Nishikawa, Osamu, Hayasaka, Yasutaka,2017, LA-ICP-MS Zircon U–Pb and Muscovite K-Ar Ages of Basement Rocks from the South Arm of Sulawesi, Indonesia, Lithos.

[6] Dana, 1977, Manual of Mineralogy, John Wiley & Sons, p.317.

IOP Conf. Series: Materials Science and Engineering **676** (2019) 012035 doi:10.1088/1757-899X/676/1/012035

Chemical Characteristics and Correlation of Heavy Metal Elements in Lumpue Beach, Parepare City

H Sirajuddin[1]* and A Tonggiroh[1]

[1]Geology Engineering Department, Faculty of Engineering, Hasanuddin University, Jalan Poros Malino Km. 6 Borongloe, Gowa 92172, Indonesia

*Email: haerany_sirajuddin@yahoo.com

Abstract. This study aims to analyze heavy metal elements in Lumpue Beach. Characterization of Lumpue Beach sediments in addition to the megaskopis and petrographic observation approach, determination of mineral chemical aspects using ICP-OES analysis was also done. This analysis done on 7 samples of very fine sand obtained from sieving and aimed to determine the level of heavy metal elements on the ppm scale. Based on ternary plotting diagrams, that the dominant Lumpue Beach is composed of sedimentary texture very fine sand - coarse sand. The ratio of Pb elements in the study area has an average level of 41.38 ppm and a trend line R2 = 0.25, Cd has an average level of 1.65 ppm and a trend line R2 = 0.12, Cu elements has an average level of 14.5 ppm and a trend line R2 = 0.02imes and Zn element has an average level of 62.125 ppm and a trend line R2 = 0.0022. They are indicated to exceed normal limits. Based on the use of Pearson method with two paired variables on metals Cu, Fe, Ni, Zn, Pb, Ti and Mn, it is known that Cu metal has a very weak correlation with Fe, Zn, Ni, Pb and Ti, while Fe metal has a strong correlation to Zn., Ti and Mn but have a weak correlation with Ni and Pb. Fe and Al metals are deposited with low values, this occurs in the concentrations of the two metals and follows the sediment fraction controlled by currents and tides.

1. Introduction

Increased levels of heavy metals in seawater occur due to the entry of waste containing heavy metals into the marine environment. Waste that contains a lot of heavy metals usually comes from industrial activities, mining, settlement and agriculture. In general, before going to the sea, heavy metals are affected by properties such as tides, sea breezes, water permeation and the geological conditions of the surrounding rocks. Lumpue Beach has quite high turbidity [1, 2].

Turbidity that occurs in the Lumpue Beach area is affected by the presence of sediment resuspension [3]. Increasing ionic strength causes the pulling force between particles to become stronger and result in the accumulation of a material which is often called a floc. If the resultant attraction is large, the size of the floc will be even greater. In addition, the particles in Lumpue Beach have the ability to adsorb heavy metals, so that the dissolved metal content of water becomes reduced, then the content of this metal is deposited in the sediment.

Lumpue Beach is the largest tourist destination beach in the City of Parepare which is a gathering place for people to spend their holidays with family (Figure 1). Therefore, this research is important to determine whether Lumpue Beach is safe against harmful heavy metal elements. Because information about the analysis and distribution of heavy metals in Indonesia is still very limited, especially in the Lumpue Beach area, Parepare City, South Sulawesi.

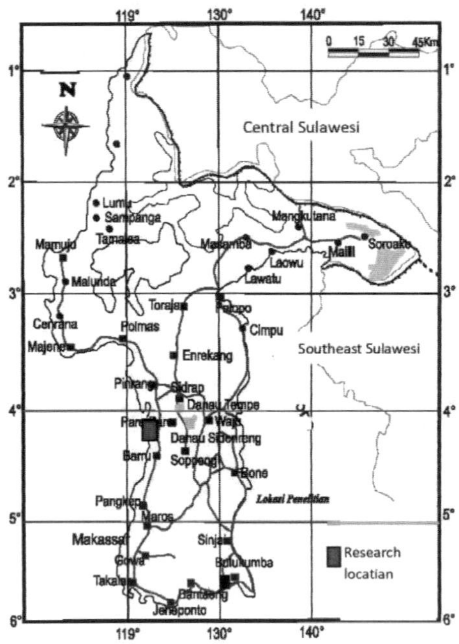

Figure 1. Map showing the location of the research area

The concentration of heavy metals will change during their time on the coast. The changes in metal concentrations are influenced by various processes in the coast such as dilution, flocculation, adsorption and desorption processes by particles. Based on geomorphological and petrography conditions of volcanic rocks regionally, it can be distinguished from south to north, namely Baturape, Lompobattang, Cindako, Camba, Soppeng and Parepare volcanic [4]. The process of heavy metal concentration occurs due to sedimentation from the geomorphological conditions of the original rock [5].

With the adsorption process, which is then followed by a flocculation process, the concentration of this metal will experience a reduction and vice versa if there is a desorption process or the dissolution of the particles, the concentration of heavy metals will increase [6]. This process can be seen by looking at the distribution of heavy metals in terms of the distribution of metal concentrations so that needs for heavy metal concentration data.The distribution of heavy metals that studied from sediment deposits in Lumpue Beach area are Pb, Cd, Cu, and Zn.

2. Geological of Study Area

The geomorphological group units in the study area consist of morphographic and morphometric approaches. Based on the morphographic approach, the dominant study area consisted of terrain, hills and mountains. This aspect of land formation is known by observing the parameters of each topography such as the shape of the peak, the shape of the slope and the shape of the valley. Based on the morphometry approach, the study area consists of land morphological units located in west of the Karajae River flow including the Kampungbaru, Watanbacukiki, Lumpue and Pekkae regions. This morphological unit occupies alluvial, tuff and intrusion trakit units (Figure 2).

Based on unofficial lithostratigraphy, the rock unit in the Lumpue Beach area can be divided into 2 rock units which are described sequentially from the youngest are alluvial and volcanic breccias units. The description of each rock unit contained in the study area will start from the oldest unit to the youngest.

Volcanic breccias units occupy about 25.1km^2 or about 28.33% from the total area of the mapping area. Lithology which composes units consists of volcanic breccias, agglomerate, lapili tuff, coarse tuff, ignimbrite and basalt. This unit occupies or around the total area of the study dominates from the southern to the north, including Latasiso, Bilalang, Jawi-Jawi, Lawalane, Lapance and Bacukiki.

Figure 2. The appearance of the morphology of the terrain in the Bacukiki area

The field appearance of the breccia unit that found in the study area has physical characteristics, namely fresh grayish brown color, brownish black weathered color, has a coarse clastic texture, grain size until coarse sand, with angular - subangular material shape, poorly sorted and open fabric. This rock is composed of components consisting of fragments from the igneous rock porphyry basalt and andesite porphyry, while the rock matrix is composed of volcanic material in the form of minerals which consists of plagioclase, orthoclase, biotite, and pyroxine, while the one that act as cement in rock is volcanic ash. Generally in difficult volcanic breccias outcrops there are a rock position or layered structure, but the layer of volcanic breccias can be generally measured if volcanic breccias are exposed together with other unit members, namely with coarse tuffs or lapilli tuff. Based on its constituent components, this rock is volcanic breccias [7].

In the study area exposed rocks in the form of porphyry trachyte which are igneous rock types with massive structures (Figure 3). Direct observation in the field that being study area shows igneous rock has a contact with porphyry trachyte rock. The distribution directionof rocks is N320°E or relatively southeast to northwest [8]. From the results of microscopic analysis, it found the brown absorption color, black interference color, hypocrystalline crystallinity, porphyritic granularity (porphiroafanitic), anhedral–euhedral mineral form, mineral size <0.02–1.20 mm, mineral composition consisting of plagioclase, oligoclase, hornblende, biotite, quartz and groundmass (Figure 4).

Based on the lithology equation and its geographical distribution close to the type location, volcanic breccias units with Parepare volcanic rocks can be correlated with rocks deposited on land environment. In the regional stratigraphy relationship, the volcanic breccias unit of the study area is the same as the Parepare volcanic rock that is Plistocene [4, 9].

Alluvial units are spread along the Bojo river, forming river terrace or as active deposits in the Bojo river, which occupy an area of approximately 2.7 km^2 or around 2.9% of the entire study area. This alluvial unit is located in the southwest part of the study area.

The alluvial unit is composed of lumps of rock that come from igneous rocks, but sometimes there are also other rocks such as volcanic breccias but in very small quantities, the rest is sand-sized to clay, but this sand material is dominated by sand-sized material very coarse. As for the comparison group is the Qac group (alluvium, lake and beach deposits) composed of clay, silt, sand and gravel components along the large river.

The geological structure that developed in the study area was in the form of a sliding fault, which is the southwest-northeast direction, where there is a fault mirror or slickenslide, which was found in the Mangimpuru village. Age of Tolong fault, where the rock unit displaced is an old tuff unit between the Middle Miocene - Lower Pliocene, it can be concluded that the age of the Tolong fault is Lower Post-Pliocene [2, 4].

IOP Conf. Series: Materials Science and Engineering **676** (2019) 012035 doi:10.1088/1757-899X/676/1/012035

Figure 3. The appearance of porphyry trachyte rock outcrops at the Lumpue coast

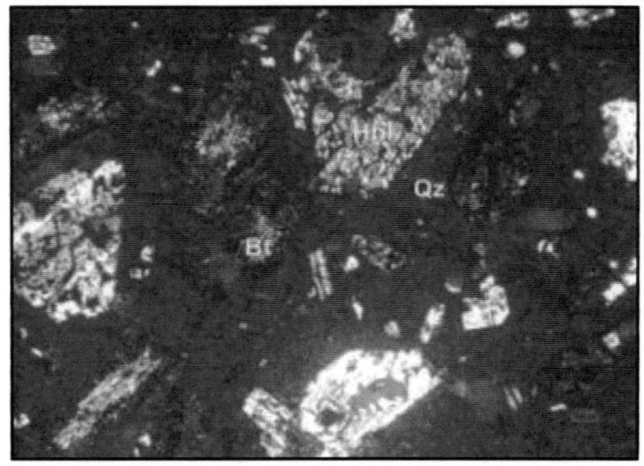

Figure 4. Photomicrograph of thin section showing the presence of plagioclase, hornblende, biotite and quartz minerals. Photographed with 50x magnification.

3. Analysis of Sediment Grain Size in Lumpue Beach

The grain size and distribution of Lumpue Beach sediment particles are identified and determined based on compilation of field observations in the form of descriptions of sediment samples in the form of wells test and results of laboratory data processing. The description of each layer of the well test is based on variations in sediment materials, color, structure or other sedimentary characteristics. The description results are then used to describe the main characteristics of each layer. The main characteristics of its components are sedimentary materials, texture and grain size of sediments, color of sediment material, thickness and slope of sediment distribution structure of sediments and layering fields [11]. Based on the results of compilation of the data, it is known that the dominant research area is composed of very coarse sand sediments to sandy silt.

Semi logarithmic curves (Figure 5) at stations 01 to stations 03 have in common the percentage of weight passes and grain size, while at station 04 there is a difference in weight passing by 20%. Sediment grain size distribution at stations 05, 06 and 07 are relatively smooth from station 4 with a grain size of ± 0.51 mm. It is interpreted that the direction of sedimentation is relative to stationary 04 or southeast. Based on ternary plotting diagram [10] in the form of abundance percentage and texture of sediment components (size of grains sediment), it is known that the dominant Lumpue Beach is composed of very fine sand sedimentary (Figure 6).

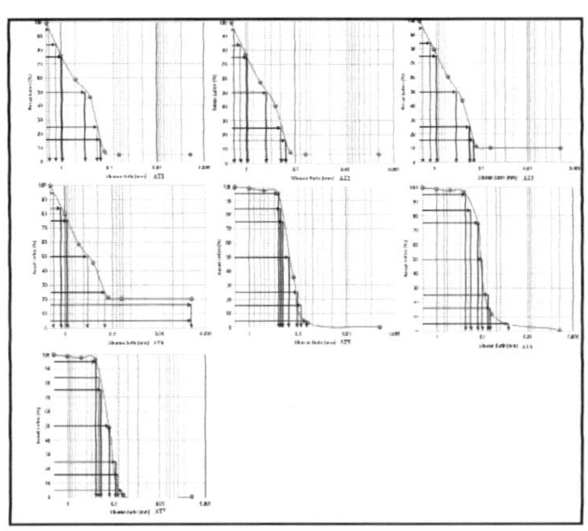

Figure 5. Semi-logarithmic curve of Lumpue Beach sediment distribution.

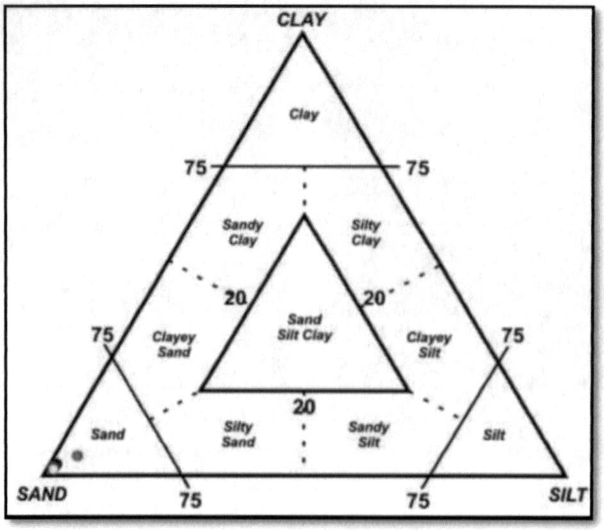

Figure 6. Distribution of Lumpue Beach sediments based on the texture and percentage of components

Based on the results of petrography analysis of mineral smear slides on sedimentary material in the study area, it is known that the composition of the constituent materials includes of rock fragments, plagioclase, hornblende, biotite and opaque minerals (Tabel 1 and Figure 7).

Table 1. Composition of the main components in Lumpue Beach sediments

Component	ST-01	ST-02	ST-03	ST-04	ST-05	ST-6	ST-7
Rock Fragment	40,00	25,00	45,00	45,00	10,00	25,00	25,00
Plagioclase	45,00	30,00	30,00	25,00	40,00	30,00	50,00
Hornblende	10,00	40,00	25,00	30,00	30,00	40,00	15,00
Biotite	5,00	0,00	0,00	0,00	5,00	0,00	5,00
Opaque Mineral	0,00	5,00	0,00	0,00	15,00	5,00	5,00

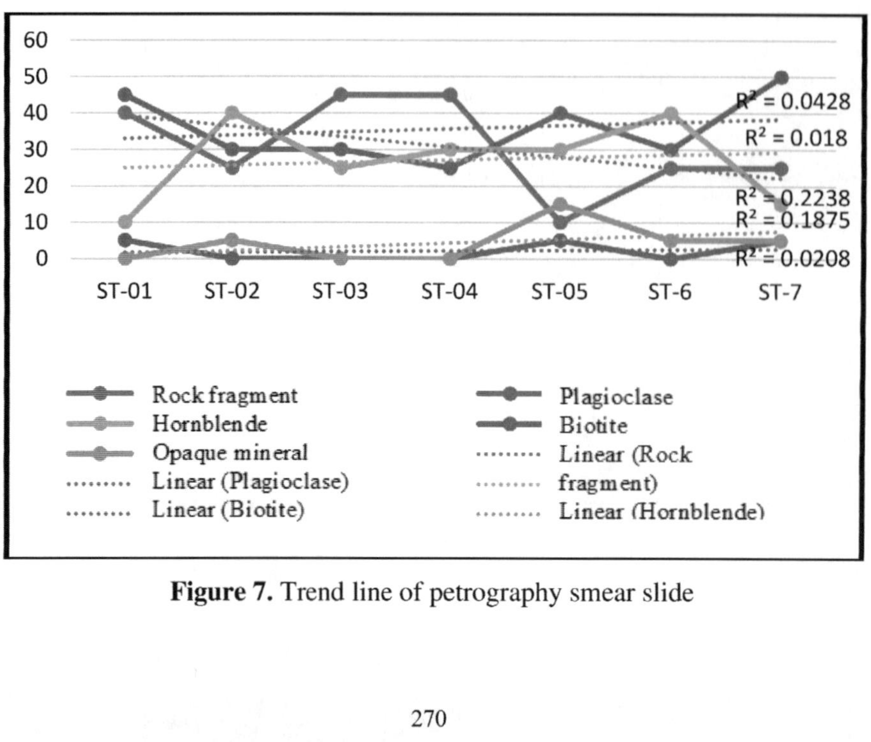

Figure 7. Trend line of petrography smear slide

IOP Conf. Series: Materials Science and Engineering **676** (2019) 012035 doi:10.1088/1757-899X/676/1/012035

4. Chemical Characteristics and Correlation of Heavy Metal Elements in Lumpue Beach

Characterization of Lumpue Beach sediments in addition to the megaskopis and petrography observation approach, determination of mineral chemical aspects using ICP-OES analysis was also done [12]. This analysis done on 7 samples of very fine sand obtained from sieving and aimed to determine the level of heavy metal elements on the ppm scale.

Based on the use of Pearson method with two paired variables on metals Cu, Fe, Ni, Zn, Pb, Ti and Mn, it is known that Cu metal has a very weak correlation with Fe, Zn, Ni, Pb and Ti, while Fe metal has a strong correlation to Zn, Ti and Mn but have a weak correlation with Ni and Pb (Table 2a).
Ni metal has a strong correlation to Ti and Mn, and has a correlation of weaknesses against Fe, Zn and Pb. Zn has a strong correlation with Fe, Ti and has a weak correlation with Ni, Pb and Mn. Metal Pb is weakly correlated to the above elements.

Ti correlates strongly with Ni, Zn and Mn, while Mn correlates strongly with Fe metal, Ni and Ti. This correlation shows that there are fairly close relationship between heavy metals with alkaline compositions to other heavy metals, a strong correlation indicates that the higher level of a heavy metals will be followed by other metals.

Table 2a. Elements correlation

Correlations

		Cu	*Fe*	*Ni*	*Zn*	*Pb*	*Ti*	*Mn*
Cu	*Pearson Correlation*	*1*	*-.414*	*-.572*	*-.240*	*-.031*	*-.467*	*-.532*
	Sig. (2-tailed)		*.308*	*.138*	*.567*	*.942*	*.243*	*.174*
	N	*8*	*8*	*8*	*8*	*8*	*8*	*8*
Fe	*Pearson Correlation*	*-.414*	*1*	*.376*	*.570*	*.213*	*.609*	*.904***
	Sig. (2-tailed)	*.308*		*.358*	*.140*	*.612*	*.109*	*.002*
	N	*8*	*8*	*8*	*8*	*8*	*8*	*8*
Ni	*Pearson Correlation*	*-.572*	*.376*	*1*	*.436*	*-.050*	*.800**	*.532*
	Sig. (2-tailed)	*.138*	*.358*		*.281*	*.906*	*.017*	*.175*
	N	*8*	*8*	*8*	*8*	*8*	*8*	*8*
Zn	*Pearson Correlation*	*-.240*	*.570*	*.436*	*1*	*.279*	*.809**	*.377*
	Sig. (2-tailed)	*.567*	*.140*	*.281*		*.503*	*.015*	*.357*
	N	*8*	*8*	*8*	*8*	*8*	*8*	*8*
Pb	*Pearson Correlation*	*-.031*	*.213*	*-.050*	*.279*	*1*	*-.055*	*-.017*
	Sig. (2-tailed)	*.942*	*.612*	*.906*	*.503*		*.897*	*.969*
	N	*8*	*8*	*8*	*8*	*8*	*8*	*8*
Ti	*Pearson Correlation*	*-.467*	*.609*	*.800**	*.809**	*-.055*	*1*	*.593*
	Sig. (2-tailed)	*.243*	*.109*	*.017*	*.015*	*.897*		*.121*
	N	*8*	*8*	*8*	*8*	*8*	*8*	*8*
Mn	*Pearson Correlation*	*-.532*	*.904***	*.532*	*.377*	*-.017*	*.593*	*1*
	Sig. (2-tailed)	*.174*	*.002*	*.175*	*.357*	*.969*	*.121*	
	N	*8*	*8*	*8*	*8*	*8*	*8*	*8*

***. Correlation is significant at the 0.01 level (2-tailed).*

**. Correlation is significant at the 0.05 level (2-tailed).*

IOP Conf. Series: Materials Science and Engineering **676** (2019) 012035 doi:10.1088/1757-899X/676/1/012035

In table 2b there is an element of Ca which strongly correlates with heavy metals Fe and Mn. This correlation shows that changes in heavy metal in the study area are influenced by Ca and Fe [13] and heavy metals Fe and Mn generally come from coastal material carried by currents. Statistical graph of the correlation of heavy metal elements on Lumpue Beach is shown in Figure 8.

Table 2b. Elements correlation

Correlations									
		Cu	*Fe*	*Ni*	*Zn*	*Pb*	*Ti*	*Mn*	*Ca*
Cu	*Pearson Correlation*	*1*	*-.414*	*-.572*	*-.240*	*-.031*	*-.467*	*-.532*	*-.336*
	Sig. (2-tailed)		*.308*	*.138*	*.567*	*.942*	*.243*	*.174*	*.416*
	N	*8*	*8*	*8*	*8*	*8*	*8*	*8*	*8*
Fe	*Pearson Correlation*	*-.414*	*1*	*.376*	*.570*	*.213*	*.609*	*.904***	*.841***
	Sig. (2-tailed)	*.308*		*.358*	*.140*	*.612*	*.109*	*.002*	*.009*
	N	*8*	*8*	*8*	*8*	*8*	*8*	*8*	*8*
Ni	*Pearson Correlation*	*-.572*	*.376*	*1*	*.436*	*-.050*	*.800**	*.532*	*-.029*
	Sig. (2-tailed)	*.138*	*.358*		*.281*	*.906*	*.017*	*.175*	*.946*
	N	*8*	*8*	*8*	*8*	*8*	*8*	*8*	*8*
Zn	*Pearson Correlation*	*-.240*	*.570*	*.436*	*1*	*.279*	*.809**	*.377*	*.176*
	Sig. (2-tailed)	*.567*	*.140*	*.281*		*.503*	*.015*	*.357*	*.677*
	N	*8*	*8*	*8*	*8*	*8*	*8*	*8*	*8*
Pb	*Pearson Correlation*	*-.031*	*.213*	*-.050*	*.279*	*1*	*-.055*	*-.017*	*.317*
	Sig. (2-tailed)	*.942*	*.612*	*.906*	*.503*		*.897*	*.969*	*.444*
	N	*8*	*8*	*8*	*8*	*8*	*8*	*8*	*8*
Ti	*Pearson Correlation*	*-.467*	*.609*	*.800**	*.809**	*-.055*	*1*	*.593*	*.124*
	Sig. (2-tailed)	*.243*	*.109*	*.017*	*.015*	*.897*		*.121*	*.769*
	N	*8*	*8*	*8*	*8*	*8*	*8*	*8*	*8*
Mn	*Pearson Correlation*	*-.532*	*.904***	*.532*	*.377*	*-.017*	*.593*	*1*	*.700*
	Sig. (2-tailed)	*.174*	*.002*	*.175*	*.357*	*.969*	*.121*		*.053*
	N	*8*	*8*	*8*	*8*	*8*	*8*	*8*	*8*
Ca	*Pearson Correlation*	*-.336*	*.841***	*-.029*	*.176*	*.317*	*.124*	*.700*	*1*
	Sig. (2-tailed)	*.416*	*.009*	*.946*	*.677*	*.444*	*.769*	*.053*	
	N	*8*	*8*	*8*	*8*	*8*	*8*	*8*	*8*
**. *Correlation is significant at the 0.01 level (2-tailed).*									
*. *Correlation is significant at the 0.05 level (2-tailed).*									

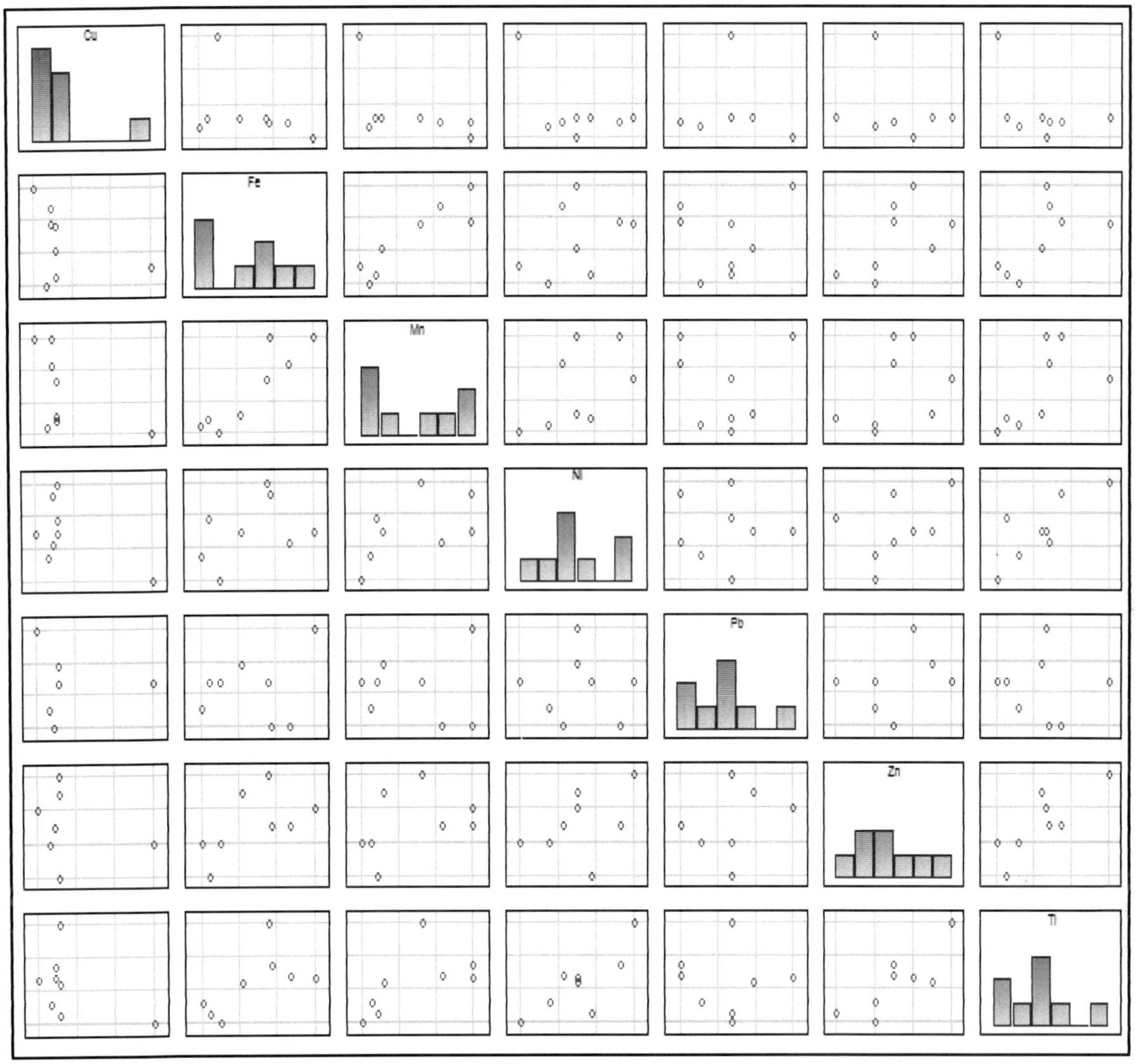

Figure 8. Statistical graph of the correlation of heavy metal elements on Lumpue Beach.

The geochemistry of Lumpue Beach sedimentary with the regression value shown Al: Fe in the linear regression graph shows that layer A has the correlation properties of Al and Fe, as follows that increased Fe and Al have similarities from the same source, Al and Fe concentrations in sediments change according to the size of the fraction and concentrated Al elements increase gradually compared to Fe.

Linear regression value B is shown by Al: Fe, where in this graph shows that the B layer of the study area has Al to Fe correlation properties and comes from different sources and deposited as part of the coastal sediment layer. Fe and Al metals are deposited with low values in layer A and slowly accumulate or element enrichment in layer B. This occurs in the concentrations of the two metals and follows the sediment fraction controlled by currents and tides [14].

The low value of Fe is seen in layer B but has a high value of Al in layer A. This shows that Fe is gradually deposited well from the sediment surface and then in the lower layer eroded by the influence of waves. Whereas Al occur the opposite, the more value increases in the lower layer (Figure 9).

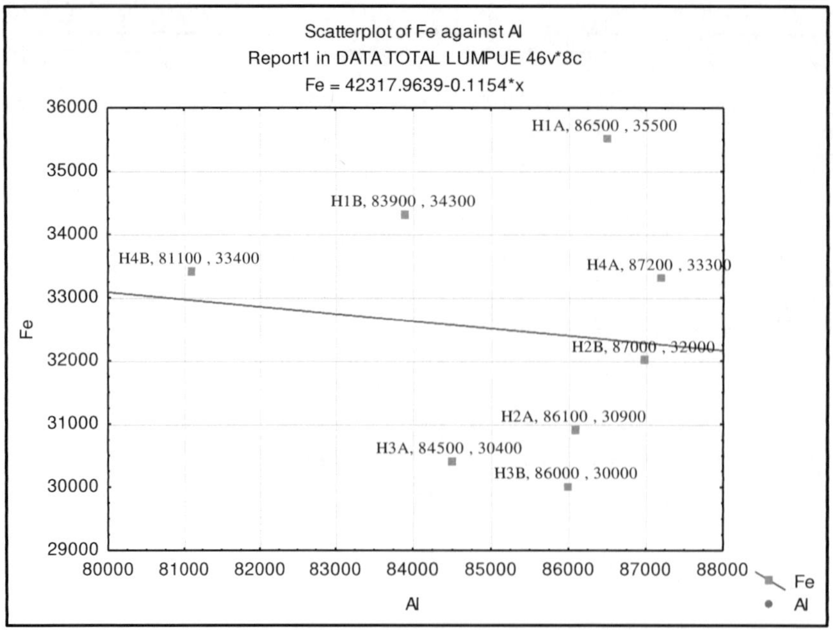

Figure 9. Total linear regression graph

Chemical characteristics of heavy metals in Lumpue Beach sediments were identified and determined based on the results of compilation of field observation data and geochemical results (ICP-OES). For the elements observed in the form of Pb, Cd, Cu, and Zn with a percentage of 1 ppm (Table 3). Based on the processing of statistical data and the distribution of heavy metal elements in the study area, there are relationship between the distribution of sediment particles, heavy metal elements to the influence of ocean wave currents on Lumpue Beach sediments.

Table 3. Lumpe Beach heavy metal elements content

Stasions	Heavy Metal Elements			
	Pb	Cd	Cu	Zn
H1A	48	2,8	8	63
H1B	37	1,4	12	62
H2A	42	1,2	34	61
H2B	44	2,0	13	64
H3A	42	1,4	13	59
H3B	39	1,0	11	61
H4A	42	1,4	13	65
H4B	37	2,0	12	62

Physical and chemical characteristics of Lumpue Beach sediment showed that the development of sedimentation in the study area is highly controlled by ocean waves, in addition changes from environmental conditions are identified to occur due to the development of geological activities in the form of sedimentation, changes in physiographic conditions and activities of local residents, especially those relating to household waste processing.

Based on the results of coastal sediment geochemical analysis (ICP-OES) which shows that the distribution of elements and changes in the composition of heavy metal elements are in the normal range and experience of element enrichment in the northern part of the research area. The distribution of Pb element at station 1 is relatively normal and tends to decrease, but at stations 2 - 4 there is an enrichment of Pb element with a trend line ratio that is R2 = 0.25. At stations 1 to 4 the level of Cd element is relatively stable where the trend line ratio is R2 = 0.12. The element Cu experienced a significant change at stations 1-2 and at stations 3-4 the elemental content was relatively stable with a trend line ratio of R2 = 0.02. At stations 1 to 2 the increase in Zn element is relatively stable and regresses at station 3 and decreases again at station 4 with the trend line ratio that is R2 = 0.0022.E (Figure 10).

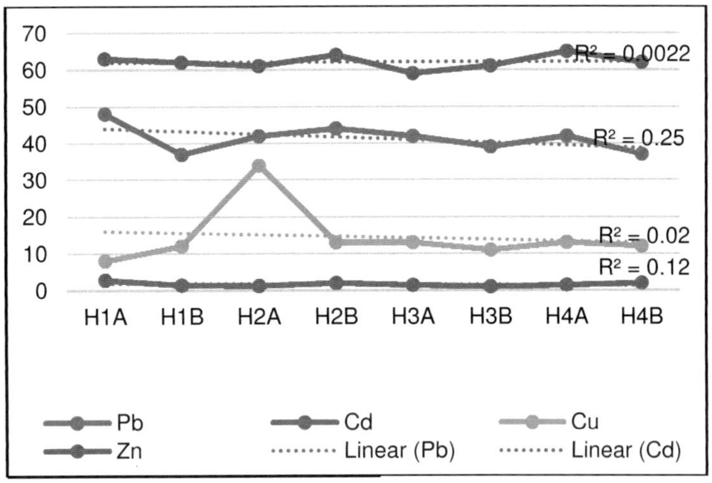

Figure 10. Trend line of Heavy Metal Elements Pb, Cd, Cu, Zn

5. Lumpue Beach Heavy Metals Distribution

The percentage of heavy metals Al and Fe are also poured in the graph below (Figure 11) where the elements of Al with a percentage of 50 ppm and the elements of Fe with a multiple of 0.01.

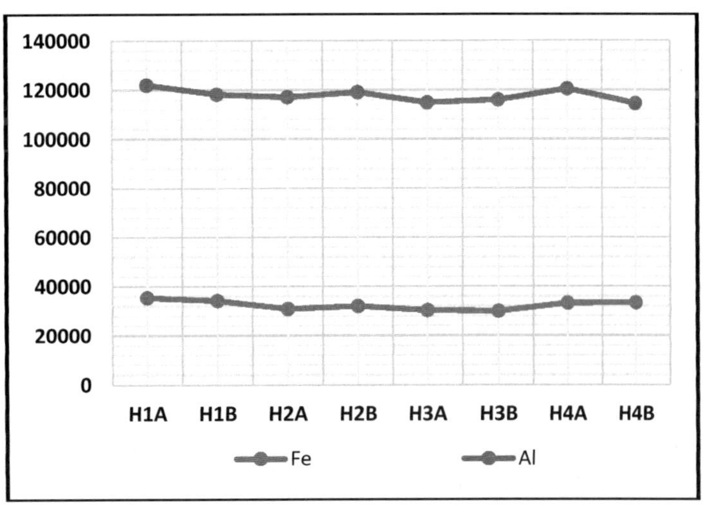

Figure 11. Trend line of heavy metal elements Fe and Al.

The distribution of heavy metals relative to the north caused by the direction of ocean currents in Lumpue Beach which is relatively south to north. The elemental abundance of Ca is found in station 1 with an average level of 53,636.96 - 60,049.89 ppm. The dominant Na element at stations 1 and 2 with an average level reached 26,108.69 ppm. The dominant Cu element is in station 2 with the element content reaching 17.17 - 23.49 ppm. The dominant Fe elements are found at station 1 and station 4 with an average level of 3.28 - 3.48%. Ti element is dominant at station 1 with the element content reaching 0.667 - 0.684 ppm. Pb elements are more dominant in stations 1 and 2 with the elements content reaching 42.24 - 42.99 ppm.

Based on the value from the maximum level of heavy metal types on the nature of conservative metals, it is known the ratio of the concentration of heavy metal elements (in ppm) in the study area are as follows:

1. The ratio of Pb elements in the study area has an average level of 41.38 ppm and a trend line R2 = 0.25. This indicates that the presence of Pb element in the study area is indicated to exceed the normal threshold. Where the normal level of reference for Pb is 0.00003 and the maximum limit is 0.05 ppm.
2. The ratio of Cd element in the study area has an average level of 1.65 ppm and a trend line R2 = 0.12. This shows that the presence of Cd element in the study area is indicated to exceed the normal threshold. Where the normal level reference for the Cd element is 0.00011 and the maximum limit is 0.01 ppm.
3. The ratio of Cu element in the study area has an average level of 14.5 ppm and a trend line R2 = 0.02. This shows that the presence of Cu element in the study area is indicated to exceed the normal threshold. Where the normal level reference for Cu elements is 0.002 and the maximum limit is 0.05 ppm.
4. The ration of Zn element in the study area has an average level of 62.125 ppm and a trend line R2 = 0.0022. This shows that the presence of Pb element in the study area is indicated by a normal threshold. Where the normal level reference for Pb is 0.002 and the maximum limit is 0.01 ppm.

6. Conclusion

1. Distribution of Pb element at station 1 and tends to decrease, but at stations 2 - 4 enrichment occurs. At stations 1 to 4 the levels of Cd were relatively stable while Cu element experienced significant changes at stations 1-2 and at stations 3-4 the levels of elements were relatively stable. At stations 1 to 2 the increase in Zn element is relatively stable and regresses at station 3 and decreases again at station 4.
2. Abundance element of Ca is found in station 1 with an average level of 53,636.96 - 60,049.89 ppm. The dominant Na element at stations 1 and 2 with an average level reached 26,108.69 ppm. The dominant Cu element is in station 2 with levels of elements reaching 17.17 - 23.49 ppm. The dominant Fe element is found at station 1 and station 4 with an average level of 3.28 - 3.48%. The Ti element is dominant at station 1 with levels of elements reaching 0.667 - 0.684 ppm. The Pb element is more dominant at stations 1 and 2 with elemental levels reaching 42.24 - 42.99 ppm.
3. The characteristics of heavy metals in Lumpue Beach have ratio for each element where Pb, Cd, Cu and Zn are indicated to exceed normal limits.

Acknowledgments

The authors are grateful to the Institute which has provided motivation and support in the preparation of this journal. We would like to say thank to Kaharuddin, Irwan Zibuka and assistance student in Department of Geology for help in the field and data processing. This study is contributed to the research in Geology Engineering, Hasanuddin University and for scientific progress.

References

[1] Sirajuddin, H., Suriamihardja, D.A., Imran, A.M., Thaha, A., 2015, Coastal Vulnerability Zonation of Pinrang District and Parepare City Based on Morphodynamic Review, International Journal of Inventive Engineering and Sciences Volume-3 Issue-12, pp. 23-26, December.

[2] Sirajuddin, H., Suriamihardja, D.A., Imran, A.M., Thaha, A.,2014, Coastal Vulnerability Based on Tectonics and Shoreline Change Along Coastal Area of Lumpue Coast, South Sulawesi, Proceedings of the 9th International Symposium on Lowland Technology, Saga, Japan, September 2014.

[3] Eliot, M. 2016, Coastal Sediments, Beaches and Other Soft Shores. Coast Adapt Information Manual 8. National Climate Change Adaptation Research Facility, Gold Coast.

[4] Sukamto. R., 1982, Geology of Pangkajane and Watampone Western, Research Centre and Development of Geology, Bandung.

[5] Sankaran, G., Jothivel, S., Govindasamy, V., Nateesan, M., 2014, Geochemical and Textural Characterization of Minerals in Thondi Coastal Sediments Along the Southeast Coast of India, Journal of Oceanography and Marine Science, Vol. 5(5), pp. 37-44, August, 2014.
[6] Herrmann, S. and N.A. Stroncik, 2013. Data report: Si, Al, Fe, Ca, and K Systematics of Volcaniclastic Sediments from Selected Cores of Hole U1347A, IODP Expedition 324. Proceedings of the Integrated Ocean Drilling Program, Volume 324.
[7] Sirajuddin, H., Suriamihardja, D.A., Imran, A.M., Thaha, A., 2016, Coastal Characteristics Control for Coastal Vulnerability Rating, Asian Academic Research Journal Of Multidisciplinary, Volume 3 Issue 12 (December 2016).
[8] Sirajuddin, H., Suriamihardja, D.A., Imran, A.M., Thaha, A., 2015, Influence of Sediment Texture on Beach Morphodynamic State, International Journal of Development Research Vol. 05, Issue 08, pp. 5249 – 5254.
[9] Arifin, F., 1985, Surface Geological Mapping at Parepare City, Geology Department, Faculty of Engineering, Hasanuddin University, unpublished.
[10] Souza, Nascimento, F.D., Freire, G.S., and Miola, B., 2010. Geochemistry of Marine Sediments of The Brazilian Northeastern Continental Shelf. Universidade Federal do Ceará - Laboratório de Geologia Marinha Aplicada (Campus Universitário do PiciBloco 912/913, 60455-760 Fortaleza, CE, Brasil).
[11] Elhabab, Adel A, A, and Adsani., I., 2013. Geochemical and Mineralogical Characters of the Coastal Plain Sediments of the Arabian Gulf, Kuwait. College of Technological Studies, Petroleum Engineering Department, Kuwait.
[12] Mounteney, et al. 2018, Heavy mineral analysis by ICP-AES a tool to aid sediment provenancing. Journal of Geochemical Exploration 184 (2018) 1–10.
[13] Fu, Baihua, et al. 2008. Influence of Particle Size on Geochemical Suspended Sediment Tracing In Australia. Sediment Dynamics in Changing Environments (Proceedings of a Conference in Christchurch, New Zealand, December 2008). IAHS Publ. 325.
[14] Panteleit, B, et al. 2001, Geochemical Characteristics of Salinization-Zones in The Coastal AquifertestField (CAT-Field) In North-Germany. First International Conference on Saltwater Intrusion and Coastal Aquifers - Monitoring, Modeling, and Management. Essaouira, Morocco.

IOP Conf. Series: Materials Science and Engineering **676** (2019) 012036 doi:10.1088/1757-899X/676/1/012036

Light Distribution Analysis on Buildings Located on the Coastal

N Jamala[1]*, R Rahim[1], S Latief[1] and H Ramli[1]

[1]Department of Architecture, Faculty of Engineering, Hasanuddin University, Indonesia

*E-mail: nuruljamala@yahoo.co.id

Abstract. Issues of global warming are important. The effects of illuminance level of natural light on the earth's surface. The use of natural light into buildings can minimize energy consumption as a source of artificial lighting, but it is important to attention to negative effects such as glare, brightness and thermal. The Aryaduta Hotel building located on the coastal area, is an object of research that needs to analyze the distribution of natural light come in to the building. The purpose of this study was to determine how the distribution of natural light in buildings on the coastal area. The results of research concluded that the farther away from the veil the building then the light distribution will decrease, orientation of the building affects the distribution of sunlight and distribution of natural light into the building affects the sun's radiation in the morning, afternoon and evening. This research can be the base for energy efficient building design.

1. Introduction

The concept of energy-efficient architecture is to optimize the lighting system by considering the integration between natural lighting (daylight) and artificial (lights). The use of energy in buildings, can be reduced if it can maximize the distribution of natural light, but by considering the negative effects include: thermal, glare and brightness.

Aryaduta hotel is located around the losari beach. Position of the hotel rooms are the view city and view losari beach. This building is one of the object in this study, the distribution of daylight is the different, because sunset and sunrise are different direction. Base on this, so the research to need to analyze how the distribution daylight in to the rooms are oriented towards the West and East.

Figure 1. Aryaduta Hotel in Makassar

The building lighting design is designed according to the illumination standard recommended by SNI 03-6575-2001. The Level of Illuminance at the hotel is lobby of 100 lux, bedroom of 150 lux, ballroom of 200 lux and cafe of 250 lux [1]. The sky conditions consist of clear sky, intermediate sky and overcast sky. Light distribution in buildings, can be influenced by sky conditions, position, wide and shape window [2]. Recommendations of illumination standard of office workspace refers to the values recommended by the CIE (Commission International de I'Eclaire) [3] and IES (Illuminating Engineers Society) [4] which is the National and International standards for lighting design [5]

Based on this recommendation, the designer can determine amount of armatures required on space, according to the activities in the space. According Nurul [6] that although it does not meet these standards, can be running well, the activities of space. Based on this, it is necessary to study the distribution of natural light on selected objects.

2. Literature Study

Classification of sky condition i.e. overcast, clear and intermediate sky, trying to define the luminance distribution of the intermediate sky Nakamura et al. [7]. Local Government Rules number 38 suggested that the building envelope is a building element that surrounds the building. They are transparent or not transparent walls or the roof [8].

Satwiko [9] suggested that Indonesia is a tropical humid region with high relative humidity characteristics (60% -80%), high radiation (80% per year), and unstable velocity (urbanized between 0-> 30 m / seconds). Newsham et al. [10]. examined quantitative guidance of illuminance, luminance and ratio values in office buildings [11]. According to Soegijanto [12], sky conditions based on the number and type of clouds can be grouped into: (a) Clear sky is the sky without clouds; (b) Overcast sky is a sky completely covered in white or gray white clouds or partial or complete partial clouds; (c) intermediate sky is a partially clouded sky with a variety of darker clouds and numbers: and (d) Uniform sky is the sky with the same luminance in all positions not dependent on the geographic latitude and the height of the sun [13].

The Phinisi tower building uses a diagonal and horizontal facade on the building's envelope, so there is no glare and brightness in space. This building is one example of building in designing high building. In terms of aesthetics, building facade is a positive value but needs to be considered the visual comfort of space users so that work productivity can increased. The study concluded the decline percentage in the value of the illuminance after the building using the building facade is 49% -74% and a mean value of 60.3%, so it can be concluded that the building facade effects on the natural lighting [14].

3. Methodology

Quantitative research method is analyzing the data of measurement results of illuminance levels statistically descriptive. The research conducted was to analyze illuminance levels in a number of rooms with orientation of the building envelope opening to the east and west. Measurements were made at a measuring point with a distance of 50 cm (measuring point A1-A5) and 150 cm (measuring point B1-B5) from the building envelope opening. This research was conducted by analyzing the data objectively in assessing the results of the measurement of illuminance levels in a number of rooms.

4. Result of Research

In Aryaduta hotel there are various functions of space including: meeting room, fitness room, swimming pool area, administration room and hotel room. The orientation of some hotel rooms leads to the city (city view / east direction) and some point to Losari beach (view beach / west direction) as shown in Figure 2.

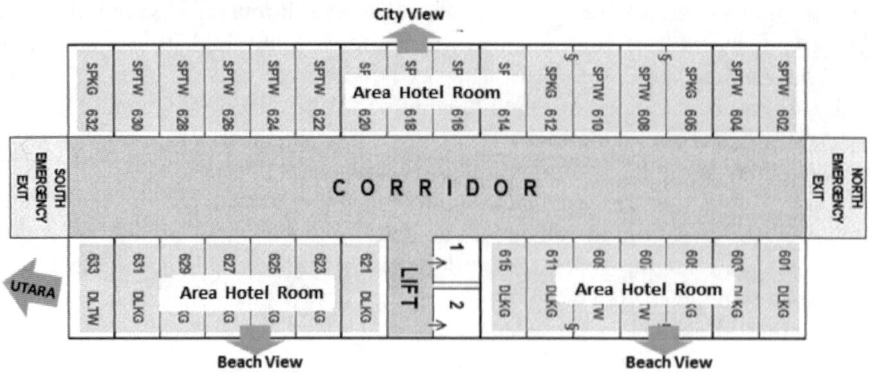

Figure 2. 6th floor plan

Research activities were measuring the illuminance level in 8 Aryaduta hotel rooms located on floors 6, 8 and 9. Measurements were made on August 18-19 2018 in the morning, afternoon and evening. This study analyzed how the distribution of natural light in some hotel rooms that east and west oriented. The measuring point uses the A-B and 1-5 notation as shown in Figure 3.

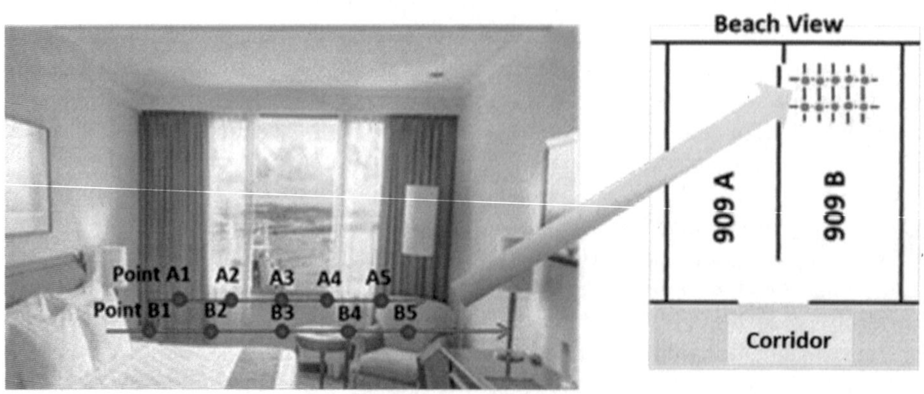

Figure 3. Layout of measurement point

4.1. Analysis of illuminance level in the morning

4.1.1. Hotel rooms oriented to the west (beach view)

Room 909 is located on the 9th floor and this room consists of two rooms namely a bedroom (909-A) and a family room (909-B), where each room is directly related to the building envelope opening. The orientation of this hotel room leads to the west (beach view) (Figure 4).

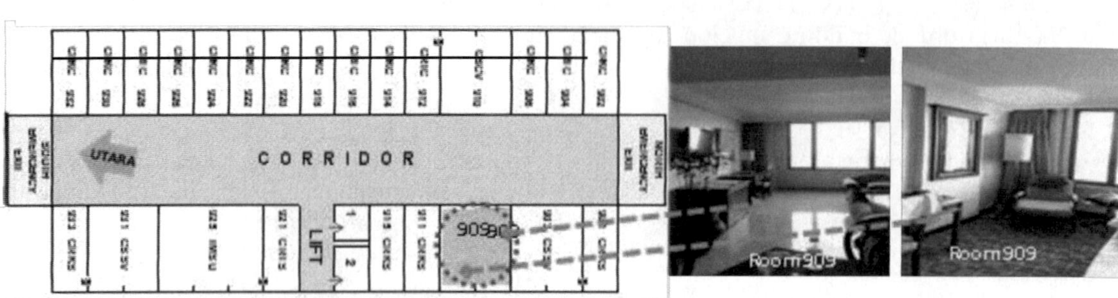

Figure 4. Room hotel 909-A and 909-B to the west orientation building

Table 1 shows the average level of illuminance for 2 measurement days in the morning. The distribution of natural light in this space is 192-372 lux 9 (measuring points A1-A5) and 157—213 lux (measuring point B1-B5). The results of the analysis show that the farther away from the building envelope the light distribution will decrease as shown in Table 1 and Figure 3 below.

Tabel 1. The mean value of illuminance level on 909 A-B Room

Room	illuminance level (lux) 909-A					Room	illuminance level (lux) 909-B				
Number	1	2	3	4	5	Number	1	2	3	4	5
909A-1	226	338	428	265	187	909A-2	158	196	208	207	211
909A-1	221	337	372	282	190	909A-2	168	207	210	214	140
909B-1	202	420	360	215	198	909B-2	117	200	215	204	192
909B-1	211	380	326	256	194	909B-2	183	218	220	190	180
Average	215	369	372	255	192	Average	157	205	213	204	181

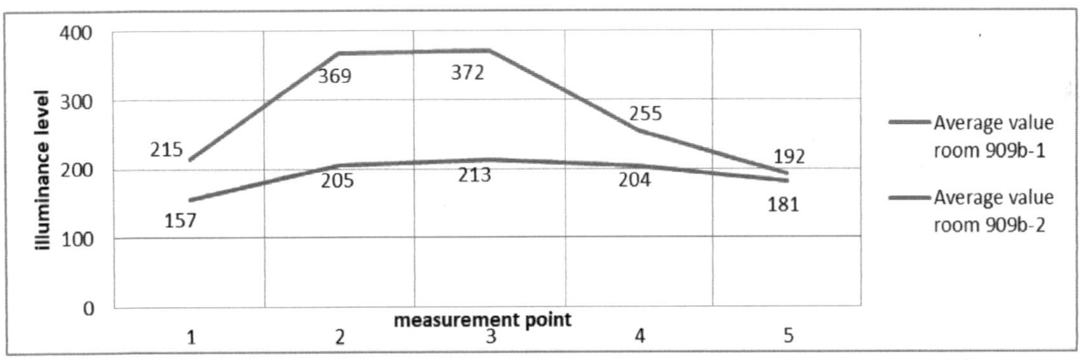

Figure 5. Graph of illuminace level at the rooms 909 A-B

4.1.2. Hotel rooms oriented to the east (city view)

Hotel rooms 616 and 624 are located on the 6th floor and are oriented towards buildings eastward (view city). This room is directly related to the building envelope opening with an area of 32 m^2 as shown in Figure 6.

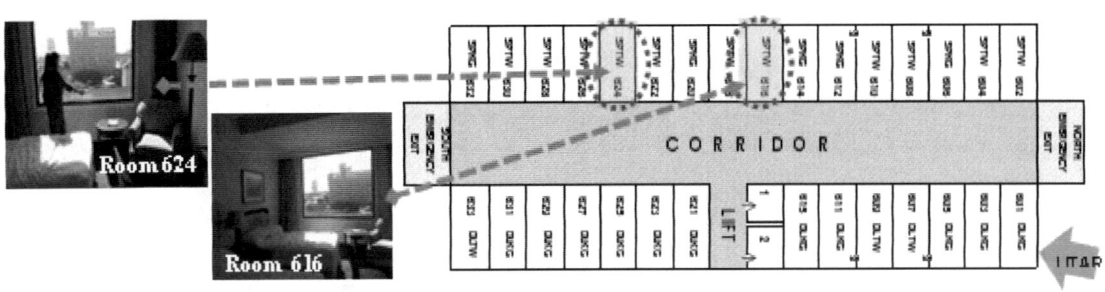

Figure 6. Position of rooms hotel 612 and 624 on west orientation building

The light distribution in these two rooms is very high, with the highest illuminance level of 7757 lux (measuring point 616-A2) and the lowest is 1415 lux (measuring point 616-A5) at an area of 50 cm from the building envelope opening. While the light distribution in the 150 cm area of the building envelope opening has the highest value of 2109 lux (measuring point 624-B3) and the lowest is 1187 lux (measuring point 616-B1) as in Table 2.

Tabel 2. The Mean value of Illuminance Level on Room 616 and 624

Room Number	Illuminance Level(lux) Room 616-624					Room Number	Illuminance Level (lux) Room 616-624				
	1	2	3	4	5		1	2	3	4	5
Point 616-A	1804	7757	6263	5700	1415	Point 616-B	1247	1532	1613	1461	1216
Point 616-A	1587	7402	6870	6520	1524	Point 616-B	1187	1530	1632	1516	1270
Point 624-A	1998	6553	7077	6545	1733	Point 624-B	1676	1789	1980	1877	1543
Point 624-A	1908	6800	7300	6788	1984	Point 624-B	1669	1898	2109	1932	1435

Table 2 shows that mean value of illuminance level on room 616 point A-B and 624 point A-B. The based on the measurement data, a graph is made to analyze the illuminance value in both rooms. (Figure 7). The results of analysis show that of illuminance level at point A is higher than point B, because point A is the closest area of the building envelope, so that the light distribution is higher.

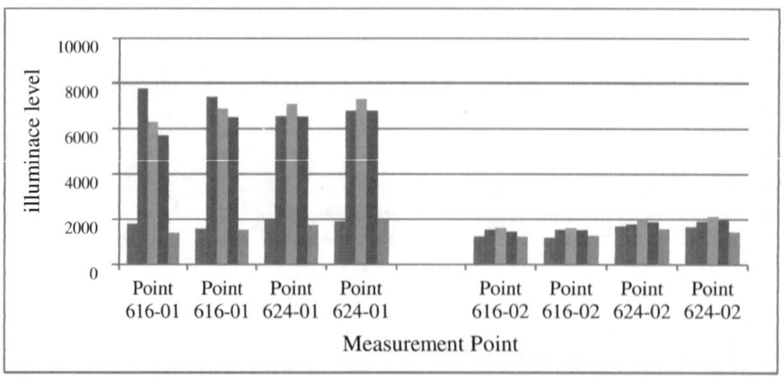

Figure 7. Level of illuminance in the rooms 6th floor

Table 3 shows that average value of illuminance level in the both rooms with point 01 and 02. This table show that the higher of illuminance level on point 01 than 02.

Tabel 3. The Level of illuminance on room 616 and 624

Mesurement Point	The Average rooms 616 and 624 point measurement 01 dan 02				
	1	2	3	4	5
Point 01	1824	7128	6878	6388	1664
Point 02	1445	1687	1834	1697	1366

The average of illuminance level on measurement point 1 and 2 show that illumination level at point 1 is higher than level 2 (Figure 8). The highest of the illuminance level at point 01 is 7128 lux and the lowest is 1664 lux, while at the highest point 02 is 1834 lux and the lowest is 1366 lux.

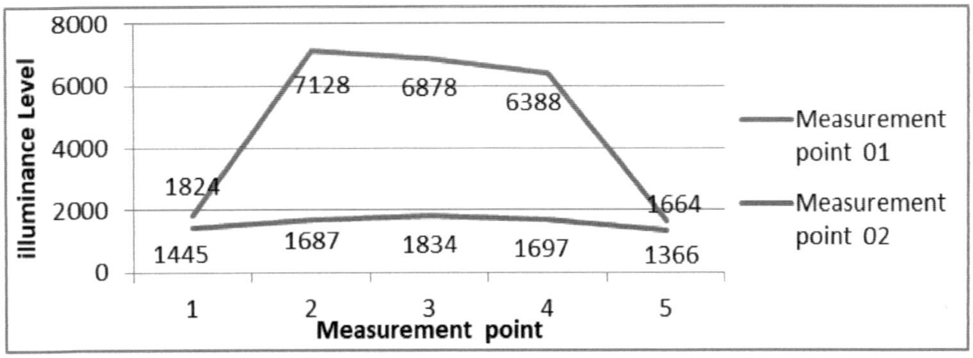

Figure 8. The average of illuminance level on measurement point 01 and 02

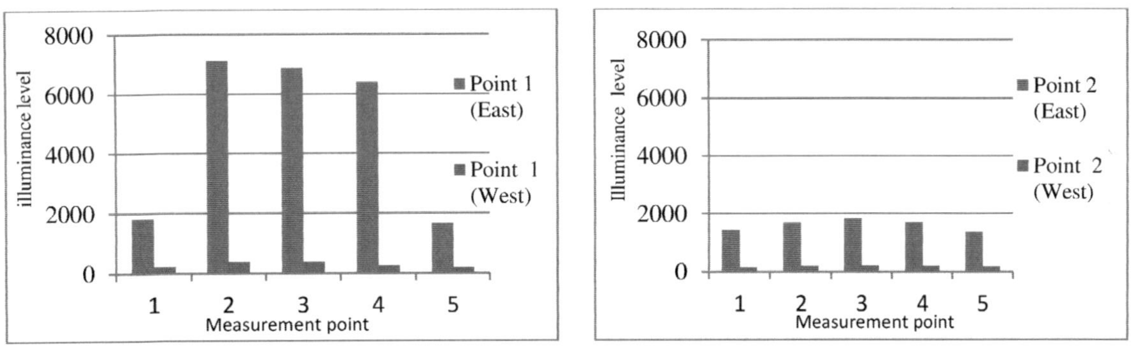

Figure 9. The average of illuminance level on 6th floor (rooms 616 and 624)

The average level of illumination shows that the distribution light in the morning in east and west oriented rooms are very different. This can be seen in the graph of illuminance levels on West orientation (city view) of 1515 lux to 4408 lux, while in the west-oriented rooms (beach view) of 186 lux up to 293 lux. Comparison of illuminance levels between east and west orientation rooms is 200% (Figure 10).

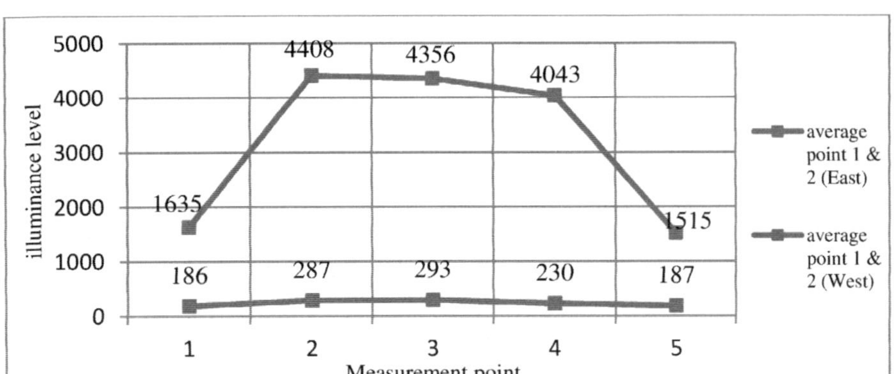

Figure 10. Comparison of illuminance level in room hotel in west and east oriented

4.2. Analysis of differences in illuminance level in building orientation west and east at noon.

Comparison of light distribution in Aryaduta hotel East and West rooms oriented during the day is analyzing illumination level in rooms 909-A and 909-B (west orientation) with rooms 808, 818, 830, 828, 630 and 604 (East orientation).

4.2.1. Hotel room oriented toward the west orientation (beach view) at noon.
The average illuminance level for 2 measurement days is as follows:

- Room 909-A at the A1-A5 measuring point of 570-2821 lux and at the measuring point B1-B5 of 398-760 lux
- Room 909-B at the A1-A5 measuring point is 672-3308 lux and at the measuring point B1-B5 is 421-677 lux.
- The average value of rooms 909-A at A1-A5 was 1778 lux and B1-B5 was 593 lux
- The average value of rooms 909-B at A1-A5 was 1829 lux and B1-B5 was 531 lux.

The analysis shows that the farther away from the building envelope the illuminance level will decrease by 67% (room 909A) and 71% (room 909B). The average illuminance levels in the rooms are oriented towards the West (rooms 909 A and B) at the measuring points A1-A5 and B1-B5 can be seen in Figure 11.

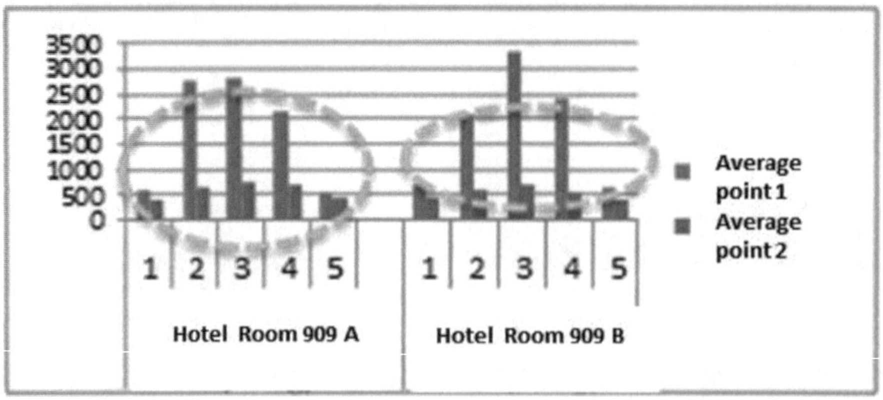

Figure 11. The average of illuminance level in west orientation rooms in the noon

The above graph shows the difference in illuminance level at a distance of 50 and 150 cm from the building envelope opening. Then analyze the average illuminance level in rooms 909-A and B for 2 days of measurement (Figure 12).

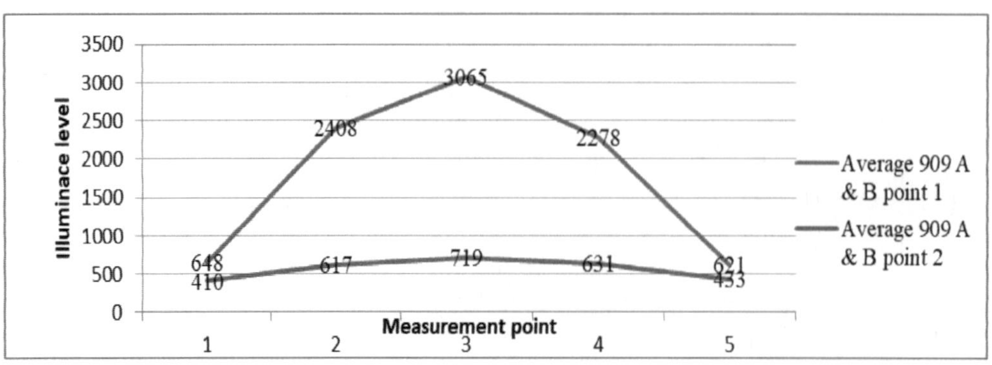

Figure 12. Comparison of illuminance level in west orientation rooms

4.2.2. Hotel room oriented toward the east (city view) at noon.
Illumination measurements are carried out in six rooms, namely rooms 808, 818, 630, 604, 830 and 828 which are oriented towards the east, as in the table below. Illuminance level fluctuated, because the position of the room is different, on the 6th and 8th floors. The average value of the six rooms at the measuring points 1 and 2 is as in the following table.

IOP Conf. Series: Materials Science and Engineering **676** (2019) 012036 doi:10.1088/1757-899X/676/1/012036

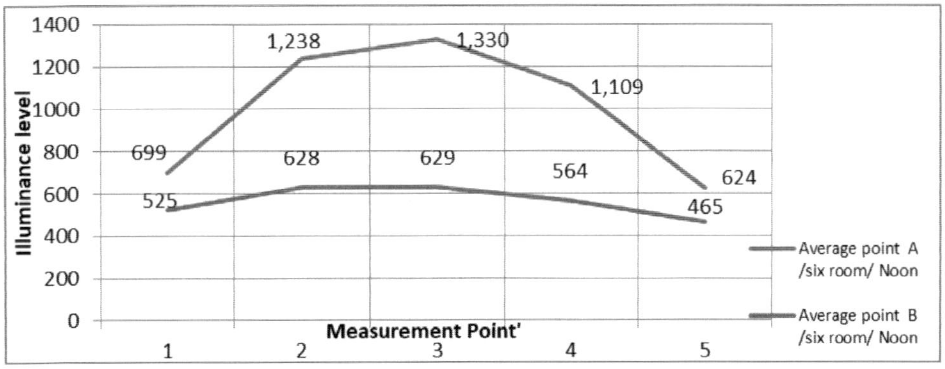

Figure 13. Average of illuminance level in east orientation rooms.

Furthermore, the comparison of illuminance level in east and west oriented rooms during the day. This graph shows that the distribution of light in the west-oriented rooms is higher at 621-3065 lux (A1-A5 measuring point) and 410-719 lux (B1-B5 measuring point). Rooms oriented towards the East are 525 lux-1330 lux (A1-A5 measuring points) and 624-629 lux (B1-B5 measuring points). This graph shows that the light distribution in the west orientation room is higher than the east orientation, both at the A1-A5 measuring point and at the B1-B5 measuring point.

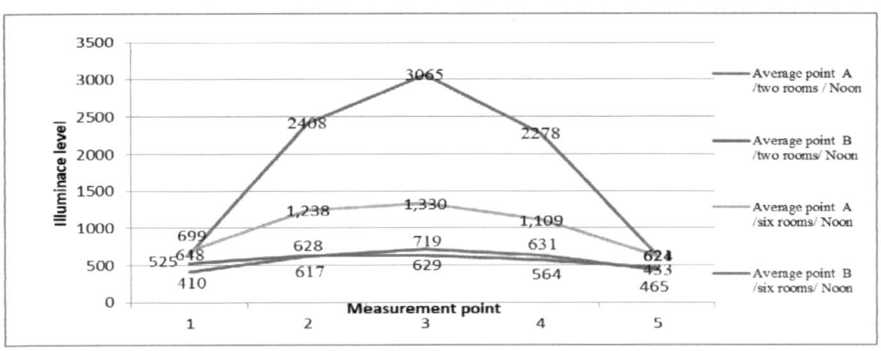

Figure 14. Average of illuminance level of measurement point A and B

Comparison of illuminance level in east and west oriented rooms during the day with mean values at the A1-A5 and B1-B5 measuring points. The graph shows that the distribution of light in the room is oriented westward (view beach) higher than east (view city). Light distribution has a west orientation of 3065 lux while the east orientation is 1330 lux as shown in the following Figure 15.

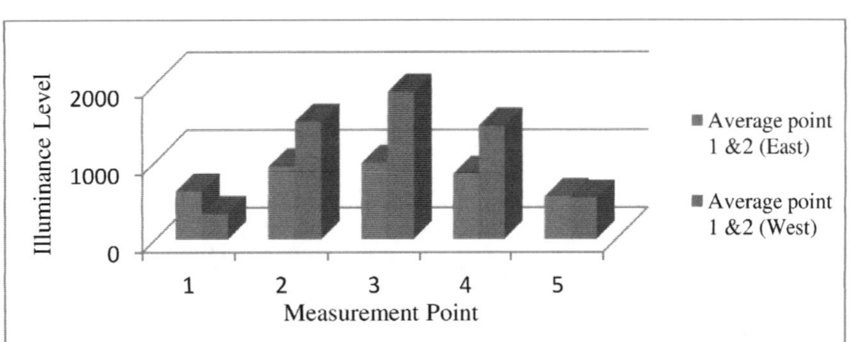

Figure 15. The average of measurement point on west and east orientation building

IOP Conf. Series: Materials Science and Engineering **676** (2019) 012036 doi:10.1088/1757-899X/676/1/012036

4.3. Analiysis differences of building orientation of illuminance level toward west and east in the afternon

Measurement of illuminance level in the afternoon is carried out in 8 rooms. Before analyzing the measurement data, an outlier was carried out to eliminate insignificant data, so that only 6 rooms would be analyzed. Furthermore, analyzing the comparison of rooms oriented to west and east. The mean value of measurement results is as in the following Table 4 and Figure 16.

Tabel 4. The average of illuminance level in the rooms 909 A and B (west orientation)

Average point A (909 A)	2466	3718	4918	4421	1498
Average point B (909 A)	2046	3423	3419	1508	1186
Average point A (909 B)	213	554	775	522	280
Average point B (09 B)	274	358	512	323	300

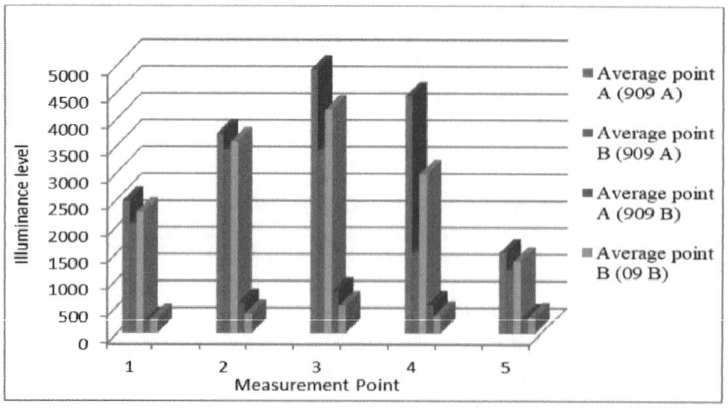

Figure 16. Average of illuminance level on room 909 A and B

The average illuminance level at the A1-A5 and B1-B5 measuring points were in the west oriented room of 3404 lux and 2316 lux, while in the east oriented rooms were 469 lux and 353 lux. The results of the analysis show the percentage decrease in light distribution at the A1-A5 measuring point to the B1-B5 measuring point of 32% (West orientation) and 25% (east orientation). The percentage difference between the two directions and the two measuring points is equal to A1-A5 and B1-B5 86% and 85%.

4.4. Comparison of illuminance level in the morning, afternoon and evening in the west and east orientated room

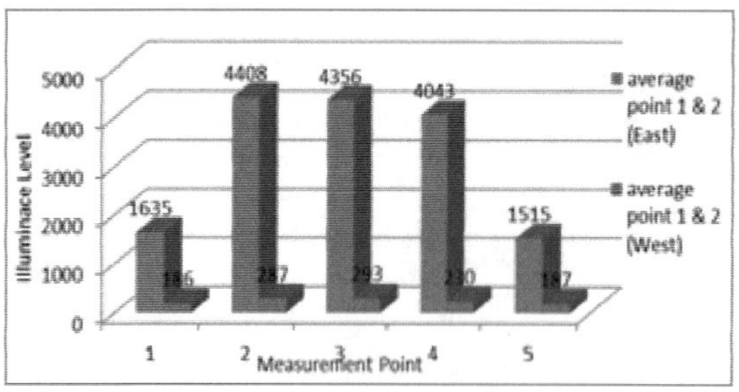

Figure 17. The average of illuminance level in west and east building orientation in the morning

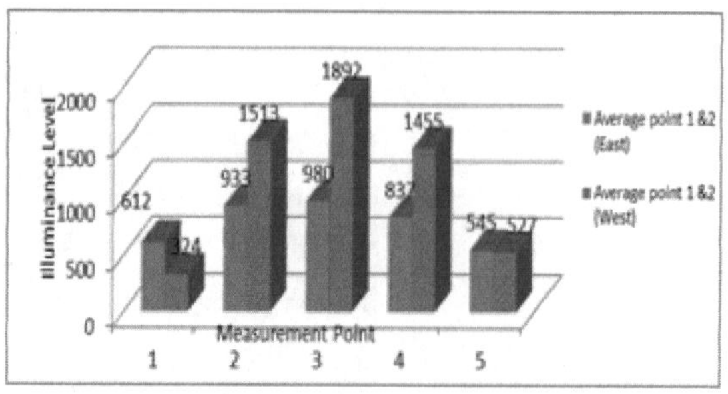

Figure 18. The average of illuminance level in west and east building orientation at noon

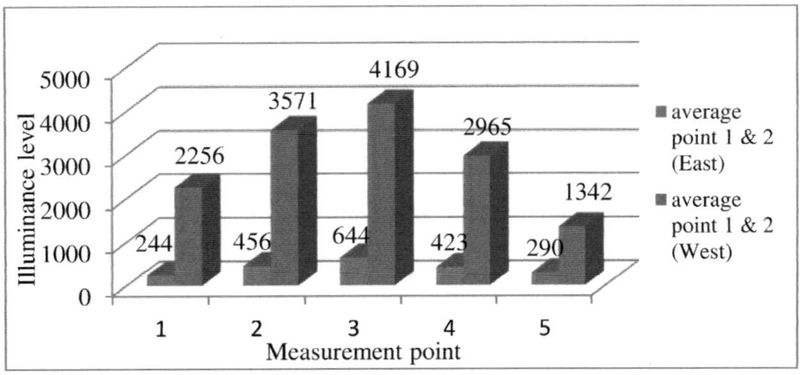

Figure 19. The average of illuminance level in west and east building orientation in the afternoon

5. Conclusion

The analysis result of research show the percentage decrease in light distribution at the distance of 50 cm to 150 cm on building envelope namely 32% (West orientation) and 25% (east orientation). Based on the results of the analysis it can be concluded that the farther away from the veil the building then the light distribution will decrease.

Distribution of natural light in the orientation of buildings westward: (1) morning: 293-186 lux; (2) noon 1892-324 lux; and (3) afternoon 4169-1342 lux. Then the orientation of the building eastward: (1) morning: 4408-1515 lux; (2) noon: 980-545 lux; and (3) afternoon: 644-244 lux. The results of the analysis show that the orientation of the building in the West direction is higher from morning to evening, while in the east the lower. The conclusion is that the distribution of natural light into the building affects the sun's radiation in the morning, afternoon and evening.

The results showed the difference in the value of light distribution in the morning to evening, namely: (1) In the morning: the distribution of light was much higher in the orientation of the eastward building, which was 4408 lux compared to the westward orientation of 293 lux; (2) noon: a slightly higher west orientation of 1892 lux compared to the east orientation of 980 lux: and (3) afternoon: a much higher orientation towards the west at 4169 lux compared to the east at 644 lux . Based on the results of the analysis, it can be concluded that the orientation of the building affects the distribution of sunlight.

References

[1] Indonesia National Standard 03-6575-2001 *The Design of The System of Artificial Light in The Building* (Original in Indonesia language)

[2] Indonesia National Standard 03-2396-2001 *The Design of The System of Natural Light in The Building* (Original in Indonesia language)

[3] Commission International de I'Eclaire (CIE) 1981 *An Analytical Model for Describing the Influence of Lighting Parameters upon Visual Performance*.

[4] Illuminating Engineering Society (IES) 1977 *code for Interior Lighting* IES Lighting Handbook London

[5] United Nations Environment Program (UNEP) 2006 *Energy Efficiency guide for industry in Asia* (India: *www.energyefficiensiasia.org*) p. 1-43.

[6] Jamala N 2012 *The Visual Comfort of the drawing Room Studio Using the Echotec Program* Gelagar Engineering Scientific Journal Vol. 26, p. 40-46

[7] Nakamura H et al 1985 *Luminance Distribution of Intermediate Sky* Journal of light and Vis. Environment 9 (1) p n6-13

[8] Jakarta Regional Regulation Number 38 of 2012 concerning *Green Building* article 1 p. 3

IOP Conf. Series: Materials Science and Engineering **676** (2019) 012036 doi:10.1088/1757-899X/676/1/012036

[9] Satwiko P 2009 *Building Phisics* Second Edition Yogyakarta

[10] Newsham G R et al 2004 *Lighting Design for Open Plan Offices* (Canada: Construction Technology) No. 62

[11] Jamala N dan Rahim R 2017 *Theory and Application of Visual Comfort* (Indonesia: the Publisher of the Makassar State University)

[12] Sugianto 1998 *Building in Indonesia with a Tropical Climate Humidity from Aspect Building Physics* Jakarta (Original in Indonesia language)

[13] Rahim R 2009 *Theory and Application of sky Luminance Distribution in Indonesia* National Library (Original in Indonesia language)

[14] Jamala N 2017 *The Effect building Façade on Natural lighting (Case Study: Building Of Phinisi Tower UNM)* Proceeding by AIP Publishing 978-0-7354-1499-0 (American Institute of Physics) Volume 1831- 02006

IOP Conf. Series: Materials Science and Engineering **676** (2019) 012037 doi:10.1088/1757-899X/676/1/012037

Glare from Windows Assessment at Offices with Three Types of Internal Solar Shadings

A Dinapradipta[1]*, E Sudarma[1], I Defiana[1] and C Erwindi[1]

[1]Architecture Department, Faculty of Architecture, Design and Planning. Institut Teknologi Sepuluh Nopember, Kampus ITS Sukolilo-Surabaya, Indonesia

*E-mail: asdina_p@arch.its.ac.id

Abstract. Successful daylighting requires trade off and optimization between competing elements of façade, space, and lighting system. It also requires reliable tool for assessing the indoor visual comfort, especially discomfort glare from windows. The objective of the study was to investigate the users' glare perception due to the use of internal solar shading systems by using laboratory experimentation with subjects and measurements. The purpose was to compare these results with a glare rating equation and Luminance Distribution Ratio (LDR) and to assess the reliability of this glare assessment tool to be used for assessing glare from window and in the context of humid tropics climate. The laboratory experimentations were conducted at three identical office rooms at the Department of Architecture of ITS in which each of the room's windows was covered with different internal solar shading types e.i. Venetian, Horizontal, and Roller Blinds. Results showed good correlation between Luminance Distributions Ratio (LDR) with the users' responses. Among the three internal shading, the Roller blind type had the highest correlation on both working activities and sitting positions to the glare perception as well as the fittest value complying the standard of luminance distribution. The research concluded that the Luminance Distribution Ratio method can be a reliable tool for assessing glare from window for many office situations in humid tropical climate.

1. Introduction

1.1. Background

Recently, energy and well-being are both the most challenging discourse in construction industries. In order to take a benefit from daylighting, whether they are related to energy, productivity, and well-being of the people, most of office buildings are tended to be built in a huge glass wall [1] (Figure 1.a.). Unfortunately, according to several researchers, it was found that the application of that wide glass wall had increased risks of both improper indoor illumination and discomfort glare from window [1, 2]. Further, empirical evidence showed that if the daylight becomes disturbing, the occupants will probably make use of the indoor solar shadings and after the daylighting condition returned to a comfortable state, most of the occupants either forget or do not willing to re-adjust the shadings [1] (Figure 1.b.). This might at the end, cancel the benefit of daylight and increase the buildings' energy for lighting.

(a) (b)

Figure 1. Some of full glass office buildings in Jakarta (a). The design has increased the risk of daylighting problems such as improper illumination and glare. Additionally, the problem of the utility of internal shadings is vivid and this can cancel the benefit of daylighting (b)

Many tasks should be accomplished in designing using daylighting. The daylight design has become a complex system integration challenge in which it is fortunate that the development of glass technology and shading devices has resulted on a wide selection of a new façade design solutions. Therefore, a successful daylighting design needs to consider a wide range of parameters including the occupants' well-being and comfort. It needs a trade-off or an optimization among all competing elements of the façade including the shadings. Besides that and the most importantly, it is also required reliable tools for assessing different aspects of comfort. For many aspects, reliable tools are available – but only a few for discomfort glare from windows. In addition, many studies revealed that these tools were appropriate for assessing discomfort glare in sub-tropics climate – but a few for assessing glare in buildings in the tropics. Some researchers on discomfort glare asserted that some factor influencing discomfort glare were still unknown [3]. In addition, in terms of glare, several researchers had doubted the use of luminance ratio method if it is used for assessing glare from windows. The objective of this study therefore, was to investigate the users' glare perception due to the use of daylighting and internal solar shading systems. The method used was a laboratory experimentation with subjects and measurements. The purpose was to compare these results with the glare rating as well as to assess the reliability of the glare assessment tools that fit to be used for assessing glare from window especially for those built in hot humid tropics region. In the study, typical office task such as reading, writing, and typing as well as two types of viewing directions were investigated in order to derive on reliable glare rating.

1.2. Predicting glare from window – luminance distribution ratio

Several very similar methods were developed in which they related luminance, position, and apparent size of the glare source, and the adaptation luminance to subjective judgments of the degree of discomfort experienced in indoor environments. Especially for artificial lighting, these had been expressed in some equations such as the British Glare Index (BGI), CIE Glare Index (CGI), Unified Glare Rating (UGR). Meanwhile, for daylighting the common use of glare equations used are Daylight Glare index (DGI), Visual Comfort Evaluation Method (VCE), and Predicted Glare Sensation Vote (PGSV). Many of these glare indices were simply measuring the luminance ratio. In addition, since disability glare is rarely happened in office space and there is still no satisfactory model to predict and evaluate this condition, many glare indices in office space are simply measuring discomfort glare [4, 5]. This is also stated in the ISO Standard 9241-6 (ISO-2000): "the most important factors for ensuring good daylighting is an acceptable surfaces' luminance and luminance contrasts in the office room" [6]. Based on this, the research simply considers luminance distribution ratios to be glare and thus do not distinguish between disability and discomfort glare as the other researchers specifically investigate luminance ratios as a critical subset of glare since there are clear impact of surface luminance and luminance distribution ratios on the level of perceived glare.

While several researchers argued that the surface luminance and luminance ratio were suitable to investigate glare due to the use of artificial light, others insisted that these methods could also be used in daylit offices. There are several sources for recommendations on acceptable luminance ratios for daylit offices (Table 1):

Table 1. Acceptable luminance distribution ratios according to several sources

Egan (1983) [7]	**IEA sourcebook (2000)** [8]	**IESNA** [9]
Task to adjacent darker surroundings 3:1	Variation in luminance across task max 2.5:1 to 3:1	Task to surroundings max 5:1
Task to remote darker surfaces 10:1	Task to background max 3:1	Normal field of view max 4:1
Lighting fixtures/ window 20:1	Task to remote surface 10:1	Windows to adjacent surfaces max 2:1
Field of view max 40:1		

Note:
Definition of task = 50 solid angle (field of view from the eye)
Near surround = 50 < 600 solid angle
Far surround (background)> 600 solid angle

These evidences show that the acceptable LDR can be predicted using some indicators i.e.: (1) For the surfaces ratios within a cone of 60° centered about the line of sight (*ergorama*): task (VDT screen) to adjacent surroundings (e.g. table, divider) max 3:1 or 1:3 (Luminance VDT/ Luminance surroundings >0.33 or <3). Task (VDT screen) and adjacent paper max 3:1 or 1:3 (Luminance VDT/Luminance paper >0.33 or<3), (2) For the surfaces ratios within a cone of 120° centered about the line of sight (*panorama*): Task (VDT screen) and immediate surroundings (Windows, window frame, and blinds) max 10:1 (Luminance VDT/Luminance surroundings >0.1 or <10). In the study, these indicators will be used as points in glare assessment.

2. Method

Since this study aimed at investigating the reliability of Luminance Distribution Ratio (LDR) used in a daylit office room in the tropical climate, the main method used in this study was to collect in-situ visual discomfort assessments. The method consists of collecting almost simultaneously measures of daylighting environments and subjective discomfort glare evaluations in several days scenarios. Despite the sitting positions, this study used also three types of shadings (i.e., the Venetian, Vertical and the Roller Blinds) in order to derive more accurately the reliability of glare rating in many office situations.

2.1. Experimental rooms and interior shadings investigated

The test rooms were six identical 4x7 m^2 private lecturers' rooms at the Department of Architecture ITS. Among these, the first three rooms had windows facing to the East and the others were facing to the West. Since luminance performance are also highly influenced by the rooms' interior surface attributes, the optical characteristics these are also considered in the measurements and discussions. The attributes of the rooms were as the followings: window glass: clear glazing (VT=90, R=5); floor: grayish ceramic (T=0, R=40); ceiling: white painted (R=80); working plans: dark brown laminated wood (T=0, R=5); interior solar shadings: cream/light brown Venetian blind (T=5, R=65); yellowish Vertical blinds (T=70, R=10); and cream / light brown blackout dotted Roller blind (T= 40, R=10). Note that T= transmittance; R=reflectance; VT=visible transmittance.

Points of measurements were at three work planes (WPs) located in zone 0-3 m from window perimeter (i.e. WP 1, WP 2, WP 3) from the nearest window distance. The luminance values were measured at these three represented working plans and at diverse sitting positions/viewing directions

(WP 1 and 3 were parallel to the windows, while WP 2 was against the windows) (Figure 2). Luminance measurements were taken during the months of July and August. For measuring these, a luminance meter used and held at 1.0 m high (the assumption of the height of the workers' eyes when sitting on their work plane) and at their chair positions. They were set to record every hour from 8 AM to 3 PM and focused to approximately 40° in altitude and 90° in azimuth as the area of the human field of view (FOV). So that, it measured luminance at some points such as the table, the computer display/video display terminal (VDT), and its surrounding surfaces such as the door and its frames, the papers at the table and the wall. The measurements were conducted at the average outdoor illuminance of 42.000 Lux and under the average of partly cloudy and overcast types of sky. The luminance measurements were taken when the window blinds (Figure 3) were executed by tilting the slats so they were opened in several degrees.

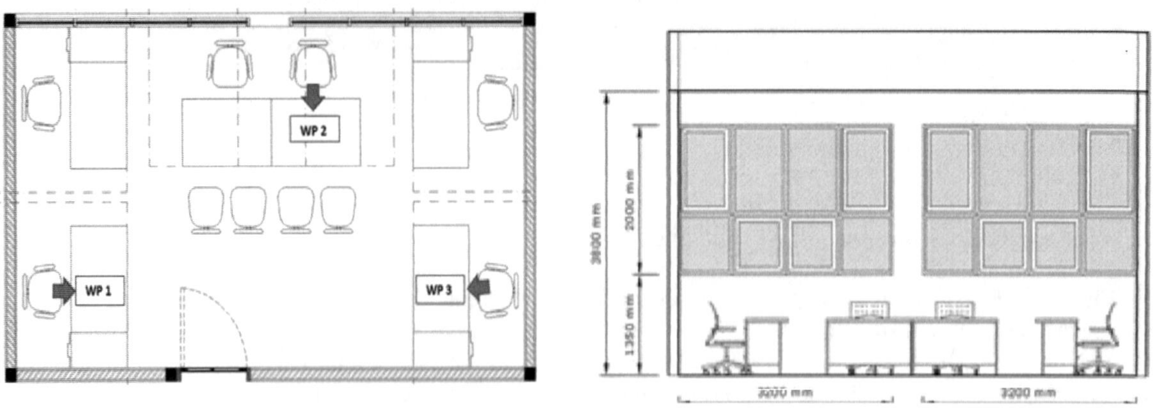

(a) **(b)**

Figure 2. The Experimental rooms: the plan and the position of work-planes (i.e WP1 and WP3 – viewing directions were parallel to the windows; WP2 – viewing direction was against the window) where the luminance was measured (a) and the section (b)

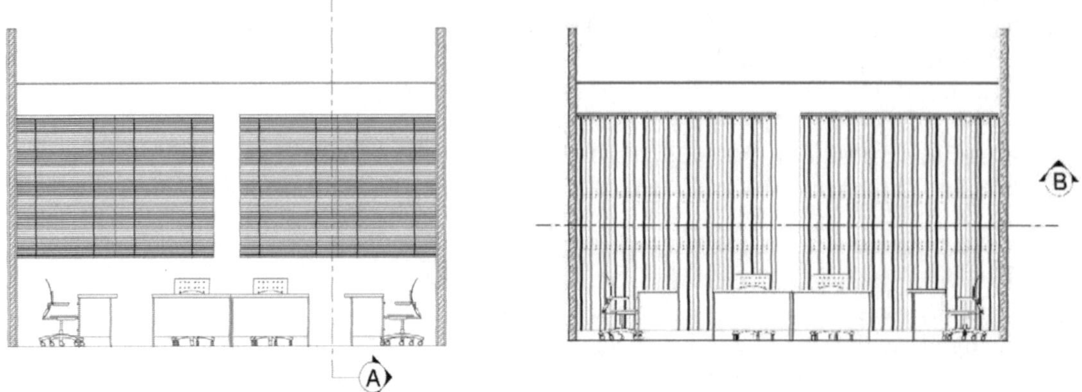

(a) **(b)**

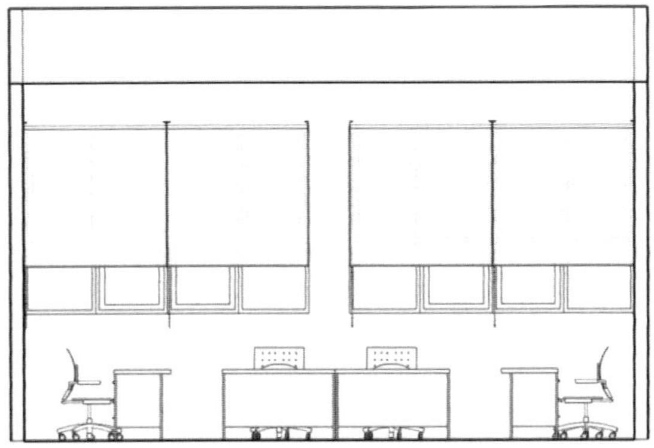

(c)

Figure 3. The section of the rooms with the venetian blind (a), vertical blind (b), roller blind (c)

2.2. Data collection methods

In the test room, the subjects were asked about their impression and opinion of the room, the windows and occurring glare problems. The subjects performed similar office tasks (typing using computer/VDT – for horizontal view and reading/writing on desk – for downward view) and the task presentation order was fixed and data on users' performance was recorded. The order of presentation of the three windows' blind types and two viewing directions was carefully controlled. Glare assessment for each of the three different windows' blinds occlusions scenarios (totally opened, 45° opened, and totally closed) and sitting positions lasted about approximately 30 minutes and measured at both morning and afternoon sessions. No artificial lighting was used in this experimentation.

The questionnaire on the daylighting conditions was divided into 4 main parts. The demographic questions (part 1) considered gender, age, the wearing of glasses or contact lens, and sensitivity to bright light. Part 2 consisted of questions related to rating the lighting conditions when typing using computer and writing on desk. The subjects described the perception of the visual environment by means of a set of line rating scales and magnitude of glare on a 4-point scale with pre-defined glare criteria (imperceptible, noticeable, disturbing and intolerable). Part 3 was subdivided into two parts where the first part concentrated on general lighting conditions within the room before the subject could change the system according to their wishes, while the second part concentrated on why they had changed the initial set-up of the solar shading systems. Part 4 focused on indoor climate conditions in the room. All collected data had been analyzed by a repeated measure of cross tabulation and dependency correlation.

3. Results and Discussions

A total of 18 subjects (6 males and 12 females) involved in this study. Demographically, they are all lecturers and sitting in their own sitting position in which the windows are covered with different type of solar blind system. They were asked to carry out several tasks in 5 days both in the morning and in the afternoon. So that for each respondent evaluated the luminance performance of his/her room 10 times and the data gathered were totally of 180. Within the groups, 8 subjects using glasses and none using contact lens. All the respondents were right-handed people.

For illuminance performance, due to the use of different solar shadings under the average of 42.207 lux of outdoor horizontal illuminance, the Vertical blinds systems provided a mean work-plane illuminance around 283 lux, while the Venetian and the roller blind systems provided 245 and 437 Lux respectively. The Roller blinds type seemed to contribute to the highest illuminance since if the blinds were opened, almost 90% of light could freely enter the room. However, in the average, the subjects reported that the Roller blinds yielded more adequately lit environment compared to the other two blinds type if the Roller blinds were in the totally closed scenario.

Subjects rated glare from windows by several rating scales (1 – not at all and 10 – very much) and four points scale for glare criteria (imperceptible, noticeable, disturbing, and intolerable). The reported glare was different and the metal Venetian blind was reported to cause higher contribution to glare than the other two blind systems. Subjects also reported a higher degree of glare perception with the increasing of blinds slat angles and occlusions. All blinds systems have clear tendency that subjects are more affected by glare looking horizontal working on VDT than looking downward reading paper on desk. Furthermore, the Venetian blinds and the large opening of slats/occlusions cause more glare problems in the lower part of the field of view, since reflections of the sun and sky on the lower slats are visible within the field of view. There was also a tendency of higher reported glare in the afternoon, when the subjects were more affected by the light reflected by the blinds. Correlations and dependencies of these working activities and sitting position are shown in Table 2. This table shows that there is a correlation between working activities types and sitting positions with glare perception. Strong correlations and dependencies are shown by the use and the operations of Roller blinds.

Table 2. Dependency analysis results between the subjects' working activities and sitting positions with glare perception

No	Shadings type & type of correlation	Critical Area		Comp. X^2 and X^2_{table}	Interpretation of dependency	The strength of dependency C-Cremer value
		x^2	$x^2_{(k-1);0,05}$			
A	**Vertical blinds type**					
1.	The subjects working activities (typing and writing) vs. glare perception	6.14	5.991	$\chi^2 > \chi^2_{table}$, = reject H0, accept H1	Working activities influenced glare perception.	Weak (0.402)
2.	The subjects' sitting position vs. glare perception.	16.5	9.48	$\chi^2 > \chi^2_{table}$, = reject H0, accept H1	Sitting position influenced glare perception	Strong enough (0.6589)
B	**Venetian blinds type**					
1.	The subjects working activities (typing and writing) vs. glare perception	13.5	9.49	$\chi^2 > \chi^2_{table}$,= reject H1, accept H0	Working activities influenced glare perception.	Weak (0.457)
2.	The subjects' sitting position vs. glare perception.	18.1	9.488	$\chi^2 > \chi^2_{table}$, = reject H0, accept H1	Sitting position influenced glare perception	Strong (0.74)
C	**Roller blinds type**					
1.	The subjects working activities (typing and writing) vs. glare perception	20.3	5.991	$\chi^2 > \chi^2_{table}$, = reject H0, accept H1	Working activities influenced glare perception.	Very strong (0.81)
2.	The subjects' sitting position vs. glare perception.	17.5	9.488	$\chi^2 > \chi^2_{table}$,= reject H0, accept H1	Sitting position influenced glare perception	Strong enough (0.6)

Note: Hypothesis: Ho=Subject proportions were the same; H1=Subject proportions were different.

Further evaluation was on the preferred blinds' slats or occlusions positions. After the subjects completed to rate the daylighting conditions, they were asked to adjust the closed blinds – as the initial position, to their preferred positions for each combination of window blinds and viewing directions. For the Venetian and Vertical blinds systems the subjects were allowed to change only the slats positions, but not to raise (for Venetian type) or to shove (for Vertical type) the blinds. It was found that more than 70% of the subjects asserted that they were either very uncomfortable or slightly uncomfortable with the initial setting of the Venetian and Vertical blinds and wanted to change the setting of the blinds to maintain comfortable daylighting in the room. Inversely, for the windows covered by Roller blinds, almost 70% of the subjects felt very comfortable with the initial setting. They asserted that in the closed position of the Roller blinds, they still can work well. Besides the illuminance performance, the subjects were also satisfied with the Roller blinds for eliminating glare. They stated that increasing the slat angle of both the Venetian and the Vertical blinds tended to cause glare since the slats became a very good reflector that reflected the excessive daylight especially in the afternoon. When asked about the controllability of the blinds, almost 80% of the subjects were agree that Roller blind systems was easier to be operated since they did not need to frequently adjust the blinds.

4. Conclusions

It is found that the using of luminance distribution can manage the evaluation of glare occurrences for office building in hot and humid climate. Using this tool also, the study revealed that there is a strong enough correlation between the glare indicator i.e. the value of luminance distribution and how subjects reported discomfort glare in an experimental set-up with three different solar shading systems and two different view directions. It can be concluded that due to the variability and uncertainties with the currently available glare assessment, the study proposes the use of a simple luminance distribution method since the evaluation of the results from the experiments shows good correlation between luminance distribution values with the user response. The study also found that glare from window assessment can be evaluated in conjunction with the use of several solar shading systems. The data set provides ample opportunities for further glare analysis tools such as Glare Index (GI), Daylight Glare Index (DGI), and other tools to be used within the context of offices in humid tropics region.

Acknowledgement

The authors gratefully acknowledge the Ministry of Research, Technology and Higher Education of The Republic of Indonesia for providing the research grants under the contract number of 863/PKS/ITS/2018. Authors would also like to thanks to all the participants who involved in this rooms' glare/visual comfort assessments.

References

[1] Dinapradipta A 2015 Office building facades for functionality and adaptability in humid tropical cities, multi-case studies of office buildings in Jakarta – Indonesia (TU/e dissertation, the Netherlands).

[2] Taylor P, Duponchel H and Bannnister P 2009 The impact of glazing selection on daylighting and energy performance for an office building in Canberra (International IBPSA Conference Proceedings, Glascow, Scotland) pp. 1678–1684.

[3] Chauvel P, Collins J B, Dogniaux R and Longmore J 1982 *Lighting Research Technology* **14**, pp. 31-46.

[4] Osterhaus W K E 2005 *Solar Energy* **79** pp.140-158 (Elsevier).

[5] Dubois M C 2001 Impact of shading device on daylighting in offices (KFS AB. Lund).

[6] ISO 2000 Ergonomic Requirements for office Work with Visual Display Terminals (VDTs)- Part 6: Guidance on the Work Environment (French European and International Norm NF EN ISO 9241-6: 1999, AFNOR, Paris, ISSN 0335-3931).

[7] Egan M D 1983 *Concepts in architectural lighting* (McGraw-Hill, New York).

[8] IEA SHC Task 12/ECBCS Annex 29 2000 *Daylight in buildings - a sourcebook on daylighting systems and components*.

[9] IESNA *Light, vision, and perception* (IESNA ED 100.2).

IOP Conf. Series: Materials Science and Engineering **676** (2019) 012038 doi:10.1088/1757-899X/676/1/012038

Climatic Significance of Colonial House Forms in Surabaya

I G N Antaryama[1]*, S N N Ekasiwi[1], Irvansyah[1], P M Dewi[2] and M S Ulum[2]

[1]Department of Architecture, Institut Teknologi Sepuluh Nopember (ITS), Kampus ITS Sukolilo, Surabaya 60111
[2]Research Assistance at the Laboratory of Architectural Science & Technology ITS, Department of Architecture Kampus ITS Sukolilo, Surabaya 60111

*Email: antaryama001@gmail.com

Abstract. The present study explores the morphological characteristics of colonial houses in Surabaya. Four houses are included as the subject, and physical features of the building are recorded through field observations. Basic drawings of the building are made and used for the morphological exploration. Formal and topological characteristics of the house are then analysed and its climatic response as well as climatic adaptation potentials are evaluated. Principles of climatic adaptation and findings from previous studies on the subjects are used as the basis of the evaluation. The study found that the colonial houses in Surabaya have a great potential in terms of climatic adaptation. Despite the compact shape, the layout of the building allows internal spaces to poses high connection to outdoors, and hence increases its degree of connectedness. The colonial houses show moderate degree of exposure and have high degree of porosity due to the configuration and size of windows and doors of the building. All of these formal characteristics contribute to the significant climatic response of the building.

1. Introduction

Formal and spatial characteristics of a building modify external environment, including climate. Building shape could determine the amount of heat gain and air flow pattern around the building. Spatial arrangement within the house could influence ventilation as well as light penetration. From the passive design point of view, these formal and spatial characteristics play a primary role in producing an energy efficient design.

Colonial houses in Indonesia have a specific morphological characteristic that embody both features of Indonesian traditional dwellings and Dutch buildings. Passchier cited in [1] considered the design as a hybrid architecture. The house appears in a relatively new form, which neither resembles Dutch nor Indonesian traditional building. Common features of colonial house are of a compact shape and has surrounded by abundant open space. The building is typically made of thick brick walls, and has large openings, big as well as steep roof.

Many studies regard the colonial house as having good climatic adaptation [2-4]. These studies, however, do not pay attention much on investigating spatial and formal characteristics of the building in relation to its environmental performance. Conversely the studies that focus on formal aspects of the design tended to explore only the formal characteristics of the building, such as shape, configuration/composition [5-7], without putting any specific reference to climatic conditions.

From a wider perspective, Dall stated all architecture is influenced by climate [8]. The reason for this is building is required to protect internal conditions and the building from negative impact of the climatic. Building shape and façade are important aspects in determining thermal condition of a building [9]. Window's properties such as size, orientation, and disposition will influence ventilation [10] and daylight condition [11] in the building.

The present study noted that aside from the fact that many studies have conducted over the last decade on investigating the climatic performance of a building, only a view of them paying attention on exploring climatic significance of colonial house form. Aim of the study, therefore, will be set in investigating morphological characteristics of the house in relation to its climatic adaptation. It will analyze spatial and formal characteristics of the building and evaluate its climatic response as well as climatic adaptation potential, especially with reference to the warm-humid climate of Surabaya.

2. Method

The present study explores the morphological characteristics of the colonial houses in Surabaya. Four houses are included as the subject, and physical features of the building are recorded through field observations. The houses considered in the study are located in urban areas, i.e. houses in Tapak Siring street (HC1-Siring), Nias street (HC2-Nias), Sutomo street (HC3-Sutomo) and Musi street (HC4-Musi). Area of the buildings is 132.4 m2, 133.9 m2, 152.2 m2 and 205.1 m2 respectively. For the study, only the original colonial house that were taken into account, and building extensions are excluded. Figure 1 shows elevation of the four houses.

2.1. Morphological measures

Basic drawings of the building, i.e., plan, and elevation, are recorded and used for the morphological exploration. Formal and topological characteristics of the house are established by using diagrams depicting wall adjacencies and accesses which exist on internal and external walls. Wall adjacency only considers walls that have a length of 1 meter and more, i.e., a dimension that enables window or opening to be incorporated on the walls. Wall access includes all existing openings/windows and doors on the walls.

Based on wall adjacencies and accesses diagrams, justified access graphs indicating relations between internal-external spaces, and degree of connectedness (DC) of the house is calculated. Other calculated measures are average number of adjacency and access of the building. The first is used to indicate degree of exposure (DE), and the second used to indicate degree of porosity (DP). The first indicator chiefly relates to thermal response of the building and the second relates to response of the building to daylight and ventilation. This method is adopted from Antaryama [12].

2.2. Analysis of climatic response of buildings

Climate analysis was first conducted to describe characteristics of the climate of Surabaya. The second is analysis on dimensional characteristics of the house. This includes window-to-wall ratio (WWR) and window-to-floor ratio (WFR) which are used to evaluate and indicate climatic response of the building. Climatic response potentials of the house were also analyzed on the basis of the possible use of wall for promoting daylight and ventilation. Basis of the climatic response evaluation is the Indonesian National standard (SNI) for the design of daylighting [13] and natural ventilation [14] as well as energy efficiency for building envelope design [15, 16]. This analysis is adopted from Ibrahim and Hayman [17].

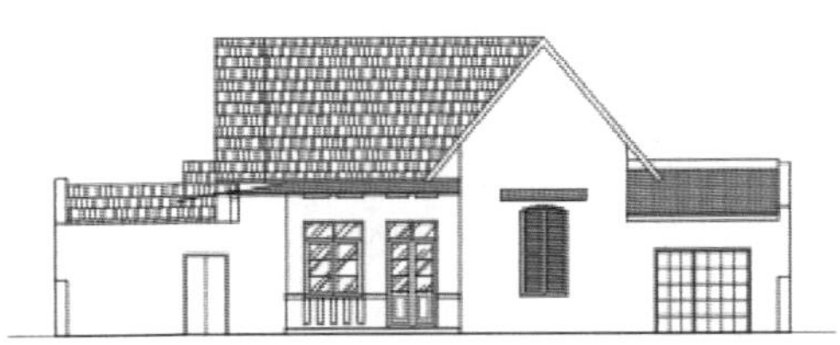

a) HC1-Siring

b) HC2-Nias

IOP Conf. Series: Materials Science and Engineering **676** (2019) 012038 doi:10.1088/1757-899X/676/1/012038

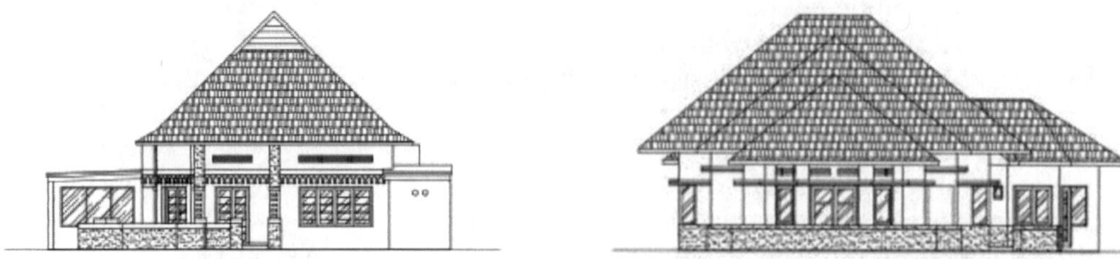

c) HC3-Sutomo d) HC4-Musi

Figure 1. Elevation of the four colonial houses under study [18]

3. Results and Discussions

Results of the analysis and discussion are arranged into three parts. The first describes climatic conditions of Surabaya. The second part explores the morphological characteristic of the house, and the last part evaluate the climatic response and the climatic response potential of the buildings.

3.1. Climatic condition of Surabaya

Average climatic conditions of Surabaya are obtained form the MET Office of Juanda, Surabaya. Figure 2 describes conditions of air temperature, air humidity, wind speed and direction, and solar radiation. The city of Surabaya is laid on 7,22o south of the equator and regarded as having warm and humid climate. Average air temperature (28-30 °C), solar radiation (up to 1200 W/m2), and relative humidity (76%) are relatively high, whereas wind speed is generally low (2.6-3.6 m/s). This condition indicates that the city experiences almost constant warm and humid conditions throughout the year. Wind, which can be utilized as a means of restoring comfort, unfortunately is not high enough for this purpose.

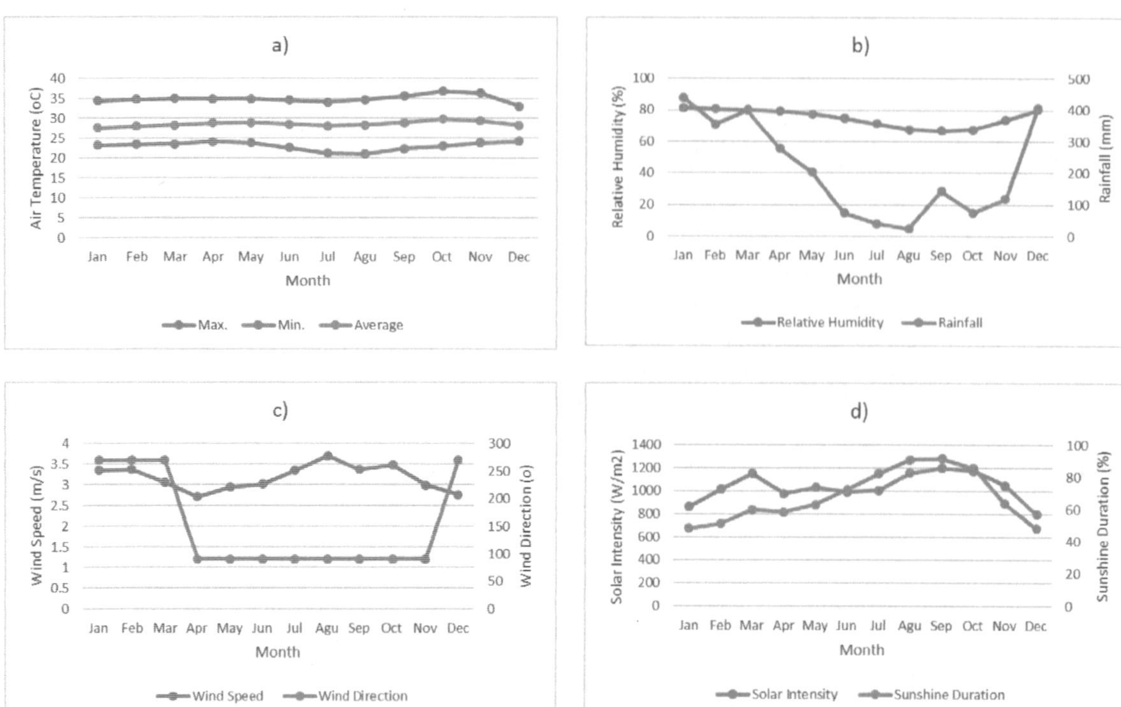

Figure 2. Climatic condition of Surabaya: air temperature (a), air humidity & rainfall (b), wind (c), solar intensity & sunshine duration (d)

The monthly rainfall is high (280-400 mm) and the sky ranges from partly cloudy to cloudy/overcast as indicated by the sunshine duration which is within a range of 42-85%. The sky characteristics could prevent heat dissipation towards the atmosphere and consequently could create uncomfortable thermal condition in the city. A care must be taken in order to prevent the building from gaining heat and excessive light, and to promote ventilation for restoring comfort in this type of climate.

3.2. Morphological characteristics of the house

Despite the compact plan shape, layout of colonial houses in Surabaya allows a great deal of internal and external accesses (figure 3a). Many accesses both internally and externally can be associated with good cross ventilation and sufficient natural light penetration. All the houses are designed in such a way that all internal spaces have at least one internal and external accesses. A two-row type of layout enables this characteristic.

Adjacency between internal spaces can be maintained into a high level in the design of the house, offering a greater opportunity for spaces to have internal connections. The same also applies to external walls where spatial configuration of the house could increase the number of external wall adjacency. Although this characteristic could increase the amount of heat gain, the availability of external walls however could also increase the number of opening/window on the building that are attached on the wall, and thus increasing airflow through the building and natural light penetration.

Observing the justified access graph of the house shown in figure 4, it can be seen that all of internal spaces in the building are apparently connected to outdoors. Some of the spaces are also internally linked. Degree of connectedness of the house is high (DC=0) indicating that all spaces are directly connected to outdoors and consequently the buildings have a great chance for external penetration.

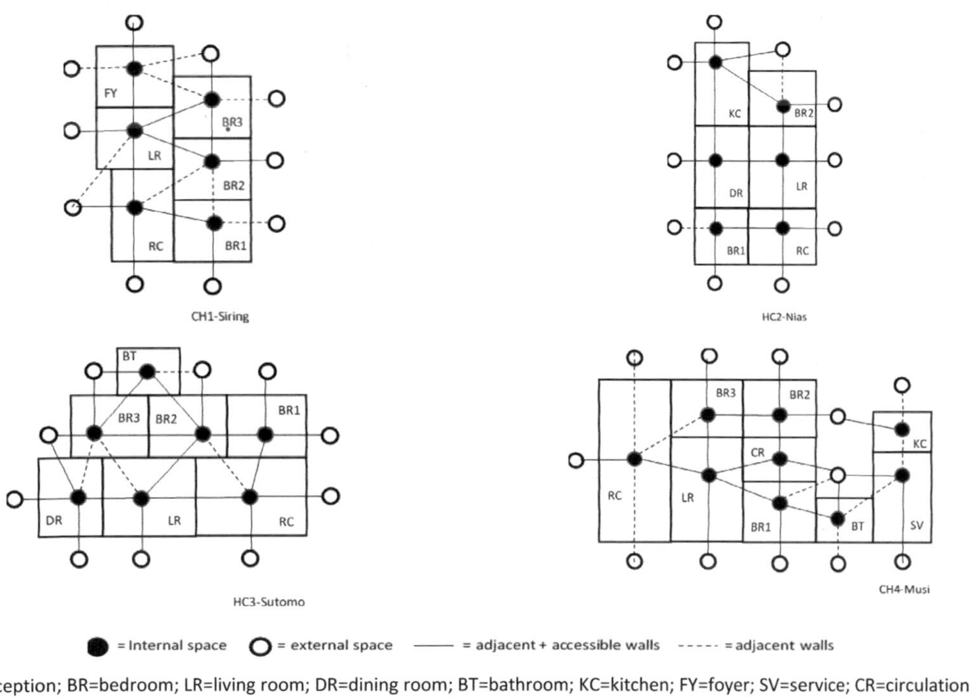

Figure 3. Diagram denoting adjacent and accessible walls in the colonial houses

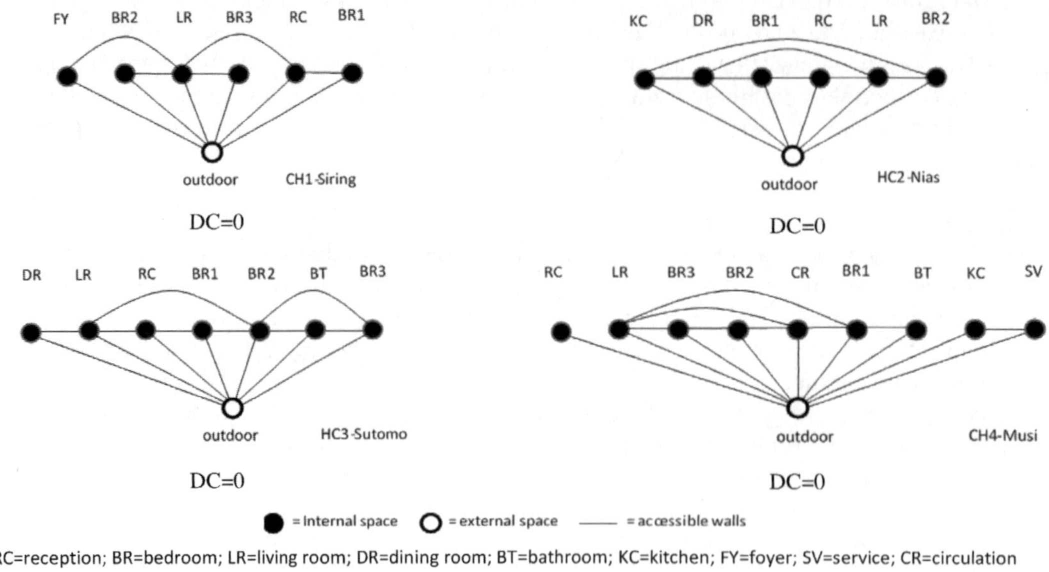

Figure 4. Justified access graphs of the four house under study

Colonial houses in Surabaya have relatively moderate degree of exposure (DE). As figure 5.a shown, the average number of external adjacency of the four houses under study is around 2. It follows that about half of the walls in each rooms are exposed to external environment. From thermal design standpoint, this characteristic could be associated with an increase in heat gain. Average number of internal adjacency ranges between 2 and 3. Two houses, i.e., HC1-Siring and HC4-Musi have low level of DE (2). Average of overall internal-external adjacency is between 4 to 5.

Degree of porosity (DP) of the house is high (figure 5.b). This can be observed from the average number of internal-external access which are between 3 and 4. This average number indicates that each space in the building has access in almost all its perimeter. Average number of external access clusters around 1 and 2, and average number of internal access is about 2 in all cases. This characteristic illustrates a great allowance for cross ventilation and light penetration.

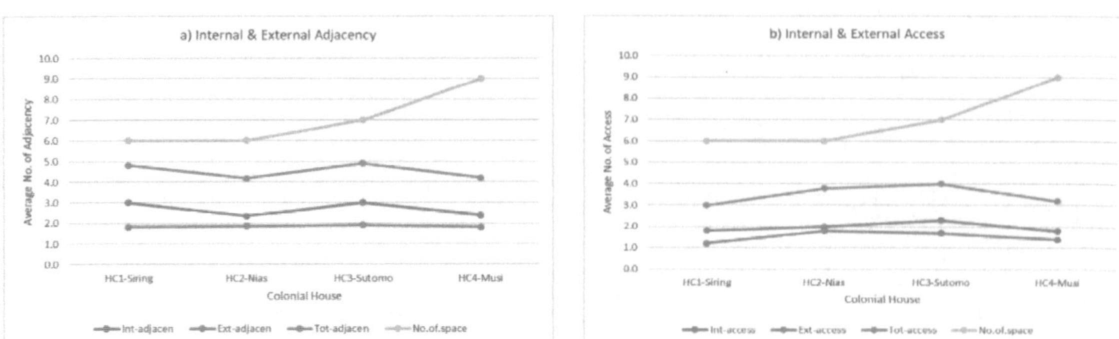

Figure 5. Degree of exposure and porosity of the building

3.3. Climatic significance of the house form

Response of building to thermal condition is evaluated on the basis of the value of WWR that meets the energy efficiency requirement, i.e., overall thermal transfer value (OTTV) of 35 Watt/m2. The method assumes that the building is air conditioned and thermal response of the building is determined by the heat gain through the building envelope. Despite the fact that the building is naturally ventilated, the method is regarded as adequate for assessing the role of building envelope in modifying the thermal conditions. For the evaluation WWR for different building orientation is estimated using charts given in [11]. Figures 6.a and 6.b give results of the analysis.

The two figures show that in general the standard WWR can be attained by all of the houses. However, there are minor cases, such as in HC1-Siring and HC2-Nias, where the values are over the standard. The result implies that the colonial houses only transfer small amount of heat into the building, and thus it is able to promote energy efficiency. WWR of HC3-Sutomo and HC4-Musi in particular are all below the recommended WWR (figure 6.b).

From promoting ventilation point of view, it can be seen in figure 6.c that the external window-to-floor area ratio (WFR) of the houses is well above the standard (5%). As illustrated in the figure, the external WFR of the building lays between 15 to 25%. It follows that natural ventilation can be induced in the building. Internal WFR that ranges between 8 to 16% increases the opportunity for the house to have better cross ventilation. Comparing all the house, it can be seen that external WFR of HC4-Musi is the lowest (around 13.4% WFR). The values are slightly increase in HC3-Sutomo (17.8%), and then followed by HC1-Siring (19.1%) and HC2-Nias (23.7%) respectively. From this result, it can be said that climatic significance of all of the houses is high.

Similar results are found when analysing the ability of the design in promoting day light. As can be seen in figure 6.c, it is observed that external WWR of the houses are well above the standard (8%). This implies that daylighting can be incorporated in the design of all colonial houses without any big efforts. Thus it can be said that all the houses have high climatic significance especially in terms of daylight provision. Among the four houses under study, HC2-Nias in this case is the best.

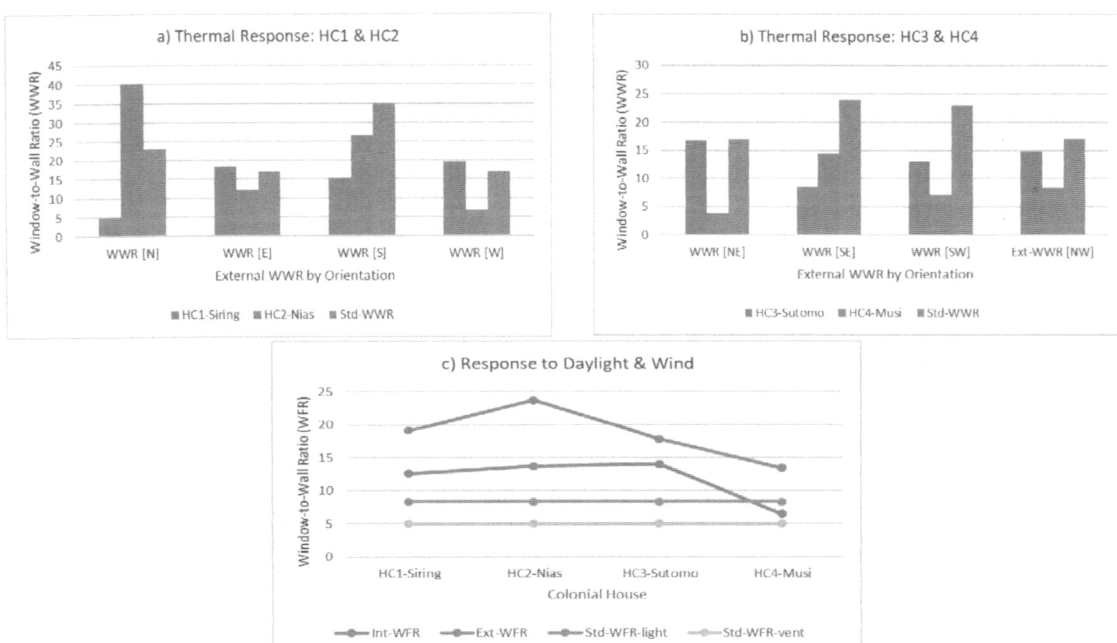

Figure 6. Climatic response of the building

Climatic response potentials of the houses are shown in figure 7. From this figure, it can be seen that number of internal and external adjacencies is generally higher than internal and external access (figure 7.a). This follows that the house still has an allowance to provide daylight and ventilation if it is required, especially when the existing coverage area for light and wind penetration is reduced by obstructions such as overhang or neighbouring buildings. In terms of external access, the opportunity is widely available in the four houses under study, however in terms of internal access, only in HC2-Nias that all allowances are already utilized.

Considering the overall access and adjacency in the building, it can be seen that similar tendency also occurs in the case of overall access and adjacency (figure 7.b). Allowance for providing daylight and ventilation in all cases is high. The results again underline the good climatic response of the building, especially for providing a more distributed airflow in the building. The condition particularly gives significant impact in restoring thermal comfort in the house.

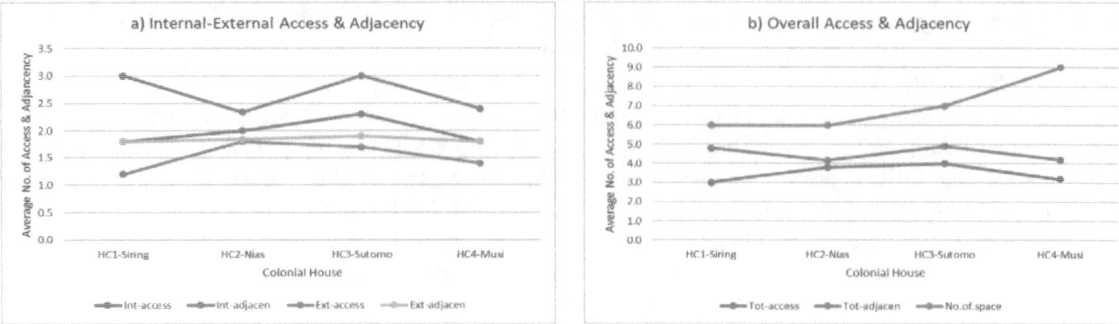

Figure 7. Climatic response potential of the building

4. Conclusions

From the above discussion it can be concluded that under the warm humid condition of Surabaya the colonial houses have a great potency in promoting comfortable indoor environment in the building. The morphological characteristics of the house show this tendency. The high degree of internal and external connectedness (DC=0) and high degree of porosity (DP>5% and 8%) are indication that the house can offer sufficient access to natural ventilation and daylighting. Moderate degree of exposure (DE=2) and relatively moderate WWR (<35%) are also indicators that the building will transfer less heat through its building envelope. An allowance for further effort in promoting daylighting and ventilation actually exists, because not all of the available walls are used for openings/windows. It should be noted, however, such an allowance for daylighting or ventilation can pose a detrimental impact to the thermal condition of the building.

Acknowledgments

The authors wish to acknowledge the contribution of the Ministry of Research and Technology and Higher Education of the Republic of Indonesia for allocating a research allowance under the scheme of Penelitian Terapan Unggulan Perguruan Tinggi (PTUPT) ITS 2018.

References

[1] Peter JMN 2009 *Masa Lalu dalam Masa Kini Arsitektur di Indonesia* (Jakarta: Penerbit Gramedia Pustaka Utama).
[2] Mas S 2001 *Dimensi Teknik Arsitektur* **29**(1) 34-42.
[3] Leonardus MFP, 2004 *Dimensi Teknik Arsitektur* **32**(2) 138-39.
[4] Antonius A, Achmad D, Ikaputra, Jatmika AS 2015 *International Journal of Scientific and Research Publication* **5**(4) 1-6.
[5] Jarwa PSH 2015 *Jurnal Inovasi dan Kewirausahaan* **4**(1) 21-30.
[6] Rina W, Meyka W 2012 *Jurnal Ilmiah Desain & Konstruksi* **2**(11) 3-14.
[7] Joko TS, Noviani S, Antariksa 2013 *Jurnal Ruas* **11**(2) 37-50.
[8] Torben D 2010 *Climate and Architecture* (Oxon: Routledge).
[9] Waraporn R. and Wathanyoo R. 2014 Energy Procedia **52**(2014) 10-17.

[10] Ardalan A, Norhayati M, Zakaria AMA, Mohamad RB, 2014 E3S Web of Conferences **03** (01025).

[11] Srikonda R 2018 Journal Architectural Engineering & Technology **7**(1).

[12] IGN Antaryama 1999 *House Form Transformation and Climate in Bali* (University of Manchester: Unpublished PhD Thesis).

[13] BSN 2001 *SNI 03-2396-2001: Tata cara perancangan sistem pencahayaan alami pada bangunan gedung* (Jakarta: BSN).

[14] BSN 2001 *SNI 03-6572-2001: Tata cara perancangan sistem ventilasi dan pengkondisian udara pada bangunan gedung* (Jakarta: BSN).

[15] BSN 2011 *SNI 6396:2011: Konservasi energi selubung bangunan pada bangunan gedung* (Jakarta: BSN).

[16] Dinas Penataan Kota Pemprov DKI Jakarta 2013 *Panduan Penggunaan Bangunan Gedung Hijau Jakarta* (Jakarta: Pemprov DKI Jakarta).

[17] NL.Nik Ibrahim, S. Hayman 2005 Daylight Design Rule of Thumb, *Proc. Int. Con. on Sustainable Building South East Asia.*

[18] I.G.N. Antaryama, Sri Nastiti NE, Andy M, 2017 *Strategy of Climatic Control of Colonial Houses in Surabaya, Part I : Thermal Conditions.*

IOP Conf. Series: Materials Science and Engineering **676** (2019) 012039 doi:10.1088/1757-899X/676/1/012039

Local Wisdom that forms of the Bajo Ethnic Settlements in Belopa Village

M Ali[1]*, A M Ramadhani[1] and A S Ekawati[1]

[1]Department of Urban and Regional Planning/Laboratory Waterfront, Engineering Faculty, Hasanuddin University

*E-mail: mukti_ali93@yahoo.com

Abstract. In general, the characteristics of the Bajo ethnic (Bajo) settlement are built on the waters. Likewise with the Bajo in Belopa Village, their settlements were formed starting above the waters. But along with the development and increasing population, their settlement pattern increasingly leads to land. So that, the arrangement of the settlements began to differ from its characteristics. This condition is feared to have an impact on local wisdom, which will gradually disappear. This study examines the extent of local wisdom that underlies the formation of Bajo settlements in Belopa Village. The analytical method used are descriptive analysis, comparative analysis, spatial analysis and photo pmpping analysis. The results of the study show that there are several forms of local wisdom that underlie the creation of Bajo, namely governance, value systems, procedures, and special provisions. In addition, it was also found that the condition of the settlement of the Bajo that currently exists has characteristics based on orientation patterns and material and type of building. The condition of the Bajo settlement facilities and infrastructure, some of its associations still cannot serve the needs of the community and its maintenance is still not optimal.

1. Introduction

The Bajo tribe, known as the nomadic tribe, is a tribe that depends entirely on the sea. In accordance with their livelihood as fishermen, the Bajo people are very familiar with their world. They are traditional fishermen who have wisdom in the use of marine resources in their environment, both regarding the types of marine resources, utilizing and preserving them. Bajo settlements have a characteristic, built on water, simple and able to adapt in extreme environments, such as waves, ocean currents, and wind, which become the pulse of their daily lives.

Local wisdom contained in a community group is the basis for regulating the growth of a city or region. Bajo is a tribe that has a local wisdom in the use and preservation of marine resources. Bajo settlement in Belopa Village has a characteristic built on the water, simple and able to adapt in its environment. The settlement area of the Bajo continues to grow in line with the increasing number of residents who live and the existing needs. The settlement conditions of the Bajo that are not neatly arranged give the impression of slums coupled with the condition of poor sanitation facilities, the absence of landfills, and the problem of other settlement infrastructure facilities.

The Bajo settlement pattern in Belopa Village is now approaching towards the land. This condition occurs because the Bajo people who began to leave the way of life in the sea to live on land. This will eliminate the characteristics of those who initially settled on the waters. Therefore, efforts need to be made to preserve the local wisdom. The initial step to be taken is to identify local wisdom that underlies formation of the Bajo settlement in Belopa Village.

IOP Conf. Series: Materials Science and Engineering **676** (2019) 012039 doi:10.1088/1757-899X/676/1/012039

2. Methodology

Research was carried out in the Bajo fishermen's settlement in Belopa Village, Luwu Regency, South Sulawesi. The data was obtained by observation, questionnaire, and interviews with the Bajo people. The analysis was carried out with qualitative descriptive analysis methods, spatial analysis, and photo mapping.

3. Result and Discussion

3.1. History of the emergence of settlement

The Bajo who settled in Belopa Village was of genuine descent and some were mixed descendants whose parents married local people. There are some perceptions about the origin of the existence of the Bajo on the coast of Luwu Regency.

The first, the Bajo is a tribe that does not have a place to live on land. They live on boats and live by the sea. Sometimes there is some time they pull over to the coast if they need something or repair a boat. From here the origin of the Bajo is on the coast of Belopa Village. Their ancestors at first only stopped a few times on this coast, then some of them chose to settle down and build wooden houses on the seafront.

The second story is told that the ancestors of the Bajo were people who were made soldiers by the kingdom of Luwu who were stationed on the coast to protect the kingdom from enemy threats. Over time, they finally built settlements on the coast and settled there until now.

In another story it is mentioned that one day a princess from the Bajo was washed away and disappeared into the sea. Then the king sent all the kingdom's people to split up looking for princesses. The king issued an order that no one should return before the princess was found. It is known that the daughter was found by the people of the Gowa kingdom and later married and settled there. Therefore, the Bajo people who had spread around the oceans and islands were unable to find a daughter and could not return because the Bajo king had issued an order, settling them on the nearest shores until now their keturan still settled at the location of the initial settlement.

3.2. Initial forms of settlements

Initially, the Bajo in Luwu District built their settlements on the water near the coast. They built their houses with very simple building materials. They built houses using large tree trunks they could find on land (Figure 1). The roof of their house is covered with some dried leaves such as dried coconut leaves and so on and their house foundation uses coral rocks. The foundation of his house is made of sea corals which are arranged and stockpiled lined up as a barrier to house poles. They live in a very simple way and occasionally go ashore just to buy basic commodities and then return to the sea. The vehicles they have and use daily are the boats they make themselves. Generally every house is inhabited by more than one family head. For mck and other activities carried out above the sea (Figure 2).

Figure 1. A house of the Bajo in Belopa Village

IOP Conf. Series: Materials Science and Engineering **676** (2019) 012039 doi:10.1088/1757-899X/676/1/012039

Figure 2. People who park their boats next to the house

The existing settlement conditions have not changed much from their original form. The foundation of the house no longer uses coral rock but has begun to change to the embankment and hardened form. Some residential buildings have changed their shape from stilts to permanent homes.

Settlement patterns spread over the sea not far from the coast. In the beginning there were only about ten houses built together. If the house does not fit in, it is occupied by existing family members, then the house is built again for the new family head and so on. Their main livelihood is as fishermen. Some also have side jobs as boat makers. They assemble their own bait and fishing gear they use. Fishermen usually use small boats to catch fish. They will usually leave at dawn on their own or bring one of their family members and go home at dusk. There are also several groups of fishermen who use boats that are rather large to catch fish in a farther place and will go home for a week. This tradition of fishing is still done until now. Their boats are parked below or beside the house. Their house poles were made rather high to prevent the tide from entering the house.

The first Bajo people who came to the coastal area of Belopa Village built settlements on the water near the beach. At first there were only a few houses filled with more than one family head. The settlement is entirely on the water. The existing settlement has a spread pattern with the orientation of several houses facing land and some facing the sea.

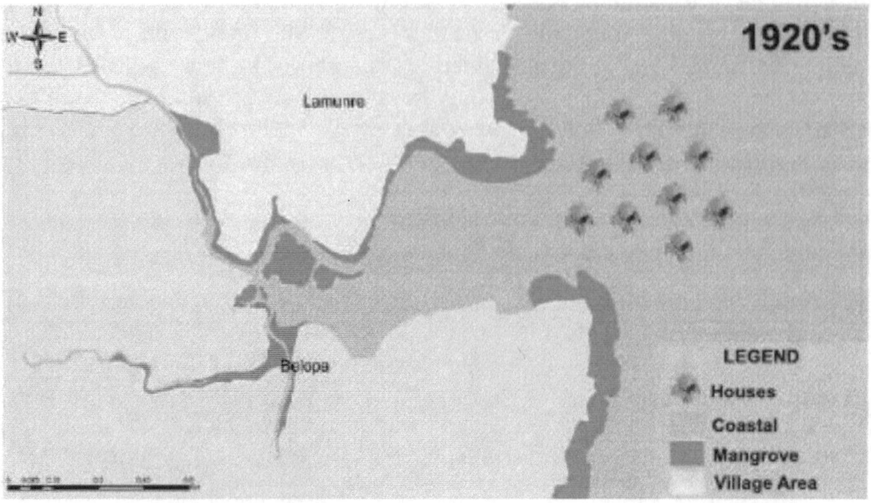

Figure 3. Illustration map of the beginning of the arrival of the Bajo

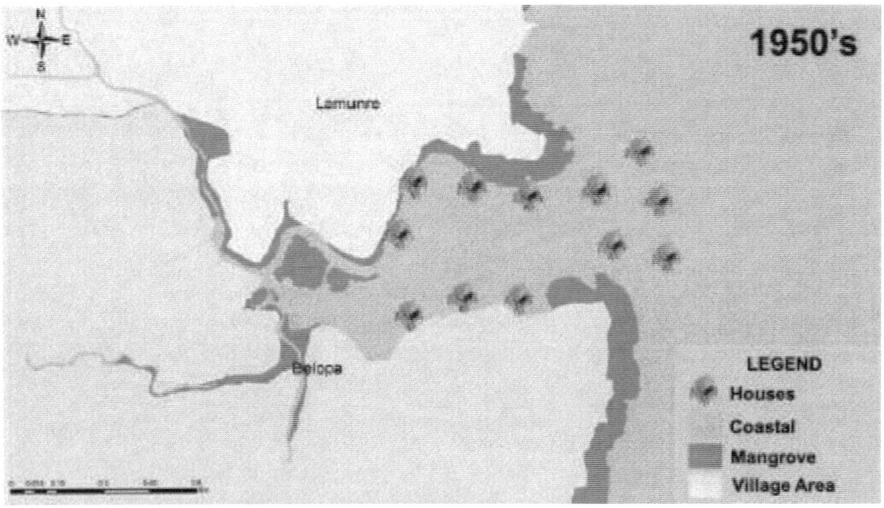

Figure 4. Illustration map of settlement conditions the 1950s

Next, the resident population tries to adjust. Some of them started to build or move their houses a little closer to the land or were right on the coast of Belopa Village. At this stage, settlements began to experience pattern changes, namely the existing houses partially linearly patterned by following the existing coastline, some still spread over the water. Houses that are linear follow the coastline oriented toward the sea, while others are unchanged in orientation (Figure 4).

As time goes by, the Bajo settlements grow, the existing houses increase. Some build closer to land and some more build still around the coast. But in this case, there are no more houses that are really above the sea. For the settlement pattern itself, it is still linear following the coastline and oriented towards the sea. Because the existing settlements continue to grow, the government finally starts to gradually build the facilities and infrastructure needed, the most basic example is building roads (Figures 5 and 6).

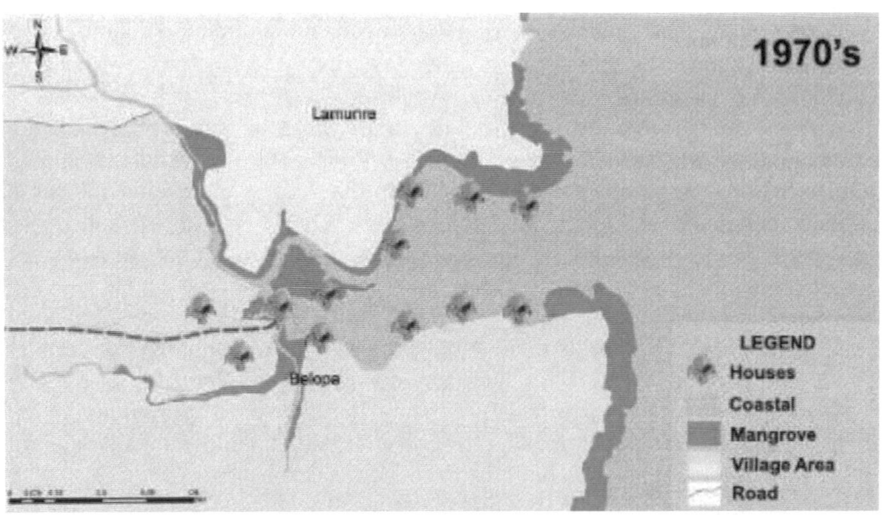

Figure 5. Illustation map of settlement conditions the 1970s

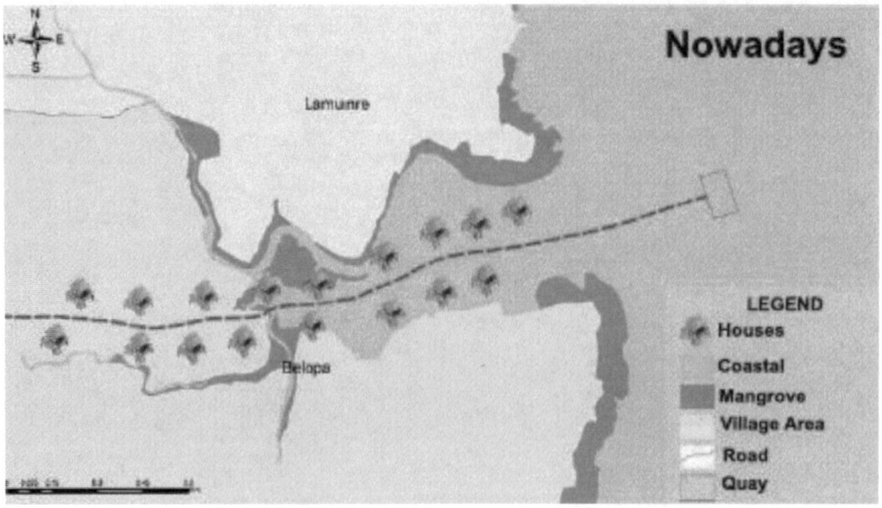

Figure 6. Illustration map of settlement conditions nowadaays

3.3. Local wisdom that forms the characteristics of the Bajo in Belopa Village

Local wisdom is local ideas that are wise, full of wisdom, good value, embedded and become tradition (steady) and followed by members of the community. The types of local wisdom included in the formation of spatial utilization include governance, value systems, procedures/procedures, and special provisions [1].

3.3.1. Governance. Governance consists of two parts, namely the social system and the social hierarchy structure. In each region there is generally a social system that regulates the social structure and linkages between existing community groups. The social hierarchy structure is also a part of governance, such as the authority of the customary leader in decision making, and the rules of sanctions and social fines for violators of certain customary laws and regulations.

The same thing with the other Bajo, the Bajo in Belopa Village also has the name of the Head of the tribe in its social system. The Chief of the tribe here himself has a similar role to the Village Head. The Head of the tribe presides over every customary program held, such as "*Macera Tasi*" or a sea party held every day or a customary event during a wedding.

For the structure of social hierarchy, descendants of the king or Bajo nobility are given the title "*Lolo Bajo*". Usually those who get the title of "*Lolo Bajo*" are native descendants where both parents are the original Bajo. For their role in society, they do not have a special position. What distinguishes them from ordinary people is when they hold a wedding. When "*Lolo Bajo*" and their descendants hold wedding events, they will put up a flag called "*Ulala*" as a sign that they are descendants of nobility.

3.3.2. Value system. The value system is a value developed by a traditional community that regulates the ethics of good and bad judgment. The values that develop in the daily life of the Bajo in Belopa Village are influenced by the values in Islamic. They highly uphold simplicity in everyday life. Starting from the place of residence, social interaction between people every day, transportation, livelihood, and others. There are no specific rules for ethics in this Bajo.

Their residence is built with simple shapes and materials. The social interaction they do everyday is to visit each other's homes in a friendly relationship, work together to do their work such as making boats or embroidering nets. They have a very strong kinship.

3.3.3. Procedure. Some customary rules in the region have provisions regarding the right time to go to sea and traditional calendar systems that can estimate the suitability of the season for various other fishing activities. Bajo is known as a tribe that depends most or even all of its life on the sea. Then it can be concluded that almost all Bajo people work as fishermen or the like. Bajo in Belopa Village also has community characteristics as explained. Almost all people have jobs as capture fishermen.

There are no specific rules that determine the right time to go to sea and a special traditional calendar system. Generally for fishermen who only have small boats will start their fishing activities at night and back at dawn or early morning. They usually go to sea on their own or are accompanied by family. Whereas for fishermen who go to sea using a much larger boat can spend two to four days in the sea and then return to land. They depart in a group consisting of five to ten fishermen.

3.3.4. Special provisions. Regarding the preservation and protection of sensitive areas related to forms of adaptation and mitigation of shelter against climate, disasters or other threats, traditional communities have also developed various forms of traditional house architecture that can provide protection and are friendly to the environment. For the Bajo people, the sea is a purified area. Because from there they come from, in the sea they live, and in the sea they also earn a living.

Table 1. Table of Local Wisdom of Bajo.

No	Type of Local Wisdom of	The Bajo Generally	Bajo in Belopa Village
1.	Govermance	- The Bajo noble descendants (*lolo Bajo*) and ordinary Bajo (*same*) people. - In the adat institution (*pamangku ade`*), the Bajo led by the chief (*punggawe*), the [2].	- Chief of the Tribe acts as the administrator of the adat. - *Lolo Bajo* is a noble title given to indigenous people of the Bajo.
2.	Value System	- Community Bajo is famous for its simple value system with a strong sense of togetherness. - The law or fine that applies in the Bajo tribe is that every Bajo citizen who makes a mistake there will be a penalty and sanctions that must be paid in the form of money depending on the amount paid to the traditional head or dukun [3].	- Values that develop in everyday life Bajo are influenced by values in Islam. - Highly upholds simplicity in daily life. Starting from the place of residence, social interaction between people every day, transportation, livelihood, and others. - There are no specific rules for ethics in this Bajo.
3.	Procedures	- Generally Bajo fishermen in Wuring are fishermen with one day fishing time. Climate change affects the certainty of fishing, changes in fishing loca-tions and reduced numbers of fishermen [4].	- There are no specific rules that determine the right time to go to sea and a special traditional calendar system. - Generally, for fishermen who only have small boats will start their fishing activities at night and back at dawn or early morning. - Whereas for fishermen who go to sea using a much larger boat can spend two to four days in the sea and then return to land.

4.	Special Provisions	- The traditional house of the Bajo is in the form of a stage with a pole support which vertically consists of three parts, namely the first *rakkeang/pammakkang*, located at the top. Here there is a ceiling where the roof is located, it also functions as a storage warehouse for rice as a symbol of the life/welfare of the owner. The second is the *ball / kale ale*, located in the middle [5].	- Bajo in Belopa Village has an unwritten rule in building houses. Their characteristics are houses made of wood or ordinary boards. Being on the water with supporting poles made of ordinary bamboo. Usually the pole is made higher than the stage house in general so that the sea water does not enter the house, especially when sea water is high. Some use coral as a home foundation.

In terms of special provisions, the Bajo in Belopa Village has an unwritten rule in building a house. Their characteristics are houses made of wood or ordinary boards. Being on the water with supporting poles made of ordinary bamboo. Usually the pole is made higher than the stage house in general so that the sea water does not enter the house, especially when sea water is high. Some use coral as a home foundation. The roof of the house is also only coated with ordinary zinc. It is known that simple stage houses are easier to build on water. But the building is more weathered.

3.4. Settlement pattern and orientation

Settlement Bajo in Belopa Village has three forms of pad pattern settlement. The first pattern is linear, growing following the existing path. Linear patterns are known to be the initial pattern in previous settlements. The second pattern is the cluster pattern, settlements grow clustering in one area. The third pattern is grid pattern. This grid pattern is the most recent pattern that arises in Bajo settlements in Belopa Village. The condition of the Bajo settlement pattern in Belopa Village is currently created because the community's land needs are increasing. The types of patterns that are formed are also created because of the adjustment to the geographical conditions that exist in building a house. The current pattern also does not have far differences with the existing initial settlement patterns.

Figure 7. Map of Bajo settlement patterns in Zone 1

In linear patterns, settlements grow along the road without any further land expansion. This is because the geographical conditions of the land are quite wet and some are still ponds or water areas. It will be easier for them to build a house alongside the road. If you build a house away from the road, you will be worried about difficulties in terms of access. Although the pattern is linear, the existing houses remain clustered and will not build away from other houses. This is because the local wisdom in the Bajo is that they live close together and group together (Figure 7).

IOP Conf. Series: Materials Science and Engineering **676** (2019) 012039 doi:10.1088/1757-899X/676/1/012039

For cluster patterns, this pattern arises in settlements whose growth direction is inland. The creation of clustering patterns that are clustered is due to the flatter geographical conditions of the land and no longer wetlands. This area is the land area that is closest to the water area there. With the aforementioned geographical conditions, houses have the opportunity to build further back. This pattern is also created because the need for land and this area provides an opportunity to build houses more easily, whether it is stage houses or non-stage houses. The negative of this settlement pattern is that settlements can grow spreading and uncontrolled. Which if it happens means that it has come out of the local wisdom system that the Bajo has there that is to stay close together and group together. Moreover, the road access created will be very irregular.

For the existing grid pattern is a new pattern in the Bajo Tribe settlement. The pattern that has the shape of the grid is a new settlement built by the government. For now, the number of buildings for the Bajo Tribe is 25 units (Figure 8).

Figure 8. Map of Bajo settlement patterns in Zone 2

The grid pattern in the Bajo settlement looks good, especially in the house setting and access. But in this new settlement, the bridge or road that was built was deemed to be inappropriate in its construction. Road access is known to have blocked access to every house by the sea. The boats of each resident can no longer enter and be parked next to their homes as usually because of roads or bridges built around existing houses.

3.5. Material and type of building

Characteristics of buildings The Bajo in the village of Belopa has similarities that are not so far from the Bajo in other places. The Bajo settlement in Belopa Village has simple building characteristics. In the building material, it is dominated by non-permanent material. For buildings with non-permanent material, usually the building is made of wood or boards with wooden or bamboo poles. For semi-permanent material, usually part of the building has changed to stone. Starting from the foundation of the house that has undergone pavement and continued with some of the board walls being replaced with stones.

For non-permanent building materials, more are found in areas closer to the mainland. People who change their houses to become non-permanent have a reason to make their homes more durable. However, this condition is contrary to the local criteria of the Bajo there, namely using a non-permanent house (Figure 9).

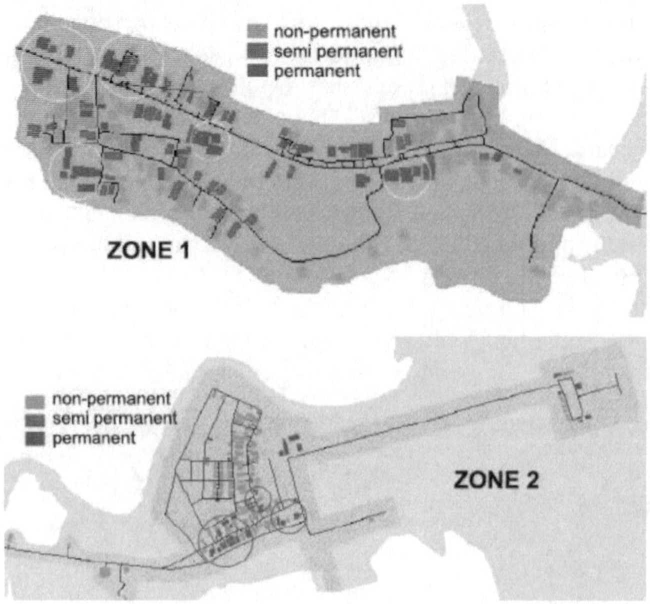

Figure 9. Map of Bajo settlement materials building

There are two types of buildings found in the Bajo in Belopa Village, stage and non-stage buildings. For building types, the stage is more dominating. This is because the type of stage house is the earliest type of house in the Bajo settlement in Belopa Village. The stage house has become part of the local wisdom of the Bajo people there. The house on stilts is considered more suitable to be used if the settlement is on water. The costs incurred are cheaper because the material is non-permanent. They too can use their house as a boat. But now the condition is no longer possible to park the boat under the house (Figure 10).

For fewer non-stage houses. Non-stage houses grow more inland. The community that changes the type of building becomes non-stage so the building is stronger.

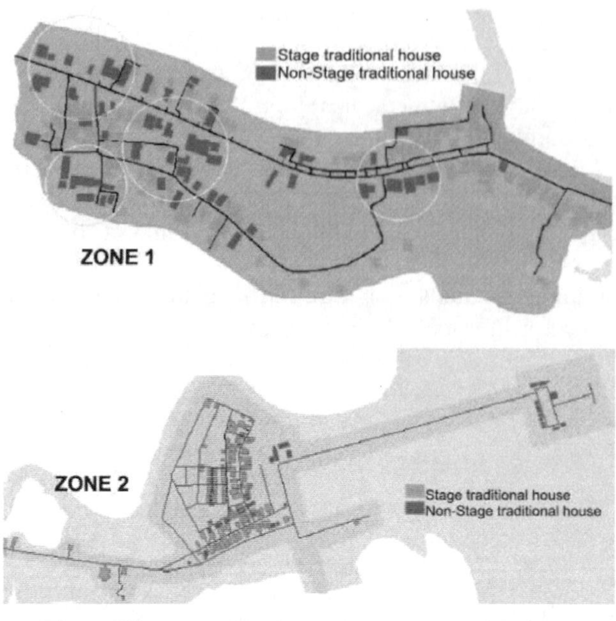

Figure 10. Map of Bajo settlement types building

Conditions that occur in materials and types of buildings that have changed to permanent materials and types of non-permanent buildings have resulted in the disappearance of local wisdom in the Bajo. From the analysis carried out, it is known that permanent settlements that have non-stage types will grow closer to the land and will further away from the sea.

3.6. The Bajo who live on land

The development of Bajo in Belopa Village has now undergone quite obvious changes. The development towards modernization has a negative impact as well as a positive impact on the community and its settlements.

In this case it is categorized as negative is the change in behavior and the order of life which is no longer in accordance with the local wisdom that the Bajo should have in Belopa Village. There is a desire from the Bajo people to make changes, because now they assume that they are no longer a society that is marginalized from other communities. The Bajo people are able to change their lives, which were traditional in nature and can now become more modern in accordance with the times. These conditions make their lives increasingly distant from the order of life in accordance with existing local wisdom. People began to build their houses with permanent materials and changed the shape of their houses to no longer form a stage. Whereas in reality the Bajo in Belopa Village has building characteristics that are appropriate to their local wisdom which is in the form of a stage with non-permanent material. The thing that is feared later is that if this continues to be allowed to continue without the arrangement that brings them to maintain local wisdom, then the true identity or value of the Bajo will be lost. Along with the times that are progressing more and more, people will continue to grow and change leaving the local criteria which they should keep in existence.

The conditions that occur at this time are one of their ways to adapt to the changing times that exist today. It does not close the possibility that in the future it will be more advanced now. This actually cannot be considered as violating the rules. Because the people themselves will continue to grow in accordance with the times. What can be done is to limit the space, which one can be accepted into the Bajo tribal settlement environment and which ones still need to be maintained so as not to lose their identity as a sea tribe.

4. Conclusion

From the discussion about the initial history, the initial form of the settlement and the form of local wisdom of the Bajo in Belopa Village, it was suggested that there were several points in the form of local wisdom that underlie the creation of Bajo settlements in Belopa Village. These points are as follows:

- The sea is designated as an area considered sacred to the Bajo people in Belopa Village. From generation to generation they have been taught to appreciate the sea because they believe that the sea they come from, they build settlements, and at sea they also earn a living.
- Bajo in Belopa Village prefer their settlements to be on the water as the initial form of their actual settlement. They prefer to stay in the water because the Bajo's main livelihood is as fishermen, so this also makes it easier for them to go to sea.
- Their settlements are above the water because their majority of livelihoods are as fishermen. They cannot live or settle too far from the sea
- The building characteristics of the Bajo people in Belopa Village are simpler. With non-permanent materialized buildings made of wood or bamboo shaped. The type of house is stage and stands tall on a foundation composed of marine corals.
- Bajo in Belopa village have a habit of living from time immemorial because they have very close kinship ties with each other. Although the pattern seen at the beginning of the formation of settlements is more linear, but when they build houses must remain close to the settlement of the tribe.

IOP Conf. Series: Materials Science and Engineering **676** (2019) 012039 doi:10.1088/1757-899X/676/1/012039

Reference

[1] Mukhtar "Mecula" and "Haroa Ano Laa": 2016, An Overview of Local Wisdom of North Buton Community in Land Use (*Suatu Tinjauan Kearifan Lokal Masyarakat Buton Utara dalam Pemanfaatan Lahan)*, Yogyakarta, Ed 1, Cet, 1.

[2] La Ode Taufik Nuryadin, Social Capital of the Bajo Community The Case Study of the Bajo Community on Baliara Island, Southeast Sulawesi Province, (*Kapital Sosial Komunitas Suku Bajo Studi Kasus Komunitas Suku Bajo Di Pulau Baliara Provinsi Sulawesi Tenggara*) University of Indonesia.

[3] Francois-Robert Zacot "BAJO PEOPLE Tribe of Sea Nomads" *("ORANG BAJO Suku Pengembara Laut"*) Forum Jakarta-Paris KPG (Popular Literature Gramedia) p. 101

[4] Nurlaili, Strategy for Adapting Bajo Fishermen to Climate Change: Bajo Fisheries Study in Sikka District, Flores, East Nusa Tenggara, Journal of Society & Culture, (*Strategi Adaptasi Nelayan Bajo Menghadapi Perubahan Iklim: Studi Nelayan Bajo Di Kabupaten Sikka, Flores, Nusa Tenggara Timur, Jurnal Masyarakat & Budaya*) Volume 14 No. 3 of 2012, Ministry of Maritime Affairs and Fisheries, Jakarta.

[5] I Made Krisna Adhi Dharma, Response of Bajo Traditional Houses to Tropical Climate, 2017 Iplbi Scientific Meeting (*Respon Rumah Tradisional Suku Bajo Terhadap Iklim Tropis, Temu Ilmiah Iplbi 2017)*, Halu Oleo University.

Development of a Shrouded Wind Turbine with Various Diffuser Type Structures

Y Klistafani[1]* and M I Mukhsen[1]

[1]Mechanical Engineering Department, State Polytechnic of Ujung Pandang, Makassar, 90425, Indonesia

*E-mail: yiyin_klistafani@poliupg.ac.id

Abstract. The Shrouded wind turbine is an innovative mean to increase power generated by wind turbine. By encompassing the rotor with a diffuser structure, it is possible to increase the wind velocity through the turbine up to 1.8 times of free stream velocity. Specifically, this study is numerical simulations that investigate the effect of wind velocity on various diffuser type structures to develop shroud wind turbine. Numerical study was conducted using Computational Fluid Dynamics (CFD) method. A reasonable agreement between the computed results, available experimental data and previous simulation is obtained. The results show that simulation results have good agreement to experimental results. However, Curved diffuser give the best increment of wind velocity (at centreline diffuser) than flat diffuser and others, which is 84.18%, with the maximum velocity was 9.21 m/s. Regarding the velocity contour results, vortex in downstream curved diffuser is bigger than others. Therefore, suction effect on curved diffuser is strongest and can generate positive impact on the wind velocity quality inside the diffuser.

1. Introduction

Alternative energy is very interesting to be studied more deeply considering the limited availability of fossil energy sources on earth. Wind power investigation is very important to address environmental issues. It is well known that wind speed fluctuates depend on natural weather. But if wind speed is very low, it will certainly be one of the inhibiting factors in the utilization of wind energy conversion technology. The power of the wind is proportional to the cubic power of the wind velocity approaching a wind turbine. This means that even small amount of its acceleration gives large increase of the energy generation [1]. Wind turbines usually operate for the rated wind speed of around 8-11 m/s [2, 3]. Therefore, wind turbine innovation is required in optimizing the utilization of wind, as have been done by many research groups that have tried to find a way to accelerate the wind velocity effectively [4-9]. Some interesting studies were reported by Yadav and Kumar [4], Kannan et al. [5], Igra [6], Gilbert and Foreman [7], and Lipian et al [8, 9]. Selection of the diffuser type structure as a good shroud for wind turbine was given by Ohya et al. [10] who employed three typical hollow structures. It was confirmed that the diffuser structure was the most effective for collecting and accelerating the wind than other nozzle and cylinder structures.

It is possible to model engineering problems using Computational Fluid Dynamics (CFD) approaches. The present simulations have been compared and validated against measurement data since then. The simulations range from the simplest 2D Reynolds-Averaged Navier-Stokes (RANS) approach. Wang et al. [11] investigated the dynamic stall behaviour of a pitching airfoil in comparison to the experimental data using the Shear-Stress-Transport (SST) k-ω [12] and Wilcox k-ω [13] turbulence model. They concluded that the Wilcox (or standard) k-ω model was too dissipative and could not deliver the prediction accurately. On the other hand, the SST model provided a better agreement against the experimental data. A good agreement of the CFD computations using the SST turbulence model was also obtained in several computations, for example Pape and Lecanu [14], Weihing et al. [15], Jost et al. [16], Klistafani [17, 18] and Klistafani et al. [19].These encourage the use of CFD for predicting the fluids engineering problems especially with the help of the Menter SST k-ω model.

Having considered the above background, the development of a wind power system with high output aims at determining how to collect wind velocity efficiently and what kind of diffuser design can generate energy effectively from the wind speed. The computed results were compared with corresponding experimental data [10] to show the fundamental capability of predicting complex turbulent flows.

2. Numerical Method

Numerical study is conducted using Computational Fluid Dynamics (CFD) method. The procedures performed on numerical research are pre-processing, solving, post-processing and validations. The first stage on numerical method is designing the diffuser type structure on two dimensional. The structures are illustrated in Figure 1, while detailed information about their dimension is given in Table 1 and Figure 2.

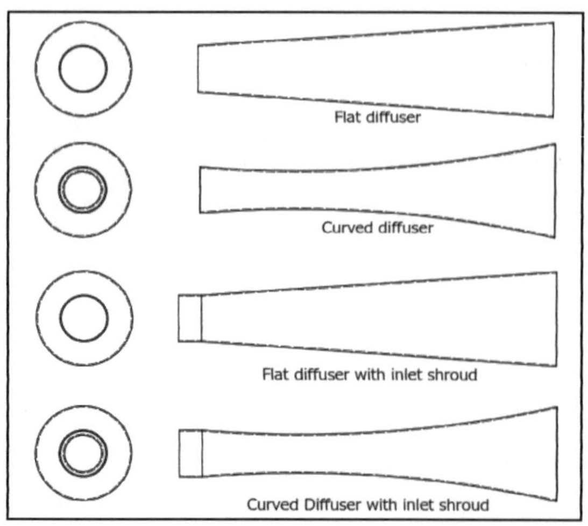

Figure 1. Diffuser structures

ANSYS Workbench 18.2 is software which used for modelling. The flat diffuser was generated according to the geometry specified in the experimental studies carried out by Ohya et al [10] for the length, diameter, and angle of divergence diffuser. Whereas design of curved shaped diffuser is adjusted flat diffuser dimensions that only interior modified into curved. We designed thickness of diffuser (12.5 mm) by considering recent studies [20] which used 10 layers of plates with each thickness of the plate is 1.25 mm.

The domain of the simulation is shown in Figure 3. The velocity inlet boundary condition was applied at inlet of the flow which is located five times inlet diameter of diffuser (D). The flow leaves the computational domain at 8.5D distance from the outlet of the diffuser with the outflow boundary condition. The side walls were set as a non-slip wall that are sufficiently far away from the area of interest to ensure the minimal effect on the flow characteristics near the diffuser. The computations were carried out using the commercial software Ansys Fluent 18.2. The flow was assumed to be steady, incompressibility and heat transfer effect was neglected. This is reasonable because wind turbines usually operate at a much smaller velocity than the speed of sound. An initial wind velocity of 5 m/s was prescribed at the velocity inlet plane as same as the velocity was employed in experimental [10]. The turbulence model is using the two-equation SST k-ω model. The pressure velocity coupling uses the SIMPLE method. All the variables were solved using the second order discretization. The computations were carried out for 10,000 iterations, otherwise convergence was achieved if the residual of the momentum reaches 1e-6.

Table 1. Diffuser type structure dimensions.

Specification	Diffuser Type Structure			
	Flat diffuser	Curved diffuser	Flat diffuser with inlet shroud	Curved diffuser with inlet shroud
Inlet diameter (*D*)	400 mm	400 mm	400mm	400mm
Diffuser length (*L*)	3080mm	3080mm	3080mm	3080mm
Diffusser thickness (*t*)	12.5 mm	12.5 mm	12.5 mm	12.5 mm
Inlet shroud length (*l*)	-	-	200mm	200mm
Diverging angle (α)	4°	4°	4°	4°
Circle radius of arch (*r*)	-	10^4mm	-	10^4mm

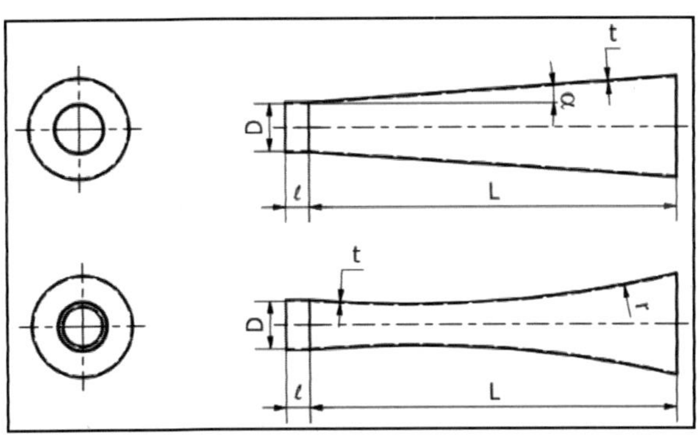

Figure 2. Detailed dimensions of diffusers with inlet shroud

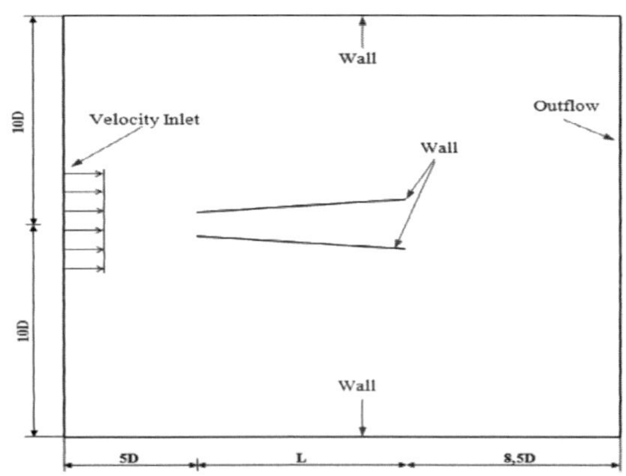

Figure 3. Computational domain and boundary conditions

The mesh was generated using ANSYS Workbench 18.2 software. Mesh parameters and controls are shown in Table 2. Grid independence studies were carried out in advance to ensure that the results are independent of the mesh resolution. The computational results are plotted in Figure 4 where the streamwise velocity ratios (u/U_∞) of flat diffuser at the centreline are compered with experimental results [10] and previous CFD Simulation conducted by Wibowo et al. [21]. It can be seen that simulation results has good agreement to experiment results with the error relative to the experiment is about 6.01%. The numerical method used greatly affects the computational results, despite using same software. This numerical simulation used second order and second order upwind for the methods and defined stricter convergency criteria than previous simulation [21]. Present simulation determine the bigger iteration limit, it is ten thousand iterations, than previous which only one thousand iterations. Therefore, those make present computations results closer to the experimental results than previous simulation.

Table 2. Mesh parameters and controls.

Mesh Parameters		Mesh Controls	
Size function	**Curvature**	**Method**	**Quadrilateral dominant**
Relevance center	Fine	Element order	Use global setting
Target skewness	0.900000	Free face mesh type	All quad
smoothing	High	Refinement	3
Inflation Option	Smooth transition	Number of divisions	90
Growth Rate	1.2	Size function	Uniform
Maximum layers	2	Bias factor	5.0

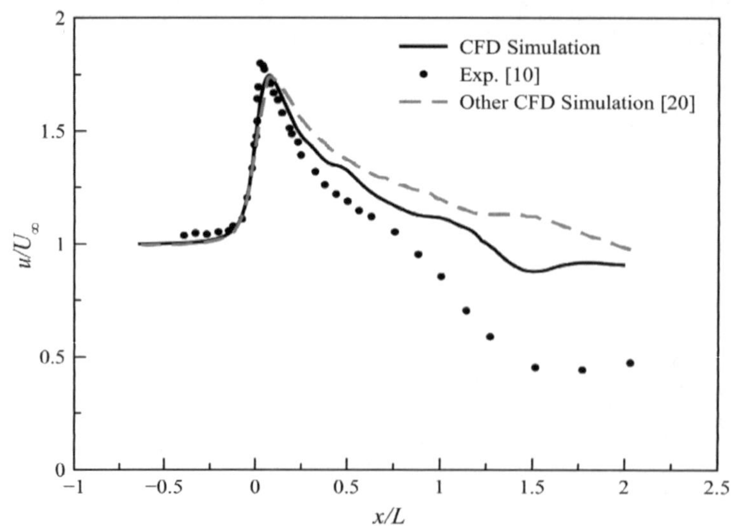

Figure 4. Streamwise flow velocity distribution for the flat diffuser.

Comparison of streamwise flow velocity distribution is intended to validate the results of numerical computations with experimental results. Wind speed increment data compared is data for flat diffuser, where the specification of the flat diffuser dimensions in the simulation refers to the diffuser specifications used by Ohya et al [10]. The highest velocity increase in experimental results reached 1,795 times of the freestream velocity (5 m/s), while in the numerical results the highest wind velocity increase reached 1,742 times. This indicates that the difference in numerical predictions with experiments in terms of the highest wind velocity increase about 2.95%.

3. Results and Discussion

The dimensionless streamwise velocity at centreline diffuser along the axial positions plots for all diffuser type structures are presented in Figure 5. Curved diffuser generates wind velocity increment highest than other diffuser type structures, which is up to 1.842 times (x/L = 0,285) of initial wind velocity set up (5 m/s). It is found that the inlet shroud has no significant impact on the augmentation of the wind velocity, but it just makes wind velocity increase occurs earlier. However, the highest wind speed between flat diffuser and curved diffuser occurs at different sections. The smallest cross-sectional area both of them is different, so that makes it happen. Flat diffuser has an inlet section as a smallest cross-sectional area, while the smallest cross-sectional area of curve diffuser has distance of several mm from the entrance section. It is noted, however, that the curved diffuser has a better performance than the flat diffuser, provided that location of the rotor is not at the near entrance but around $x/L \approx 0.36$. Detailed information regarding the comparison of the wind velocity at centreline of all diffuser type structures is shown in Table 3.

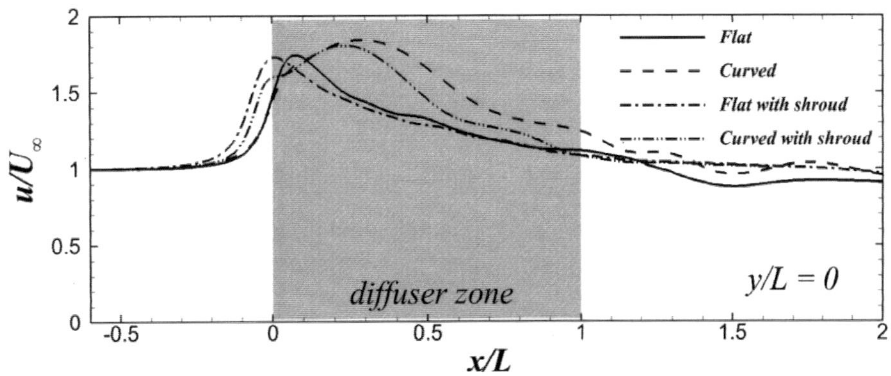

Figure 5. Streamwise flow velocity distribution at centreline diffuser for four diffuser type structures

Table 3. Wind velocity at centreline for all diffuser type structures

Value	Flat diffuser	Curved Diffuser	Flat diffuser with inlet shroud	Curved diffuser with inlet shroud
Wind velocity at inlet diffusser (*m/s*)	7.34	7.20	8.63	7.98
Maximum wind velocity (*m/s*)	8.71	9.21	8.64	9.02
Location of maximum wind velocity (x/L)	0.074	0.285	0.009	0.233
Increment	74.23%	84.18%	72.80%	80.39%

The velocity contour for all diffuser type structures is shown on Figure 6. As seen in Figure 6, the wind flows into the diffuser as it is inhaled. The vortex formed in the downstream area can give the effect of suction fluid into the diffuser. The greater the flow vorticity that occurs, the better the increase in wind velocity enters the diffuser. The flow vortices that occur in the downstream area for all four diffusers vary depending on the type of diffuser structure. Vortex size that formed in the downstremcurved diffuser area is greater than the flat diffuser's, therefore curved diffuser is better in increasing wind velocity.

As seen on Figure 6, inlet shroud that attached on flat diffuser instead the size of the vortex in the downstream area becomes smaller than flat diffuser without inlet shroud. The addition of inlet shroud makes the suction flow effect decrease. Therefore, as seen in Table 3, the increment of wind velocity generated by the flat diffuser is 1.43% larger than the flat diffuser with inlet shroud. The same thing happened to the curved interior diffuser, where the curved diffuser has better flow suction power compared to curved diffuser with inlet shroud which the difference of wind velocity increment about 3.97%.

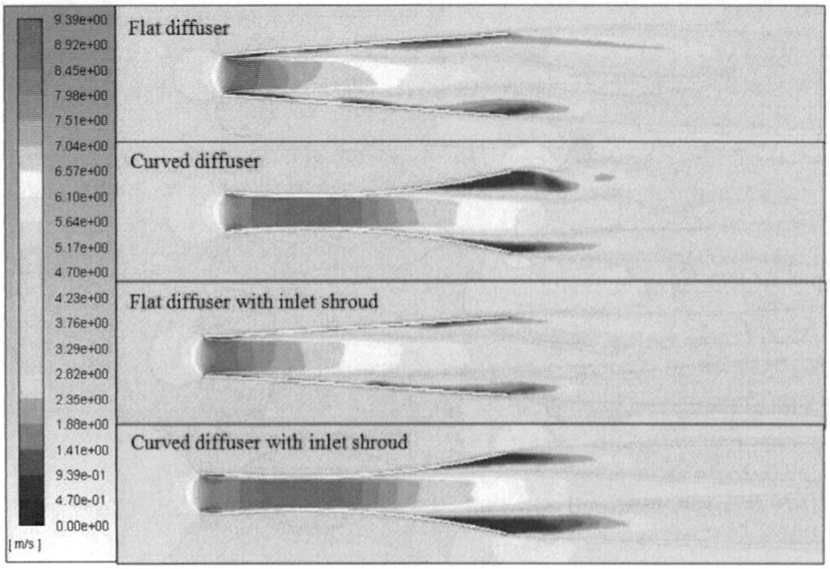

Figure 6. Velocity contour of all diffuser type structures (velocity magnitude, *m/s*)

4. Conclusions

Numerical simulations were carried out for flow fields around various diffusser type structures to develop shrouded wind turbine. The main conclusions derived from the study are as follows:

- Computational results (present study) are in good agreement with the corresponding experimental data. It has been confirmed from those validation that present computational procedure is very useful to investigate flow field of the diffuser type structure.
- The curved diffuser gave the best increment of wind velocity than flat diffuser and others, which was 84.18%, with the maximum velocity was 9.21 m/s.
- The vortex in downstream curved diffuser is bigger than others. Therefore, suction effect on curved diffuser is strongest and can generate positive impact on the wind velocity quality inside the diffuser.

Acknowledgments

The research leading to these results has received funding from the directorate of research and community service, directorate general of strengthening research and development, ministry of research, technology and higher education of Indonesia.

References

[1] Abe K and Ohya Y 2004 An investigation of flow fields around flanged diffusers using CFD *J. Wind Engineering and industrial aerodynamics* **92** 315.

[2] Bangga G, Hutomo G, Wiranegara R and Sasongko H 2017 Numerical study on a single bladed vertical axis wind turbine under dynamic stall *J. Mech. Science and Tech.* **31** 261.

[3] Bangga G, Lutz T, Jost E and Krämer E 2017 CFD studies on rotational augmentation at the inboard sections of a 10 MW wind turbine rotor *J. Renewable and Sustainable Energy* **9** 023304.

[4] Yadav A J and Kumar D 2017 Review of a shrouded wind turbine for low wind speeds International digital library of technologi & Research **1.**

[5] Kannan S A and Lau Y H K 2013 Design and flow velocity simulation of diffuser augmented wind turbine using CFD *J. Eng. Science and Tech.* **8** 372.

[6] Igra O 1981 Research and development for shrouded wind turbines *Energy Convers. Manage.* **21** 13.

[7] Gilbert B L and Foreman K M 1983 Experiments with a diffuser-augmented model wind turbine *J. Energy Resour.* **21** 13.
[8] Lipian M, Karczewski M and Olasek K 2015 Sensitivity study of diffuser angle and brim height parameters for the design of 3kW diffuser augmented wind turbine *Open Eng.* **5** 280.
[9] Lipian M, Karczewski M and Molinski J 2016 Numerical simulation methodologies for design and development of diffuser-augmented wind turbines – analysis and comparison *Open Eng.* **6** 235.
[10] Ohya Y, Karasudani T, Sakurai A, Abe K and Inoue M 2008 Development of a shrouded wind turbine with flanged diffuser *J. Wind Eng. and Industrial Aerodynamics* **96** 524.
[11] Wang S, Ingham D B, Ma L, Pourkashanian M and Tao Z 2010 Numerical investigations on dynamic stall of low reynolds number flow around oscillating airfoils *Computers & Fluids* **39** 1529.
[12] Menter F R 1994 Two-equation eddy-viscosity turbulence models for engineering applications *AIAA Journal* **32** 1598.
[13] Wilcox D C 1993 Turbulance modelling for CFD *DCW Industries.*
[14] Pape A L and Lecanu J 2004 3D Navier–stokes computations of a stall regulated wind turbine *Wind Energy* **7** 309.
[15] Weihing P, Letzgus J, Bangga G, Lutz T and Krämer E 2016 Hybrid RANS/LES capabilities of the flow solver FLOWer—Application to flow around wind turbines *Proc. Symposium on Hybrid RANS-LES Methods Springer Champ* 369.
[16] Jost E, Fischer A, Bangga G, Lutz T, and Krämer E 2017 An investigation of unsteady 3-D effects on trailing edge flaps *Wind Energy Science* **2** 241.
[17] Klistafani Y 2017 Studi numerik steady RANS aliran fluida di dalam asymmetric diffuser *J. INTEK* **4** 20.
[18] Klistafani Y 2018 Karakteristik aliran fluida di dalam asymmetric diffuser dengan penambahan vortex generator *J. INTEK* **5** 21.
[19] Klistafani Y, Mukhsen M I, and Bangga G 2018 Assessment of various diffuser structures to improve the power production of a wind turbine rotor *Technische Mechanik* **38** 256.
[20] Hu J and Wang W 2015 Upgrading a shrouded wind turbine with a self adaptive flanged diffuser *Energies* **8** 5319.
[21] Wibowo A T H, Wahyuono R A and Nugroho G 2013 Studi numerik pengaruh geometri dan desain *diffuser* untuk peningkatan kinerja DAWT (*diffuser augmented wind turbine*) *J. Teknik Mesin* **14** 90.

IOP Conf. Series: Materials Science and Engineering **676** (2019) 012041 doi:10.1088/1757-899X/676/1/012041

Optimization of Pouring Temperatures and Stirrer Speed Parameters on a Semi-Solid Slurry of ADC12 Al Alloy Prepared by Mechanical Stirring

S Rasyid[1,2*], I Renreng[1], H Arsyad[1] and M Syahid[1]

[1] Department of Mechanical Engineering, Faculty of Engineering, Hasanuddin University, Makassar, 90245, Indonesia
[2]Department of Mechanical Engineering, Politeknik Negeri Ujung Pandang, Makassar, 90245, Indonesia.

*E-mail: syaharuddinrasyid@poliupg.ac.id

Abstract. This paper presents an experimental design approach for the optimization process parameters for the aluminum solid alloy casting of ADC12 alloys. To achieve this goal, the pouring temperatures and stirring speed are selected and two levels of this parameter are considered. Design of expert (DOE) of tests was used for experimental design and analysis of results. The aluminum ADC12 slurry is stirred by a mechanical stirrer (round rods stirrer models) with a variation of speed 250, 300, 350 rpm for 20 seconds. Furthermore, the aluminum slurry of ADC12 is poured on a metal mold with a temperature of 620 °C, 600 °C, and 580 °C. The microstructure characteristics were examined by direct observation using optical microscopy, secondary α-Al phase dendrite arm spacing and shape factor were identified. The mechanical properties were investigated by the hardness test and tensile test. The mechanical properties and microstructure of aluminum alloys ADC12 made with semi-solid rheocasting casting technology using several parameters pouring temperature and stirring speed have been studied, and the results obtained can be synergized as follows. ADC12 aluminum alloy slurry preparation using a round rod mechanical stirrer on semi-solid casting can improve mechanical properties. Low pouring temperature and high stirring speed are reliable techniques for producing the right mechanical and microstructure properties for a semi-solid casting process (Rheo casting). The pouring temperature and stirring speed are recommended to produce optimal mechanical properties and the round microstructure is 580 °C and 350 rpm.

1. Introduction

One of the methods for fabricating a near net shaped castings is semi-solid metal processing (SSM). It takes advantage of the flow behavior of raw materials prepared in the mushy zone. Bilateral combinations like this show solid and slurry-like behavior. This means that as solid, the material maintains structural integrity and stand-alone characteristics in the form of temporary billets by applying shear strength, the material flows easily as a slurry and fills the mold cavity progressively. The advantages of this process are lower porosity and shrinkage voids, lower processing temperatures, less mold erosion, and higher mold life [1-4].

The final microstructure is very sensitive to superheat the melting and casting temperatures low in promoting spheroidal morphology of a phase-Al in the aluminum alloy prepared by the new processing rheocasting [5]. To get the semi-solid slurry in the formation of rheo, pouring temperature, and cooling rate of the molten alloy [6-8] are strictly controlled. The liquefaction is carried out in the solid-liquid phase area [9] to ensure the primary phase morphology is in the solid-liquid phase [10].

Aluminum alloy is one of the most popular alloys for the SSM process. This alloy has a wide solidification range, a relatively low melting point, superior formability, and better mechanical properties. The desired structure in all SSM processes is defined as a structure free of dendrites, spherical, minimum eutectic particles or without being trapped. Therefore, globular morphology is the most noteworthy problem [11].

Dendritic transformation into spheroidal morphology with a stirring mechanism has been explained by several previous researchers. The decrease in surface energy due to shear velocity causes dendrites to bend and break [12, 13]. The change in dendritic transformation becomes spheroidal morphology because of the stirring of dissolved convection [14]. A number of researchers have investigated the formation of semi-solid metals with mechanical stirring.

The effects of pouring temperature on the slurry manufactured by weak electromagnetic stirring were researched. The results indicate that it is feasible to manufacture the slurry with particle-like primary phases by low superheat pouring and weak electromagnetic stirring, and there is an important effect of the pouring temperature (superheat temperature) on the morphology and the size of primary 0-A1 in A356 Al alloy [15].

Rasyid et al. [16] have investigated the effects of pouring temperature on microstructure and mechanical properties of ADC12 Aluminum Alloys. The aluminum ADC12 slurry is stirred by a mechanical stirrer at 300 rpm for 60 seconds. Furthermore, the aluminum slurry of ADC12 is poured on a metal mold with a starting temperature of 580-680 °C. The pouring temperature had an effect on the change of mechanical properties and microstructure of aluminum alloy of ADC12.

Furthermore, Rasyid et.al. [17] have investigated the effects of stirring parameters on the rheocast microstructure and mechanical properties of Aluminum Alloy ADC12. A short period of stirring below the liquidus temperature to form a non-dendritic structure. The final morphology of the primary particles after a short period of stirring time has little impact as the stirring time increases. The optimal mechanical properties (hardness and tensile strength) were obtained at 20 seconds of stirring time and 300 rpm of stirring rate.

Design of Expert (DOE) has been used by several researchers to optimize various types of manufacturing processes [18-21]. In this study, a model has been developed to predict the optimization of cast temperature and stirring speed in the ADC12 aluminum alloy semi-solid casting process.

In this work by using a mechanical stirrer (round rods model) slurry maker constructed by the authors, several experiments have been conducted on the effects of pouring temperatures (580-620 °C) and stirring speed (250-350 rpm) with stirring times 20 second. For all these experiments, investigated the microstructure and mechanical properties of aluminum alloys ADC12 in the semi-solid casting.

2. Experimental Procedures

In this study, the secondary ADC12 alloy was used. The chemical composition of this alloy is shown in Table 1. The liquidus temperature of this alloy is 582 °C. The chemical composition of the aluminum alloy ADC12 is seen in Table 1.

Table 1. The composition of ADC12 aluminum alloys

ADC12 Aluminum Alloys	Weight %											
	Si	Cu	Fe	Mn	Mg	Zn	Ti	Cr	Ni	Pb	Sn	Al
	9.55	2.01	0.91	0.16	0.22	1.31	0.03	0.02	0.14	0.11	0.02	85.49

2.1. Preparation of SSR slurry

Metal molds are prepared and heated to a temperature of $\pm 300^{\circ C}$. Aluminum alloy material ADC12 is prepared (±300 gram). The alloy material of aluminum ADC12 is melted to 650 °C temperature using a gas furnace. Metal mold temperature measurements using infrared temperature gauges and aluminum molten measurements using a thermocouple gauge. In a temperature of 630 °C, 610 °C, and 590 °C, the aluminum alloy molten or slurry of ADC12 is stirred with a mechanical stirrer round rod stirrer models (Fig.1) of stirring speed of 250, 300, and 350 rpm. The liquid or aluminum alloy of ADC12 aluminum is poured into the metal mold. The specimens of casting result are made of the tensile specimen, hardness, and microstructure. In this study, the effect of pouring temperature and stirring speed on size, the main feature of ADC12 rheocast, and recognizing stirring requirements for significant changes in dendritic microstructure was examined.

Figure 1. Mechanical stirrer round rod model

2.2. Mechanical testing and microstructure analysis

The mechanical properties of the foundry are investigated experimentally, including the nature of hardness and tensile properties. Hardness is evaluated by a Brinnel hardness tester, where steel ball indenter is used at 613 N load for 5 s. Tensile properties are checked at room temperature using a universal screw driven type screw machine with a capacity of 100 kN. The test specimens were designed based on ASTM B557 (Fig 2). Characteristics of microstructures are examined by optical microscopy (MO). The grain size of the α-al phase and the size of the eutectic base phase Si were measured using image analysis.

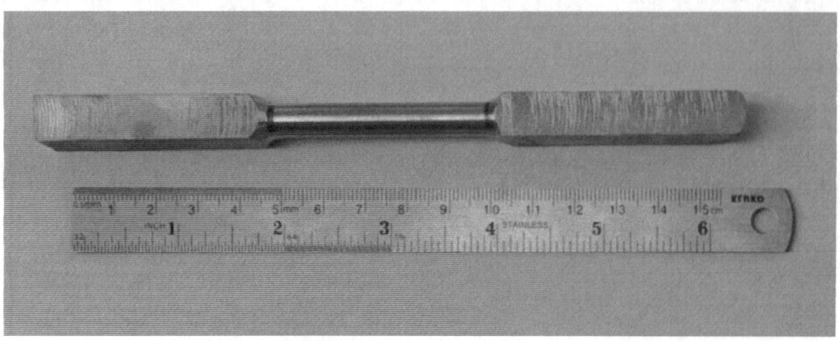

Figure 2. Tensile specimens (ASTM B557)

Figure 3 shows the surface microstructure of aluminum alloy products ADC12 stirred slurry aluminum by mechanical stirrer round rod model. Optical micrographs showing the effect of stirring speed rate on the microstructure in semi solid casting at pouring temperature 580 °C.

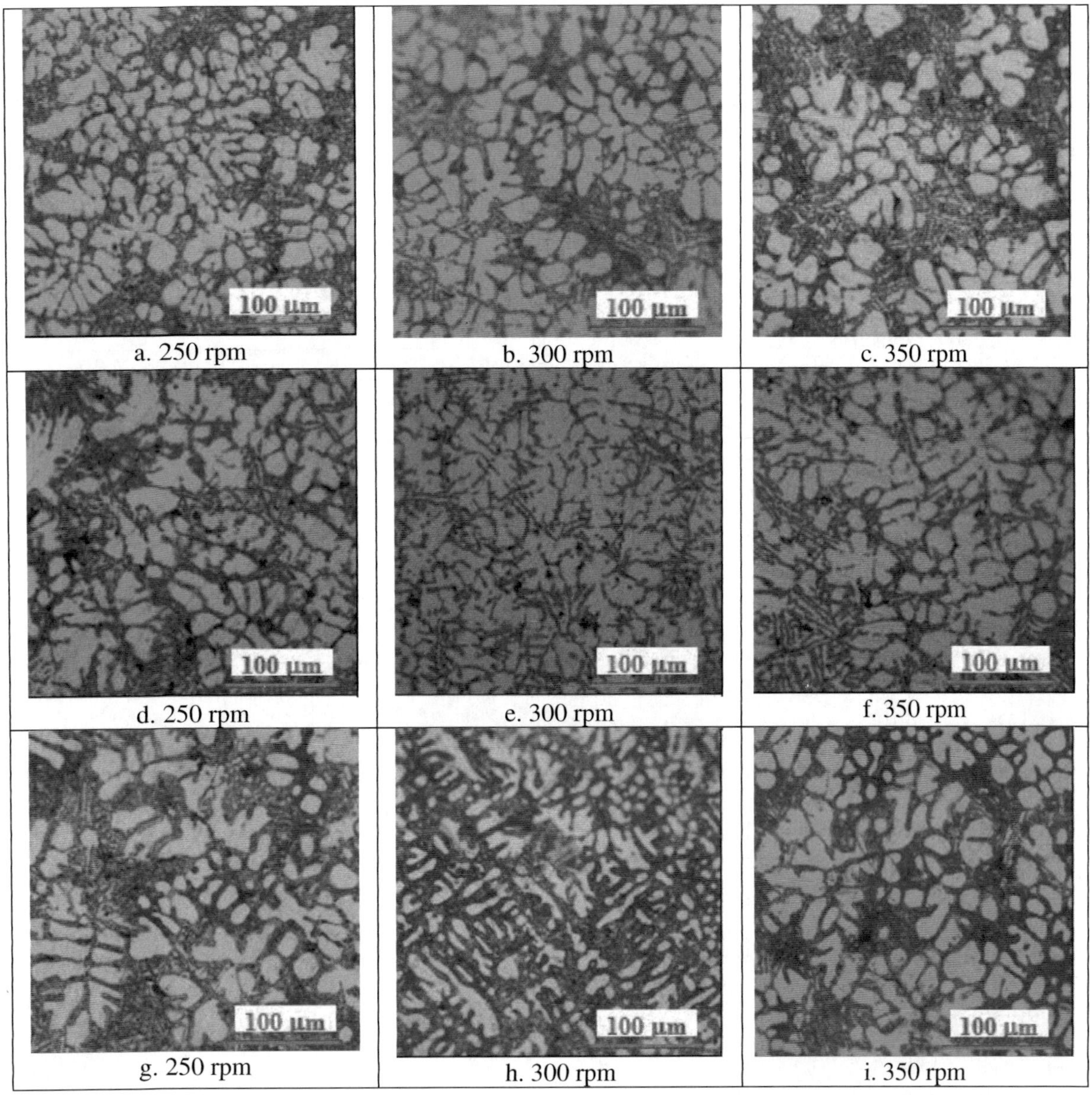

Figure 3. Optical micrographs showing the effect of stirring speed rate on the microstructure in semi solid casting at pouring temperature 580 °C (a, b, and c), 600 °C (d, e, and f), 620 °C (g, h, and i)

Image analysis techniques were used to investigate the microstructure of rheocast samples following standard metallographic procedures and average grain size was measured using a linear intercept method. The equation used to calculate the particle shape factor (SF) of the particle is [22]:

$$SF = \frac{4.\pi.A}{P^2} \quad (1)$$

Where P and A are the perimeter and area of individual particles respectively. A perfectly spherical particle would have a shape factor value of unity, and an infinitely long needle-like particle would have a shape factor equal to zero.

IOP Conf. Series: Materials Science and Engineering **676** (2019) 012041 doi:10.1088/1757-899X/676/1/012041

2.3. Experimental design and statistical analysis

To explore the effect of the operational factors on the response in the region of investigation, a DOE at two levels was performed. Pouring temperatures (C, A) and Stirring speed (rpm, B) were selected as independent factors. The range of values and coded levels of the factors are given in Table 2.

A polynomial equation (Eq. 2) was used to predict the response as a function of independent factors and their interactions. Interaction is the failure of the one factor to produce the same effect on the response at different levels of another factor [23]. In this work, there were four independent factors; therefore, the response for the quadratic polynomials becomes:

Table 2. Independent factors and their levels for DOE of semi solid casting process

Independent Factors	Unit	Level		
		-1	0	1
Pouring temperatures (A)	(°C)	580	600	620
Speed stirrer (B)	(rpm)	250	300	350

$$Y = \beta_0 + \sum \beta_i x_i + \sum \beta_i x_i^2 + \sum\sum \beta_{ii} x_i x_j \qquad (2)$$

where b_0, b_i, b_{ii}, b_{ij} are the constant, linear, square and interaction regression coefficient terms, respectively, and xi and xj are the independent factors (A and B). Design-Expert 6 software was used for multiple regression analysis, analysis of variance (ANOVA), and analysis of ridge maximum of data in the response surface regression (RSREG) procedure. The goodness of the model was evaluated by the coefficient of determination R^2 and its statistical significance was checked by the F-test.

3. Result and Discussion

This study demonstrates the effect of pouring temperature and stirring speed for the optimization of the semi solid casting route. The design is used to obtain 13 design points within the whole range of two factors for experiments. The designs and the response are given in Table 3. Following the experiments, the response surface is approximated by the DOE.

Table 3. Design layout and experimental results

Std	Pouring temperatures (rpm)	Stirring speed (rpm)	Coded		Hardness (HBN)	Tensile Strength (N/mm^2)	Grain Size (µm)	SF
			A	B				
1	580	250	-1	-1	80.4	237.8	31	0.61
2	600	250	0	-1	81.0	213.6	33	0.50
3	620	250	1	-1	80.8	205.5	45	0.59
4	580	300	-1	0	81.3	234.1	29	0.64
5	600	300	0	0	81.0	222.4	30	0.49
6	600	300	0	0	80.8	222.2	29	0.50
7	600	300	0	0	81.2	222.6	31	0.48
8	600	300	0	0	80.8	222.6	30	0.50
9	600	300	0	0	81.2	222.4	29	0.48
10	620	300	1	0	79.3	213.8	35	0.62
11	580	350	-1	1	82.4	220.5	22	0.72
12	600	350	0	1	81.5	226.7	24	0.66
13	620	350	1	1	80.8	225.0	31	0.67

3.1. Hardness

Results for hardness at pouring temperature and stirring speeds show that it fits the linear model. The ANOVA for the hardness data is given in Table 4. Having its Prob>F of much less than 0.01, the linear model is valid. As for the coefficients, the pouring temperature and stirring speeds were considered as a significant factor.

Table 4. ANOVA with CI = 95% for model and factors of the hardness

Source	Sum of Squares	DF	Mean Square	F Value	Prob > F	
Model	11.38	2	5.69	32.86	<0.0001	significant
A	4.34	1	4.34	25.04	0.0005	
B	7.04	1	7.04	40.68	<0.0001	
Residual	1.73	10	0.17			
Cor Total	13.10	12				

The obtained empirical equation of the hardness in the form of an actual factor is as stated in equation (3),

$$Hardness = 80.67 - 0.85A + 1.08B \quad (3)$$

where A is pouring temperature ($^{\circ}$C) and B is speed stirrer (rpm).

For convenience, the equation can be displayed as response surface contour as well as three-dimensional surfaces, as shown in Figure 4.

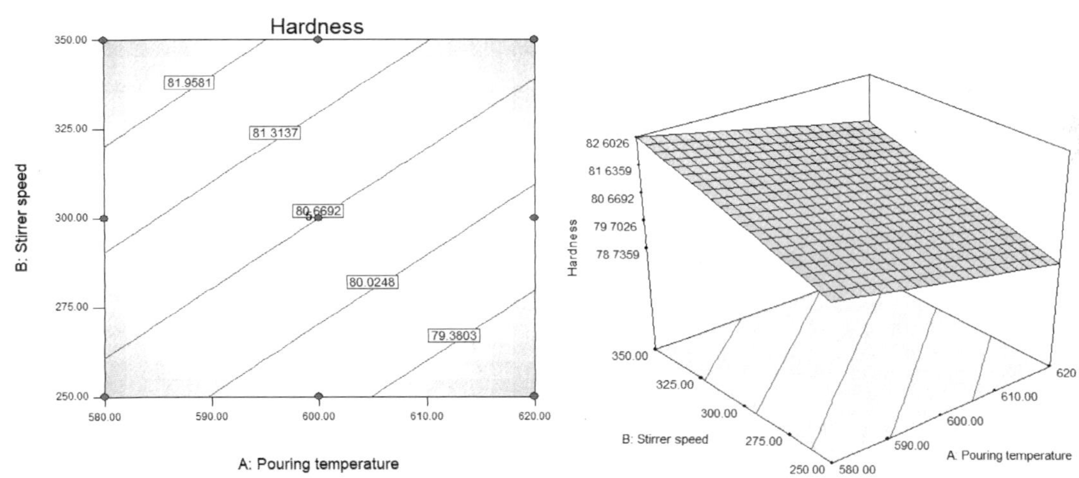

Figure 4. Response surface graph of (a) contours and (b) 3D surface for hardness

Figure 1 shows the hardness of casting results with variations of pouring temperature (580 $^{\circ}$C, 600 $^{\circ}$C, and 620 $^{\circ}$C) and stirring speed (250 rpm, 300 rpm, and 350 rpm). The higher the pouring temperature the hardness value decreases, but on the increase of stirrer speed, the hardness value increases. The highest hardness is 82.4 HBN at a pouring temperature of 580 $^{\circ}$C and a stirring speed of 350 rpm.

3.2. Tensile strength

Results for tensile strength at a various pouring temperature and stirring speeds show that it fits the quadratic model. The ANOVA for the tensile strength data is given in Table 5. Having its Prob>F of

IOP Conf. Series: Materials Science and Engineering **676** (2019) 012041 doi:10.1088/1757-899X/676/1/012041

much less than 0.01, the quadratic model is valid. As for the coefficients, both of the pouring temperature and stirring speeds was considered a significant factor. Tensile strength was insensitive to the change in pouring temperature and stirring speeds.

Table 5. ANOVA with CI = 95% for model and factors of the tensile strength

Source	Sum of Squares	DF	Mean Square	F Value	Prob > F	
Model	832.15	5	166.43	427.23	<0.0001	significant
A	556.81	1	556.81	1430.04	<0.0001	
B	244.48	1	244.48	627.90	<0.0001	
A^2	6.95	1	6.95	17.85	0.0039	
B^2	27.65	1	27.65	71.00	<0.0001	
AB	2.72	1	2.72	6.99	0.0332	
Residual	2.73	7	0.39			
Cor Total	834.85	12				

The obtained empirical equation of tensile strength in the form of an actual factor is as stated in equation (4),

$$Tensile\ Strength = 222.39 - 9.63A + 6.39B + 1.59A^2 - 3.16B^2 - 0.82AB \quad (4)$$

Where A is pouring temperature (°C) and B is speed stirrer (rpm).

For convenience, the equation can be displayed as response surface contour as well as three-dimensional surfaces, as shown in Figure 5.

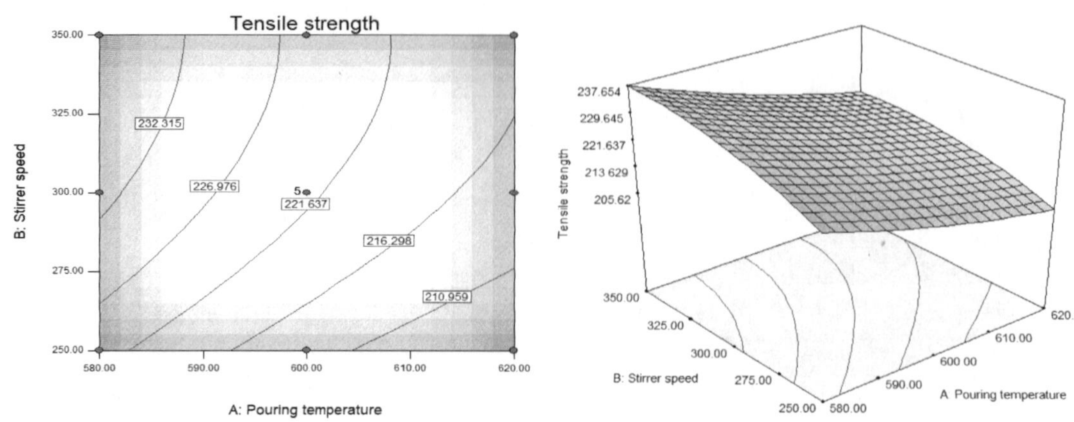

Figure 5. Response surface graph of (a) contours and (b) 3D surface for tensile strength

Figure 5 shows the tensile stress values of casting results with pouring temperature (580 °C, 600 °C, and 620 °C) and stirring speed (250 rpm, 300 rpm, and 350 rpm). The tensile stress of aluminum alloy ADC12 increases with increasing stirrer speed. However, the tensile stress of aluminum alloy ADC12 decreases with increasing pouring temperature. The maximum tensile stress is 237.8 N / mm2 at a pouring temperature of 580 °C and a stirring speed of 350 rpm.

3.3. Grain size

Results for grain size at a various pouring temperature and stirring speeds show that it fits the quadratic model. The ANOVA for the grain size data is given in Table 6. Having its Prob>F of much less than 0.01, the quadratic model is valid. As for the coefficients, both of the stirring speeds and stirring time was considered a significant factor. Grain sizes were insensitive to the change in pouring temperature and stirring speeds.

Table 6. ANOVA with CI = 95% for model and factors of grain size

Source	Sum of Squares	DF	Mean Square	F Value	Prob > F	
Model	347.16	5	69.43	29.40	0.0001	significant
A	121.50	1	121.50	51.45	0.0002	
B	192.60	1	192.60	81.58	<0.0001	
A^2	28.10	1	28.10	11.90	0.0107	
B^2	0.27	1	0.27	0.11	0.7470	
AB	2.25	1	2.25	0.95	0.3615	
Residual	16.53	7	2.36			
Cor Total	363.69	12				

The obtained empirical equation of grain size in the form of an actual factor is as stated in equation (5),

$$Grand\ Size = 29.52 - 4.5A - 5.67B + 3.19A^2 - 0.31B^2 - 0.75AB \qquad (5)$$

Where A is pouring temperature (°C) and B is speed stirrer (rpm).

For convenience, the equation can be displayed as response surface contour as well as three-dimensional surfaces, as shown in Figure 6.

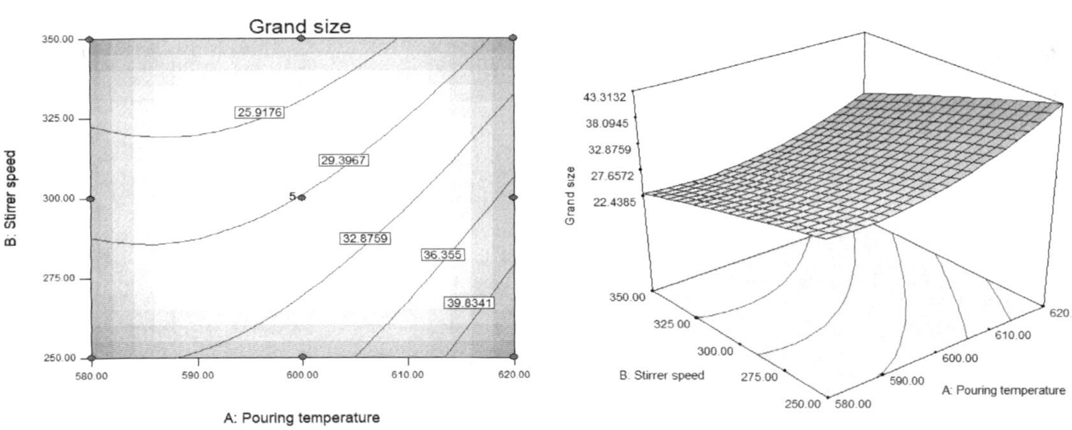

Figure 6. Response surface graph of (a) contours and (b) 3D surface for grain size

Figure 6 shows the grade of grain size with variations of pouring temperature (580 °C, 600 °C, and 620 °C) and stirring speeds (250 rpm, 300 rpm, and 350 rpm). The grain size of ADC12 aluminum alloy decreases with increasing stirrer speed, the grain size of aluminum alloy ADC12 increases with increasing pouring temperature. The minimum grain size is 20 μm at a pouring temperature of 580 °C and a stirring speed of 350 rpm. The grain size of ADC 12 aluminum alloy strengthens the results of mechanical properties testing. The mechanical properties of the material are inversely proportional to grain size.

3.4. Shape Factor (SF)

Results for shape factor at a various pouring temperature and stirring speeds show that it fits the quadratic model. The ANOVA for the shape factor data is given in Table 7. Having its Prob>F of much less than 0.01, the quadratic model is valid. As for the coefficients, both of the temperature and stirring speeds was considered a significant factor. Shape factor was insensitive to the change in temperature pouring and stirring speeds.

Table 7. ANOVA with CI = 95% for model and factors of shape factor

Source	Sum of Squares	DF	Mean Square	F Value	Prob > F	
Model	0.083	5	0.017	21.20	0.0004	significant
A	1.350E-003	1	1.350E-003	1.72	0.2310	
B	0.022	1	0.022	27.53	0.0012	
A^2	28.10	1	28.10	40.61	0.0004	
B^2	7.588E-003	1	7.588E-003	9.67	0.0171	
AB	2.50E-004	1	2.50E-004	0.29	0.608	
Residual	5.492E-003	7	7.846E-004			
Cor Total	0.089	12				

The obtained empirical equation of shape factor in the form of an actual factor is as stated in equation (6),

$$SF = 0.50 - 0.015A - 0.060B + 0.11A^2 + 0.052B^2 - 7.500E - 003AB \quad (6)$$

here A is pouring temperature (°C) and B is speed stirrer (rpm).

For convenience, the equation can be displayed as response surface contour as well as three-dimensional surfaces, as shown in Figure 7.

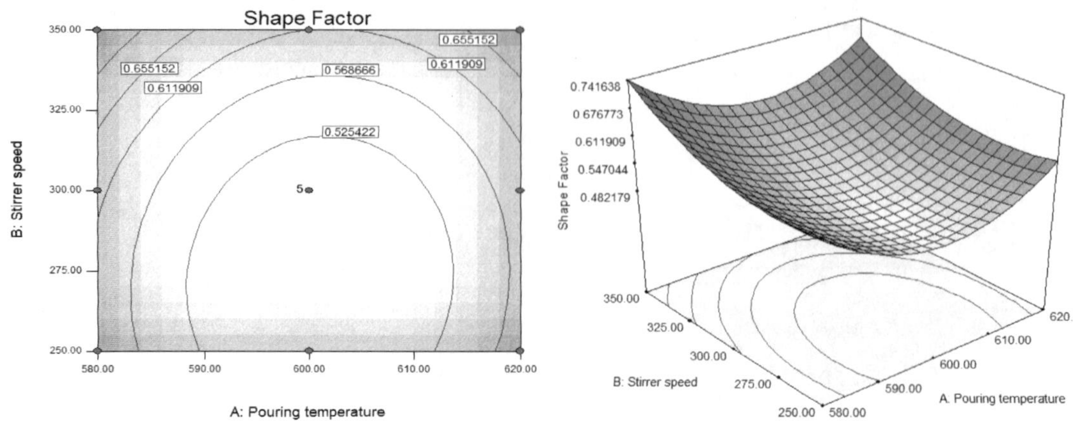

Figure 7. Response surface graph of (a) contours and (b) 3D surface for shape factor

Figure 7 shows the grade of grain size with variations of pouring temperature (580 °C, 600 °C, and 620 °C) and stirring speeds (250 rpm, 300 rpm, and 350 rpm). The shape factor aluminum alloy ADC12 increases with increasing stirrer speed, whereas Shape factor aluminum alloy ADC12 decreases with increasing pouring temperature. The maximum shape factor is 0.72 at a pouring temperature of 580 °C and a stirring speed of 350 rpm. Roundness or form factor is inversely proportional to grain size. The smaller the grain size, the greater the form factor and the increasing mechanical properties.

3.5. Optimization

Now that the empirical model for all casting responses as a function of pouring temperature and stirring speed has been obtained, the selection of the optimal casting parameter setting can be performed. One can adjust the expected range of each casting response and the range of pouring temperature and stirring speed in line with expectations for all foundry responses can be determined. For example, that in order to obtain optimal mechanical properties, minimum grain size, and optimum shape factor, the stirring parameters should be carried out at 350-325 rpm rotation range and 580-590 °C pouring temperature. To achieve this criterion, the pouring temperature range and the speed stirring must be within the yellow plot of the overlay (Figure 8) of all casting responses.

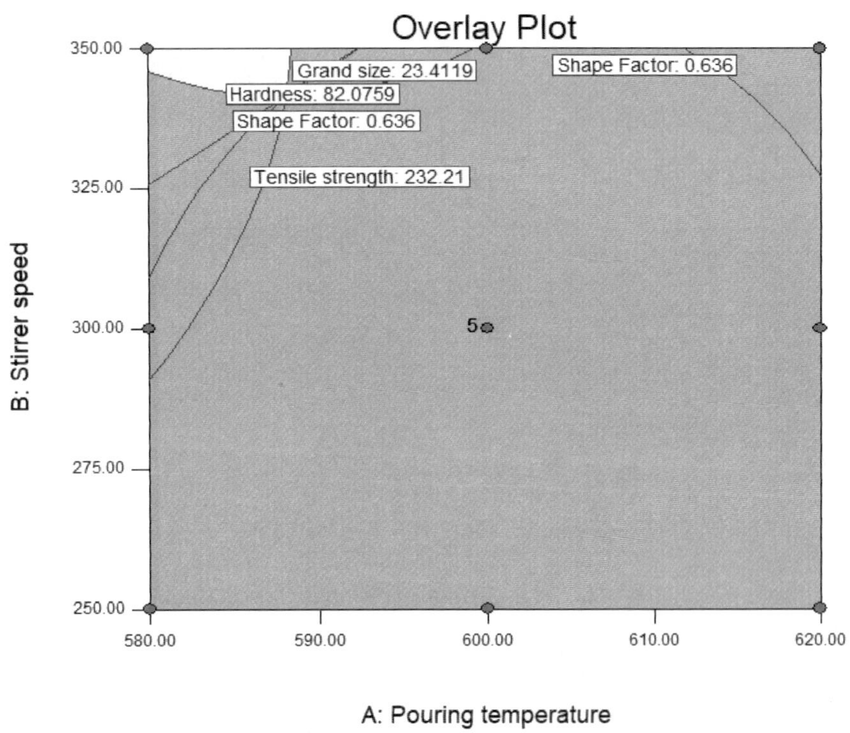

Figure 8. Overlay plot of the input factors for the predetermined response criteria of a minimum of 82.0759 HBN hardness, 232.21 N/mm^2 tensile strength, 0.636 shape factor, and 23.119 μm grin size

4. Conclusion

The mechanical properties and microstructure of aluminum alloys ADC12 made with semi-solid rheocasting casting technology using several parameters pouring temperature and stirring speed have been studied, and the results obtained can be synergized as follows. ADC12 aluminum alloy slurry preparation using a round rod mechanical stirrer on semi-solid casting can improve mechanical properties. Low pouring temperature and high stirring speed are reliable techniques for producing the right mechanical and microstructure properties for a semi-solid casting process (Rheo casting). The pouring temperature and stirring speed are recommended to produce optimal mechanical properties and the round microstructure is 580 °C and 350 rpm.

Acknowledgments
This article can be completed in cooperation between the lead author and lecturer of the Department of Mechanical Engineering, Hasanuddin University. Thanks to the Department of Mechanical Engineering, Ujung Pandang State Polytechnic, and Geology Department of Hasanuddin University for the provision of research facilities.

References
[1] Flemings M C 1991 Met. Trans. A 22A 952–981.
[2] Kirkwood D H 1994 Int. Mater. Rev. 39 173–189.
[3] Fan Z, 2002 Int. Mater. Rev. 47 49–85.
[4] Nafisi S, Emadi D, Shehata M T, Ghomashchi R 2006 Mate. Scie. and Engi. A 432 71–83.
[5] Cardoso E, Atkinson H V, Jones H 2004 Proc of the 8" Int Conf on Sem. Sol. Proc. of Allo. and Compo. 296-307.
[6] Masao K, Osamu N 2004 J. Appl. of Nano Cast Metho. 34(11): 21-26.
[7] Pan Y, Aoyama S, Liu C 1997 Proc. of the 5" Asia. Found. Cong. 443-451.
[8] Vieira EA, Junior B A, Ferrante M. 2004 Proc. of the 8" Int Conf on Semi-solid Proc. of Allo. and Compo. 243-251.
[9] Zhao J X, Zhu M F, Kim J M 2004 J. PTCA (Part A: Phys. Test. 40(9): 433-438.
[10] Wang P, Gui M L U, Zhong J C 2004 J.Mate. Revie. 6(5): 72-74.
[11] Nafisi S, Emadi D, Shehata M T, Ghomashchi R 2006 Mate. Scie. and Engi. A 432 71–83.
[12] Vogel, Zhang K 2003 J. of Mate. Proce. Tech. 137, p. 195-200.
[13] Vogel A, Doherty R D, Cantor B 1979 Proce. of Inter. Confe. on Solid. Met. Soc. 518.
[14] Mullis A M 1999 J. Acta Mater, 47:1783.
[15] Zheng L, Wei MM, Zheng Z 2006 Trans. Nonf. Met. Soc. 16 71-76.
[16] Rasyid S, Arif E, Arsyad H, and Syahid M 2018 J. of Eng. and Appl. Scie. 13, p. 2032-2037, (2018).
[17] Rasyid S, Arif E, Arsyad H, and Syahid M 2018 Mat. Web of Conf. 197, 12004.
[18] Khosravi H, Eslami F R, Askari PM 2014 T. Nonferr. Metal. Soc. 24 961-968.
[19] Khosravi H, Eslami F R, Askari PM 2017 J. of Part. Scie. and Tech. 3 145-154.
[20] Sultan A Z, Sharif S and Kurniawan D 2015 The 4th Inter. Confe. on Adva. Manu. Tech. (ICAMT).
[21] Qing Z C, Simard A 2010 Adva. in Light Weight Mate.– Alum. Cast. Mate. and Mag. Tech. SP-2294.
[22] Lee HH and Chen Y PP 2014 Patt. Recog. Lett. vol.49, pp. 155-161.
[23] Montgomery D.C. 1997 Desi. and Anal. of the Expe.

IOP Conf. Series: Materials Science and Engineering **676** (2019) 012042 doi:10.1088/1757-899X/676/1/012042

Effects of Work on Shock Absorber and Spiral Springs Against Vertical Loads of Vehicles Burdening the Road Structure

S Ka'ka[1]*, S Himran[2], I Renreng[2] and O Sutresman[2]

[1]Mechanical Engineering Department, State Polytechnic of Ujung Pandang, South Sulawesi, Indonesia
[2]Mechanical Engineering Department, Engineering Faculty, Hasanuddin University, South Sulawesi, Indonesia

E-mail: *simon_kaka@poliupg.ac.id

Abstract. This research aims to examine the effect of vertical dynamic load of vehicles and changes in dimensional barriers on the road surface in its path. Experimentally this fluctuating load is replaced by a pneumatic force change based on the regulation of air pressure on the regulator. The deviations generated by the varying load work are measured by placing a proximity sensor along the spring movement. The amount of vertical load transformation reaches the road surface is measured by using Load cell. Characteristics of vertical dynamic vibration occurring due to several dimensional barriers, U (cm) obtained using mathematical modelling method with 2 DOF suspension system transfer function. The results showed a condition on the body and wheels of vehicles experienced a brief overshot for 0.14 seconds with deviation of 0.178 m. From the graph shows that the rate of deviation that occurs is large enough that Y2d = 1.03 m / s caused by a sudden shock that occurred on the wheels of the vehicle. This condition does not last long that is only duration t = 0.22 s, because the spring reaction force and shock absorber can absorb 45% vibration against the sprung and un-sprung vertical load of the vehicle.

1. Introduction

Increasing the volume of vehicles, especially the four-wheeled vehicles that cross the road will further increase the load, while the road capacity to support the recurrent load of vehicles decreases with increasing age of use. Vibrations originating from vertical dynamic loads of vehicles often fluctuate due to unstable drivers and passengers. Such conditions will further weaken the ability of the road structure to accept the fluctuating load.

The spring and shock absorber mounted on each of the wheels of the vehicle is expected to be able to overcome and reduce the vertical dynamic load of vehicles overloading the road structure. The vehicle suspension system according to [1, 2] is composed of a spring and a shock absorber arranged in parallel. The main function of the suspension system is to support the weight of the vehicle, to provide comfort for the rider, to keep traction of the wheel on the road surface condition, and to maintain the alignment of the front wheel and rear wheel [2-4]. If the suspense spring is very rigid, then the shock absorber will not efficiently absorb the shock that comes from the form of road surface resistance [4].

IOP Conf. Series: Materials Science and Engineering **676** (2019) 012042 doi:10.1088/1757-899X/676/1/012042

The shape and mechanism of the suspension system for quarter vehicles is shown in Figure-1 (a), with sprung mass (m_2), un-sprung mass (m_1), suspension spring (k_2), shock absorbers (c) and tire elastic constants on wheels (k_1), [5, 6].

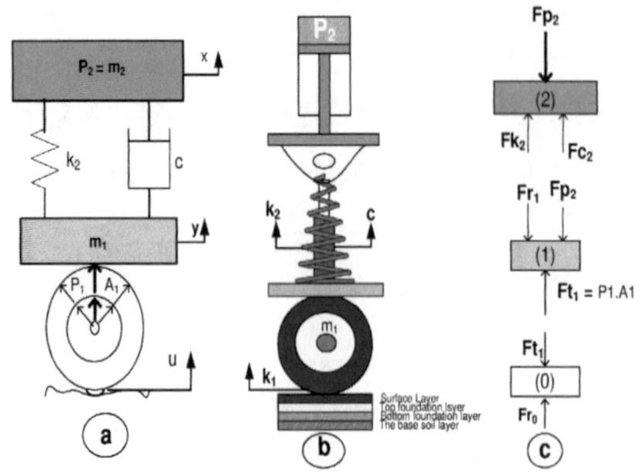

Figure 1. Pneumatic system experimental loading mechanism

The magnitude of the pneumatic cylinder thrust force generated by the air pressure P_2 (bar) according to [5, 7] is formulated as;

$$F_k = \frac{\pi}{4} D^2 P_2 \left(N\right) \tag{1}$$

If the dimensions of the cylinder, D = 100 mm (0.100 m) are substituted to equation (1) then the compression force of the suspension mechanism is obtained:

$$F_k = 785 \times P_2 \text{ and } F_{p2} = 706.5 P_2 = 707 P_2 \left(N\right) \tag{2}$$

Based on the free-body diagram of Figure 1(c), the equation of the wheel loading the road structure is:

$$\left.\begin{aligned} F_{t1} &= F_{p2} + F_{r1} \\ F_{t1} &= \left(k_2 + c\right) x_2 + \left(m_1 \cdot g\right) \end{aligned}\right\} \left(N\right) \tag{3}$$

If in equation (3), the spring constant k_2 (N / mm), the coefficient of damping c (Ns / mm), the weight of the wheel m_1 (kg) and the gravitational acceleration g (m / s^2) are obtained:

$$F_{t1} = F_{ro} = 707 P_2 + \left(m_1 \cdot g\right)\left(N\right) \tag{4}$$

The mass sprung of m_2 supported by the spring and the shock absorber with the spring stiffness of k_2 and the damping coefficient c will overload the vehicle wheel as the mass unsprung of m_1 further gives the action force to the contour of the road surface.

The reaction force of the road contour to the tire will be distributed in the direction of u through the tire elasticity constant of k_1 Thus the transfer function of a quarter vehicle suspension system for a front wheel with two degrees of freedom (2DOF) is obtained by comparing the output and input as follows:

$$\frac{Y(s)}{U(s)} = \frac{k_1\left(cs + k_2\right)}{m_1 m_2 s_4 + \left(m_1 + m_2\right) cs^3 + \left[k_1 m_2 + \left(m_1 + m_2\right) k_2\right] s2 + k_1 cs + k_1 k_2} \tag{5}$$

Calculation of spring constant value, k_2 by [8, 9] respectively in the book "Machinery's Handbook, 29th Edition and the journal Mechanism and Machine Theory are formulated as follows:

$$k_2 = \frac{Gd^4}{8n_a D^3} \tag{6}$$

G is the modulus of stiffness (N / cm^2), d is the diameter of the wire (cm), n_a is the number of active coils and D is the mean coil diameter (cm) obtained from the difference between the outer diameter of the coil and the diameter of the wire.

The number of active coils, n_a is usually less than the total coil, n whereas the spring compress area is between the minimum compression of 20% and the maximum compression of 80%. If the number of active coils, $n_a = 80\% \times n$ and the mean coil diameter, $D = D_0 - d$ then the value of the spring constant, k_2 can be obtained by using equation (6). The energy analysis and efficiency of shock absorber can produced by the suspension work on the wheels of the vehicle uses a linear dynamic equation [10, 11].

2. Research Method and Material

2.1. Pneumatic cylinder and direction control valve

Experimental tests conducted on the mechanical work of suspension on the wheels of the vehicle is by adjusting the working pressure P_2 (bar) on the regulator ranging from 1 to 8 bar [5]. Furthermore, the loading of the spring and the shock absorber by the pneumatic actuator begins to occur when the 5/2 directional control valve is operated.

(a)

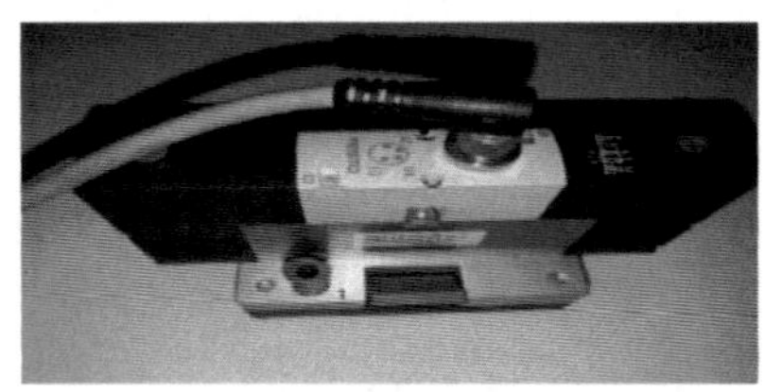

(b)

Figure 2. (a) Pneumatic actuator (b) Memory valve with 5/2-way

The dimensions of the Pneumatic actuator and the 5/2 directional control valve used at the working pressure, P = 6 (bar) are as follows; The type of pneumatic cylinder selected is, DNU-100-300-PVA and sourced from [12] with specifications; Diameter of piston D = 100 mm, stroke L = 300 mm, Weight piston W = 3.864 kg = 38.64 N, and Weight piston rod / 10 mm w = 0.090 kg . The forward and reverse thrust force of cylinder piston at a working pressure of 6 bar are respectively F_k = 4496 N and F_m = 4221 N, piston weight, W = 3.864 kg = 38.64 N Weight / 10 mm and piston rod, w = 0.090 kg / 10mm.

2.2. Experimental set up

The change in the position of compression motion on the spring and the shock absorber is measured by placing the proximity sensor along the spring movement. The amount of deviation detected by the sensor can be transferred to the LCD in digital form. It should be noted that the change in the size of the readings on the LCD screen is correlated with the regulatory pressure settings. By placing a Load cell gauge just below the wheel of the vehicle, the measurement of the vertical dynamic load of the vehicle successfully transformed to the road surface can be obtained through a direct reading on the LCD screen.

Optimization of physical models, using pneumatic cylinders as dynamic vertical drive test simulators on wheel drive suspension work has been discussed by [5, 13]. Mathematical model optimization is done by making the weight data of wheel axis, m_1 (kg) and body weight, m_2 (kg) as

IOP Conf. Series: Materials Science and Engineering **676** (2019) 012042 doi:10.1088/1757-899X/676/1/012042

INPUT as shown in Figure 3 (b). Similarly, the constant values of k_1, k_2, and c_2 are the input data to create the program in Mat Lab.

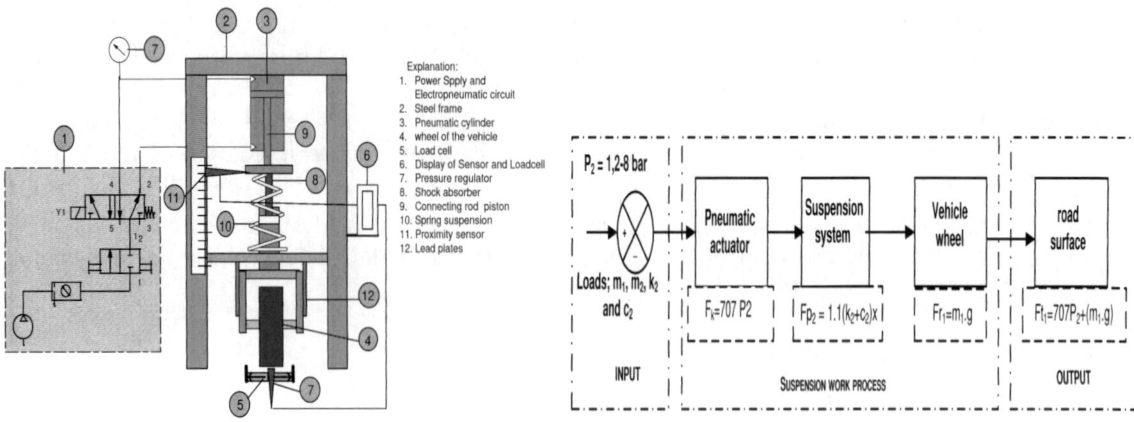

(a) (b)

Figure 3. (a) Loading experiments with pneumatic actuators (b) Block diagram of vehicle suspension loading process

The test spring used in the suspension system mechanism is a kind of "Helical" with the following specifications: inside and outside coil diameters are Di = 12.985 cm and Do = 15.815 cm, diameter of spring wire, d = 1.415 cm, and number of coils, n = 5 pieces. The spring material used is from Chrome Vanadium, ASTM A231 with Stiffness Modulus, G = 7.929E + 10 Pa = 7.929×10^6 N / cm^2.

2.3. Population and sample

The population of the targeted vehicle is the medium category vehicles including Terios, New Rush, Suzuki R3, Grand Max, Suzuki APV, Isuzu D-Max pickup until Fortuner. The sample selection of medium weight vehicles is shown in Table 1, each of which is randomly performed.

A number of vehicles belonging to the medium weight category, Terios is one medium-weighted vehicle with a net weight of m_2 = 1665 kg and axis weight, m_1 = 600 kg, then selected as research sampling, representing the population of other vehicle types.

Table 1. Vehicle data with medium weight categories

No.	Vehicles	Vehicle Weight (kg) *) Body m_2	Total/Net m_2	Maximum m_2	wheel axis m_1
1	Terios	1235	1665		600
2	New Rush	1950	2510		600
3	PickUp, Grandmax	1950	2510		600
4	Suzuki R3	1185	1745		480
5	Suzuki APV	1950	2510		600
6	Isuzu D-max, Pick up	1990	2550	2550	600

**) PT. Astra Internasional, Tbk. Indonesia*

2.4. Data collection technique

Procedures and steps taken on primary data retrieval activities are: *First*; Collect medium-weighted vehicle data (M_2) which has axle weight, m_1 (kg) and body weight m_2 (kg) as in Table 1, relevant journals, books, and other references. *Second*; Calculating the spring stiffness factor k_2, damping value, c_2 and wheel specifications in each vehicle type either obtained from reference journal or through calculation. *Third;* Conducting experimental and observation activities directly by operating the motion simulation test equipment as follows; Prepares the pressurized air on the air service unit while adjusting the working pressure on the regulator starting from P2 = 1 bar to 8 bar. *Fourth*; Operate the actuator lever on the 5/2 direction control valve by turning the lever towards the cylinder piston movement, up and down. When the cylinder piston moves down, the loading process will occur in the spring and shock absorber in the compression direction.

The drift, x (mm) that occurs due to loading, the magnitude can be read on the available digital display ranging from 0 to 60 mm by the proximity sensor. It should be noted that the magnitude of the deviation is highly correlated with the pressure setting on the Manometer.
Moving time, t (s) can be measured using a stopwatch when there is a drift from x_0 to x (mm). The amount of vertical load that reaches the road surface can be measured by placing "Load cell" just below the wheel of the vehicle.

3. Result and Discussion

3.1. Load and programming analysis

Programming with Mat Lab software referring to equation (2.3) will produce the suspension vibration characteristics with small and large deviations as shown in Figure 4 (a), (b) and (c). The execution result of the program has obtained the characteristic graph of vibration on the suspension as the impact of vertical dynamic load. In Figure 4 (a) if the INPUT resistance is U = 0.1 m, then the suspension on the wheels will experience vibration with maximum deviation, Y = 0.188 m with a vibration time of 0.339 s. then two seconds later, the minimum deviation occurs at Y = 0.101 m. Proportionally author [14] has obtained 0.12 m deviation on the measured vehicle wheel as a strain using the WIM sensor that placed below the tire.

The characteristic shape of the vibration in the wheel suspension as in Figure 4 (a) is caused by a mound as high as U = 0.1 (m) on the road surface. Similarly, the reaction to the total dynamic load of the vehicle can result in a vertical overshot deviation of Y = 0.188 (m) with a duration of 0.1 seconds. Furthermore, within a time interval of 2.34 seconds later, the vibrations begin to shrink and stabilize at 0.66 seconds.

The result of the research in Figure 4 (b) shows that the largest deviation rate occurs at Y2d = 1.03 m / s when t = 0.22 s, while the lowest deviation rate is Y2d = 0.02 m / s when the vibration time t = 2.65 s. From the graph shows that the rate of deviation that occurs largely enough caused by a sudden shock that occurs on the wheel. Such conditions can not last long because of the resistance of the spring and shock absorber.

The change in the drift velocity shown in Figure 4 (c) has a significant slowdown of Y2dd = (- 10.9 + 8.41) m / s = - 2.59 m / s^2 at t = 0.301 s. It should be noted that the perceived comforts of the vehicle are often disrupted by the uneven, hollow, bumpy surface of the road surface and are filled with mound-shaped materials. Comfort can also be felt by passengers and drivers when the mechanism of shock absorber damper is able to reduce the vertical dynamic load of vehicle well.

Given the circulation of liquid or gaseous fluid determined by the regulation of the flow control valve in the cylinder chamber, the performance of the absorbent shock is getting better. The amount of deviation experienced by the body and the vehicle's axle while experiencing vertical dynamic loads of experimental results and modelling using Mat Lab is shown in Figure 4.

IOP Conf. Series: Materials Science and Engineering **676** (2019) 012042 doi:10.1088/1757-899X/676/1/012042

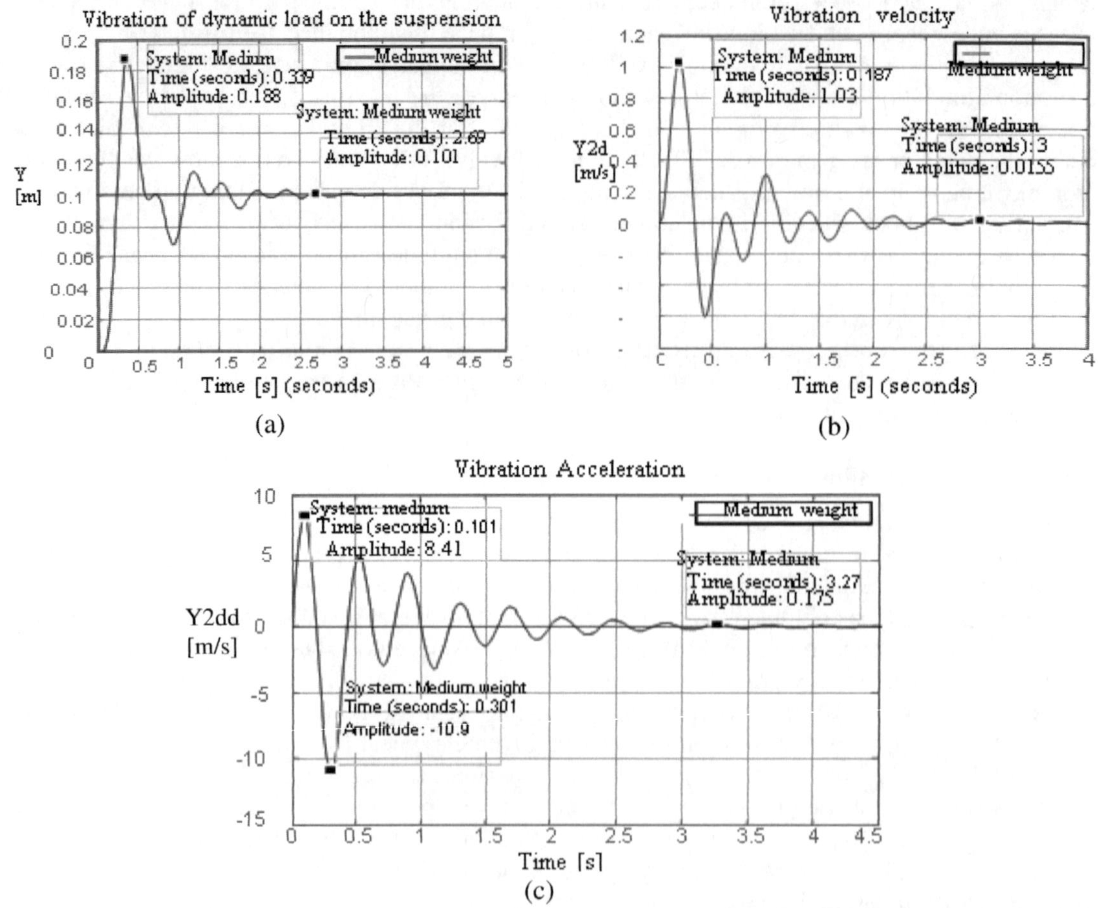

Figure 4. (a) Characteristics of vibration on the body and wheels with deviation Y (m), (b)The characteristics of vibration velocity on the body and wheels, V (m/s), (c) The characteristics of vibration acceleration on the body and wheels, a (m / s^2).

3.2. Pneumatic and road structure force

Work of loads in the vertical direction generated by pneumatic actuators based on equation (1.2) get reactions from the spring force, shock absorber, and air pressure in the tire [2, 5]. The magnitude of the vertical dynamic load of vehicles that overload the structure of the road according to [1, 7] is referring to the equation (1.3). Some variable and constant data such as pressure, P_2 (bar), m_1 axis load (kg), vehicle body load, m_2 (kg), spring constant, k2 (N / m), and shock absorber c_2 (Ns / m) will be substituted into equations (1.2) and (1.3).

The relationship between the pneumatic action force and the force that overlies the road structure is shown in Figure 5. From the graph showing the damping characteristics by shock absorber gives a very big influence on the dynamic load of the vehicle which overload the road structure. Such a condition causes the dynamic load of F_{t1} (N) to overload the structure of the road according to Figure 5 to be around 10000 (N) or equivalent to 1000 kg. If the medium weighted vehicle being tested has a wheel axis weight of m_1 = 600 kg, body weight m_2 = 1665 kg and total weight, m_t = 2265 kg then obtained the dynamic load of the reduced vehicle average is $m_r = m_t$ - 1000 kg = 2265-1000 = 1265 kg. The dynamic load drop presentation that weighs the structure of the path by the attenuation of the shock absorber is 55.85 % [5, 15].

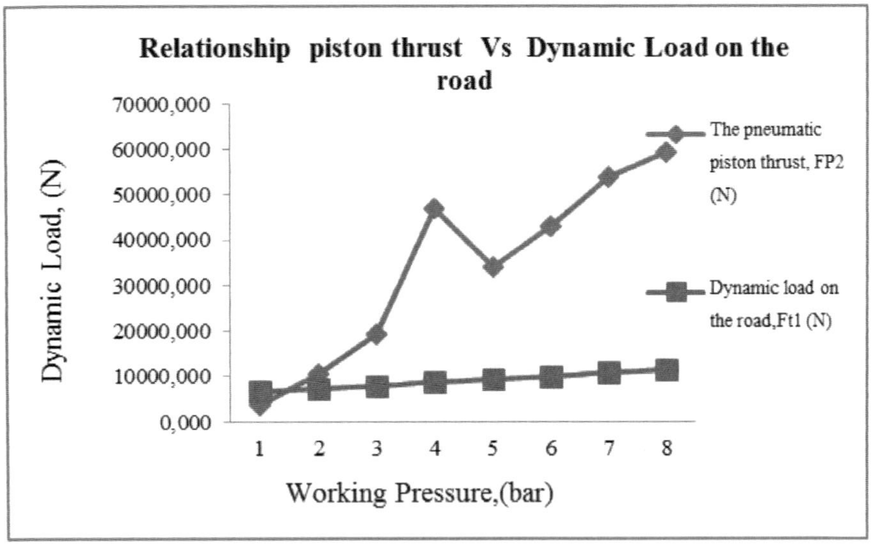

Figure 5. The relationship between piston force, F_{P2} and dynamic load, F_{t1} which overloads the road structure

3.3. Comparison between experiment and theoretical

The comparison of deviation values between the experimental results and the programming with Mat-Lab is shown in Table 2.

Table 2. Characteristics of Experimental Results and Theoretical Deviation

Pressure	Experimental		Theoretical	
P1(bar)	t(s)	Y(m)	t (s)	Y(m)
1	0.5	0.059	0.5	0.11
2	1.0	0.067	1	0.087
3	1.6	0.078	1.5	0.11
4	2.1	0.086	2	0.1
5	2.7	0.088	2.5	0.098
6	3.3	0.092	3	0.1
7	3.9	0.096	3.5	0.1
8	4.6	0.098	4	0.099

The experimental vibration characteristics and theoretical studies show a trend that starting at the working pressure of 4 bars vibration deviation is increasingly blessing each other and finally at the working pressure of 8 bars the deviation is at Y = 0.1 m.

The mean deviation of the experimental results is Y = 0.083 m, with a mean duration of 2.5 s, while the average deviation of the programming results is Y = 0.1 (m) in the mean duration, t = 4.5 s. The rate of deviation and the acceleration of movement of the vehicle body that weighs 300 kg by [2] are 0.03 m and 1.8 m/s^2 respectively and stabilized at t = 3.5 (s).

The shape of the deviation graph shows that at the time of vibration, t = 0.5 s there is a deviation gap of Y = 0.051 m, when pressure P = 1 bar, but after reaching the vibration time t = 2.1 s when pressure P = 4 bar , then the deviation gap starts to decrease to Y = 0.014 m.

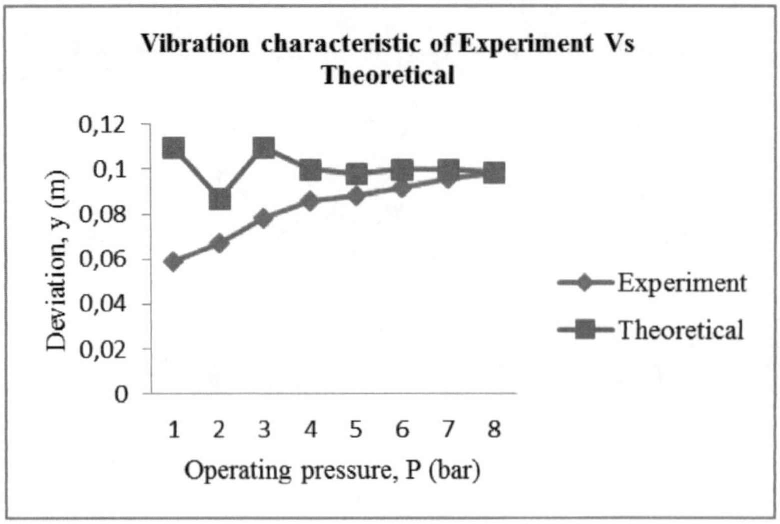

Figure 6. Relationship of vibration characteristics between experimental and theoretical studies.

4. Conclusion

Based on the results of experimental studies and analyzes that have been done on the vehicle suspension work performance, it can be concluded as follows:

1. Characteristics of the working mechanism of the suspension on the body and wheels of the vehicle for the type of medium weight vehicle if experiencing the disturbance U = 0.1 m it will happen the largest deviation Y = 0.18 m with a duration of 0.1 seconds. Within a span of 2.34 seconds later, the vibrations begin to shrink and eventually stabilize at 0.66 seconds.
2. Characteristics of suspension vibration rate on the largest vehicle wheels occur at Y2d = 1.03 m/s when t = 0.22 seconds, while the lowest deviation rate is Y2d = 0.02 m/s when the vibration time t = 2.65 seconds.

Acknowledgement

The author would like to thank to Kemenristekdikti, team supervising commission, Mechatronics Study Program of State Polytechnics of Ujung Pandang, and the Department of Mechanical Engineering of Hasanuddin University have much support and facilitate the authors in the form of grants, guidance, and support of laboratory equipment and workshops.

References

[1] M. F.-D. J. Brembeck, "LPV Control of Full-Vehicle Vertical Dynamics using Semi-Active Dampers," *Science Direct,* vol. 49-11, p. 8, 2016.

[2] M.[a,] TongyiXu[a,] ChuanLi[b,*,] ShuaiYang[a], "Design and analysis of a shock absorber with variable moment of inertia for passive vehicle suspensions" *Journal of Sound and Vibration,* vol. 355, p. 20, 2015.

[3] E. A. Finney, "Dynamic Aspect Of Vehicle Size and Weight," p. 23, 2010.

[4] R. J. G. R. S.Gopinath*, J.Dineshkumar "Design and fabrication of magnetic shock absorber," *International Journal of Engineering & Technology,* vol. 3, pp. 208-211, 2014.

[5] Simon Ka'ka[1,] ,Syukri Himran[2], Ilyas Renreng[2] and Onny Sutresman[2], "Pneumatic Actuator As Vertical Dynamic Load Simulator On The Suspension Mechanism Of A Quarter Vehicle Wheels," *ARPN Journal of Engineering and Applied Sciences,* vol. 12, 2017.

[6] X. J. Y. Z. W. Ma, "Nonlinear dynamic characteristics of a micro-vibration fluid viscous damper," *Springer Science+Business Media, B.V.,* vol. 92, p. 10, 2018.

[7] Simon Ka'ka[1,] , S.Himran, Ilyas Renreng[2] and Onny Sutresman[2], "The Pneumatic Actuators As Vertical Dynamic Load Simulators On Medium Weighted Wheel Suspension Mechanism," *Journal of Physics,* vol. 962, p. 10, 2018.

[8] E. Edge, "Helical Compression Spring Design Equation and Calculator," *Became An Engineers Edge Contributor,* 2017.

[9] P. A. Y. Hassen Trabelsi,*, Jamel Louati, Mohamed Haddar "Interval computation and constraint propagation for the optimal design of a compression spring for a linear vehicle suspension system," *Mechanism and Machine Theory , journal homepage: www.elsevier.com/locate/mechmt,* vol. 48, pp. 67–89, 2015.

[10] K. V. C. I.M. Ryabova, A.V. Pozdeeva*, "Energy Analysis of Vehicle Suspension Oscillation Cycle," *International Conference on Industrial Engineering, ICIE 2016,* vol. 150, pp. 384 – 392, 2016.

[11] V. V. N. I. V. Ryabova, A.V. Pozdeeva,*, "Efficiency of Shock Absorber in Vehicle Suspension," *International Conference on Industrial Engineering, ICIE 2016,* vol. 150, pp. 354 – 362, 2016.

[12] K. Festo, "Automation With Pneumatic," in *Festo Didactic, KG* vol. NR.23099-01/1, ed. Esslingen: Festo Pneumatic, 2015.

[13] L. D. Y. A. Asst. prof. Munaf F. Badr[1,] , Engineer Ahmed Kadhiam Jaliel3, "Position Control of the Pneumatic Actuator Employing ON/OFF Solenoids Valve," *International Journal of Mechanical & Mechatronics Engineering IJMME-IJENS,* vol. Vol.17 No.02, p. 9, 2016.

[14] Wenbin Zhang 1, Chunguang Suo 2 and Qi Wang 1, "A Novel Sensor System for Measuring Wheel Loads of Vehicles on Highways," *Sensors www.mdpi.com/journal/sensors,* vol. 8, p. 19, 2008.

[15] D. B. Manjunatha.T.R[1,] H.V[2], Prabhunandan.G.S[3], "Design and Static Structural Analysis of Cylinder and Piston of Two Stage Reciprocating Compressors Using ANSYS" *International Journal of Innovative Research in Science, Engineering and Technology (An ISO 3297: 2007 Certified Organization),* Vol. 2, p. 7, 2013.

Design and Analysis of Fixture for Welding Casing Components of Cross-Flow Turbine

M Muchtar[1]*, S Rasyid[1] and L Sonda[1]

Department of Mechanical Engineering, Politeknik Negeri Ujung Pandang, Makassar, 90245, Indonesia

*E-mail: syaharuddinrasyid@poliupg.ac.id

Abstract. The purpose of this research is to improve the quality of the turbine casing product by using the welding fixture. The method used was to design and make welding fixture of turbine casing welding, test the straightness, alignment, straightness, and analyzes the deviation of measurement results. Based on the design result of welding fixture assembling of cross-flow turbine casing component, it had been obtained prototype welding fixture which can facilitate the assembling process of cross-flow turbine casing component. The conclusions of this research are the quality of cross-flow turbine casing products can be increased along with the small geometry deviations in the cross-flow turbine runners by using a fastening tool. The maximum deviation of the straightness of the turbine wall is 1.6 (wall A) and 0.85 (wall B). The process of assembling the components of the turbine runner is shorter by using a faster welding assembly tool with a better level of accuracy.

1. Introduction

The use of a fixture has a direct impact on welding quality, productivity, and costs. The purpose of the fixture is to reduce distortion so as to reduce production losses and also produce lead times for welding, positioning, and containment of parts. Various residual stresses generated when welding is responsible for distortion. Preheating, peening, post-welding heat treatment, relieving stress with natural aging, eliminating vibration tension are a number of ways to control residual stress.

The residual stress of welding affects the integrity of the structure and is one of the greatest unknown pressure. Residual stresses are difficult to measure and estimate theoretically but are often significant when compared to the pressure of the service they put. Loss of performance in corrosion, fatigue, and fracture are the causes of high tensile stresses [1, 2].

Jig and Fixture is a tool used in machining processes in order to produce more accurate parts duplication. Jig and Fixture are selected and designed in accordance with the shape of parts and machining processes to be processed. One type of Fixture used for welding process is Welding Fixture. This type of tool served to support the workpiece when the welding process operates [3, 4].

The manufacture and assembly of micro-hydro turbine components are carried out using conventional machining processes such as electric welding processes, lathe processes, drilling processes, grinding processes etc. One component of the micro-hydro turbine is a turbine casing. The manufacture of turbine casing components is a phase that focuses on the tolerance of shapes that must be fulfilled on a single element. Form tolerances related to the nature of components are straightness, alignment, straightness, etc. While the turbine casing component assembly is a phase that focuses on orientation tolerance, location tolerance, and alignment tolerance, which must be met in the related elements. Related elements must take into account their attributes for a given tolerance such as alignment tolerance, stability tolerance, etc [3].

The use of improper tools can make the welding process is inhibited. Operators will have difficulty installing two components to be welded so that the product produced is not accurate. The welding equipment used today requires the operator to hold the tool during the welding process. This results in a product defect because the workpiece will easily slide so that welding is not correct. This has an impact on the long adjustment time and increased production costs. One way that can be done to reduce this is to do the design and selection of the right product equipment [3].

Several studies report that jigs are widely used to assist the welding process to avoid rotation distortion in the front of the welding heat source. A methodological approach to the design of welding equipment for robotic cells in spot welding of car bodies based on the principle of modularity. Regarding welding deformation control, there are several reports relating to external constraints and burdens [5]. Park et al. [6] investigate angular distortion and residual stress under various pre-tension conditions by changing the direction and magnitude of the pre-stress voltage. Schenk et al. [7] studied the effects of clamping on buckling distortion and angular distortion for overlapping joints and T-shape fillet joints. They found that residual stresses and welding distortion were strongly influenced by clamping conditions. Shateryana et al. [8] investigated the constraint effects on welding deformation and residual stress on aluminum alloy lap joints under three local U-shaped equipment types by performing 3D finite element analysis. Ziaee et al. [9] studied the effect of boundary conditions on buckling mode during thin plate welding. It was found that external constraints can increase buckling resistance but cannot eliminate buckling.

Welding the turbine casing with a simple tool causes difficulty in controlling the dimensions or uniformity of the shape, and requires high costs and a longtime during assembly. To get a uniform product and in accordance with the basic size, accurate and precise assembly process with an accelerator system is needed to avoid mistakes in assembling these components [3]. The quality of the cross-flow turbine runner product can be improved along with the small deviation of geometry in the turbine runner by using tools. The assembling process of turbine runner components has a shorter time by using welding assembly tools faster with better accuracy. Based on the results achieved in this study it is advisable to optimize this tool in the manufacture of cross-flow turbine runner components [3]. Welding equipment can be used as a tool for the production of the front engine mounting component welding process. The use of welding equipment can help to provide welding production process solutions from the time side, ease of work and production process speed [4].

The effect of the jig position is limited and the distance to the welding deformation can be reduced by the fastening device in the plate welding, the three-way jig constraint has a large influence on the component deformation. Normal jig direction constraints effectively reduce the distortion angle of the compressive force and shrinkage. In general, angular distortion can be reduced by jig position and smaller distances [10]. The control style can increase flexibility in the car bodywork process when compared to conventional law use [11]. The Joint Jig Assembling System can stabilize the connecting process because the AJJ system uses many components such as angular joints, adaptive PC bolts, expanding mandrel, and modular search. With the AJJ system, it can be operated efficiently and comprehensively, so as to reduce production costs [12].

Based on the results that have been carried out by several researchers, it can be concluded that fixture (production equipment) can improve the efficiency of the component assembly process. The purpose of this research is to improve the quality of the turbine casing products using welding fixture equipment.

2. Experimental Procedures

2.1. Materials and equipment

The materials and components used to make the turbine welding tool are: 1). Steel Streep plate 10x50x600 mm, 2). Steel plate 400x500x10 mm, 3). steel elbow 40x40x6000 mm, 4). steel shaft 50x1000 mm, and 5). Bolts, M10x60 mm. The equipment used to make the accelerator and turbine

IOP Conf. Series: Materials Science and Engineering **676** (2019) 012043 doi:10.1088/1757-899X/676/1/012043

casing components is 1). Lathe, 2). CNC milling machine, 3). Drilling machine, 4). The bending machine, 5). Cutting machine, 6). A grinding machine, 7). Electric welding machine, 8). A dial indicator, and 9). Chuck.

2.2. Design method
Jig and fixture for welding is a series consisting of standard taping, fastening pen, retaining and all integrated into a functioning unity. Some things to consider in the design of tools such as the operator to take a good welding position, fixture should be easy and fast in its use, the design should be easy and cheap, fixture must be able to provide leeway on the workpiece to move in one direction, and grafting should be prepared for various welding operations [3]. Based on several considerations in planning the tool system, it has been designed welding fastening system as shown in Figure 1.

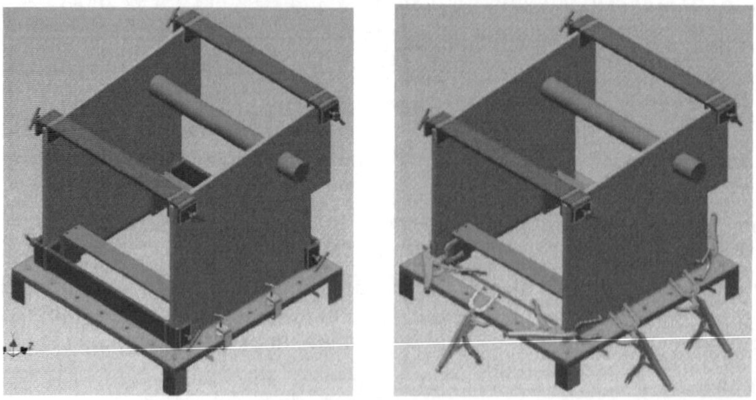

(a) (b)

Figure 1. Cross-flow turbine casing tools with a) first design and b) second design.

Based on the above two designs, the first design was chosen with the consideration of easy and cheap made, flexible in use, and welding operators easy in welding position taking.

2.3. Assembly process
The turbine casing assembly procedure is carried out in two stages: the assembly of the base frame and the installation of the casing wall. The following is an illustration of the following sequence of assembly stages with the numbering. The assembly phase of the base frame is (Figure 2); a). The base frame plate is conditioned on the pins on the fixture table, b). Block both base frame components close to the connection, c). Welding is done on the connection between components. This procedure is performed on each connection that is on the cross-flow turbine casing base frame.

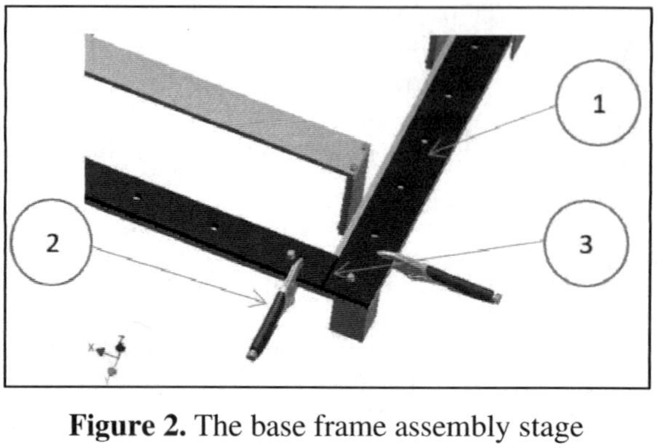

Figure 2. The base frame assembly stage

Stage of casing wall installation. The assembly procedure that needs to be done in the casing wall installation process is (Figure 3): a). Install the A casing wall on the fixture table then grip it with the clamp on both the front and back ends of the wall, b). Install the casing wall B on the fixture table and grip it like the casing wall A, c). Inserting a shaft in a hole in the casing walls to adjust the alignment of the holes between walls, d). Installing the clamp of the upper wall at the upper end of the wall of the casing, and e). Calm down the base frame next to the casing wall, then welding between the base frame and the casing walls.

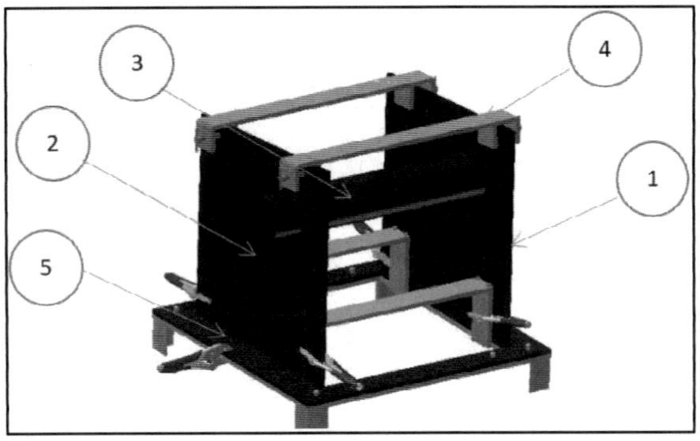

Figure 3. Stage of the turbine casing assembly

2.4. Measurement process

Turbine casing measurements are carried out to determine the straightness of the casing wall to the base frame casing, measurements are taken during assembly before welding and after welding. Measurements before welding are carried out by measuring the distance between walls using calipers and measuring the distance between walls at some point (Figure 4). Turbine casing measurements are carried out to determine the straightness of the casing wall to the base frame casing, measurements are taken during assembly before welding and after welding. Measures before welding are carried out by walls at some point (Figure 4).

Figure 4. Measurement before welded

The measurement after welding is done by two methods, namely measuring the distance between walls and calipers and measuring the straightness of the lines and alignment of the walls using a dial indicator on a CNC mill machine. Measurement of the distance between walls after welding is done at the points that have been measured before welding. Follows the measurement process after the welding process (Figure 5).

IOP Conf. Series: Materials Science and Engineering **676** (2019) 012043 doi:10.1088/1757-899X/676/1/012043

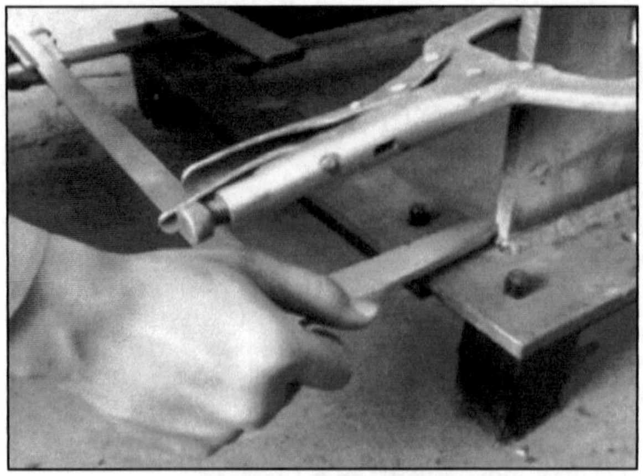

Figure 5. Measurement after welded

Straightness measurement and alignment of the casing after welding is carried out on both walls according to the design demands that are to be achieved (Figure 6).

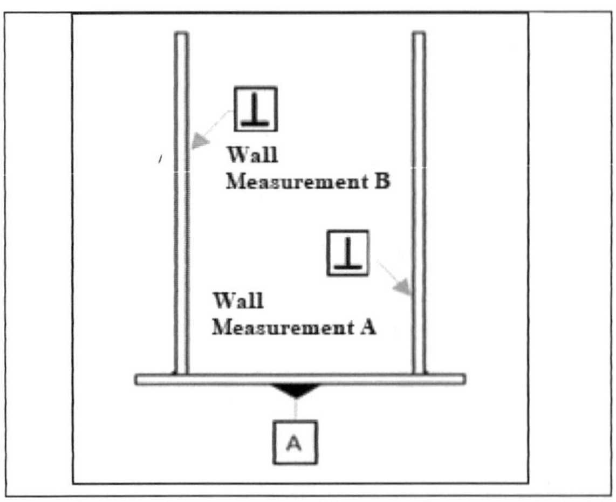

Figure 6. The point of measurement of the straightness of the wall

Measurements were made on the casing walls with instruments measuring instruments in the form of dial indicators and carried out at points A, B, and C to the base frame (Figure 7). Figure 8 shows the process of measuring casing walls in CNC mill machines.

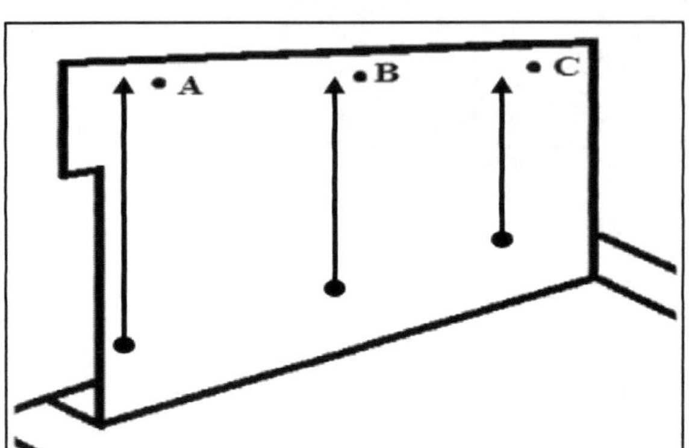

Figure 7. The point of measurement for the straightness of the casing wall

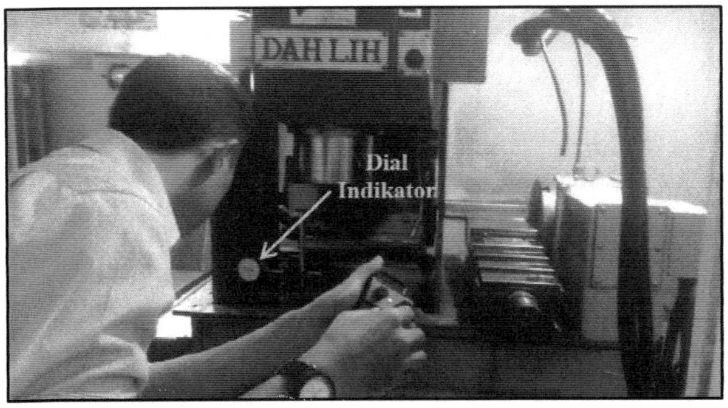

Figure 8. Turbine casing measurement process

3. Result and Discussion

The testing phase of the cross-flow turbine casing assembly aids has been carried out and can produce turbine casings that have been determined. The specimen is measured in accordance with the specified measurement procedures. Then the data obtained will be used to determine whether the chassis assembly fixture is successful in meeting the design target to be achieved.

The turbine casing measurement phase has been determined in advance where measurements are made before welding or after welding is carried out according to the specified assembly reference. The measurement before welding is only the measurement of the distance between the walls while the measurement after welding is the measurement of the distance between the walls and the measurement of the straightness of the casing walls. Measurement of distance between walls both before welding and after welding is done at points in accordance with the sketches that have been made (Figure 9).

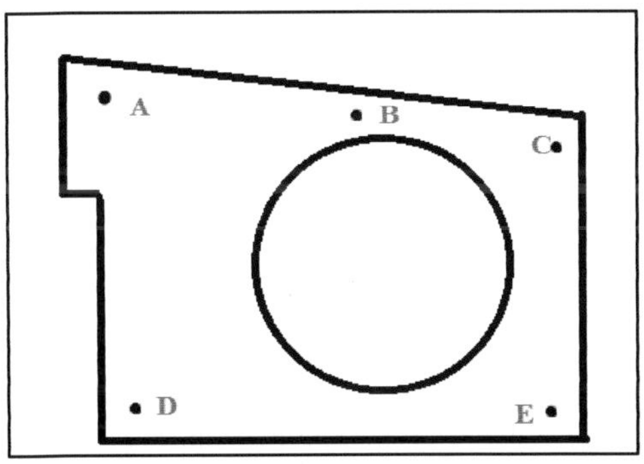

Figure 9. The point of measurement of the distance between walls

Table 1 presents the data from the measurement of the distance between the discs using calipers (in mm) carried out before and after welding.

Table 1. Data from measurements of the distance between walls

Point of measuring area	Before welded	After welded	
		Before the fixture is released	After the Fixture is released
A	447.9	447.9	450.4
B	448.1	448.1	450.5
C	448.05	447.7	449.1
D	448.2	449.0	449.7
E	447.5	447.85	448.25

IOP Conf. Series: Materials Science and Engineering **676** (2019) 012043 doi:10.1088/1757-899X/676/1/012043

Based on the data obtained indicate that there is a change in the size of the distance between walls when the fixture is removed after welding, the biggest distance between walls is approximately 2 mm from the initial size (448 mm), this deviation is caused by the wall clamp at points D and E it has not been rigid to resist the deformation of the welding process so that the wall shifts and causes large deviations to occur at points A, B, and C.

Based on the data obtained indicate that there was a change in casing firmness measurements carried out on a CNC Mill machine to find out the perpendicular deviations that occurred in the casing walls after welding, and to find out whether the geometry deviations from the casing walls had met the design demands target what you want to achieve is 1.0 mm. Measurements were made at points A, B and C on each wall shown in Figure 9. Table 2 shows the geometry deviation of the casing wall against the base frame.

Table 2. Data on measurements of turbine casing firmness.

Measurement Side	Measurement point	Minimum deviation (mm)	Actual Deviation (mm)
A wall	A	1.0 mm	1.25
	B		1.20
	C		1.60
B wall	A	1.0 mm	0.45
	B		0.73
	C		0.85

Based on the measurement results above, the deviation values that occur on the two walls are different where wall A produces the biggest deviation of 1.60 mm from the allowable deviation of 1.0 mm, while the deviation generated from wall B is greatest at point C, which is 0.85 mm below the tolerance limit. which has been specified. The value of deviations that occur in wall A that are out of the tolerance limit that has been determined because the process of welding walls is not concurrently carried out on both sides which causes deformation that occurs on the wall being welded also affects the other wall.

Based on the results of testing using a casing assembly fixture and measurements made on the results of cross-flow turbine casing assembly using a casing assembly fixture, it can be concluded that the casing assembly fixture has not been able to meet the target geometric tolerance deviation to be achieved (1.0 mm).

4. Conclusion

The conclusion of the research on the design of welding fixture assembly of cross-flow type micro-hydro turbine components is 1). The quality of cross-flow turbine casing products can be increased along with the small geometry deviations in the cross-flow turbine runner by using an auxiliary tool, 2). The maximum deviation of the turbine wall alignment is 1.6 (wall A) and 0.85 (wall B), and 3). The process of assembling the components of the turbine runner is shorter by using a faster welding assembly tool with a better level of accuracy.

Acknowledgments

This paper was composed based on work supported by RISTEKDIKTI grant. I would like to thank The Ministry of Research, Technology and Higher Education of the Republic of Indonesia and Politeknik Negeri Ujung Pandang for funding and support.

References

[1] Jigar DS, Patel KM, and Sanjay G L 2013. Design and analysis of fixture for welding an exhaust impeller. Proc. Engi. 51 514 – 519.

[2] John W.H. Price, Anna Z P, Suraj J, 2008. Comparison of experimental and theoretical residual stresses in welds. Int. J. of Mech. Sci. 50 5 13-521

[3] Muas M, Syaharuddin R, and Rusdi N. 2018. Welding fixture design for the production of cross-flow turbine runner. AIP Conf. Proc. 1977, 030020.

[4] Tricot, R., 1989. Principal metal working characteristics. Matr. Et Tech.

[5] Hajduk, M., Semjon, J., Vagas, ˇ M., 2009. Design of the welding fixture for the robotic station for spot welding based on the modular concept. Acta Mech. Slovaca 13 (3), 30–35.

[6] Park, J.U., An, G.B., Lee, H.W., 2012. Effect of external load on angular distortion in fillet welding. Mater 42, 403–410.

[7] Schenk, T., Richardson, I.M., Kraska, M., Ohnimus, S., 2009. A study on the influence of clamping on welding distortion. Mater. Sci. 45 (4), 999–1005.

[8] Shateryana M, Rezaee BA, Sabzevara M.H, Naderi B 2012. Numerical investigation of the effects of constraint design on 5XXX Al Alloy GTA Welding. Proc. of Iran Int. Al. Con. 15–16.

[9] Ziaee S, Kadivar M H, Jafarpur K .2009. Experimental evaluation of the effect of external restraint on buckling behavior of thin welded shells. J. Sci. Technol.Trans. B: Eng. 33 (B5), 397–413.

[10] Prasetyo H and Dewi P 2015. Design of Welding Fixture Manufacture of Front Engine Mounting for Suzuki Baleno. J Rek. Sist Ind. 4(2).

[11] Ma N, Huang H, and Murakawa H 2015. Effect of jig constraint position and pitch on welding deformation. J Mat. Proc. Tech. 221:154–62.

[12] Keller C and Putz M. 2016. Force-controlled adjustment of car body fixtures–verification and performance of the new approach. Proc. CIRP. Els. 44:359–64.

AUTHOR INDEX

Institute of Physics
Dirac House, Temple Back
Bristol BS1 6BE UK

ISSN: 1757-8981
ISBN 978-1-7138-0928-9

9 781713 809289

CURRAN ASSOCIATES INC.
proceedings.com

All-Russian Conference with the School for Young Scientists Thermophysics and Physical Hydrodynamics 2016

Journal of Physics: Conference Series Volume 754

Crimea
19-25 September 2016

IOP
Institute of Physics